Monographs in Computer Science

Monographs in Computer Science

Abadi and Cardelli, **A Theory of Objects**

Benosman and Kang [editors], **Panoramic Vision: Sensors, Theory, and Applications**

Bhanu, Lin, Krawiec, **Evolutionary Synthesis of Pattern Recognition Systems**

Broy and Stølen, **Specification and Development of Interactive Systems: FOCUS on Streams, Interfaces, and Refinement**

Brzozowski and Seger, **Asynchronous Circuits**

Burgin, **Super-Recursive Algorithms**

Cantone, Omodeo, and Policriti, Set Theory for Computing: From Decision Procedures to Declarative Programming with Sets

Castillo, Gutiérrez, and Hadi, **Expert Systems and Probabilistic Network Models**

Downey and Fellows, **Parameterized Complexity**

Feijen and van Gasteren, **On a Method of Multiprogramming**

Herbert and Spärck Jones [editors], **Computer Systems: Theory, Technology, and Applications**

Leiss, **Language Equations**

Levin, Heydon, and Mann, **Software Configuration Management with VESTA**

McIver and Morgan [editors], **Programming Methodology**

McIver and Morgan [editors], **Abstraction, Refinement and Proof for Probabilistic Systems**

Misra, **A Discipline of Multiprogramming: Programming Theory for Distributed Applications**

Nielson [editor], **ML with Concurrency**

Paton [editor], **Active Rules in Database Systems**

Selig, **Geometrical Methods in Robotics**

Selig, **Geometric Fundamentals of Robotics, Second Edition**

Shasha and Zhu, **High Performance Discovery in Time Series: Techniques and Case Studies**

Tonella and Potrich, **Reverse Engineering of Object Oriented Code**

Iman Hafiz Poernomo
John Newsome Crossley
Martin Wirsing

Adapting Proofs-as-Programs

The Curry–Howard Protocol

Iman Hafiz Poernomo
Department of Computer Science
King's College London
Strand, London, WC2R 2LS
United Kingdom
iman@dcs.kcl.ac.uk

John Newsome Crossley
School of Computer Science and Software Engineering
Monash University
Victoria, Australia
jnc@cs.monash.edu.au

Martin Wirsing
Institut für Informatik
Ludwig-Maximilians Universitat München
Oettingenstr. 67
D-80538 München, Germany
wirsing@informatik.uni-muenchen.de

Series Editors:
David Gries
Cornell University
Department of Computer Science
Ithaca, NY 14853
U.S.A.

Fred B. Schneider
Cornell University
Department of Computer Science
Ithaca, NY 14853
U.S.A.

Cover illustration: "Creating Seven" by Soerodipoero Paul Poernomo.

Mathematics Subject Classification (2000): (Primary): 03B15, 68N18, 68Q65 (Secondary): 03B20, 68N30

Library of Congress Cataloging-in-Publication Data
Poernomo, Iman Hafiz, 1976–
Adapting proof-as-programs : the Curry-Howard protocol / Iman Hafiz Poernomo, John Newsome Crossley, Martin Wirsing.
p. cm. — (Monographs in computer science)
Includes bibliographical references and index.

1. Curry-Howard isomorphism. 2. Proof theory. 3. Logic, Symbolic and mathematical. 4. Functional programming (Computer science) 5. Lambda calculus. 6. Abstract data types (Computer science) I. Crossley, John N. II. Wirsing, M. (Martin) III. Title. IV. Series.
QA9.54.P64 2005
511.3′6—dc22 2005046411

Printed on acid-free paper.
ISBN 978-1-4419-2014-0 e-ISBN 978-0-387-28183-4

Softcover reprint of the hardcover 1st edition 2005

9 8 7 6 5 4 3 2 1

springeronline.com

Preface

This book finds new things to do with an old idea. The proofs-as-programs paradigm constitutes a set of approaches to developing programs from proofs in constructive logic. It has been over thirty years since the paradigm was first conceived. At that time, there was a belief that proofs-as-programs had the potential for practical application to semi-automated software development. Initial applications were mostly concerned with fine-grain, mathematical program synthesis. For various reasons, research interest in the area eventually tended toward more theoretic issues of constructive logic and type theory. However, in recent years, the situation has become more balanced, and there is increasingly active research in applying constructive techniques to industrial-scale, complex software engineering problems.

This monograph details several important advances in this direction of practical proofs-as-programs.

One of the central themes of the book is a general, abstract framework for developing new systems of program synthesis by adapting proofs-as-programs to new contexts. Framework-oriented approaches that facilitate analogous approaches to building systems for solving particular problems have been popular and successful. These methods are helpful as they provide a formal toolbox that enables a "roll-your-own" approach to developing solutions. It is hoped that our framework will have a similar impact.

The framework is demonstrated by example. We will give two novel applications of proofs-as-programs to large-scale, coarse-grain software engineering problems: contractual imperative program synthesis and structured program synthesis. These applications constitute an exemplary justification of the framework. Also, in and of themselves, these approaches to synthesis should be interesting for researchers working in the target problem domains.

The monograph serves a dual purpose of providing a state-of-the-art overview of the field and detailing tools and techniques to stimulate further research. The intended audience is graduate students in computer science or mathematics, the proofs-as-programs research community, and the wider computational logic, formal methods, and software engineering communities.

The ideas presented in this monograph originate from research conducted over the past five years by Iman Poernomo and John Crossley at Monash University in collaboration with Martin Wirsing at Ludwig–Maximilians Universität. A significant portion of the monograph is based on the PhD thesis of Iman Poernomo [Poe03b]. Some of the results were presented previously in conference papers and journal articles. Part II of the book presents ideas that resulted from work of Poernomo and Crossley [PC01], [CP01] and [JPBC03]. Part III ellaborates and extends the work done in [Poe99], [PC03] and [Poe03a]. Part IV extends work that was first presented in [CPW00] and [PCW02] by Crossley, Poernomo and Wirsing.

Several people have helped us in the preparation of this book. We are particularly grateful to John Shepherdson and Masahiko Sato for their valuable comments and suggestions for improvement. We also thank John Jeavons, Bolis Basit, Helmut Schwichtenberg, Dirk Pattinson, Stuart Allen and the Nuprl seminar group for fruitful discussions.

Iman Hafiz Poernomo
Caulfield East
May 2004

Overview

The Curry–Howard isomorphism says that intuitionistic logic can be presented as a constructive type theory in which proofs correspond to terms, formulae to types, logical rules to type inference and proof normalization to term simplification. In order to represent intuitionistic proofs, terms of the constructive type theory contain constructive information used to prove formulae. This information can be used to synthesize correct, error-free programs from proofs. Such approaches to program synthesis, based upon the Curry–Howard isomorphism, constitute the area referred to as the proofs-as-programs paradigm.

The advantage of proofs-as-programs techniques is that the task of programming a function is reduced to reasoning with domain knowledge. After more than three decades of research, proofs-as-programs constitutes a mature field with an established theory and set of best practices. State-of-the-art approaches to proofs-as-programs usually involve some form of optimization and extraction strategy, transforming intuitionistic proofs to a commonly used functional programming language that can encode a simply typed lambda calculus, such as *SML*, *Scheme* or *Haskell*.

Work has been done in providing analogous results to the Curry–Howard isomorphism and proofs-as-programs for other logical systems and programming languages. However, little work has been done in identifying a general framework that generalizes the form such analogies should take over arbitrary logical calculi and programming languages. Such a framework would serve as a guide to go about adapting proofs-as-programs to new contexts.

This book defines such a framework, which we call the Curry–Howard protocol. It requires an analogous property to the Curry–Howard isomorphism to hold between a given logic and type theory. However, generalizing state-of-the-art approaches to proofs-as-programs, the protocol requires an optimization and extraction strategy from proofs represented in the logical type theory to programs in a separate programming language. While program synthesis methods have been developed that conform to our protocol, such a framework has not been explicitly identified previously.

We then use the protocol to show how proofs-as-programs can be adapted to two different contexts.

- *Proofs-as-imperative-programs.* The Hoare logic provides a method for the simultaneous development of imperative programs and proofs of their properties. We adapt proofs-as-programs to the Hoare logic for the purpose of extending it to developing imperative programs with side-effect-free return values and views on state.
- *Structured proofs-as-programs.* Structured algebraic specifications are an approach to the compositional design of software systems based on the development of data types. There are proof systems that enable us to reason about structured specifications. We develop such a system and use proofs-as-programs–style techniques for the synthesis of programs from proofs about specifications, and the eventual refinement of specifications into structured code.

These adaptations constitute an exemplary justification for the applicability of the protocol to different contexts.

Contents

Part III Imperative Proofs-as-Programs

Part IV Structured Proofs-as-Programs

Part I

Prologue

1

Introduction

Ultimately, software developers would like to solve problems by building well-structured, comprehensible, correct programs, solely through the application of domain knowledge. The proofs-as-programs paradigm has been proposed as a means of achieving this goal. These methods use constructive, intuitionistic logic and the Curry–Howard isomorphism to generate correct programs from proofs of their specifications [CMH86, HKPM97, CS93, PC01]. The programs generated are pure functions (stateless programs written in languages such as *SML*, *Haskell*, *Scheme* or *LISP*). The advantage of these techniques is that the task of programming a function is reduced to reasoning with domain knowledge.

After more than thirty years of research, proofs-as-programs constitutes a mature field with an established theory and set of best practices.

An active area of research concerns analogous results to the Curry–Howard isomorphism and proofs-as-programs for other programming paradigms and logical systems. Such adaptations leverage the successes of proofs-as-programs in correct, domain-knowledge–oriented development, for a wider range of programming and reasoning contexts.

The usefulness of these results lead us to question if there exists a general framework for adapting proofs-as-programs. A framework would serve as a guide to go about adapting proofs-as-programs to new contexts, abstracting properties and constraints required of a logic and programming language for program synthesis to be achieved. This book defines such a framework, which we call the Curry–Howard protocol.

We shall use the protocol to adapt proofs-as-programs for two different contexts:

- *Proofs-as-imperative-programs.* The Hoare logic of [Hoa69] provides a method for the simultaneous development of imperative programs and proofs of their properties. We adapt proofs-as-programs to the Hoare logic for the purpose of extending it to the synthesis of imperative programs with side-effect-free return values and views on state.

- *Structured proofs-as-programs and structured program synthesis.* Structured algebraic specifications are an approach to the compositional design of software systems based on the development of data types [Wir90, CoF01]. One important area of research is proof systems that enable us to reason about these specifications. We develop such a system and use proofs-as-programs–style techniques for the synthesis of programs from proofs about specifications, and the eventual refinement of specifications into structured code.

These adaptations are significant results in themselves and therefore constitute an exemplary justification for the applicability of the protocol to different contexts.

1.1 Proofs-as-programs

The term proofs-as-programs originates from the title of Bates and Constable's paper [BC85]. Proofs-as-programs methods enable the synthesis of programs using constructive type theory and the Curry–Howard isomorphism. These methods may be more widely classified as *deductive methods* of program synthesis, following Tyugu [Tyu88, p. 8]. In deductive synthesis, a program that solves a problem is derived from the deduction of solvability of the problem. This is in contrast to, for example, refinement-based methods of synthesis, where programs are derived by means of verified transformation steps from an abstract model.

In proofs-as-programs, correct functional programs are synthesized from intuitionistic proofs of specifications. For instance, if a proof of $\forall x : t \bullet \exists y : s \bullet A(x, y)$ is given, then a computable function f can be synthesized from the proof. The function is a "correct" program in the sense that $\forall x : t \bullet A(x, f(x))$ is satisfied [How80].

This sense of correctness corresponds to a form of constructive realizability in the sense defined by Kreisel [Kre59, BS95a, Dil80]. (See Appendix A for an overview of intuitionistic logic, constructive type theory and realizability.) The correctness property is guaranteed to hold because f is synthesized from the constructive content of the intuitionistic proof. From the perspective of the designer, implementation details are hidden, i.e., encapsulated as the constructive content of the proof. Programs are developed solely through reasoning with domain knowledge encoded as axioms and theorems of intuitionistic logic.

By the property known as the Curry–Howard isomorphism, a form of type theory can be used to represent intuitionistic proofs, storing constructive content from which it is possible to synthesize correct, realizing programs.

The proofs-as-programs paradigm has evolved over time, from what we classify as naïve approaches, to sophisticated, state-of-the-art approaches. The former approaches use constructive type theory as a programming language itself, and realizability to define program correctness. In the latter approaches, a proof is transformed and optimized into a program of a commonly used functional programming language (such as *Scheme*, *SML* or *Haskell*). A different notion of correctness is employed, which is obtained by modifying the concept of realizability to apply between functional programs and formulae.

In this book, we will be concerned with the adaptation of state-of-the-art approaches to different logical and programming contexts.

1.1.1 The Curry–Howard isomorphism

Proofs-as-programs is based on the the Curry–Howard isomorphism. This property tells us that intuitionistic logic can be represented by a kind of type theory where proofs correspond to terms, formulae to types, logical rules to type inference, and proof normalization to term simplification. The original idea was first described by Curry [Cur34] and extended to intuitionistic first order logic by Howard [How80].

Essentially, a constructive type theory corresponding to intuitionistic predicate logic is a typed lambda calculus with dependent product and sum types, and disjoint unions. The rules of natural deduction then have corresponding type formation rules.

Example 1.1. The formula $(A \vee B)$ in intuitionistic logic can be considered as a disjoint union type of constructive type theory. The $(\vee\text{-I}_1)$ rule of intuitionistic natural deduction corresponds to a typing rule

$$\frac{\Gamma \vdash p^{A}}{\Gamma \vdash \mathsf{inl}(p)^{(A \vee B)}} \ (\vee\text{-I}_1)$$

The rule tells us that the term $\mathsf{inl}(p)$ is correctly typed by $(A \vee B)$, provided that p is typed by A.

Example 1.2. The $(\forall\text{-I})$ rule of natural deduction for first order intuitionistic logic with arithmetic corresponds to a typing rule

$$\frac{\Gamma \vdash p^{A[y/x]}}{\Gamma \vdash \lambda x.p^{\forall x \bullet A}} \ (\forall\text{-I})$$

The rule tells us that $\lambda x.p$ is correctly typed with $\forall x \bullet A$, provided that p is typed by $A[y/x]$. The formula $\forall x \bullet A$ is taken as a dependent product type, by virtue of the type inference rule corresponding to $(\forall\text{-E})$:

$$\frac{\Gamma \vdash p^{\forall x \bullet A}}{\Gamma \vdash (pa)^{A[a/x]}} \ (\forall\text{-E})$$

This is the elimination rule for dependent product types, showing that $\forall x \bullet A$ parametrizes the type A over possible instantiation by a term a.

First order, and many-sorted, logics have straightforward type theories — see, for instance, the type theory of Schwichtenberg in [Sch99b, pp. 1–13]. Crossley and Shepherdson [CS93] provide a constructive type theory that was extended by Crossley and Poernomo in [PC01, CP01] to be modular over sorts (with datatypes such as natural numbers, booleans, and lists).

The Curry–Howard isomorphism can also be applied to a range of fully higher-order constructive type theories, each corresponding to a different form of intuitionistic logic that permits predication over logical formulae. The predicative type theories of Martin–Löf [ML75, ML84] restrict quantification according to hierarchies of type universes. In contrast, the impredicative type theories of, for instance, Girard [Gir72], Reynolds [Rey74] and Coquand [MLM90] permit quantification over types to form a type itself. (See Appendix A for more background details.)

1.1.2 Naïve proofs-as-programs

The earliest proofs-as-programs approaches directly invoke the Curry–Howard isomorphism to extract programs by identifying terms of constructive type theory with programs. We classify these approaches as naïve, because they directly take constructive type theory as a programming language, rather than synthesizing programs in a conventional language.

The idea is simple, following directly from the properties of type theory. The terms of a constructive type theory constitute a lambda calculus that is equipped with reduction rules. By the isomorphism, the closure of these reduction rules corresponds to proof normalization. Also, these rules, considered as operational semantics, permit us to regard the type theory as an executable functional programming language. A formula is considered as a specification of input/output behavior, and realizability defines how a program satisfies a specification. By constructing a realizing proof we simultaneously provide an algorithm that satisfies the specification.

Thus a proof can be viewed as an executable, functional program.

This style of proofs-as-programs has been defined for higher-order predicative type theories by Martin–Löf [ML85] and by Constable and Mendler who implemented it in the *Prl* and *Nuprl* systems [BC85, CMH86]. Program synthesis of this form has also been defined for the impredicative Calculus of Constructions by Coquand and Huet in [CH88], implemented in the *Coq* system. (See Appendix A for an overview of these higher-order type theories.)

Example 1.3. For example, given an intuitionistic proof of $\forall x : t \bullet \exists y : s \bullet A(x, y)$, we can form a corresponding term p in a Martin–Löf type theory of the form

$$\lambda x : t.(g_1(x) : s, g_2(x) : A(x, y)) : \forall x : t \bullet \exists y : s \bullet A(x, y)$$

If we define

$$f = \lambda x : t.\pi_1(p\ x)$$

(where π_1 is the first projection) then f is considered a program such that, on every input $x : t$, (fx) terminates and $\forall x : t \bullet A(x, (fx))$ is satisfied, and is consequently a correct program corresponding to a proof of the specification. The former item holds because it is possible to show that the terms are strongly normalizing, while the latter item is true because the terms form realizers for types.

1.1.3 State-of-the-art proofs-as-programs

There are practical limitations to the use of the constructive type theories used in naïve approaches. These concern the efficiency and usability of resulting programs: terms of a constructive type theory generally contain computationally irrelevant information, and have types that are only representable in experimental programming languages.

Commonly used functional programming languages, such as *SML*, *Haskell* or *Scheme* do not have dependent sum and product type constructors. But these constructors are essential for defining types that correspond to first-order and many-sorted formulae such as $\forall x : t \bullet \exists y : s \bullet A(x, y)$. Consequently, to execute a realizing, inhabiting term, a custom-built compiler or interpreter for the type theory must be written. This is the situation for the *Nuprl* and *Coq* systems. Currently these implementations are not bytecode compilers but are rather interpreters encoded within a conventional functional programming language, which is, in turn, interpreted rather than compiled. For larger scale practical programming problems, this can result in inefficient code that is not reusable or maintainable.

A further problem is that lambda terms corresponding to intuitionistic proofs often encode irrelevant, non-constructive information. Such irrelevant information is introduced when proving Harrop formulae [Har60].

Example 1.4. The atomic formula $y = 2 * x$ is Harrop. Given an intuitionistic proof of $\forall x : int \bullet \exists y : int \bullet y = 2 * x$, we might form a corresponding term p in a Martin–Löf type theory of the form

$$\lambda x : int.(2 * x, q) : \forall x : int \bullet \exists y : int \bullet A(x, y)$$

The number $2 * x$ is the witness term for the y in the existential statement. The term p denotes the proof that $2 * x$ can stand for y in $y = 2 * x$ and give a true statement. The witness term is the constructive information in the proof and, consequently, is of interest to us. The term q is irrelevant from a computational perspective (but, of course, relevant from a logical view).

To solve these problems, later proofs-as-programs approaches distinguish between proofs of specifications and the programs that are ultimately obtained. We refer to these approaches as state-of-the-art (SOA) proofs-as-programs.

These approaches still use the Curry–Howard isomorphism for representing proofs within a constructive type theory. However, they do not treat the constructive type theory as a programming language. Term simplification is not identified with program execution but, instead, only with simplification of the corresponding proof.

SOA approaches synthesize programs of a commonly used functional programming language, such as *SML*, *Scheme* or *Haskell*. This is done by means of an extraction map from proofs (terms of the constructive type theory) to programs (of the functional language). The resulting programs satisfy the proved formulae (type of the term) according to a specialized notion of realizability that takes into account redundant non-constructive information and the fact that reasoning and synthesis are done in two separate languages.

In [PC01], Crossley and Poernomo defined a specialized notion of realizability to hold between *SML* programs and many-sorted formulae. This book will use and extend this notion of realizability.

Example 1.5. According to this definition of realizability, the *SML* program

$$\texttt{fn x : int => } 2 * \texttt{x}$$

is a realizer of

$$\forall x : int \bullet \exists y : int \bullet y = 2 * x$$

This program is more optimal than the corresponding proof-term in Martin–Löf type theory (it is smaller), and contains no redundant non-constructive information (no information corresponding to proofs of Harrop formulae).

Example 1.6. Under this notion of realizability, a *SML* program p is a realizer of

$$\exists x : int \bullet Prime(x)$$

with $Prime(x)$ a Harrop formula, provided that p can be executed to give an answer a that can be represented as a witness a with $A[a/x]$ being provable.

The relation between constructive proofs and extracted programs obey the following diagram (based on an observation by Anderson [And93, p. 36] and modified by Iman Poernomo and John Crossley in [PC01]). Let L denote a constructive type theory, C a target programming language and extract be the extraction map between the two languages.

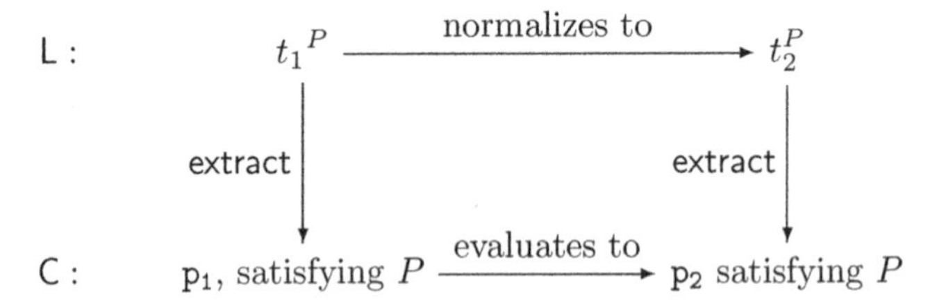

where satisfaction of a specification P is defined by the specialized notion of realizability.

The advantage of a SOA approach over a naïve approach is twofold: programs are optimized for execution and readability, and are implementable in a commonly used functional programming language. Consequently these methods produce optimized programs that are easier to understand and use by programmers who have no knowledge of constructive type theory but who require programs that are correct for a given specification.

While the basic idea is the same between authors, the target programming language, and the application and definition of extraction and specialized realizability differ between authors. Nordström and Petersson were among the first authors to advocate a separation between constructive type theory and programs for the purposes of optimal extraction [NP83]. In [Sch82, Sch85], Schwichtenberg defines an optimizing extraction map from intuitionistic proofs to functions in a simply typed lambda calculus which can then be transformed into *Scheme* programs. Sasaki [Sas86] provides an optimizing method of extraction for the *Nuprl* system (based on Martin–Löf predicative type theory), mapping intuitionistic proofs to programs in the subset of *Nuprl* corresponding to the simply typed lambda calculus. A different optimizing extraction map is given by Paulin–Mohring in [PM89] for transforming proofs of the Calculus of Constructions into terms of Girard's F_ω (a superset of the simply typed lambda calculus [Gir72]). In [PMW93], Paulin-Mohring and Werner showed how this method can be adapted to synthesize programs in *ML*.

The work at Monash University by John Crossley and Iman Poernomo has defined an SOA approach that uses a constructive type theory for many-sorted logic with extraction into *SML* [PC01, CP01, JPBC03]. In Chapter 2 of Part II, we will provide the details of this approach. In later chapters, we will generalize and adapt this SOA approach to different logics and other programming languages. So, when we claim to adapt the proofs-as-programs paradigm, we mean that we adapt this SOA approach. However, we argue that this approach is typical and our results therefore give a fair generalization of the other SOA methods mentioned here.

1.1.4 Related methods

There are several important related deductive program synthesis methods based in intuitionistic logic. Similar to SOA proofs-as-programs, these methods differentiate between programs and proofs, and synthesize programs by ignoring redundant, non-constructive proof information.

Hayashi and Nakano's system *PX* — described in [HN88, Hay90] — is based on Feferman's theory of functions and classes [Fef79]. The *PX* system can be used as a constructive logic, and is equipped with an optimizing extraction map from proofs of specifications to untyped *LISP* programs. The system itself is untyped, but can be used as a foundational framework for constructive type theories.

The deductive synthesis methods developed by Manna and Waldinger are based in constructive logic — see, e.g., [MW91]. These methods use a special tableaux style presentation to develop proofs and optimized programs in tandem. This is in contrast to the SOA approaches, where the program is extracted after a proof is complete. The method uses the same notion of a program's correctness with respect to a formula as a specialized notion of realizability. This work was implemented at NASA with the *Amphion* system. The work of Bundy in proof-plans [KBB93] is a development of this work, offering a similar means of program synthesis with some improvements to the tableaux style reasoning.

In a wider context, there are many logic-based approaches to program synthesis. The proofs-as-programs approaches we have described are all interactive (semi-automated), by virtue of the fact that they involve proof goals in predicate logic (which can never be fully automatically derived, by Gödel's incompleteness theorem).

However, automated deductive synthesis is what occurs in high-level logic programming languages such as *Prolog* and automatic theorem provers such as the Boyer–Moore prover [BM79] or *OTTER* [McC92].

Tyugu devised an automatic approach to synthesis based in constructive logic, implemented in the *NUTS* system [Tyu88, MT98]. This uses a type theory corresponding to propositional intuitionistic logic — essentially the simply typed lambda calculus with disjoint unions. It has the advantage that, under certain constraints, proofs are decidable, and so the inhabitation of a type is decidable. The usefulness of the type theory comes from the idea that simple types can correspond to a more detailed specification of term (program) behavior (such as the primality of an integer), in contrast to the usual typing by a sort name (such as being an integer). Because of the decidability of the subset of propositional proofs considered, this kind of program synthesis is automatic.

Finally, we note that deductive synthesis contrasts with two alternative approaches to logic-based synthesis: the transformational and inductive. In transformational synthesis, a program is derived stepwise from a specification by means of transformations or refinements. Refinement calculi (see, e.g., [Dij76, MV93, Mor94, Bac80]) achieve transformational synthesis through languages that mix non-executable specifications and programs. Related techniques (see, e.g., [HHS85]) have been employed to obtain structured programs from both model-oriented specifications (such as *B* specifications [Abr96, pp. 501–550]), and from structured algebraic specifications (such as *OBJ* [FD88, GWM$^+$00] or *CASL* [CoF01]). In inductive synthesis, a program is built on the basis of a declaration of input-output requirements or examples of input-output pairs. Examples of methods that fall into this category include inductive logic programming [Plo71, Mug92] and neural and belief networks [RN95, pp. 563–597].

1.2 Generalizing constructive synthesis

We have surveyed proofs-as-programs approaches to the synthesis of functional programs from constructive proofs.

An interesting and largely unexplored area of research concerns how proofs-as-programs can be adapted to different contexts (programming paradigms and logics other than functional programming and constructive intuitionistic logic).

1.2.1 Research on adaptation

Throughout the 1990s, research on adapting proofs-as-programs largely focused on two areas: synthesis of functional programs from classical proofs and synthesis of functional languages with catch-and-throw exception mechanisms.

Schwichtenberg and Berger developed a method for synthesis of functional *Scheme* programs from classical proofs [BS93, BS95b, BS95a, Sch99b]. Their method is an extension of a SOA method for extracting programs from intuitionistic proofs about functionals. This work involves a translation of the classical derivation to an intuitionistic proof, followed by extraction of an optimized program that can be executed in *Scheme*. A comparable approach is that of Murthy [Mur91], which uses a less elegant refinement translation and yields less optimal programs.

These methods develop functional programs by adapting SOA methods. Also, some interesting work has been done in adapting *naïve* proofs-as-programs to classical logic. These results show that certain logics, given a type theoretic presentation, correspond directly a kind of functional programming language with catch-and-throw control mechanisms. Classical logic is used to achieve this in [Gri90, Par93].

Interesting work was done in [Nak94], where Nakano defines a new logic with connectives that enables explicit reasoning about catch-and-throw mechanisms. This work is similar in philosophy to this monograph — defining new logics for new programming paradigms, with an adaptation of the Curry–Howard isomorphism and proofs-as-programs for program extraction from proofs. In related work, Sato examines the relation of catch-and-throw-mechanisms to classical and intuitionistic deduction in [Sat97].

1.2.2 Generalizing constructive synthesis

Little work has been done on defining what *should* constitute an adaptation to general cases of logics and programming languages. An important part of our project is therefore concerned with identifying a framework that generalizes a specific constructive program synthesis approach over a range of possible logics and programming language paradigms.

Our framework, which we call the *Curry–Howard protocol*, generalizes SOA approaches to proofs-as-programs. It assumes a Curry–Howard style isomorphism to hold between a given logic and type theory. Following an important property of SOA approaches, our protocol requires an optimization and extraction strategy from proofs to programs. Programs are elements of a separate programming language, not part of the logical type theory.

The rest of this monograph will then concern applying this protocol to adapt proofs-as-programs for imperative program synthesis, and to reasoning with structured algebraic specifications and structured program synthesis.

1.3 Imperative programs

Imperative programs produce results by manipulating values stored in a computer's memory: producing *side-effects*. This is done by executing sequences of individual statements that are determined by iterative and conditional commands. These programs are in contrast to pure functional programs, which do not involve changes of state.

For both historical and practical reasons, imperative programming dominates industry development. It is rare to see purely functional programs employed in industrial applications.

Also, many functional languages, such as *LISP* and *SML*, are not pure and offer imperative constructs. In many imperative and object-oriented languages currently used in industry, such as C++, Java, C# or Visual Basic, it is possible to program in the functional style with simulated higher-order anonymous functions. These languages sometimes utilize side-effect-free functionals to access data and provide views of state [Mey00].

Pure functional and imperative programming styles can be, and often are, mixed. It is therefore of value to provide a formal means for reasoning about, and synthesizing, correct programs that involve both imperative and pure functional aspects. Hoare logic is an established approach to reasoning about, and synthesizing, imperative programs. By considering a constructive variant we can use proofs-as-programs techniques to synthesize imperative *SML* programs that involve complicated pure, functional return values. This will form one of the main concerns of this book: proofs-as-imperative-programs.

1.3.1 Hoare logic

One of the most important ideas in formal software development is the Hoare calculus, first described in [Hoa69]. This is a deduction system for simultaneously constructing and reasoning about imperative programs, based on a semantics for programs due to Floyd [Flo67].

The idea is to specify a program in terms of its side-effects via pre- and post-conditions. Both conditions are usually formulae with special variables that denote the state of a computer's memory. The post-condition explains how the program execution should affect the state of memory, assuming the pre-condition was true prior to execution.

The Hoare calculus involves rules that show how to obtain imperative programs that satisfy such specifications. An example of a theorem is

$$\vdash \{Even(g)\}\mathtt{g} := !\mathtt{g} + \mathtt{1}\{Odd(g)\}$$

The middle term is a *SML* program, while the left-hand bracketed term is a pre-condition and the right-hand term is a post-condition. In this example, the theorem tells us that, assuming that the state **g** is an even number, after executing the addition program, we have an odd number.

Hoare logic has had successful application in the development of imperative programs. Hoare logic has been extended to the specification and construction of nondeterministic, parallel and distributed programs [Hoa85, Har84, HHH$^+$87] and object-oriented programs [AL97]. The state-based specification of an imperative program via pre- and post-conditions forms the basis of model-based specification languages such as the *B* method [Abr96]. The object constraint language (*OCL*) part of *UML* [WK98] is based on the notion of pre- and post-condition specifications. These specifications also form the basis of the real-time assertion checking system of the Eiffel programming language [Mey97].

We will limit ourselves to a version of the basic Hoare logic for imperative *SML* programs with while loops.

1.3.2 Synthesis of imperative programs with functional return values

The presence of side-effects is what distinguishes the imperative paradigm from the functional one. However, side-effect-free functions are also important in imperative programs because they enable access to data, obtaining views of state and producing return values. For instance, the *SML* program

$$\mathtt{s} := 10; !\mathtt{s} * 2$$

involves a side-effect producing assignment statement, $\mathtt{s} := 10$, followed by a side-effect-free term $!\mathtt{s} * 2$, which will evaluate to a return value.

Hoare logic is good for reasoning about and developing side-effect producing aspects of imperative programs. However, there are some inadequacies in the traditional means of using logic for developing side-effect-free aspects of imperative programs. Commonly, Hoare logic relies on the user directly constructing a required side-effect-free function along with a proof of its required property. Return values and state views are specified by a post-condition that associates the return value itself with a special designated variable (see, for instance, [Abr96, pp. 240–241]).

Example 1.7. We can associate the variable *return* with a required return value. When the variable is mentioned in the pre- and post-condition of a Hoare triple, it denotes the return value of a program's side-effect is equivalent to the program of the triple. So, the triple

$$\{\}\mathtt{s} := 10\{return = s * 2 \wedge Even(return)\}$$

describes a *SML* program $\mathtt{s} := 10; !\mathtt{s} * 2$ which returns an even number.

The problem with this approach is that the user is required to explicitly define a side-effect-free function while proving properties about it.

As we have mentioned, in many imperative languages, return values can potentially take the form of complex functional programs that are difficult to synthesize using the usual approach in Hoare logic. Proofs-as-programs has had success in the synthesis of such programs. Constructive methods hide implementation details from the designer, permitting a functional program to be developed solely through reasoning about domain knowledge.

We would prefer to be able to hide the details about the definition of a return value so the designer need not think about the way the return value or view of state is to be coded, and focus instead on manipulation of domain knowledge. It is therefore of interest to see how to adapt proofs-as-programs to the Hoare logic, and to combine imperative construction and the functional synthesis of return values.

In constructive program synthesis, a proof of a statement can be used to synthesize a realizer of the statement. The realizer is a functional program that satisfies the statement as a specification. For example, an existential statement

$$\exists x : int \bullet Even(x)$$

can be used to synthesize a functional program that returns a witness value p such that $Even(p)$ is provable. Because the realizer is synthesized from a proof, the details about its definition are hidden from the prover. The prover need only be concerned with using logic to reason about a problem, not with the definition of a program.

We will adapt this property to the specification of return values.

Example 1.8. For instance, given a constructive proof of the theorem

$$\{\}\texttt{s} := \texttt{10}\{\exists x : int \bullet Even(x)\}$$

we want to synthesize a program of the form

$$\texttt{s} := \texttt{10};\texttt{f}$$

with a side-effect-free function `f` that realizes the existential post-condition

$$\exists x : int \bullet Even(x)$$

with a value that is a witness for the x. An example of such a realizing return value function might be $!\texttt{s} * 2$.

In our methods the user does not need to manually code the return value, but instead the Hoare logic is used to prove the theorem from which the return value is extracted, and incorporated with the side-effect producing part of the theorem to give a final program.

The adaptation is not trivial, as, unlike functional program synthesis, our specifications involve initial and final values of state and our extracted side-effect-free functions can involve state references.

1.3.3 Alternative approaches to synthesis

We identify four distinct attempts at using proofs-as-programs notions in imperative program synthesis.

The first three involved encodings within ordinary constructive logic.

Filliâtre developed a denotational semantics of imperative programs within *Coq*, providing a means of transforming constructive content of proofs into monadic representations of imperative programs, and then finally into executable imperative programs [Fil99, Fil03].

A similar approach was taken separately by Manna and Waldinger for the synthesis of imperative *LISP* code [MW87], by encoding a semantics of imperative programs within their deductive synthesis system [MW91].

Stark and Ireland use a straightforward metalogical encoding of a Hoare-like logic within constructive logic that then facilitates interactive theorem proving, using tools based in constructive type theory [SI98a, SI98b].

Finally, Bellot developed a logic based on Girard's linear logic, for defining requirements of imperative programs [BR90, BCR+99]. This work might be characterized as an adaptation of naïve proofs-as-programs because it defines a logical type theory that can also be understood as an imperative programming language, in the same way that naïve methods view constructive type theory as a functional programming language. The novelty of his work is that the logic, represented as a type theory, corresponds to a programming language with imperative qualities. In particular the normalization of a proof — and the corresponding operational semantics of the proof-term — is ordered and sequential in an imperative sense. A new form of realizability is defined to hold between specifications and imperative programs, corresponding to inhabitability of types in the type theory. A problem with this logic and programming language is that they are nonstandard and difficult to understand and use for a new user.

None of these methods can be said to adapt state-of-the-art proofs-as-programs techniques. In particular our work is unique amongst these approaches, such that we use Hoare logic to develop side-effects of imperative programs and constructive methods to develop side-effect-free return values.

Popular alternatives to the Hoare logic for imperative program synthesis are refinement calculi. Refinement calculi achieve synthesis through languages that mix non-executable specifications and programs — see, e.g., [Dij76, MV93, Mor94, Bac80]. These calculi provide rules for refining non-executable specifications into executable terms that satisfy the given specification. Repeated recursive application of rules over a term with non-executable subterms will eventually yield an executable term. Related techniques (see, e.g., [HHS85]) have been employed to obtain structured programs from both model-oriented specifications (such as specifications in the *B* language [Abr96, pp. 501–550]). Refinement calculi have the advantage over Hoare logic of being scalable, and of appearing more like programming languages.

However, the problem of specifying and synthesizing return values is usually dealt with in a similar way to that of the Hoare logic, using specially designated return value symbols in specifications. The developer is required to explicitly define a side-effect-free function during the refinement process. It is an open question as to whether proofs-as-programs ideas could be incorporated into refinement calculi for return value synthesis.

1.4 Structured specifications and programs

An important research area in software engineering is that of scalable methods for modular program specification and correct development. The need for provably correct, hierarchical designs and implementations of programs was voiced in the program of the 1968 NATO conference on Software Engineering [NR68, pp. 45–55, 181–186]. Since that time, various formal abstractions have been studied to provide modular programming and specification languages, methods of proving correctness of modular programs against their specifications, and methods for the synthesis of structured programs from specifications. A popular example of such a formal abstraction is structured algebraic specifications — see, e.g., [Wir90].

Structured algebraic specifications provide a data-centric view of a software system. They can be considered as a hierarchical means of defining abstract data types, enabling us to specify systematically the required functionality in terms of an algebraic theory. A basic theory consists of a signature — type, function, and predicate symbols — together with axioms that define the required behavior of the algebraic entities represented by the signature. A new theory can be built from an existing theory via structuring operators: for instance, by renaming its types and constants, by abstraction (forgetting some types and constants and perhaps renaming the rest), combining two theories, or parametrizing and instantiating. These theory-building operations allow large theories to be built in a flexible and well-structured fashion. Structuring operators facilitate specification according to compositional, divide-and-conquer principles.

There is a variety of specification languages available. The earliest specification language was *Clear* developed by Burstall and Goguen [BG77, BG80]. Significant and well-developed work is the provided by *OBJ2* and *OBJ3* systems [FGJM85, GWM$^+$00], which are based on order-sorted algebras. Sanella, Tarlecki and Wirsing developed *ASL* as a core language for developing specifications [Wir82, SW83, Wir86, ST88a].

Over the past few years, the CoFI group has defined a standard for algebraic specification, called *CASL* [CoF01], which incorporates ideas from previous work. *CASL* specifications are based on many-sorted, partial, first-order algebras. Many of the ideas from *ASL* and other specification languages have been incorporated into the *CASL* standard. Our work shall be concerned with *CASL*. However, because *CASL* resembles other systems, many of our results could easily be adapted to other systems.

We will be interested in *reasoning about* and *refining and extending* structured specifications. We will use program extraction to aid refinement and extension using proofs about specifications.

1.4.1 Reasoning about specifications

Reasoning about algebraic specifications is important for the purposes of understanding the consequences of a specification and ensuring that a specification meets the requirements of the domain being modelled.

It is possible to reason about a structured specification simply by using many-sorted logic. This is true because structured specifications can be collapsed into an equivalent single set of axioms with a signature [Wir91].[1]

To prove properties about the specification, we need only equip ordinary logic with the axioms, and reason using the signature. However, this has the disadvantage of not being compositional. A compositional proof system constructs a proof about a structured specification in a modular fashion, using knowledge about sub-specifications to derive knowledge about the composed specification. This promotes the desirable features of a divide-and-conquer approach and proof reuse.

Our project is concerned with extensions to the compositional proof systems for structured specifications defined by Martin Wirsing in [Wir91]. Similar systems have been investigated in [FC92, BCH99, HWB97]. The original idea has roots in work done on modular reasoning by Sannella [SB83]. This used an extension to the Edinburgh *LCF* theorem-proving system that permitted the construction of *Clear*-like specifications and provided inference rules and strategies for compositional proofs about structured theories.

In this approach, proofs are conducted in a fashion that mirrors the structuring of specifications. The user derives statements of the form

$$\textsc{Sp} \diamond P$$

where Sp is a structured specification, and P is a known truth about the specification. The system is compositional in that it enables the simultaneous composition of old specifications to form new specifications and the derivation of new truths from known truths.

For example, the translation operation $(\rho \bullet \textsc{Sp})$ permits us to rename the signature and axioms of a specification Sp using a signature morphism ρ to give a new specification with renamed symbols. If we consider a specification as specifying component requirements, the renamed specification can be considered as a means of wrapping the component requirements with a new interface. In Wirsing's system, a renaming rule permits the formation of a renamed specification:

$$\frac{\textsc{Sp} \diamond \rho^{-1}(P)}{(\rho \bullet \textsc{Sp}) \diamond P}$$

Besides constructing the new specification, the rule shows how to derive a truth P about the renamed specification from the previously known truth $\rho^{-1}(P)$. In the sense that the logic enables two things to be done — the construction of

[1] This follows from the normal form theorem which states that a specification is equivalent to a normal, non-structured, specification of an algebra.

new entities from old and the reasoning about the result — it resembles Hoare logic's treatment of programs and theorems.

The overall proof system is parametrized with respect to a first-order logic. Usually this is classical logic, but, in this book, we replace it with intuitionistic logic, to obtain a constructive system for compositional reasoning about specifications. Our motivation for this is to adapt the Curry–Howard isomorphism and state-of-the-art proofs-as-programs results in order to extract provably correct functions from proofs about specifications: *structured proofs-as-programs.*

1.4.2 Refinement of specifications

One of the main aims of algebraic specification is to provide a formal basis to support the systematic development of correct programs from specifications by means of verified refinement steps. Refinement is the process of transforming abstract specifications into more concrete ones. If the concrete specification is executable, then we consider it as a program that implements the specification.

Sannella and Tarlecki argue that refinement should proceed in a stepwise fashion, from abstraction to implementation, gradually enriching the original specification with more detail and incorporating design, architecture, and implementation decisions [ST88b, ST97, SST92]. Such decisions include choosing between alternative behaviors of functions, data representation, and structure. Stepwise refinement is important for large specifications, because it permits the designer to decide upon the implementation of different aspects of a specification at separate points in the development process.

1.4.3 Proofs-as-programs for function extraction and refinement

A proof system for reasoning about *ASL* specifications was developed by Wirsing, Peterreins and Crossley in [WCP98, Pet96]. That work developed the ideas of Wirsing's calculus of [Wir91], but was based in natural deduction, over which Curry–Howard terms could be provided to encode proofs. The calculus used classical deduction.

In this book, we will develop an intuitionistic version of that natural deduction calculus, using the *CASL* standard for expressing specifications. Curry–Howard terms are defined in a similar fashion to the classical calculus. However, because we use constructive deduction, we can adapt proofs-as-programs to enable the extraction of correct functions from proofs about specifications.

Example 1.9. For instance, we could derive a proof of the theorem

$$\textsc{Int} \diamond \forall x : int \bullet \exists x : int \bullet y > x \wedge Prime(y)$$

where $\textsc{Int}$ is a specification of the natural numbers, with the predicates $>$ and *Prime* given appropriate axioms. Using our methods, we could extract a realizing lambda term t such that

$$\text{I}_{\text{NT}} \diamond \forall x : int \bullet t > x \wedge Prime(t\ x)$$

is true. In this way, the term t is a function that satisfies the original theorem as a specification.

By virtue of our methods, these programs can then be consistently added back to a specification for correct extension. For example, the term t can be associated with a function f, with the equational axiom $\{f = t\}$ consistently extending the specification INT. This provides a formal means of designing structured specifications by consistent extension. This is another important result: structured proofs-as-programs.

Finally, we show how our techniques can be used to define processes for the synthesis of structured programs and the refinement for specifications. By deriving constructive proofs of the axioms for a function, we can extract an executable definition of the function. By repeating this process, we can achieve a stepwise development of full executable structured programs from a structured specification.

We are not the first to propose developing correct *SML* programs from structured specifications. Sannella and Tarlecki proposed a stepwise development process and designed *Extended ML* (see [KST97]) as a language for expressing specifications and *SML* programs by one single syntax. However, those techniques did not involve program extraction techniques.

Also, the technique used by Smith in the *SpecWare* system bears some similarity to ours [Smi93]. There he uses similar rule-based techniques to construct specification morphisms. Our technique differs from Smith's in both the specification-building operations and in the approach to program synthesis. Even though he uses program synthesis techniques, he does not involve constructive proofs-as-programs methods.

1.5 Overview

This book deals with the following concepts.

1.5.1 The Curry–Howard protocol

Some work has been done in providing analogous results to the Curry–Howard isomorphism and proofs-as-programs for other logical systems and programming languages, often in the domain of classical logic. We contribute a novel result to the field by identifying a general framework that generalizes the form such analogies should take over arbitrary logical calculi and programming languages. The Curry–Howard protocol provides the framework. It is useful because it can then be used as a guide for adapting proofs-as-programs to new contexts, such as imperative program synthesis.

1.5.2 Proofs-as-imperative-programs

For the most part, even when tackling logics other than intuitionistic logic, proofs-as-programs research has been concerned with the synthesis of side-effect-free programs. Little work has been done in adapting proofs-as-programs for imperative program synthesis.

Our method involves the synthesis of imperative programs from proofs of specifications. Specifications concern side-effects and side-effect-free return values at the same time, but with different treatments. Specifications of side-effects are given as assertions about initial and final states as in standard Hoare logic. Specifications of side-effect-free aspects are given by considering assertions as constructive specifications, with constructive realizers as return values. By conforming to the Curry–Howard protocol, we adapt proofs-as-programs (defining an extraction map from proofs in the Hoare logic) to imperative programs of *SML*. The resulting programs satisfy specifications of both side-effects and side-effect-free return values.

Also, by defining a constructive version of Hoare logic including proof-terms, we provide a type-theoretic description of Hoare logic that is useful for theorem-proving implementations.

1.5.3 Structured proofs-as-programs and structured program synthesis

We present a method, using a version of the logical system of [WCP98, Pet96], for obtaining *SML* programs from specifications written in the algebraic specification language *CASL*. These programs are provably correct.

The logical calculus adds structural rules corresponding to the standard ways of creating structured specifications as presented in *CASL*: translating, hiding signatures, taking unions of specifications and building structured and parametrized specifications.

We then adapt proofs-as-programs to this logic, applying the Curry–Howard protocol to extract programs from proofs in our logic. We show that these techniques lead to consistent extensions of specifications, and the stepwise development of structured code.

1.5.4 Book organization

The monograph is organized into parts, corresponding to the main contributions above.

- Part I, Chapter 1 is this introductory chapter.
- Part II introduces the Curry–Howard protocol:
  - Chapter 2 provides an example of SOA proofs-as-programs for the synthesis of functional programs.
  - In Chapter 3, we define the Curry–Howard protocol. We illustrate how it generalizes our SOA approach.

- Part III defines a method for imperative program synthesis by the application of the Curry–Howard protocol.
  - Chapter 4 presents our logic for reasoning about side-effect relations.
  - Chapter 5 provides some necessary properties of the logic. We show soundness and completeness and define the proof-theory necessary for applications of the protocol. We present a logical type theory, and show an analogous result to the Curry–Howard isomorphism.
  - Chapter 6 describes how the Curry–Howard protocol is applied to achieve the synthesis of imperative programs from proofs in our logic.
- Part IV develops a method for reasoning about structured specifications, the synthesis of functions by application of the Curry–Howard protocol and refinement of specifications.
  - Chapter 7 presents the logic for reasoning about structured specifications.
  - Chapter 8 provides proof-theory necessary for the application of the protocol. We present a logical type theory for our logic, and show an analogous result to the Curry–Howard isomorphism. We establish the Church–Rosser and strong normalization theorems.
  - Chapter 9 describes how the Curry–Howard protocol is applied to achieve the synthesis of correct functions from proofs of our logic and how specifications can be consistently extended by these functions.
  - Chapter 10 extends our results to generic, parametrized specifications as they are treated in the algebraic specification language *CASL* [CoF01].
  - Chapter 11 provides a methodology for using our results for the refinement of structured specifications into structured executable code.
- Part V, Chapter 12, offers concluding remarks and suggests directions for future research.

Part II

Generalizing Proofs-as-Programs

2

Functional Program Synthesis

We are now going to show how to synthesize *SML* functional programs from constructive proofs. The approach is a state-of-the-art (SOA) proofs-as-programs approach in the sense described in the previous chapter.

In order to do this we first define a constructive logic where formulae assert truths about a problem domain and specify required *SML* programs as modified realizers, following [Kre59, BS95a, Dil80]. We first give a simple indication of how we are going to do this. We shall use Skolem functions for the Skolem form of a formula.

Example 2.1. For example, the formula $A = \forall x : int \bullet \exists y : int \bullet x + 10 = y$ asserts that, for every integer x there is an integer y equal to x plus 10.

The Skolem form of A is written as

$$Sk(A) = \forall x : int \bullet x + 10 = (f_A\ x)$$

Any function that can be substituted for f_A is called a *modified realizer* of A.

So, besides being a statement about the integers, the formula A can also be considered as a specification of a modified realizer, a functional program **a** that evaluates to the Skolem function f_A in a proof

$$\vdash_{\mathsf{Int}} Sk(\forall x : int \bullet \exists y : int \bullet x + 10 = y)[a/f_A]$$

The *SML* program

```
fn x : int => x + 10
```

satisfies this specification, because, given an appropriate axiomatization of *SML* programs it is possible to derive

$$\vdash_{\mathsf{Int}} \forall x : int \bullet x + 10 = (fn\ x : int => x + 10)\ x$$

The major innovation of this presentation of proofs-as-programs can be stated as follows. In contrast to naïve approaches to proofs-as-programs, we differentiate between proofs and programs, using different languages for each.

Proofs are represented in a logical type theory: a constructive type theory whose type inference rules reflect the rules of the constructive logic according to the the Curry–Howard isomorphism [Cur34, How80]. Programs are from a different language: the simply typed lambda terms with disjoint unions and product types, written in a subset of *SML*. The two languages are related via a common signature so program values can be represented as terms of the logic and vice versa.

Synthesis of correct programs from proofs is done via an extraction map between proofs in the type theory and programs of *SML*, producing *modified realizers* for proved specifications.

In the next chapter we provide a framework for generalizing the approach given here to other programming paradigms and logics. We claim that the method of this chapter is essentially a simplification of alternative SOA approaches that often involve additional features that are not relevant to our purposes here. For instance, unlike some of the SOA approaches mentioned in the previous chapter, we do not use full higher-order logic. We could, however, easily extend our work to full higher-order logic. Consequently, we claim that our framework, as a generalization of this chapter, is also an adequate generalization of all SOA approaches.

This chapter proceeds as follows:

- Section 2.1 discusses abstract data types, signatures and well-formed many-sorted formulae to be used in our logic.
- Section 2.2 provides a summary of many-sorted intuitionistic logic.
- In Section 2.3 we describe the logical type theory for representing proofs in our logic, according to the principles of the Curry–Howard isomorphism.
- Section 2.4 discusses the subset of *SML* that we use to extract programs.
- Section 2.5 presents our notion of realizability and describes the extraction process.
- We illustrate our methods with a password checking system example in Section 2.6.
- Section 2.7 provides a discussion of our results.

Notation 2.1 *(List notation).* We will use the following notation throughout the text.

Lists of elements are represented as follows: $[a_1; \dots; a_n]$. We will use a bar above a symbol to denote a list — e.g., we can define $\bar{a}$ to be a list $[a_1; \dots; a_n]$.

Concatenation of lists is given by the :: operator. For example, if $\bar{a}$ is $[a_1; \dots; a_n]$ and $\bar{b}$ is $[b_1; \dots; b_m]$, then $\bar{a} :: \bar{b}$ is $[a_1; \dots; a_n; b_1; \dots; b_m]$.

2.1 Abstract data types

We shall be reasoning with a many-sorted predicate logic, parametrized with respect to an assumed specification of abstract data types. We need the concept of signatures to define this specification.

2.1.1 Signatures and lambda terms

Signatures define sorts and function and predicate symbols.

We use the standard definition of first-order signatures (we use that given in [CoF01, p. 3]) extended to include sorts closed under functional, disjoint union and product sorts and also a unit (single element) sort.

Definition 2.1.1 (Many-sorted signature with total functions). A *many-sorted signature* $\Sigma = \langle S, TF, P \rangle$ consists of:

- A set, S, of sorts. Sorts are generated from a set of *basic sorts*, $B(S)$ according to the following inductive definition. First, $B(S) \subseteq S$. Also, if s_1 and s_2 are in S, then so are
  — the function sort $(s_1 \to s_2)$
  — the product sort $(s_1 * s_2)$
  — the disjoint union $(s_1 | s_2)$
  We assume that $B(S)$ includes a special sort, called $Unit$.
- Sets $TF_{w,s}$ of total function symbols, for each *function profile* (w, s). A function profile (w, s) is a pair of words, consisting of a sequence of argument sorts $w \in S^*$ and a result sort $s \in S$. Constants are treated as functions with no arguments. The length of w is called the *arity* of function symbols in $TF_{w,s}$. We assume that $TF_{\emptyset,Unit}$ contains a unit symbol, written $()$ (this denotes the single inhabitant of the sort $Unit \in B(S)$).
- Sets P_w of predicate symbols, for each *predicate profile* w. A predicate profile consists of a sequence of argument sorts $w \in S^*$. The length of w is called the *arity* of predicate symbols in P_w. For each basic sort $s \in B(S)$, there is a distinguished equality predicate $=_s \in P_{ss}$.

Constants and functions are also referred to as *operations*. The symbols that identify operations and predicates may be overloaded, occurring in more than one of the above sets. Whenever there is ambiguity in sentences, function symbols f and predicate symbols P should be qualified by profiles, written $f_{w,s}$ and p_w respectively. When no ambiguity is present, these profiles can be omitted.

We define the *terms* for a signature $\Sigma = \langle S, TF, P \rangle$, $Term(\Sigma)$, as in Fig. 2.1. This includes the usual definition of terms for a signature freely generated over a set of term variables Var. However, we have extended the definition to include a lambda calculus (with lambda terms written in an *SML* style syntax). Note that the variables Var are assumed to be disjoint from the constants in TF.

Notation 2.2. In later sections, we will require another lambda calculus for representing proofs, distinct from the lambda calculus for a signature. To distinguish between the terms of the two calculi, we will refer to the terms for a signature as *individual terms*. However, when there is no confusion, we will simply refer to them as terms.

$a, b, c ::=$	elements of $Term(\Sigma)$
$f(a_1, \ldots, a_n)$	$f \in TF_{w,s}$, w of arity n and $(a_1, \ldots, a_n)$ is a (possibly empty) list of elements of $Term(\Sigma)$
x	a variable $x \in Var$
$Inl(a)$	in left
$Inr(a)$	in right
$match\ a\ with$ $Inl(x) => b$ $\mid Inr(y) => c$	match case, $x, y \in Var$
$fn\ x : s\ => b$	lambda abstraction, s in S
$(a\ b)$	application
(a, b)	pair
$fst(a)$	first projection
$snd(a)$	second projection

Fig. 2.1. Syntax terms of $Term(\Sigma)$.

We have the usual notions of free and bound variables of the lambda terms of $Term(\Sigma)$.

Definition 2.1.2 (Free and bound variables of $Term(\Sigma)$). Let x be any variable of Var, and t a term of $Term(\Sigma)$.

Then, x is *bound* in t if there is a subterm of t of the form

$$fn\ x : s => b$$

or

$$match\ a\ with\ Inl(x) => b \mid Inr(y) => c$$

or

$$match\ a\ with\ Inl(y) => b \mid Inr(x) => c$$

If x is not bound in t, then x is free in t. We write $BV(t)$ for the set of all bound variables of t, and $FV(t)$ for the set of all free variables of t. A program with no free variables is called *closed*.

We write *Closed*($Term(\Sigma)$) for the set of closed terms from $Term(\Sigma)$.

Terms of a signature Σ are associated with sorts according to the sort inference rules provided in Fig. 2.2. These are the standard rules for inferring sorts of the lambda calculus. They involve a sorting relation (:) between terms and sorts. An inference takes the form

$$\Gamma \vdash_{\Sigma} a : s \tag{2.1}$$

where Γ is a *context*, consisting of variables associated with sorts, of the form $\{x_1 : s_1, \ldots, x_n : s_n\}$. The inference's intended meaning is that the term a has the sort s, when its free variables $x_1, \ldots, x_n$ denote possible terms of sorts $s_1, \ldots, s_n$. If an inference of the form (2.1) can be made for a term a and sort s, we say that a is *well-sorted* with sort s *for context* Γ. If the context Γ can be determined with no ambiguity from examining a, we simply say a is well-sorted with sort s.

$$\frac{}{\Gamma, x : s \vdash_{\Sigma} x : s} \text{(Ass)}$$

$$\frac{f \in TF_{(s_1 \ldots s_n),s} \quad \Gamma_1 \vdash a_1 : s_1 \ldots \Gamma_n \vdash a_n : s_n}{\Gamma, \Gamma_1, \ldots, \Gamma_n \vdash_{\Sigma} f(a_1, \ldots, a_n) : s} \text{(Fn)}$$

$$\frac{\Gamma \vdash_{\Sigma} a : s_1}{\Gamma \vdash_{\Sigma} \mathit{Inl}(a) : (s_1 | s_2)} \text{(Union}_1\text{)} \qquad \frac{\Gamma \vdash_{\Sigma} a : s_2}{\Gamma \vdash_{\Sigma} \mathit{Inr}(a) : (s_1 | s_2)} \text{(Union}_2\text{)}$$

$$\frac{\Gamma_1 \vdash_{\Sigma} a : s_1 \quad \Gamma_2 \vdash_{\Sigma} b : s_2}{\Gamma_1, \Gamma_2 \vdash_{\Sigma} (a, b) : (s_1 * s_2)} \text{(Prod)}$$

$$\frac{\Gamma \vdash_{\Sigma} a : (s_1 * s_2)}{\Gamma \vdash_{\Sigma} \mathit{fst}(a) : s_1} \text{(Proj}_1\text{)} \qquad \frac{\Gamma \vdash_{\Sigma} a : (s_1 * s_2)}{\Gamma \vdash_{\Sigma} \mathit{snd}(a) : s_2} \text{(Proj}_2\text{)}$$

$$\frac{\Gamma, x : s_1 \vdash_{\Sigma} a : s_2}{\Gamma \vdash_{\Sigma} \mathit{fn}\, x : s_1 \Rightarrow a : s_1 \to s_2} \text{(Abs)}$$

$$\frac{\Gamma_1 \vdash_{\Sigma} a : s_1 \quad \Gamma_2 \vdash_{\Sigma} b : (s_1 \to s_2)}{\Gamma_1, \Gamma_2 \vdash_{\Sigma} (b\ a) : s_2} \text{(App)}$$

$$\frac{\Gamma_1 \vdash_{\Sigma} a : (s_1 | s_2) \quad \Gamma_2, x : s_1 \vdash_{\Sigma} b : s \quad \Gamma_3, y : s_2 \vdash_{\Sigma} c : s}{\Gamma_1, \Gamma_2, \Gamma_3 \vdash_{\Sigma} \mathit{match}\ a\ \mathit{with}\ \mathit{Inl}(x) \Rightarrow b \mid \mathit{Inr}(y) \Rightarrow c : s} \text{(Case)}$$

Fig. 2.2. Sort inference rules for terms of Σ.

2.1.2 Formulae

Many-sorted formulae, $WFF(\Sigma)$, for a signature $\Sigma = \langle S, TF, P \rangle$ are constructed according to the following definition, given with respect to the denumerable set of term variables, Var.

Definition 2.1.3 (Well-formed formulae of a signature). Let $\Sigma = \langle S, TF, P \rangle$ be a signature. The set of well-formed formulae for a signature, $WFF(\Sigma)$ is the least set containing

- every $Q(t_1, \ldots, t_n)$ where $Q \in P_{s_1 \ldots s_n}$ is a predicate symbol in Q and every t_i $(i = 1, \ldots, n)$ is a well-sorted lambda term of sort s_i,
- every formula $(A \wedge B)$ for $A, B \in WFF(\Sigma)$,
- every formula $(A \vee B)$ for $A, B \in WFF(\Sigma)$,
- every formula $(A \Rightarrow B)$ for $A, B \in WFF(\Sigma)$,
- every formula $\forall x : s \bullet F$ where $x \in Var_s$ and $F \in WFF(\Sigma)$,
- every formula $\exists x : s \bullet F$ where $x \in Var_s$ and $F \in WFF(\Sigma)$,
- the formula $\perp$.

We often write $\neg A$ for $(A \Rightarrow \perp)$.

Remark 2.1. Observe that our formulae $WFF(\Sigma)$ can involve lambda terms from $Term(\Sigma)$.

2.1.3 Specification of abstract data types

The results of this chapter are parametrized with respect to a specification of abstract data types,

$$\mathcal{ADT} = \langle \Sigma, Ax \rangle,$$

where Σ is a signature and $\mathcal{AX}$ is a set of formulae from $WFF(\Sigma)$.

Remark 2.2. Signatures are associated with a semantic *structure* through an interpretation function, over which, formulae can be determined to be true or false, in the usual sense. Thus our specification corresponds to a set of models — structures that are constrained to satisfy the axioms. We do not provide the details of semantics in this chapter as this is well-known and not essential to understanding our program extraction method. Instead, we shall simply refer informally to the "models of $\mathcal{ADT}$." In Part IV (Section 7.1 of Chapter 7), a more general treatment of signatures and structures will be given that will describe the semantics for specifications such as $\mathcal{ADT}$.

Remark 2.3. As we shall see in Section 2.4, in order to obtain correct programs from proofs, we assume that the signature and axioms correspond to a *SML* library, which must be loaded before executing our synthesized programs, of given functions that satisfy the specification $\mathcal{ADT}$.

2.2 Intuitionistic logic

We now introduce our many-sorted intuitionistic logic, in order to reason about $\mathcal{ADT}$. The calculus, Int, is given in a natural deduction presentation. The inference rules of the calculus may be divided into *basic rules* and *axioms and schemata*. The basic rules are the standard rules of intuitionistic logic for introducing and eliminating the connectives and quantifiers of many-sorted formulae. Axioms and schemata are used to assert extra-logical properties of $\mathcal{ADT}$ in proofs.

2.2.1 Judgements

We deal with *judgements* which we write in sequent form as

$$\Gamma \vdash A$$

where A is a formula and the *context*, Γ, is a set of assumption formulae. The intended meaning of the judgement is that *assuming Γ are true then A is also true.*

2.2.2 Basic rules

The basic, core rules of the deductive system are presented in Fig. 2.3.

Remark 2.4 *(Proof-tree notation).* The sequent format presentation of proofs is equivalent to a "tree" format presentation. The former preferred when space needs to be conserved, the latter preferred when the steps of a deduction need to be displayed clearly. A sequent $\vdash_{\mathsf{Int}} F$ is equivalent to the following tree format presentation:

We use the usual natural deduction notation for discarding assumptions for the ($\Rightarrow$-I), ($\exists$-E) and ($\vee$-E) rules. So, the ($\Rightarrow$-I) rule tells us that, given a proof tree for B from A, we may discharge A to give a proof tree for $(A \Rightarrow B)$:

$$\dfrac{\begin{matrix}[A]\\ \vdots \\ B\end{matrix}}{(A \Rightarrow B)}\ (\Rightarrow\text{-I})$$

We denote discharging the assumption A by square brackets, $[A]$. Similar remarks hold for the ($\exists$-E), ($\vee$-I_1) and ($\vee$-E) rules.

Remark 2.5 *(Substitution for individual variables).* As usual $A[t/x]$ denotes the result of substituting t for all free occurrences of x in A subject to avoiding clashes of variables, where t and x share the same sort.

Remark 2.6 *(Eigenvariable restrictions).* The conditions on the rules ($\forall$-I) and ($\exists$-E) are the usual eigenvariable restrictions.

Motivation for the rules of intuitionistic logic is well known. We merely provide motivation for several important rules as an illustration.

Assume that x, y are arbitrary variables of sort s from signature Σ, and that a and c are well-sorted terms of sort s.

$$\frac{}{A \vdash_{\mathsf{Int}} A} \text{(Ass-I)}$$

$$\frac{\Delta, A \vdash_{\mathsf{Int}} B}{\Delta \vdash_{\mathsf{Int}} (A \Rightarrow B)} (\Rightarrow\text{-I}) \qquad \frac{\Delta \vdash_{\mathsf{Int}} A \quad \Delta' \vdash_{\mathsf{Int}} (A \Rightarrow B)}{\Delta, \Delta' \vdash_{\mathsf{Int}} B} (\Rightarrow\text{-E})$$

$$\frac{\Delta \vdash_{\mathsf{Int}} A}{\Delta \vdash_{\mathsf{Int}} \forall x : s \bullet A} (\forall\text{-I}) \qquad \frac{\Delta \vdash_{\mathsf{Int}} \forall x : s \bullet A}{\Delta \vdash_{\mathsf{Int}} A[c/x]} (\forall\text{-E})$$

provided x is free in A

$$\frac{\Delta \vdash_{\mathsf{Int}} P[a/y]}{\Delta \vdash_{\mathsf{Int}} \exists y : s \bullet P} (\exists\text{-I}) \qquad \frac{\Delta_1 \vdash_{\mathsf{Int}} \exists y : s \bullet P \quad \Delta_2, P[x/y] \vdash_{\mathsf{Int}} C}{\Delta_1, \Delta_2 \vdash_{\mathsf{Int}} C} (\exists\text{-E})$$

where x is not free in C

$$\frac{\Delta \vdash_{\mathsf{Int}} A \quad \Delta' \vdash_{\mathsf{Int}} B}{\Delta, \Delta' \vdash_{\mathsf{Int}} (A \wedge B)} (\wedge\text{-I})$$

$$\frac{\Delta \vdash_{\mathsf{Int}} (A_1 \wedge A_2)}{\Delta \vdash_{\mathsf{Int}} A_1} (\wedge\text{-E}_1) \qquad \frac{\Delta \vdash_{\mathsf{Int}} (A_1 \wedge A_2)}{\Delta \vdash_{\mathsf{Int}} A_2} (\wedge\text{-E}_2)$$

$$\frac{\Delta \vdash_{\mathsf{Int}} A_1}{\Delta \vdash_{\mathsf{Int}} (A_1 \vee A_2)} (\vee\text{-I}_1) \qquad \frac{\Delta \vdash_{\mathsf{Int}} A_2}{\Delta \vdash_{\mathsf{Int}} (A_1 \vee A_2)} (\vee\text{-I}_2)$$

$$\frac{\Delta \vdash_{\mathsf{Int}} A \vee B \quad \Delta_1, A \vdash_{\mathsf{Int}} C \quad \Delta_2, B \vdash_{\mathsf{Int}} C}{\Delta_1, \Delta_2, \Delta \vdash_{\mathsf{Int}} C} (\vee\text{-E})$$

$$\frac{\Delta \vdash_{\mathsf{Int}} \bot}{\Delta \vdash_{\mathsf{Int}} A} (\bot\text{-E})$$

provided A is Harrop

Fig. 2.3. The basic rules of many-sorted intuitionistic logic, Int.

Remark 2.7. Rules ($\vee$-I$_1$) and ($\vee$-I$_2$) are understood as follows.

Consider first the rule for $\vee$ introduction on the left:

$$\frac{\Gamma \vdash_{\mathsf{Int}} A}{\Gamma \vdash_{\mathsf{Int}} (A \vee B)} (\vee\text{-I}_1)$$

This means that from a sequent $\Gamma \vdash_{\mathsf{Int}} A$ we may infer the sequent $\Gamma \vdash_{\mathsf{Int}} (A \vee B)$. Here we are weakening the conclusion to $(A \vee B)$.

Example 2.2. The rule ($\vee$-E) is most easily understood by its analogy to proof by cases. If we have a proof of C from A and also a proof of C from B then we get a proof of C from $A \vee B$.

Likewise, for the ($\exists$-E) rule, if we have a proof of $\exists x : s \bullet A$ and a proof of C from a proof of A with free variable y, then we can get a proof of C.

The ($\bot$-E) rule requires the following definition of *Harrop formulae*, which will also be used to define realizability in Section 2.5.

Definition 2.2.1 (Harrop). A formula F is a *Harrop formula* if it is

1. an atomic formula,
2. of the form $(A \wedge B)$ where A and B are Harrop formulae,
3. of the form $(A \Rightarrow B)$ where B (but not necessarily A) is a Harrop formula, or
4. of the form $(\forall x : s \bullet A)$ where A is a Harrop formula.

We write $H(F)$ if F is a Harrop formula, and $\neg H(F)$ if F is not a Harrop formula.

Remark 2.8. Note that we restrict the premise formula of ($\bot$-E) to Harrop formulae, for reasons to do with program extraction, described in Section 2.5 (Theorem 2.5.3 and Corollary 2.5.1).

However, this restriction does not affect the intuitionistic power of our calculus, by the next lemma.

Lemma 2.2.1. *The calculus* Int *with the rule*

$$\frac{\Gamma \vdash_{\mathsf{Int}} \bot}{\Gamma \vdash_{\mathsf{Int}} A} \ (\bot\text{-E})$$

provided A is Harrop, can be extended conservatively to include the usual rule ($\bot$-E) rule*

$$\frac{\Gamma \vdash_{\mathsf{Int}} \bot}{\Gamma \vdash_{\mathsf{Int}} A} \ (\bot\text{-E}^*)$$

for all formulae A.

Proof. We assume $\Gamma \vdash_{\mathsf{Int}} \bot$. We then proceed by induction on the construction of the formula A, to obtain the inference

$$\frac{\Gamma \vdash_{\mathsf{Int}} \bot}{\Gamma \vdash_{\mathsf{Int}} A} \ (\bot\text{-E}^*) \tag{2.2}$$

from the basic rules of Int.

If A is atomic then A is Harrop and we achieve 2.2 by an application of ($\bot$-E).

Suppose A is of the form $(B \wedge C)$ then, by the induction hypothesis, we have proofs

$$\begin{array}{c} \Gamma \vdash_{\mathsf{Int}} \bot \\ \vdots \\ \Gamma \vdash_{\mathsf{Int}} B \end{array} \quad \text{and} \quad \begin{array}{c} \Gamma \vdash_{\mathsf{Int}} \bot \\ \vdots \\ \Gamma \vdash_{\mathsf{Int}} C \end{array}$$

So, using ($\wedge$-I) we have

$$\frac{\begin{array}{c} \Gamma \vdash_{\mathsf{Int}} \bot \\ \vdots \\ \Gamma \vdash_{\mathsf{Int}} B \end{array} \quad \begin{array}{c} \Gamma \vdash_{\mathsf{Int}} \bot \\ \vdots \\ \Gamma \vdash_{\mathsf{Int}} C \end{array}}{\Gamma \vdash_{\mathsf{Int}} (B \wedge C)}$$

The remaining cases are similar. □

2.2.3 Axioms and schemata

We assume the presence of axioms and schemata that define knowledge about a problem domain and provide extra-logical constraints about the behaviour of signature terms.

Recall that the axioms of $\mathcal{ADT}$ are given by a set of $WFF(\Sigma)$ formulae, $\mathcal{AX}$. To use these axioms, we use an introduction rule of the form

$$\boxed{\frac{A \in \mathcal{AX}}{\vdash_{\mathsf{Int}} A}\ (\text{Ax-I})}$$

We permit a potentially infinite number of axioms in $\mathcal{AX}$ to be generated by schemata.

Definition 2.2.2 (General form of schema). A *schema* R parametrized over lists of predicates $\bar{X}$, terms $\bar{y}$ and sorts $\bar{Z}$, has the form

$$\frac{\Gamma_1 \vdash_{\mathsf{Int}} F_1 \quad \ldots \quad \Gamma_n \vdash_{\mathsf{Int}} F_n}{\Gamma \vdash_{\mathsf{Int}} F}\ (R[\bar{X};\bar{y};\bar{Z}])$$

When applying a schema, we must substitute actual predicates $\bar{F}$, terms $\bar{t}$ and sorts $\bar{S}$, to form a rule of the form

$$\frac{\Gamma_1 \vdash_{\mathsf{Int}} F_1[\bar{F}/\bar{X}][\bar{t}/\bar{y}][\bar{S}/\bar{Z}] \quad \ldots \Gamma_n \vdash_{\mathsf{Int}} F_n[\bar{F}/\bar{X}][\bar{t}/\bar{y}][\bar{S}/\bar{Z}]}{\Gamma \vdash_{\mathsf{Int}} F[\bar{F}/\bar{X}][\bar{t}/\bar{y}][\bar{S}/\bar{Z}]}\ (R[\bar{F};\bar{t};\bar{S}])$$

Remark 2.9. The schemata rules are to be considered as a metalogical device for generating axioms in $\mathcal{AX}$. This is possible when we consider each schema application of the form

$$\frac{\Gamma_1 \vdash_{\mathsf{Int}} F_1 \quad \ldots \quad \Gamma_n \vdash_{\mathsf{Int}} F_n}{\Gamma \vdash_{\mathsf{Int}} F}$$

to generate an axiom $((\Gamma_1 \Rightarrow F_1) \wedge \ldots \wedge (\Gamma_n \Rightarrow F_n)) \Rightarrow F$ in $\mathcal{AX}$. The generated axiom and the schema application are equivalent, because repeated application of ($\Rightarrow$-E) on the former simulates satisfaction of premises in the latter.

Induction schemata may be provided for the data types of Σ that can be generated by constructor functions. These schemata are defined in the usual fashion for a data type: to prove a statement over all elements of a type we show that the statement holds over the generation of the sort.

Example 2.3. For instance, assuming Σ has a sort of integers int, with all elements generated from the constant 0 by the operation $suc : int \to int$, then we have the induction schema

$$\frac{V[0/x] \wedge \forall y : int \bullet V[y/x] \Rightarrow V[suc(y)/x]}{\forall x : int \bullet V}\ (IndInt[[V]])$$

Remark 2.10. It is possible to treat induction more generally. This is easy in higher-order logic — see, e.g., Hayashi and Nakano [HN88] or Paulin-Mohrin [PM89, PM93] for two different approaches. In Chapter 7 of Part IV (pp. 243–244), we show how to treat induction generally for a range of sorts when reasoning with algebraic specifications. There we show how to generate induction schemata for sorts with constructors. Such techniques can also be employed for this logic, but we defer them until Part IV, where the use of structured specifications makes for a more systematic treatment and aids intelligibility.

Remark 2.11. Our notion of schemata provide a limited way of simulating second order logic — see, e.g., [Lei94, 279–285].

We give several standard schemata for reasoning about equality and disjoint unions in lambda terms. These are provided in Fig. 2.4. These schemata are to be considered as a means of generating axioms that will be assumed to be included in $\mathcal{AX}$. Because they define the usual notions of equality and properties of the lambda calculus, these schemata do not affect the consistency of $\mathcal{AX}$ and the fact that there is a model for them.

$$\frac{}{\vdash_{\mathsf{Int}} u =_s r \Rightarrow r =_s u} \text{(ref)}$$

where s is a basic sort

$$\frac{P[r/y] \wedge u =_s r}{P[u/y]} \text{(subst)}[[P];[u;r];[s]]$$

where u and r are well-sorted of *basic sort* s and y is the only free variable in P

$$\frac{\vdash_{\mathsf{Int}} \forall y_1 : s_1 \bullet P[\mathit{Inl}(y_1)/x] \wedge \forall y_2 : s_2 \bullet P[\mathit{Inr}(y_2)/x]}{\vdash_{\mathsf{Int}} \forall x : s_1|s_2 \bullet P} \text{(disj-ind)}[P;[s_1;s_2]]$$

$$\frac{\vdash_{\mathsf{Int}} \mathit{Inl}(u) = \mathit{Inl}(r)}{\vdash_{\mathsf{Int}} u =_{s_1} r} \text{(union=}_1\text{)}[[u;r];[s_1;s_2]]$$

where $\mathit{Inl}(u)$ and $\mathit{Inl}(r)$ are well-sorted terms of sort $(s_1|s_2)$

$$\frac{\vdash_{\mathsf{Int}} \mathit{Inr}(u) = \mathit{Inr}(r)}{\vdash_{\mathsf{Int}} u =_{s_1} r} \text{(union=}_2\text{)}[[u;r];[s_1;s_2]]$$

where $\mathit{Inr}(u)$ and $\mathit{Inr}(r)$ are well-sorted terms of sort $(s_1|s_2)$

$$\frac{}{\vdash_{\mathsf{Int}} \mathit{Inl}(u) = \mathit{Inr}(r) \Rightarrow \bot} \text{(union}\neq\text{)}[[u;r];[s_1;s_2]]$$

where u and r are well-sorted terms of sorts s_1 and s_2 respectively

Fig. 2.4. Equality schemata and schemata for reasoning about disjoint unions.

2.3 Logical Type Theory

Our intuitionistic calculus corresponds to a type theory, *LTT*, essentially a lambda calculus with dependent sum and product types. This correspondence is known as the Curry–Howard isomorphism. The idea is that proofs formed using the calculus can be represented as lambda calculus terms (called proof-terms) with formulae considered as types of terms. The rules of Fig. 2.3 and the axiom and schemata then correspond to type inference rules.

Remark 2.12. Note that these proof-terms and types define a theory *distinct and separate from* the terms and sorts of $Term(\Sigma)$. This is an important feature of our presentation of the isomorphism — we use different terms for different tasks. Proof-terms, $PT(\mathsf{Int})$, are used to represent proofs, while terms of $Term(\Sigma)$ are used to denote elements of a problem domain for reasoning about within the logic.

In fact, we will see that terms of $Term(\Sigma)$ also serve another purpose — to denote *SML* programs. Proof-terms, however, may not be used in this way. This is in contrast to some naïve proofs-as-programs approaches, such as that of Martin-Löf [ML84, ML85], where a single type theory is used to denote terms for predication in the logic, and proofs of the logic. (Appendix A, Section A.2, provides a brief overview of the type theories for which the Curry–Howard isomorphism holds. A comparison is made between the use of a single type theory to multiple type theories to denote terms for predication in the logic, proofs of the logic, and programs.)

2.3.1 Proof-terms

The proof-terms of the type theory, $PT(\mathsf{Int})$, are given in Fig. 2.5. The grammar uses a denumerable set of proof-term variables, $Var_{PT(\mathsf{Int})}$.

2.3.2 Basic type inference rules

The basic type inference rules of *LTT* are presented in Fig. 2.6. These correspond to the basic logical rules of Int (rather than the schemata or axioms).

Remark 2.13. For the sake of clarity, we equip $PT(\mathsf{Int})$ with two forms of lambda abstraction: abstraction over proof-term variables and abstraction over Σ term variables. This necessitates two forms of application. Abstraction ($\mathsf{abstract}\ x.\ a$) and application ($\mathsf{app}(a, b)$) correspond to applications of ($\Rightarrow$-I) and ($\Rightarrow$-E), respectively. In contrast, term abstraction ($\mathsf{use}\ i : t.\ a$) and term application ($\mathsf{specific}(a, v)$) correspond to applications of ($\forall$-I) and ($\forall$-E). It is possible to reformulate the Curry–Howard isomorphism to use a *single* lambda abstraction and application to correspond to introduction and elimination for both kinds of connective. This is, in fact, what is done in much of the proofs-as-programs literature — see, e.g., Martin-Löf [ML84, ML85], the *Nuprl* system [CMH86] or the *Coq* system [CH88]. However, to do this, the distinction between proof-terms and terms is not so obvious, and the resulting theories can be difficult

$a, b, c ::=$	$PT(\mathsf{Int})$, proof-terms of Int
x^F	proof-term with type superscript, $x \in Var_{PT(\mathsf{Int})}$, $F \in WFF(\Sigma)$
$\mathsf{Axiom}(F)$	axiom, $F \in WFF(\Sigma)$
$\mathsf{Schema}(N, [\bar{e}; \bar{F}; \bar{t}; \bar{s}])$	schema application, N the name of the schema, $\bar{F}$ a list of formulae from $WFF(\Sigma)$, $\bar{t}$ a list of terms from $Term(\Sigma)$, and $\bar{s}$ a list of basic sorts
$\mathsf{abstract}\ x.\ a$	abstraction
$\mathsf{app}(a, b)$	application
$\mathsf{use}\ i : t.\ a$	term abstraction, $i \in Var$, t a sort of Σ
$\mathsf{specific}(a, v)$	term application, $v \in Term(\Sigma)$
$\langle a, b \rangle$	pair
$\mathsf{fst}(a)$	first projection
$\mathsf{snd}(b)$	second projection
$\mathsf{inl}(a)$	in left
$\mathsf{inr}(b)$	in right
$\mathsf{case}\ a\ \mathsf{of}\ \mathsf{inl}(x).b,\ \mathsf{inr}(y).c$	case
$\mathsf{abort}(a)$	abort
$\mathsf{show}(v, a)$	witness, $v \in Term(\Sigma)$
$\mathsf{select}\ (a)\ \mathsf{in}\ y : t.x.b$	select, $y \in Var$, t a sort of Σ, $x \in Var_{PT(\mathsf{Int})}$

Fig. 2.5. Syntax of the proof-terms $PT(\mathsf{Int})$ for the calculus Int.

to understand and use for a novice. We employ two forms of abstraction and application to highlight the distinction between terms and proof-terms and to aid intelligibility.

2.3.3 Axioms and schemata

We use special proof-terms to designate application of axioms and schemata. We require that there be type inference rules for all axioms and schemata of Int.

Definition 2.3.1 (General form of type inference rules for axioms). Recall the axiom application

$$\frac{F \in \mathcal{AX}}{\vdash_{\mathsf{Int}} F}\ (\mathrm{Ax})$$

We use the proof-term $\mathsf{Axiom}(F)$ to denote an application of this rule in the logical type theory, with corresponding type formation rule

$$\frac{F \in \mathcal{AX}}{\vdash_{\mathsf{Int}} \mathsf{Axiom}(F)^F}\ (\mathrm{Ax})$$

x, y are arbitrary Σ variables of some sort s, and a is a term of arbitrary sort s.

$$\frac{}{x^A \vdash_{\mathsf{Int}} x^A} \text{(Ass-I)}$$

$$\frac{\Delta, x^A \vdash_{\mathsf{Int}} b^B}{\Delta \vdash_{\mathsf{Int}} \mathsf{abstract}\ x.\ b^{(A \Rightarrow B)}} (\Rightarrow\text{-I}) \qquad \frac{\Delta \vdash_{\mathsf{Int}} a^A \quad \Delta' \vdash_{\mathsf{Int}} p^{(A \Rightarrow B)}}{\Delta, \Delta' \vdash_{\mathsf{Int}} \mathsf{app}(p, a)^B} (\Rightarrow\text{-E})$$

$$\frac{\Delta \vdash_{\mathsf{Int}} p^A}{\Delta \vdash_{\mathsf{Int}} \mathsf{use}\ x : s.\ p^{\forall x:s \bullet A}} (\forall\text{-I}) \qquad \frac{\Delta \vdash_{\mathsf{Int}} p^{\forall x:s \bullet A}}{\Delta \vdash_{\mathsf{Int}} \mathsf{specific}(p, c)^{A[c/x]}} (\forall\text{-E})$$

$$\frac{\Delta \vdash_{\mathsf{Int}} p^{P[a/y]}}{\Delta \vdash_{\mathsf{Int}} \mathsf{show}(a, p)^{\exists y:s \bullet P}} (\exists\text{-I}) \qquad \frac{\Delta_1 \vdash_{\mathsf{Int}} p^{\exists y:s \bullet P} \quad \Delta_2, x^{P[z/y]} \vdash_{\mathsf{Int}} q^C}{\Delta_1, \Delta_2 \vdash_{\mathsf{Int}} \mathsf{select}\ (p)\ \mathsf{in}\ z.x.q^C} (\exists\text{-E})$$

$$\frac{\Delta \vdash_{\mathsf{Int}} a^A \quad \Delta' \vdash_{\mathsf{Int}} b^B}{\Delta, \Delta' \vdash_{\mathsf{Int}} \langle a, b \rangle^{(A \wedge B)}} (\wedge\text{-I})$$

$$\frac{\Delta \vdash_{\mathsf{Int}} p^{(A_1 \wedge A_2)}}{\Delta \vdash_{\mathsf{Int}} \mathsf{fst}(p)^{A_1}} (\wedge\text{-E}_1) \qquad \frac{\Delta \vdash_{\mathsf{Int}} p^{(A_1 \wedge A_2)}}{\Delta \vdash_{\mathsf{Int}} \mathsf{snd}(p)^{A_2}} (\wedge\text{-E}_2)$$

$$\frac{\Delta \vdash_{\mathsf{Int}} p^{A_1}}{\Delta \vdash_{\mathsf{Int}} \mathsf{inl}(p)^{(A_1 \vee A_2)}} (\vee\text{-I}_1) \qquad \frac{\Delta \vdash_{\mathsf{Int}} p^{A_2}}{\Delta \vdash_{\mathsf{Int}} \mathsf{inl}(p)^{(A_1 \vee A_2)}} (\vee\text{-I}_2)$$

$$\frac{\Delta \vdash_{\mathsf{Int}} p^{A \vee B} \quad \Delta_1, x^A \vdash_{\mathsf{Int}} a^C \quad \Delta_2, y^B \vdash_{\mathsf{Int}} b^C}{\Delta_1, \Delta_2, \Delta \vdash_{\mathsf{Int}} \mathsf{case}\ p\ \mathsf{of}\ \mathsf{inl}(x).a,\ \mathsf{inr}(y).b^C} (\vee\text{-E})$$

$$\frac{\Delta \vdash_{\mathsf{Int}} a^{\perp}}{\Delta \vdash_{\mathsf{Int}} \mathsf{abort}(a)^A} (\perp\text{-E})$$

The type inference rules require the same conditions for application as their corresponding logical rules given Fig. 2.3.

Fig. 2.6. The logical rules of our calculus presented as type inference rules.

Definition 2.3.2 (General form of type inference rules for schemata). Given a schema rule $R[\bar{X}; \bar{y}; \bar{Z}]$ from Int, where $\bar{X}$, $\bar{y}$ and $\bar{Z}$ are lists of variables ranging over formulae, terms and sorts, respectively:

$$\frac{\Gamma_1 \vdash_{\mathsf{Int}} F_1 \quad \ldots \quad \Gamma_n \vdash_{\mathsf{Int}} F_n}{\vdash_{\mathsf{Int}} F} R[\bar{X}; \bar{y}; \bar{Z}]$$

we define corresponding type formation schemata for proof-terms of the form

$$\mathsf{Schema}(R, [[q_1; \ldots; q_n]; \bar{X}; \bar{y}; \bar{Z}])$$

written

$$\frac{\vdash_{\mathsf{Int}} q_1^{F_1} \quad \ldots \quad \vdash_{\mathsf{Int}} q_n^{F_n}}{\vdash_{\mathsf{Int}} \mathsf{Schema}(R, [[q_1; \ldots; q_n]; \bar{X}; \bar{y}; \bar{Z}])^F} R[\bar{X}; \bar{y}; \bar{Z}]$$

Example 2.4. The (subst) schema of Fig. 2.4 corresponds to the following type inference schema

$$\frac{q_1[r/y]^{P[r/y]} \quad q_2^{u=_s r}}{\mathsf{Schema}(subst, [[q_1; q_2]; P; \bar{y}; \bar{Z}])^{P[u/y]}} \text{(subst)}[[P]; [u; r]; [s]]$$

2.3.4 The Curry–Howard isomorphism

Every proof-term that is well-typed according to the inference rules corresponds to an intuitionistic proof. This fact is known as the *Curry–Howard isomorphism*, formalized according to the following theorem.

Theorem 2.3.3 (Curry–Howard isomorphism). *Let $\Gamma = \{G_1, \ldots, G_n\}$ be a set of premises. Let $\Gamma' = \{x_1{}^{G_1}, \ldots, x_n{}^{G_n}\}$ be a corresponding set of typed proof-term variables.*

Then,

1. *Given a natural deduction proof of*

$$\Gamma \vdash_{\mathsf{Int}} A$$

   we can use the type inference rules to construct a well-typed proof-term p^A whose free proof-term variables are Γ'.
2. *Given a well-typed proof-term p^A whose free term variables are Γ', we can construct a natural deduction proof of $\Gamma \vdash_{\mathsf{Int}} A$.*

Proof. The proof of item 1 follows easily by induction on the structure of the deduction, D, and the definition of the typing rules and Int. The proof of item 2 follows similarly by induction on the structure of the deduction, p. □

2.3.5 Reduction rules

Because proof-terms are terms in a form of lambda calculus, they have *reduction rules* whose application corresponds to proof normalization by the Curry–Howard isomorphism.

1. $\mathsf{app}(\mathsf{abstract}\ X.\ a^{(A \Rightarrow B)}, b^A)$	$\rhd_{\mathsf{Int}}\ a[b/X]^B$
2. $\mathsf{specific}(\mathsf{use}\ x : s.\ a^{\forall x:s \bullet A}, v : s)$	$\rhd_{\mathsf{Int}}\ a[v/x]^{A[v/x]}$
3. $\mathsf{fst}(\langle a, b\rangle^{(A \wedge B)})$	$\rhd_{\mathsf{Int}}\ a^A$
4. $\mathsf{snd}(\langle a, b\rangle^{(A \wedge B)})$	$\rhd_{\mathsf{Int}}\ b^B$
5. $\mathsf{case}\ \mathsf{inl}(a)^{A \vee B}\ \mathsf{of}\ \mathsf{inl}(x^A).b^C,\ \mathsf{inr}(y^B).c^C$	$\rhd_{\mathsf{Int}}\ b[a/x]^C$
6. $\mathsf{case}\ \mathsf{inr}(a)^{A \vee B}\ \mathsf{of}\ \mathsf{inl}(x^A).b^C,\ \mathsf{inr}(y^B).c^C$	$\rhd_{\mathsf{Int}}\ c[a/y]^C$
7. $\mathsf{select}\ (\mathsf{show}(v, a)^{\exists y:s \bullet P})\ \mathsf{in}\ z.x^P[z/y].b^C$	$\rhd_{\mathsf{Int}}\ b[a/x][v/z]^C$

Fig. 2.7. The seven reduction rules that define $\rhd_{\mathsf{Int}}$.

There are seven rules that define the normalization process over proof-terms, which are given in Fig. 2.7. Each rule of Fig. 2.7 represents a possible proof simplification. These may be obtained by matching redundant applications of elimination and introduction rules.

For example, reduction 1 of Fig. 2.7 corresponds to deleting a (⇒-I) followed by a (⇒-E):

$$\cfrac{\cfrac{\begin{array}{c}\vdots\, x\\ A\\ \vdots\, a\\ B\end{array}}{(A \Rightarrow B)}\,(\Rightarrow\text{-I}) \qquad \begin{array}{c}\vdots\, b\\ A\end{array}}{B}\,(\Rightarrow\text{-E}) \quad \text{which reduces to} \quad \begin{array}{c}\vdots\, b\\ A\\ \vdots\, a[b/x]\\ B\end{array}$$

Similarly, reduction 7 of Fig. 2.7 corresponds to deleting a (∃-I) followed by a matching (∃-E):

$$\cfrac{\cfrac{\begin{array}{c}\vdots\, a\\ P[v/y]\end{array}}{\exists y : s \bullet P}\,(\exists\text{-I}) \qquad \begin{array}{c}P[z/y]\\ \vdots\, b\\ C\end{array}}{C}\,(\exists\text{-E}) \quad \text{which reduces to} \quad \begin{array}{c}\vdots\, b\\ P[z/y]\\ \vdots\, b[a/x][v/z]\\ C\end{array}$$

The left hand side of a reduction rule is called *redex* and right hand side of a rule is called the *reduct*.

We write

$$p \,\hat{\rhd}_{\mathsf{Int}}\, p'$$

when p' may be obtained from p by the transitive closure of $\rhd_{\mathsf{Int}}$. When $p \,\hat{\rhd}_{\mathsf{Int}}\, p'$ holds, then p' is obtainable from p by a sequence of replacements of subterms using the rules of Fig. 5.4. In this case, we say that p is *reducible* to p'.

Remark 2.14. Recall that we treat terms $Term(\Sigma)$ and proof-terms $PT(\mathsf{Int})$ as serving different purposes. This is reflected in the fact that the $Term(\Sigma)$ terms that are used in term application and in witness proof-terms, are not reduced by these reduction rules. For instance, the (one-step) normalization chain:

$$\mathsf{specific}(\mathsf{use}\ x.\ \mathsf{show}(x + x, p)^{\forall x:int\bullet\exists y:int\bullet 2*x=y}, 3)^{\exists y:int\bullet 2*3=y}$$
$$\rhd\ \mathsf{show}(3 + 3, p)^{\exists y:int\bullet 2*3=y}$$

continues no further. The term $(3 + 3)$ is treated as a constant. As we shall see, the term $(3 + 3)$ can only be "reduced" when evaluated in a separate programming language.

This is a point of deviation from naïve proofs-as-programs approaches, which usually treat proof-normalization and program evaluation as one and the same. For example, a Martin-Löf type theory has reduction rules for normalization and for evaluating data types. The equivalent of the above proof-term in a

Martin-Löf type theory would reduce the witness term to give a proof-term of the form $(6, p)$ — considered to be the return value of the program.

The state-of-the-art approach of this chapter is that reduction rules only correspond to normalization of proofs, while, separately, we employ extraction to obtain optimized *SML* programs from proofs. So, in contrast to naïve-proofs-as-programs, there is a clear separation of proofs from programs. This point will be important in the next chapter where we shall take such a separation as important to the correct generalization of our approach.

2.3.6 Strong normalization

The strong normalization property tells us that the normalization process over a calculus will always terminate. To show that this property holds over our calculus, we need to show that the proof-terms of $\mathsf{LTT}(\mathsf{Int})$ are strongly normalizable, according to the following definition.

Definition 2.3.4 (Strongly normalizable proof-terms). We say that a proof-term is *normal* if it contains no redex — that is to say, it is irreducible.

Given a proof-term t, we let $\mathcal{N}(t)$ denote the least upper bound of lengths of reduction sequences for the term t. We say that t is *strongly normalizable* if all reduction sequences are finite.

Remark 2.15. Clearly, if $\mathcal{N}(t)$ is finite, then t is strongly normalizable and, by König's Lemma, conversely.

Theorem 2.3.5. *Any term t of $PT(\mathsf{Int})$ is strongly normalizable.*

Proof. This is the proof of strong normalization for many-sorted first-order intuitionistic logic and the fact that proof-terms for axiom introductions are irreducible. For a proof, see [CS93]. □

Lemma 2.3.1. *Take any proof-terms a^A and b^B. If $a^A \rhd_{\mathsf{Int}} b^B$, then the type of a is the same as the type of B.*

Proof. By induction on the length of the proof-term, using the fact that if a proof-term a reduces to b by one of the rules of Fig. 5.4, then the type of a has the same type as b. □

Remark 2.16. By the Curry–Howard isomorphism of Theorem 2.3.3 and Lemma 2.3.1, if proof-term a is reducible to b, then both represent proofs of the same formula. In this sense, reducibility is a form of equivalence between proofs.

2.3.7 The Church–Rosser Property

The Church–Rosser property says that if a proof's normalization sequence can diverge, then eventually the divergent sequences will converge to yield the same proof.

Usual proofs of the property involve showing that the transitive closure of $\rhd_{\mathsf{Int}}$ satisfies the so-called diamond property.

Definition 2.3.6 (Diamond property). A relation # over a set S satisfies the *diamond property* when

$$\text{for all } x, x_1, x_2 \text{ in } S(x\#x_1 \text{ and } x\#x_2 \Rightarrow \text{ there exists a } x_3 \text{ such that } (x_1\#x_3 \text{ and } x_2\#x_3))$$

Theorem 2.3.7 (Church–Rosser property). *The relation* $\rhd_{\mathsf{Int}}$ *satisfies the diamond property (and therefore* $\hat{\rhd}_{\mathsf{Int}}$*, as the transitive closure of* $\rhd_{\mathsf{Int}}$*, satisfies the diamond property).*

Proof. Because our proof-terms are terms in a lambda calculus, with $\rhd_{\mathsf{Int}}$ a reduction relation over the lambda calculus, we can show this by the proof given in [Bar84, pp. 59–62], a proof due to Tait and Martin-Löf. □

2.4 Programs in *SML*

We will be extracting terms from *SML* programs from proofs represented in the *LTT*. For simplicity of presentation, instead of dealing with the full *SML* language specification [MTH90], we consider the subset corresponding to the simply typed lambda calculus. This enables us to consider a simple operational semantics for our programs.

Our subset of *SML* is exactly the terms, $Term(\Sigma)$, of Fig. 2.1. We use `typewriter` font to distinguish terms of $Term(\Sigma)$ when used as programs, as opposed to terms in formulae. These terms can be considered well-formed and well-typed *SML* programs, where

- We assume all programs are evaluated with respect to a *preamble* — a library consisting of data type and terminating function declarations.
- All the functions and constants from the signature Σ are defined in the preamble.
- Basic sorts correspond to assumed *SML* types that have been defined in the preamble.
- Functional and product sorts $t \to u$ and sorts $t * u$ are taken as functional and product *SML* types `t->u` and `t * u` respectively. For the purposes of clarity and to provide a relation to terms of $\mathcal{ADT}$, we continue to write disjoint unions in the form `t|u`. However, this is to be taken as shorthand for the correct syntax in *SML*,

$$(\mathtt{t}, \mathtt{u})\ \mathtt{disjointUnion}$$

an instantiation of the following parametrized *SML* data type

```
datatype ('a,'b) disjointUnion=Inl of 'a | Inr of 'b;;
```

We assume the parametrized data type is defined in the preamble, so that our disjoint unions are available to all programs that use the preamble.

Remark 2.17. The `match` construct of $Term(\Sigma)$ forms a valid *SML* matching over the disjoint union data type defined using `Inl` and `Inr` constructors.

The terms form a lambda calculus with the usual reduction rules. We consider these rules to provide a simple operational semantics for determining program evaluation. The semantics is given by the reduction relation $\triangleright_{SML}$, defined in Fig. 2.8.

In addition to the usual r reduction rule, we provide rules for projections and cases.

Also, we provide rules for applying function symbols to arguments. We assume that all function symbols of Σ correspond to functions in the *SML* preamble. When a function is applied to arguments of appropriate arities and types, *SML* should always evaluate the result to an answer value, which can be represented as another term of $Term(\Sigma)$. This assumption is formalized by assuming a mapping, *Eval*, that gives the return value for function applications. Given a function symbol $f \in TF_{s_1 \ldots s_n, s}$ and arguments $(a_1, \ldots, a_n)$ of sort $(s_1 * \ldots * s_n)$, $Eval(f(a_1, \ldots, a_n))$ returns a term from $Term(\Sigma)$ of sort s. The term $Eval(f(a_1, \ldots, a_n))$ is exactly the return value obtained by evaluating $\mathtt{f}(\mathtt{a_1}, \ldots, \mathtt{a_n})$ in *SML*.

Assumption 2.1. For the purposes of generality, we do not explicitly define *Eval* for the function symbols that occur in lambda terms. Instead, we assume that *Eval* is always defined to represent the definition of the function symbols in a loaded *SML* preamble. We assume that, because the preamble consists of terminating programs, the definition of *Eval* is such that repeated applications of $\triangleright_{SML}$ always terminate.

We write

$$\mathtt{a} \triangleright^*_{SML} \mathtt{b}$$

if `b` can be obtained from `a` by one or more applications of the rules. We write $\hat{\triangleright}_{SML}$ for the transitive closure of $\triangleright_{SML}$, and say that `a` *evaluates* to `b` if

$$\mathtt{a}\ \hat{\triangleright}_{SML}\ \mathtt{b}$$

Remark 2.18. A standard *SML* compiler is equipped with a denotational semantics that is compatible with our rules [MTH90]. This is true because *SML* was designed to incorporate the simply typed lambda calculus.

$$
\begin{array}{r l}
(\texttt{fn x : s => p})\ \texttt{a} \rhd_{SML} & \texttt{p[a/x]} \\
\texttt{match Inl(a) with inl(x) => b | inr(y) => c} \rhd_{SML} & \texttt{b[a/x]} \\
\texttt{match Inr(a) with inl(x) => b | inr(y) => c} \rhd_{SML} & \texttt{c[a/y]} \\
\texttt{fst((a,b))} \rhd_{SML} & \texttt{a} \\
\texttt{snd((a,b))} \rhd_{SML} & \texttt{b} \\
\texttt{f}(\texttt{a}_1,\ldots,\texttt{a}_n) \rhd_{SML} & Eval(f(a_1,\ldots,a_n))
\end{array}
$$

Fig. 2.8. The operational semantics of our fragment of SML.

Because *SML* terms can be used both as programs and as terms in the logic, it is desirable to be able to reason about program evaluation using the logic. This is achieved by adding the following schema to our calculus

$$\frac{\texttt{a} \rhd^*_{SML} \texttt{b}}{\vdash_{\mathsf{Int}} a =_s b}\ (\text{red=})[[\text{a,b}];[\text{s}]]$$

where $a : s$ and $b : s$ are well-typed (well-sorted) terms of both *SML* and $\mathcal{ADT}$ (note the change in font to denote the respective uses). This schema permits us to treat reducible terms as equivalent according to an equality $=_s$ relation. In this way the logic can correctly reason about the evaluation of program terms.

Assumption 2.2. We make the following assumption. When considered as *SML* programs, functions always evaluate in a way that is consistent with their specification given by $\mathcal{AX}$. That is to say, we require that the definitions of *Eval*, $\rhd_{SML}$ and the addition of the (red=) rule to $\mathcal{AX}$ still yield consistent models for $\mathcal{ADT}$. In particular we consider our subset of the *SML* language together with the preamble to form a model of $\mathcal{ADT}$. This model is of main interest to us, because we are primarily interested in using $\mathcal{ADT}$ to reason about our *SML* programs.

2.5 Program synthesis

We are now able to show how to synthesize correct *SML* programs from intuitionistic proofs. Our method follows the principles of state-of-the-art (SOA) proofs-as-programs. We define an optimising extraction map. Given a proof-term in the *LTT* corresponding to a proof of a specification, the map produces a program in *SML* that satisfies the specification.

The way by which a formula specifies a program is formalized by a notion of *modified realizability* between *SML* programs and formulae. Following

F	$\mathsf{etype}(F)$
$P(\bar{a})$	$Unit$
$(A \wedge B)$	$\begin{cases} \mathsf{etype}(A) & \text{if not } H(B) \\ \mathsf{etype}(B) & \text{if not } H(A) \\ \mathsf{etype}(A) * \mathsf{etype}(B) & \text{otherwise} \end{cases}$
$(A \vee B)$	$\mathsf{etype}(A) \vert \mathsf{etype}(B)$
$(A \Rightarrow B)$	$\begin{cases} \mathsf{etype}(B) & \text{if not } H(B) \\ \mathsf{etype}(A) \to \mathsf{etype}(B) & \text{otherwise} \end{cases}$
$\forall x : s \bullet A$	$s \to \mathsf{etype}(A)$
$\exists x : s \bullet A$	$\begin{cases} s & \text{if } H(A) \\ s * \mathsf{etype}(A) & \text{otherwise} \end{cases}$
$\perp$	$Unit$

P is an atomic predicate.

Fig. 2.9. Definition of etype.

similar SOA approaches, our extraction map involves removal of nonconstructive information from proof-terms, and type simplification, to transform logical proof-terms into realizing *SML* lambda terms with simple types.

2.5.1 Modified realizability

We use a notion of modified realizability between our formulae and programs of *SML*, based on that given in [Kre59, BS95a, Dil80]. Essentially, an *SML* program is a modified realizer of a formula if it can act as a required *Skolem function* for the *Skolem form* of a formula. We define these concepts now.

We first need to define a sort extraction map xsort from formulae to sorts of Σ. This is given by Fig. 2.9.

Then we define the Skolem form of formulae.

Definition 2.5.1 (Skolem form and Skolem functions). Given a closed formula A, we define the *Skolem form* of A to be the Harrop formula $Sk(A) = Sk'(A, \emptyset)$, where $Sk'(A, AV)$ is defined as follows.

A unique function letter f_A, called the *Skolem function*, is associated with each such formula A, of sort $\mathsf{etype}(A)$. AV represents a list of application variables for A (that is, the variables that will be arguments of f_A). If AV is $\{x_1 : s_1, \ldots, x_n : s_n\}$ then $f(AV)$ stands for the function application $app(f, (x_1, \ldots, x_n))$.

1. If A is Harrop, then $Sk'(A, AV) = A$.
2. If $A = (B \vee C)$, then

$$\begin{aligned} Sk'(A, AV) = {} & (\forall x : \mathsf{etype}(B) \bullet f_A(AV) = Inl(x) \Rightarrow Sk'(B, AV)[x/f_B]) \\ & \wedge(\forall y : \mathsf{etype}(C) \bullet f_A(AV) = Inr(y) \Rightarrow Sk'(C, AV)[y/f_C]) \end{aligned}$$

3. If $A = (B \wedge C)$, then
   a) If B is Harrop and C is not Harrop,
   $$Sk'(A, AV) = B \wedge Sk'(C, AV)[snd(f_A)/f_C]$$
   b) If B is not Harrop and C is Harrop,
   $$Sk'(A, AV) = (Sk'(B, AV)[fst(f_A)/f_B] \wedge C)$$
   c) If B and C are not Harrop,
   $$Sk'(A, AV) = (Sk'(B, AV)[fst(f_A)/f_B] \wedge Sk'(C, AV)[snd(f_A)/f_C])$$
4. If $A = (B \Rightarrow C)$, then
   a) If B is Harrop,
   $$Sk'(A, AV) = (B \Rightarrow Sk'(C, AV)[f_A/f_C])$$
   b) If B is not Harrop and C is not Harrop,
   $$Sk'(A, AV) = \forall x : \mathsf{etype}(B) \bullet (Sk'(B, AV)[x/f_B] \Rightarrow Sk'(C, AV)[(f_A x)/f_C])$$
5. If $A = \exists y : s \bullet P$, then
   a) when P is Harrop, $Sk'(A, AV) = Sk'(P, AV)[f_A(AV)/y]$.
   b) when P is not Harrop,
   $$Sk'(A, AV) = Sk'(P, AV)[fst(f_A(AV))/y][snd(f_A(AV))/f_P]$$
6. If $A = \forall x : s \bullet P$, then $Sk'(A, AV) = \forall x : s \bullet Sk'(P, AV)[(f_A x)/f_P]$.

Example 2.5. Given a formula A defined as $\exists y : int \bullet y \geq s(s(s(0)))$, the Skolem form $Sk(A)$ is $f_A \geq s(s(s(0)))$.

Recall that our *SML* programs may be represented in formulae. We define an *SML* program to be a *modified realizer* when, treated as a term of $Term(\Sigma)$, it can be proved to be a Skolem function for the Skolem form of the formula.

Definition 2.5.2 (Modified realizability). A program p is a modified realizer of a formula F if, and only if,

$$\vdash_{\mathsf{Int}} Sk(F)[p/f_F]$$

is provable (where p is the representation of p as a term of $Term(\Sigma)$). In this case, we write

$$\mathsf{p}\ \mathsf{mr}\ F$$

We will need the following lemma

Lemma 2.5.1. *If there is a proof* $\Gamma \vdash_{\mathsf{Int}} Sk(A)[a/f_A]$ *then* $\Gamma \vdash_{\mathsf{Int}} A$.

Proof. By induction on the form of A. □

2.5.2 Extraction map

The extraction map, $\mathsf{extract}_{\mathsf{Int}}$, from proof-terms to *SML* programs, is given in Fig. 2.10.

The map presumes a set of variables in Var, each corresponding to a proof-term variable from $Var_{PT(\mathsf{Int})}$,

$$\{x_u \mid u \in Var_{PT(\mathsf{Int})}\}$$

The principle goal of our work is to produce correct code from proofs of specifications.

Theorem 2.5.3 and Corollary 2.5.1 tells us that $\mathsf{extract}_{\mathsf{Int}}$ produces modified realizers. *Together, these results provide us with the fundamental result of our SOA approach, telling us that the map extracts correct code from proofs of specifications.*

p^P	$\mathsf{extract}_{\mathsf{Int}}(p^P)$
any proof-term where $H(P)$	()
u^A	x_u not $H(A)$ () $H(A)$
abstract $u^A.\ a^B$	`fn` $\mathtt{x_u}$ => $\mathsf{extract}_{\mathsf{Int}}(a)$ not $H(A)$ $\mathsf{extract}_{\mathsf{Int}}(a)$ $H(A)$
$\mathsf{app}(c^{A\Rightarrow B}, a^A)$	$\mathsf{extract}_{\mathsf{Int}}(c)$ $H(A)$ ($\mathsf{extract}_{\mathsf{Int}}(c)$ $\mathsf{extract}_{\mathsf{Int}}(a)$) not $H(A)$
use $x : s.\ a^A$	`fn x : s` => $\mathsf{extract}_{\mathsf{Int}}(a)$
$\mathsf{specific}(a^{\forall x:s\bullet A}, v)$	($\mathsf{extract}_{\mathsf{Int}}(a)$ `v`)
$\langle a^A, b^B\rangle$	($\mathsf{extract}_{\mathsf{Int}}(a)$, $\mathsf{extract}_{\mathsf{Int}}(b)$)
case $a^{A\vee B}$ of $\mathsf{inl}(t^A).b^C$, $\mathsf{inr}(u^B).c^C$	`match` $\mathsf{extract}_{\mathsf{Int}}(a)$ `with` `Inl`($\mathtt{x_t}$) => $\mathsf{extract}_{\mathsf{Int}}(b)$, `Inr`($\mathtt{x_u}$) => $\mathsf{extract}_{\mathsf{Int}}(c)$
$\mathsf{show}(v, a^A)$	`v` $H(A)$ (`v`, $\mathsf{extract}_{\mathsf{Int}}(a)$) not $H(A)$
select $(a^{\exists y\bullet A})$ in $x.u^{A[x/y]}.b^B$	(`fn x` => $\mathsf{extract}_{\mathsf{Int}}(b)$) $\mathsf{extract}_{\mathsf{Int}}(a)$ } $H(A)$ (`fn x` => `fn` $\mathtt{x_u}$ => $\mathsf{extract}_{\mathsf{Int}}(b)$) `fst`($\mathsf{extract}_{\mathsf{Int}}(a)$) `snd`($\mathsf{extract}_{\mathsf{Int}}(a)$) } not $H(A)$
$\mathsf{inl}(a)$	`Inl`($\mathsf{extract}_{\mathsf{Int}}(a)$)
$\mathsf{inr}(a)$	`Inr`($\mathsf{extract}_{\mathsf{Int}}(a)$)
$\mathsf{fst}(a)$	`fst`($\mathsf{extract}_{\mathsf{Int}}(a)$)
$\mathsf{snd}(a)$	`snd`($\mathsf{extract}_{\mathsf{Int}}(a)$)
$\mathsf{abort}(a^\perp)$	()

We write $H(A)$ for "A is Harrop".

Fig. 2.10. Extraction map $\mathsf{extract}_{\mathsf{Int}}$ defined over the intuitionistic proof-terms.

Theorem 2.5.3. *Let* $\Gamma = \{u_1^{G_1}, \ldots, u_1^{G_n}\}$.

Let $\Gamma' = \{Sk(G_1)[x_{u_1}/f_{G_1}], \ldots, Sk(G_n)[x_{u_n}/f_{G_n}]\}$.

Take any intuitionistic proof that does not use axioms or schemata, represented in the LTT as

$$\Gamma \vdash_{\mathsf{Int}} p^P.$$

Then $\mathsf{extract}_{\mathsf{Int}}(p)$ *will produce a modified realizer of* P*, assuming* Γ'*:*

$$\Gamma' \vdash_{\mathsf{Int}} Sk(P)[\mathsf{extract}_{\mathsf{Int}}(p)/f_P]$$

Proof. First, if P is Harrop then $Sk(P) = P$, and we are done.

So, we assume that P is not Harrop and proceed by induction on the length of the proof. (Note that, in the following, we use both sequent and proof-tree style notation, depending on convenience.)

Case: $(Ass\text{-I})$. Assume that p^P is of the form A obtained by an application of $(Ass\text{-I})$:

$$\frac{}{u^A \vdash_{\mathsf{Int}} u^A}\ (\text{Ass-I})$$

So $\Gamma' = \{Sk(A)[x_u/f_A]\}$ and we can prove

$$\frac{}{\Gamma' \vdash_{\mathsf{Int}} Sk(A)[x_u/f_A]}\ (\text{Ass-I})$$

as required.

Case: $(\wedge\text{-I})$. Assume that p^P is of the form

$$\langle a, b\rangle^{(A \wedge B)}$$

obtained by an application of $(\wedge\text{-I})$:

$$\frac{\Gamma_1 \vdash_{\mathsf{Int}} a^A \quad \Gamma_2 \vdash_{\mathsf{Int}} b^A}{\Gamma_1, \Gamma_2 \vdash_{\mathsf{Int}} \langle a, b\rangle^{(A \wedge B)}}\ (\wedge\text{-I})$$

so that $\Gamma' = \Gamma_1' \cup \Gamma_2'$.

Because we assume that P is not Harrop, either

1. A and B are both non-Harrop.
2. A is Harrop and B is non-Harrop.
3. A is non-Harrop and B is Harrop.

We deal only with the first case, as the other two cases are similar. Here,

$$\mathsf{extract}_{\mathsf{Int}}(p) = (\mathsf{extract}_{\mathsf{Int}}(a), \mathsf{extract}_{\mathsf{Int}}(b))$$

and

$$Sk(P)[\mathsf{extract}_{\mathsf{Int}}(p)/f_P] =$$
$$Sk(A)[\mathit{fst}(\mathsf{extract}_{\mathsf{Int}}(p))/f_A] \wedge Sk(B)[\mathit{snd}(\mathsf{extract}_{\mathsf{Int}}(p))/f_A]$$

So, by the IH, we know there are proofs a' and b' such that

$$\begin{array}{c} \Gamma_1' \\ \vdots \\ Sk(A)[\mathsf{extract}_{\mathsf{Int}}(a)/f_A] \end{array} \tag{2.3}$$

and

$$\begin{array}{c} \Gamma_2' \\ \vdots \\ Sk(B)[\mathsf{extract}_{\mathsf{Int}}(b)/f_B] \end{array} \tag{2.4}$$

The conjunction of these two conclusions is proved by ($\wedge$-I):

$$\dfrac{\begin{array}{c} \Gamma_1' \\ \vdots\ (2.3) \\ Sk(A)[\mathsf{extract}_{\mathsf{Int}}(a)/f_A] \end{array} \quad \begin{array}{c} \Gamma_2' \\ \vdots\ (2.4) \\ Sk(B)[\mathsf{extract}_{\mathsf{Int}}(b)/f_B] \end{array}}{(Sk(A)[\mathsf{extract}_{\mathsf{Int}}(a)/f_A] \wedge Sk(B)[\mathsf{extract}_{\mathsf{Int}}(b)/f_B])}\ (\wedge\text{-I}) \tag{2.5}$$

We are required to prove

$$\begin{array}{c} \Gamma_1', \Gamma_2' \\ \vdots \\ Sk(A)[\mathit{fst}(\mathsf{extract}_{\mathsf{Int}}(a), \mathsf{extract}_{\mathsf{Int}}(b))] \wedge Sk(B)[\mathit{snd}(\mathsf{extract}_{\mathsf{Int}}(a), \mathsf{extract}_{\mathsf{Int}}(b))/f_A] \end{array}$$

Our schemata let us take reducible terms as equal. Consequently we can prove $\mathsf{extract}_{\mathsf{Int}}(a)$ is $\mathit{fst}(\mathsf{extract}_{\mathsf{Int}}(a), \mathsf{extract}_{\mathsf{Int}}(b))$ and $\mathsf{extract}_{\mathsf{Int}}(b)$ is $\mathit{snd}(\mathsf{extract}_{\mathsf{Int}}(a), \mathsf{extract}_{\mathsf{Int}}(b))$. The required conclusion follows from this and (2.5).

More formally, we proceed as follows. First we show that

$$\vdash_{\mathsf{Int}} \mathit{fst}(\mathsf{extract}_{\mathsf{Int}}(a), \mathsf{extract}_{\mathsf{Int}}(b)) = \mathsf{extract}_{\mathsf{Int}}(a)$$

This can be seen by applying the (red=) schema of p. 44:

$$\dfrac{\mathsf{d} \rhd^*_{SML} \mathsf{e}}{\vdash_{\mathsf{Int}} d =_s e}\ (\text{red=})[[\text{d,e}];[\text{s}]]$$

Taking d as $\mathit{fst}(\mathsf{extract}_{\mathsf{Int}}(a), \mathsf{extract}_{\mathsf{Int}}(b))$, e as $\mathsf{extract}_{\mathsf{Int}}(a)$ and s as $\mathsf{etype}(A)$, we have

$$\dfrac{\mathit{fst}(\mathsf{extract}_{\mathsf{Int}}(a), \mathsf{extract}_{\mathsf{Int}}(b)) \rhd^*_{SML} \mathsf{extract}_{\mathsf{Int}}(a)}{\mathit{fst}(\mathsf{extract}_{\mathsf{Int}}(a), \mathsf{extract}_{\mathsf{Int}}(b)) = \mathsf{extract}_{\mathsf{Int}}(a)} \tag{2.6}$$

with $\mathit{fst}(\mathsf{extract}_{\mathsf{Int}}(a), \mathsf{extract}_{\mathsf{Int}}(b))$ of sort $\mathsf{etype}(A)$, and so R_1 is the schema application name (red=)$[[\mathit{fst}(\mathsf{extract}_{\mathsf{Int}}(a); \mathsf{extract}_{\mathsf{Int}}(b)); \mathsf{extract}_{\mathsf{Int}}(a)]; [\mathsf{etype}(A)]]$.

Similarly, we obtain $\vdash_{\mathsf{Int}} \mathit{snd}(\mathsf{extract}_{\mathsf{Int}}(a), \mathsf{extract}_{\mathsf{Int}}(b)) = \mathsf{extract}_{\mathsf{Int}}(b)$.

$$\dfrac{\mathit{snd}(\mathsf{extract}_{\mathsf{Int}}(a), \mathsf{extract}_{\mathsf{Int}}(b)) \rhd^*_{SML} \mathsf{extract}_{\mathsf{Int}}(b)}{\mathit{snd}(\mathsf{extract}_{\mathsf{Int}}(a), \mathsf{extract}_{\mathsf{Int}}(b)) = \mathsf{extract}_{\mathsf{Int}}(b)}\ R_2 \tag{2.7}$$

where R_2 is the schema application

$$(\text{red=})[[snd(\mathsf{extract}_{\mathsf{Int}}(a); \mathsf{extract}_{\mathsf{Int}}(b)); \mathsf{extract}_{\mathsf{Int}}(b)]; [\mathsf{etype}(B)]]$$

Recall the (subst) schema of Fig. 2.4.

We apply this schema, setting u to $fst(\mathsf{extract}_{\mathsf{Int}}(a), \mathsf{extract}_{\mathsf{Int}}(b))$, r to $\mathsf{extract}_{\mathsf{Int}}(a)$ and P to $Sk(A)[y/f_A] \wedge Sk(B)[\mathsf{extract}_{\mathsf{Int}}(b)/f_B]$

$$\dfrac{\dfrac{\begin{array}{c}\Gamma_1', \Gamma_2' \\ \vdots\ (2.5) \\ Sk(A)[\mathsf{extract}_{\mathsf{Int}}(a)/f_A] \wedge \\ Sk(B)[\mathsf{extract}_{\mathsf{Int}}(b)/f_B]\end{array} \qquad \begin{array}{c}\vdots\ (2.6) \\ fst(\mathsf{extract}_{\mathsf{Int}}(a), \mathsf{extract}_{\mathsf{Int}}(b)) = \mathsf{extract}_{\mathsf{Int}}(a)\end{array}}{\begin{array}{c}(Sk(A)[\mathsf{extract}_{\mathsf{Int}}(a)/f_A] \wedge Sk(B)[\mathsf{extract}_{\mathsf{Int}}(b)/f_B]) \wedge \\ (fst(\mathsf{extract}_{\mathsf{Int}}(a), \mathsf{extract}_{\mathsf{Int}}(b)) = \mathsf{extract}_{\mathsf{Int}}(a))\end{array}}\,(\wedge\text{-I})}{Sk(A)[fst(\mathsf{extract}_{\mathsf{Int}}(a), \mathsf{extract}_{\mathsf{Int}}(b))/f_A] \wedge Sk(B)[\mathsf{extract}_{\mathsf{Int}}(b)/f_B]}\,R_3 \tag{2.8}$$

where R_3 is

$$\begin{array}{l}(\text{subst})[[Sk(A)[y/f_A] \wedge Sk(B)[\mathsf{extract}_{\mathsf{Int}}(b)/f_B]]; \\ \qquad [fst(\mathsf{extract}_{\mathsf{Int}}(a), \mathsf{extract}_{\mathsf{Int}}(b)); \mathsf{extract}_{\mathsf{Int}}(a)]; [\mathsf{etype}(A)]]\end{array}$$

We apply this schema again, setting u to $snd(\mathsf{extract}_{\mathsf{Int}}(a), \mathsf{extract}_{\mathsf{Int}}(b))$, r to $\mathsf{extract}_{\mathsf{Int}}(b)$ and P to $(Sk(A)[fst(\mathsf{extract}_{\mathsf{Int}}(a), \mathsf{extract}_{\mathsf{Int}}(b))/f_A] \wedge Sk(B)[y/f_B])$

$$\dfrac{\dfrac{\begin{array}{c}\Gamma_1', \Gamma_2' \\ \vdots\ (2.8) \\ (Sk(A)[fst(\mathsf{extract}_{\mathsf{Int}}(a), \mathsf{extract}_{\mathsf{Int}}(b))/f_A] \wedge \\ Sk(B)[\mathsf{extract}_{\mathsf{Int}}(b)/f_B])\end{array} \qquad \begin{array}{c}\vdots\ (2.7) \\ snd(\mathsf{extract}_{\mathsf{Int}}(a), \mathsf{extract}_{\mathsf{Int}}(b)) = \\ \mathsf{extract}_{\mathsf{Int}}(b)\end{array}}{\begin{array}{c}(Sk(A)[fst(\mathsf{extract}_{\mathsf{Int}}(a), \mathsf{extract}_{\mathsf{Int}}(b))/f_A] \wedge Sk(B)[\mathsf{extract}_{\mathsf{Int}}(b)/f_B] \wedge \\ snd(\mathsf{extract}_{\mathsf{Int}}(a), \mathsf{extract}_{\mathsf{Int}}(b)) = \mathsf{extract}_{\mathsf{Int}}(b))\end{array}}\,(\wedge\text{-I})}{\begin{array}{c}(Sk(A)[fst(\mathsf{extract}_{\mathsf{Int}}(a), \mathsf{extract}_{\mathsf{Int}}(b))/f_A] \wedge \\ Sk(B)[snd(\mathsf{extract}_{\mathsf{Int}}(a), \mathsf{extract}_{\mathsf{Int}}(b))/f_B])\end{array}}\,R_4$$

where R_4 is

$$\begin{array}{l}(\text{subst})[[(Sk(A)[fst(\mathsf{extract}_{\mathsf{Int}}(a), \mathsf{extract}_{\mathsf{Int}}(b))/f_A] \wedge Sk(B)[y/f_B])]; \\ \qquad [snd(\mathsf{extract}_{\mathsf{Int}}(a), \mathsf{extract}_{\mathsf{Int}}(b)); \mathsf{extract}_{\mathsf{Int}}(b)]; [\mathsf{etype}(A)]]\end{array}$$

This is the required proof, as the the conclusion is the same as writing

$$(Sk(A)[fst(\mathsf{extract}_{\mathsf{Int}}(p))/f_A] \wedge Sk(B)[snd(\mathsf{extract}_{\mathsf{Int}}(p))/f_B])$$

Case: ($\wedge$-E$_1$). Assume that p^P is of the form

$$\mathsf{fst}(q)^A$$

obtained by an application of ($\wedge$-E$_1$):

$$\frac{\Gamma \vdash_{\mathsf{Int}} q^{(A \wedge B)}}{\Gamma \vdash_{\mathsf{Int}} \mathsf{fst}(q)^{A}} \ (\wedge\text{-E}_1)$$

We are required to prove $\Gamma' \vdash_{\mathsf{Int}} Sk(A)[\mathsf{extract}_{\mathsf{Int}}(p)/f_A]$.

There are two possible cases: either B is Harrop or B is not Harrop. We reason over these cases.

1. Assume that B is Harrop, so that $Sk(B) = B$. Then,

$$\mathsf{extract}_{\mathsf{Int}}(p) = \mathsf{extract}_{\mathsf{Int}}(q)$$

and we are required to prove $Sk(A)[\mathsf{extract}_{\mathsf{Int}}(q)/f_A]$.
By the IH and the fact that $Sk(B) = B$, we know that there is a proof of the form

$$\begin{array}{c} \Gamma' \\ \vdots \\ (Sk(A)[\mathsf{extract}_{\mathsf{Int}}(q)/f_A] \wedge B) \end{array}$$

From this, we can derive

$$\begin{array}{c} \Gamma' \\ \vdots \end{array}$$
$$\frac{(Sk(A)[\mathsf{extract}_{\mathsf{Int}}(q)/f_A] \wedge B)}{Sk(A)[\mathsf{extract}_{\mathsf{Int}}(q)/f_A]} \ (\wedge\text{-E}_1)$$

as required.

2. Assume that B is not Harrop. Then,

$$\mathsf{extract}_{\mathsf{Int}}(p) = \mathit{fst}(\mathsf{extract}_{\mathsf{Int}}(q))$$

and we are required to prove $\Gamma' \vdash_{\mathsf{Int}} Sk(A)[\mathit{fst}(\mathsf{extract}_{\mathsf{Int}}(q))/f_A]$.
By the IH, we know that there is a proof of the form

$$\begin{array}{c} \Gamma' \\ \vdots \\ (Sk(A)[\mathit{fst}(\mathsf{extract}_{\mathsf{Int}}(q))/f_A] \wedge Sk(B)[\mathit{snd}(\mathsf{extract}_{\mathsf{Int}}(q))/f_B]) \end{array}$$

We apply $(\wedge\text{-E}_1)$ this to obtain the required proof

$$\begin{array}{c} \Gamma' \\ \vdots \end{array}$$
$$\frac{(Sk(A)[\mathit{fst}(\mathsf{extract}_{\mathsf{Int}}(q))/f_A] \wedge Sk(B)[\mathit{snd}(\mathsf{extract}_{\mathsf{Int}}(q))/f_B])}{Sk(A)[\mathit{fst}(\mathsf{extract}_{\mathsf{Int}}(q))/f_A]} \ (\wedge\text{-E}_1)$$

Case: $(\wedge\text{-E}_2)$. Similar to the case $(\wedge\text{-E}_1)$ above.
Case: $(\vee\text{-I}_1)$. Assume that p^P is of the form

$$\mathsf{inl}(a)^{A \vee B}$$

obtained by an application of ($\vee$-I$_1$)

$$\frac{\Gamma \vdash_{\mathsf{Int}} a^A}{\Gamma \vdash_{\mathsf{Int}} inla^{A \vee B}} \ (\vee\text{-I}_1)$$

so that $\mathsf{extract}_{\mathsf{Int}}(p) = \mathtt{Inl}(\mathsf{extract}_{\mathsf{Int}}(a))$.

We are required to show that $\Gamma' \vdash_{\mathsf{Int}} Sk(P)[\mathsf{extract}_{\mathsf{Int}}(p)/f_P]$. That is to say, we must prove

$$\Gamma' \vdash_{\mathsf{Int}} (\forall x : \mathsf{etype}(A) \bullet Inl(\mathsf{extract}_{\mathsf{Int}}(a)) = Inl(x) \Rightarrow Sk(A)[x/f_A]) \wedge$$
$$(\forall y : \mathsf{etype}(B) \bullet Inl(\mathsf{extract}_{\mathsf{Int}}(a)) = Inr(y) \Rightarrow Sk(B)[y/f_B])$$

To show this, we use the following assumptions

$$Inl(\mathsf{extract}_{\mathsf{Int}}(a)) = Inl(x) \tag{2.9}$$

and

$$Inl(\mathsf{extract}_{\mathsf{Int}}(a)) = Inr(y) \tag{2.10}$$

By the IH, there is a proof of the form

$$\begin{array}{c} \Gamma' \\ \vdots \\ Sk(A)[\mathsf{extract}_{\mathsf{Int}}(a)/f_A] \end{array} \tag{2.11}$$

Recall the (union$=_1$) schema: for any u and r of the same type,

$$\frac{Inl(u) = Inl(r)}{u =_{s_1} r} \ (\text{union}=_1)[[u; r]; [s_1; s_2]]$$

Letting u be $\mathsf{extract}_{\mathsf{Int}}(a)$ and r be x, then using assumption (2.9), we have

$$\frac{Inl(\mathsf{extract}_{\mathsf{Int}}(a)) = Inl(x)}{\mathsf{extract}_{\mathsf{Int}}(a) = x} \ R_1 \tag{2.12}$$

where R_1 is (union$=_1$)$[[\mathsf{extract}_{\mathsf{Int}}(a); x]; [\mathsf{etype}(A); \mathsf{etype}(B)]]$.

Recall the reflexivity (ref) schema of Fig. 2.4. Letting u be $\mathsf{extract}_{\mathsf{Int}}(a)$, r be x and s be $\mathsf{etype}(A)$ in this schema, then using (2.12), we have

$$\frac{\mathsf{extract}_{\mathsf{Int}}(a) = x}{x = \mathsf{extract}_{\mathsf{Int}}(a)} \ R_2 \tag{2.13}$$

where R_2 is $ref[[\mathsf{extract}_{\mathsf{Int}}(a); x]; \mathsf{etype}(A)]$.

Applying ($\wedge$-I) to the proofs (2.13) and the IH (2.11) will give us

$$\frac{\begin{array}{c} \Gamma' \\ \vdots \\ Sk(A)[\mathsf{extract}_{\mathsf{Int}}(a)/y] \end{array} \qquad \begin{array}{c} Inl(\mathsf{extract}_{\mathsf{Int}}(a)) = Inl(x) \\ \vdots \\ x = \mathsf{extract}_{\mathsf{Int}}(a) \end{array}}{(Sk(A)[\mathsf{extract}_{\mathsf{Int}}(a)/y] \wedge x = \mathsf{extract}_{\mathsf{Int}}(a))} \ (\wedge\text{-I})$$

By the (subst) schema, with P set to $Sk(A)[y/f_A]$, r set to $\mathsf{extract_{Int}}(a)$ and u set to x, we obtain

$$\dfrac{\begin{array}{c}\Gamma', Inl(\mathsf{extract_{Int}}(a)) = Inl(x)\\ \vdots \\ (Sk(A)[\mathsf{extract_{Int}}(a)/y] \wedge x = \mathsf{extract_{Int}}(a))\end{array}}{Sk(A)[x/f_A]}\ R_3 \tag{2.14}$$

where R_3 is (subst)$[[Sk(A)[y/f_A]]; [\mathsf{extract_{Int}}(a); x]; [\mathsf{etype}(A)]]$.

We apply ($\Rightarrow$-I) on (2.14) introducing assumption (2.9), and then perform ($\forall$-I) abstracting over x:

$$\dfrac{\dfrac{\begin{array}{c}\Gamma', [Inl(\mathsf{extract_{Int}}(a)) = Inl(x)]\\ \vdots \\ Sk(A)[x/f_A]\end{array}}{Inl(\mathsf{extract_{Int}}(a)) = Inl(x) \Rightarrow Sk(A)[x/f_A]}\ (\Rightarrow\text{-I})}{\forall x : \mathsf{etype}(A) \bullet Inl(\mathsf{extract_{Int}}(a)) = Inl(x) \Rightarrow Sk(A)[x/f_A]}\ (\forall\text{-I}) \tag{2.15}$$

This gives us the left hand side of the required conjunction.

Recall the (union$\neq$) schema of Fig. 2.4: for any u and r of types s_1 and s_2 respectively,

$$\dfrac{}{Inl(u) = Inr(r) \Rightarrow \bot}\ (\text{union}\neq)[[u; r]; [s_1; s_2]]$$

Letting u be $\mathsf{extract_{Int}}(a)$ and r be y, we have

$$\dfrac{}{Inl(\mathsf{extract_{Int}}(a)) = Inr(y) \Rightarrow \bot}\ R_4 \tag{2.16}$$

where R_4 is (union$\neq$)$[[\mathsf{extract_{Int}}(a); y]; [\mathsf{etype}(A); \mathsf{etype}(B)]]$.

So, the assumption (2.10) and (2.16) give a contradiction, from which we may conclude $Sk(B)[y/f_B]$ by the absurdity rule:

$$\dfrac{\dfrac{\dfrac{}{Inl(\mathsf{extract_{Int}}(a)) = Inr(y) \Rightarrow \bot} \quad Inl(\mathsf{extract_{Int}}(a)) = Inr(y)}{\bot}\ (\Rightarrow\text{-E})}{Sk(B)[y/f_B]}\ \bot - E \tag{2.17}$$

Finally, applying ($\Rightarrow$-I) to (2.17) introducing assumption (2.10), and then performing ($\forall$-I) abstracting over y will give

$$\dfrac{\dfrac{\begin{array}{c}[Inl(\mathsf{extract_{Int}}(a)) = Inr(y)]\\ \vdots \\ Sk(B)[y/f_B]\end{array}}{\mathsf{extract_{Int}}(a) = Inr(y) \Rightarrow Sk(B)[y/f_B]}\ (\Rightarrow\text{-I})}{\forall y : \mathsf{etype}(B) \bullet \mathsf{extract_{Int}}(a) = Inr(y) \Rightarrow Sk(B)[y/f_B]}\ (\forall\text{-I}) \tag{2.18}$$

Then, applying ($\wedge$-I) over (2.15) and (2.18), we obtain the required proof:

$$\begin{array}{c} \Gamma' \\ \vdots \\ (\forall x : \mathsf{etype}(A) \bullet \mathsf{extract}_{\mathsf{Int}}(p) = \mathit{Inl}(x) \Rightarrow \mathit{Sk}(A)[x/f_A]) \wedge \\ (\forall y : \mathsf{etype}(B) \bullet \mathsf{extract}_{\mathsf{Int}}(p) = \mathit{Inr}(y) \Rightarrow \mathit{Sk}(B)[y/f_B]) \end{array}$$

Case: ($\vee$-I$_2$). Similar to the ($\vee$-I$_1$) case above.
Case: ($\vee$-E). Assume that p^P is of the form

$$\mathsf{case}\ e\ \mathsf{of}\ \mathsf{inl}(x).a,\ \mathsf{inr}(y).b^C$$

obtained by an application of ($\vee$-E)

$$\frac{\Gamma_1 \vdash_{\mathsf{Int}} e^{A \vee B} \quad \Gamma_2, u^A \vdash_{\mathsf{Int}} a^C \quad \Gamma_3, v^B \vdash_{\mathsf{Int}} b^C}{\Gamma_1, \Gamma_2, \Gamma_3 \vdash_{\mathsf{Int}} \mathsf{case}\ e\ \mathsf{of}\ \mathsf{inl}(u).a,\ \mathsf{inr}(v).b^C} \ (\vee\text{-E})$$

so that $\Gamma' = \Gamma_1' \cup \Gamma_2 \cup \Gamma_3$, and $\mathsf{extract}_{\mathsf{Int}}(p)$ is defined

$$\begin{aligned} \mathsf{extract}_{\mathsf{Int}}(p) = \ & \mathtt{match}\ \mathsf{extract}_{\mathsf{Int}}(e)\ \mathtt{with} \\ & \quad \mathtt{Inl(x_u)} \Rightarrow \mathsf{extract}_{\mathsf{Int}}(a), \\ & \quad \mathtt{Inr(x_v)} \Rightarrow \mathsf{extract}_{\mathsf{Int}}(b) \end{aligned} \tag{2.19}$$

By the IH,

$$\Gamma_1' \vdash_{\mathsf{Int}} \mathit{Sk}(A \vee B)[(\mathsf{extract}_{\mathsf{Int}}(e))/f_{A \vee B}] \tag{2.20}$$

$$\Gamma_2', \mathit{Sk}(A)[x_u/f_A] \vdash_{\mathsf{Int}} \mathit{Sk}(C)[(\mathsf{extract}_{\mathsf{Int}}(a))/f_C] \tag{2.21}$$

$$\Gamma_3', \mathit{Sk}(B)[x_v/f_B] \vdash_{\mathsf{Int}} \mathit{Sk}(C)[(\mathsf{extract}_{\mathsf{Int}}(b))/f_C] \tag{2.22}$$

By ($\Rightarrow$-I) and ($\forall$-I) on (2.21) and (2.22) respectively, we obtain

$$\Gamma_2' \vdash_{\mathsf{Int}} \forall x_u : \mathsf{etype}(A) \bullet \mathit{Sk}(A)[x_u/f_A] \Rightarrow \mathit{Sk}(C)[(\mathsf{extract}_{\mathsf{Int}}(a))/f_C] \tag{2.23}$$

$$\Gamma_3' \vdash_{\mathsf{Int}} \forall x_v : \mathsf{etype}(B) \bullet \mathit{Sk}(B)[x_v/f_B] \Rightarrow \mathit{Sk}(C)[(\mathsf{extract}_{\mathsf{Int}}(b))/f_C] \tag{2.24}$$

Also, by the definition of $\mathit{Sk}(A \vee B)$, (2.20) may be rewritten:

$$\Gamma_1' \vdash_{\mathsf{Int}} M \tag{2.25}$$

where M is

$$(\forall x : \mathsf{etype}(A) \bullet \mathsf{extract}_{\mathsf{Int}}(e) = \mathit{Inl}(x) \Rightarrow \mathit{Sk}(A)[x/f_A]) \wedge \\ (\forall y : \mathsf{etype}(B) \bullet \mathsf{extract}_{\mathsf{Int}}(e) = \mathit{Inr}(y) \Rightarrow \mathit{Sk}(B)[y/f_B])$$

Using the (disj-ind) schema, we claim it is possible to prove

$$\vdash_{\mathsf{Int}} (\exists v_l : \mathsf{etype}(A) \bullet \mathsf{extract}_{\mathsf{Int}}(e) = \mathit{Inl}(v_l)) \vee \\ (\exists v_r : \mathsf{etype}(A) \bullet \mathsf{extract}_{\mathsf{Int}}(e) = \mathit{Inr}(v_r)) \tag{2.26}$$

Proof of (2.26). To see this, first recall the (disj-ind) schema of Fig. 2.4 for terms of type $(s_1|s_2)$ and arbitrary predicate Q:

$$\frac{\forall y_1 : s_1 \bullet (Q[Inl(y_1)/x] \wedge \forall y_2 : s_2 \bullet Q[Inr(y_2)/x])}{\forall x : s_1|s_2 \bullet Q} \text{(disj-ind)}[[Q]; [s_1; s_2]]$$

We use this schema with s_1 set to $\mathsf{etype}(A)$, s_2 set to $\mathsf{etype}(B)$ and Q set to the statement

$$(\exists v_l : \mathsf{etype}(A) \bullet x = Inl(v_l)) \vee (\exists v_r : \mathsf{etype}(B) \bullet x = Inr(v_r))$$

in the following proof

$$\dfrac{\dfrac{\dfrac{\dfrac{\dfrac{}{Inl(y_1) = Inl(y_1)} R_2}{\exists v_l : \mathsf{etype}(A) \bullet Inl(y_1) = Inl(v_l)} (\exists\text{-I})}{Q[Inl(y_1)/x]} (\vee\text{-I}_1)}{\forall y_1 : \mathsf{etype}(A) \bullet Q[Inl(y_1)/x]} (\forall\text{-I}) \quad \dfrac{\dfrac{\dfrac{\dfrac{}{Inr(y_2) = Inr(y_2)} R_3}{\exists v_2 : \mathsf{etype}(B) \bullet Inr(y_2) = Inr(v_2)} (\exists\text{-I})}{Q[Inr(y_2)/x]} \vee - I_l}{\forall y_2 : \mathsf{etype}(B) \bullet Q[Inr(y_2)/x]} (\forall\text{-I})}{\dfrac{(\forall y_1 : \mathsf{etype}(A) \bullet Q[Inl(y_1)/x]) \wedge (\forall y_2 : \mathsf{etype}(B) \bullet Q[Inr(y_2)/x])}{\forall x : \mathsf{etype}(A)|\mathsf{etype}(B) \bullet Q} R_1} (\wedge\text{-I}) \tag{2.27}$$

where R_1 is

$$\text{(disj-ind)}[[(\exists v_l : \mathsf{etype}(A) \bullet x = Inl(v_l)) \vee (\exists v_r : \mathsf{etype}(B) \bullet x = Inr(v_r))], [\mathsf{etype}(A); \mathsf{etype}(B)]]$$

and R_2 is

$$\text{(red=)}[[Inl(y_1); Inl(y_1)]; [\mathsf{etype}(A)|\mathsf{etype}(B)]]$$

and R_3 is

$$\text{red=}[[Inr(y_2); Inr(y_2)]; [\mathsf{etype}(A)|\mathsf{etype}(B)]]$$

both the names of applications of the schema (red=).

By setting $\mathsf{extract}_{\mathsf{Int}}(e)$ for x in (2.27), we have the required proof of (2.26). *End of proof of (2.26).*

We reason over the two possible cases given by (2.26).

Left case of (2.26). We assume the left case holds:

$$\exists v_l : \mathsf{etype}(A) \bullet \mathsf{extract}_{\mathsf{Int}}(e) = Inl(v_l) \tag{2.28}$$

We will establish

$$\Gamma', \exists v_l : \mathsf{etype}(A) \bullet \mathsf{extract}_{\mathsf{Int}}(e) = Inl(v_l) \vdash_{\mathsf{Int}} Sk(C)[\mathsf{extract}_{\mathsf{Int}}(p)/f_C] \tag{2.29}$$

We first assume that there is a $k_l : \mathsf{etype}(A)$ such that

$$\mathsf{extract}_{\mathsf{Int}}(e) = Inl(k_l) \tag{2.30}$$

Also, we observe that it is possible to derive

$$\begin{array}{c}\mathsf{extract}_{\mathsf{Int}}(e) = \mathit{Inl}(k_l) \\ \vdots \\ (Q'[\mathsf{extract}_{\mathsf{Int}}(b)/y] \wedge \mathit{Inl}(k_l) = \mathsf{extract}_{\mathsf{Int}}(e))\end{array} \tag{2.31}$$

with Q' defined to be

$$\mathsf{extract}_{\mathsf{Int}}(p) = \mathit{match}\ y\ \mathit{with}\ \mathit{Inl}(x_u) \Rightarrow \mathsf{extract}_{\mathsf{Int}}(a) \mid \mathit{Inr}(x_v) \Rightarrow \mathsf{extract}_{\mathsf{Int}}(b)$$

This follows from taking the conjunction of $P[\mathsf{extract}_{\mathsf{Int}}(b)/y]$ (that is, of the identity $\mathsf{extract}_{\mathsf{Int}}(p) = \mathsf{extract}_{\mathsf{Int}}(p)$) obtained using the (red=) schema, and of $\mathit{Inl}(k_l) = \mathsf{extract}_{\mathsf{Int}}(e))$, obtained from the assumption (2.30) $\mathsf{extract}_{\mathsf{Int}}(e) = \mathit{Inl}(k_l)$ and the reflexivity (ref=) schema.

We apply the substitution schema:

$$\frac{\begin{array}{c}\mathsf{extract}_{\mathsf{Int}}(e) = \mathit{Inl}(k_l) \\ \vdots\ (2.31) \\ (Q'[\mathsf{extract}_{\mathsf{Int}}(b)/y] \wedge \mathit{Inl}(k_l) = \mathsf{extract}_{\mathsf{Int}}(e))\end{array}}{\begin{array}{l}\mathsf{extract}_{\mathsf{Int}}(p) = \\ \quad \mathit{match}\ \mathit{Inl}(k_l)\ \mathit{with}\ \mathit{Inl}(x_u) \Rightarrow \mathsf{extract}_{\mathsf{Int}}(a) \mid \mathit{Inr}(x_v) \Rightarrow \mathsf{extract}_{\mathsf{Int}}(b)\end{array}}\ R_4 \tag{2.32}$$

where R_4 is

$$\begin{array}{l}(\text{subst})[[\mathsf{extract}_{\mathsf{Int}}(p) = \\ \quad \mathit{match}\ y\ \mathit{with}\ \mathit{Inl}(x_u) \Rightarrow \mathsf{extract}_{\mathsf{Int}}(a) \mid \mathit{Inr}(x_v) \Rightarrow \mathsf{extract}_{\mathsf{Int}}(b)]; \\ \qquad\qquad [\mathsf{extract}_{\mathsf{Int}}(e); \mathit{Inl}(k_1)]; [\mathsf{etype}(A \vee B)]]\end{array}$$

Also, using the (red=) schema we can obtain

$$\frac{t \rhd^*_{SML} \mathsf{extract}_{\mathsf{Int}}(a)[k_l/x_u]}{\begin{array}{l}\mathit{match}\ \mathit{Inl}(k_l)\ \mathit{with}\ \mathit{Inl}(x_u) \Rightarrow \mathsf{extract}_{\mathsf{Int}}(a) \mid \mathit{Inr}(x_v) \Rightarrow \mathsf{extract}_{\mathsf{Int}}(b) = \\ \quad \mathsf{extract}_{\mathsf{Int}}(a)[k_l/x_u]\end{array}}\ R_5 \tag{2.33}$$

where t stands for

$$\mathit{match}\ \mathit{Inl}(k_l)\ \mathit{with}\ \mathit{Inl}(x_u) \Rightarrow \mathsf{extract}_{\mathsf{Int}}(a) \mid \mathit{Inr}(x_v) \Rightarrow \mathsf{extract}_{\mathsf{Int}}(b)$$

and R_5 is (red=)$[[t; \mathsf{extract}_{\mathsf{Int}}(a)[k_l/x_u]]; [\mathsf{etype}(C)]]$.

Then, by application of the substitution schema, taking r to be t, u to be $\mathsf{extract}_{\mathsf{Int}}(a)[k_l/x_u]$ and P to be $\mathsf{extract}_{\mathsf{Int}}(p) = y$

$$\frac{\begin{array}{c}\mathsf{extract}_{\mathsf{Int}}(e) = \mathit{Inl}(k_l) \\ \vdots \\ \langle p_2, p_3\rangle^{(\mathsf{extract}_{\mathsf{Int}}(b) = t \wedge t = \mathsf{extract}_{\mathsf{Int}}(a)[k_l/x_u])}\end{array}}{\mathsf{extract}_{\mathsf{Int}}(p) = \mathsf{extract}_{\mathsf{Int}}(a)[k_l/x_u]}\ R_6 \tag{2.34}$$

where R_6 is $(\mathrm{subst})[[P]; [\mathsf{extract_{Int}}(a)[k_l/x_u]; t]; [\mathsf{etype}(C)]]$.

Next we prove

$$\Gamma_1', \Gamma_2, \mathsf{extract_{Int}}(e) = \mathit{Inl}(k_l) \vdash_{\mathsf{Int}} Sk(C)[(\mathsf{extract_{Int}}(a)[k_l/x_u])/f_C] \quad (2.35)$$

Proof of (2.35). Using (2.25):

$$\dfrac{\mathsf{extract_{Int}}(e) = \mathit{Inl}(k_l) \qquad \dfrac{\dfrac{\begin{array}{c}\Gamma' \\ \vdots \\ M\end{array}}{\forall x : \mathsf{etype}(A) \bullet (\mathsf{extract_{Int}}(e) = \mathit{Inl}(x) \Rightarrow Sk(A)[x/f_A])}\,(\wedge\text{-E}_1)}{\mathsf{extract_{Int}}(e) = \mathit{Inl}(k_l) \Rightarrow Sk(A)[k_l/f_A]}\,(\forall\text{-E})}{Sk(A)[k_l/f_A]}\,(\Rightarrow\text{-E}) \quad (2.36)$$

We set the x_u of (2.23) to be k_l,

$$\dfrac{\begin{array}{c}\Gamma_2' \\ \vdots\ (2.23) \\ \forall x_u : \mathsf{etype}(A) \bullet Sk(A)[x_u/f_A] \Rightarrow Sk(C)[(\mathsf{extract_{Int}}(a))/f_C]\end{array}}{Sk(A)[k_l/f_A] \Rightarrow Sk(C)[(\mathsf{extract_{Int}}(a)[k_l/x_u])/f_C]}\,(\forall\text{-E}) \quad (2.37)$$

Then, we instantiate (2.37) with with the conclusion of (2.36) to obtain (2.35):

$$\dfrac{\begin{array}{c}\Gamma_1', \mathsf{extract_{Int}}(e) = \mathit{Inl}(k_l) \\ \vdots\ (2.36) \\ Sk(A)[k_l/f_A]\end{array} \qquad \begin{array}{c}\Gamma_2' \\ \vdots\ (2.37) \\ Sk(A)[k_l/f_A] \Rightarrow \\ Sk(C)[(\mathsf{extract_{Int}}(a)[k_l/x_u])/f_C]\end{array}}{Sk(C)[(\mathsf{extract_{Int}}(a)[k_l/x_u])/f_C]}\,(\Rightarrow\text{-E})$$

End of proof of (2.35). So, using (2.34), (2.35) and (subst), we can derive

$$\dfrac{\begin{array}{c}\Gamma_1', \Gamma_2, \mathsf{extract_{Int}}(e) = \mathit{Inl}(k_l) \\ \vdots\ (2.35) \\ Sk(C)[(\mathsf{extract_{Int}}(a)[k_l/x_u])/f_C]\end{array} \quad \begin{array}{c}\mathsf{extract_{Int}}(e) = \mathit{Inl}(k_l) \\ \vdots\ (2.34) \\ \mathsf{extract_{Int}}(p) = \mathsf{extract_{Int}}(a)[k_l/x_u]\end{array}}{Sk(C)[\mathsf{extract_{Int}}(p)/f_C]}\,R_6 \quad (2.38)$$

where R_6 is the rule formed from the schema application

$$(\mathrm{subst})[[Sk(C)[y/f_C]]; [\mathsf{extract_{Int}}(e); (\mathsf{extract_{Int}}(a)[k_l/x_u])]; [\mathsf{etype}(A \vee B)]]$$

Observe that k_l does not occur in $Sk(C)[(\mathsf{extract_{Int}}(p))/f_C]$. So, we can apply (∃-E) to assumption (2.28) and (2.38):

$$\dfrac{\dfrac{}{\exists v_l : \mathsf{etype}(A) \bullet [\mathsf{extract_{Int}}(e) = \mathit{Inl}(v_l)]}\,(\text{Ass-I}) \qquad \begin{array}{c}\Gamma_1', \Gamma_2', [\mathsf{extract_{Int}}(e) = \mathit{Inl}(k_l)] \\ \vdots \\ Sk(C)[\mathsf{extract_{Int}}(p)/f_C]\end{array}}{Sk(C)[\mathsf{extract_{Int}}(p)/f_C]}\,(\exists\text{-E}) \quad (2.39)$$

This concludes our subproof using the left case of (2.26).

Right case of (2.26). We assume the right case holds:

$$\exists v_r : \mathsf{etype}(B) \bullet \mathsf{extract}_{\mathsf{Int}}(e) = Inr(v_r) \tag{2.40}$$

By symmetric reasoning to the previous case, we obtain a proof of the form

$$\begin{array}{c} \Gamma_1', \Gamma_3', \exists v_r : \mathsf{etype}(B) \bullet \mathsf{extract}_{\mathsf{Int}}(e) = Inr(v_r) \\ \vdots \\ Sk(C)[(\mathsf{extract}_{\mathsf{Int}}(p))/f_C] \end{array} \tag{2.41}$$

concluding our subproof using the right case of (2.26).

It then remains to apply ($\vee$-E) to the proofs (2.26), (2.41) and (2.39), obtaining the required conclusion

$$\frac{A \qquad \begin{array}{c} \Gamma_1', \Gamma_2', \\ [\exists v_l : \mathsf{etype}(A) \bullet \mathsf{extract}_{\mathsf{Int}}(e) \\ = Inl(v_l)] \\ \vdots\ (2.39) \\ Sk(C)[(\mathsf{extract}_{\mathsf{Int}}(p))/f_C] \end{array} \qquad \begin{array}{c} \Gamma_1', \Gamma_3', \\ [\exists v_r : \mathsf{etype}(B) \bullet \mathsf{extract}_{\mathsf{Int}}(e) \\ = Inr(v_r)] \\ \vdots\ (2.41) \\ Sk(C)[(\mathsf{extract}_{\mathsf{Int}}(p))/f_C] \end{array}}{Sk(C)[(\mathsf{extract}_{\mathsf{Int}}(p))/f_C]} \ (\vee\text{-E})$$

where A is

$$\begin{array}{c} \vdots\ (2.26) \\ (\exists v_l : \mathsf{etype}(A) \bullet \mathsf{extract}_{\mathsf{Int}}(e) = Inl(v_l)) \vee (\exists v_l : \mathsf{etype}(A) \bullet \mathsf{extract}_{\mathsf{Int}}(e) = Inr(v_l)) \end{array}$$

Case: ($\exists$-I). Assume that p^P is of the form $\exists x : s \bullet A$ obtained by an application of ($\exists$-I)

$$\frac{\Gamma \vdash_{\mathsf{Int}} a^{A[v/x]}}{\Gamma \vdash_{\mathsf{Int}} \mathsf{show}(v, a)^{\exists x : s \bullet A}} \ (\exists\text{-I}) \tag{2.42}$$

There are two cases, dependent on whether A is Harrop or not.

Case 1: Assume A is Harrop. Then $\mathsf{extract}_{\mathsf{Int}}(p)$ is defined to be v.

Also, because A is Harrop, $Sk(\exists x : s \bullet A)$ is A, and so $Sk(\exists x : s \bullet A)[\mathsf{extract}_{\mathsf{Int}}(p)/f_P]$ is $A[v/f_A]$. This means that

$$Sk(\exists x : s \bullet A)[f_A/x][\mathsf{extract}_{\mathsf{Int}}(p)/f_A] = A[v/x]$$

and, by the premise of (2.42),

$$\begin{array}{c} \Gamma \\ \vdots \\ Sk(\exists x : s \bullet A)[\mathsf{extract}_{\mathsf{Int}}(p)/f_{\exists x : s \bullet A}] \end{array}$$

By repeated application of Lemma 2.5.1, for each $G_i' \in \Gamma'$, we can derive $G_i \in \Gamma$.

So, we have

$$\begin{array}{c} \Gamma' \\ \vdots \\ \Gamma \\ \vdots \\ Sk(\exists x : s \bullet A)[\mathsf{extract_{Int}}(p)/f_{\exists x:s \bullet A}] \end{array}$$

This concludes case 1.

Case 2: Assume A is not Harrop. Then $\mathsf{extract_{Int}}(p)$ is defined as

$$(\mathsf{v}, \mathsf{extract_{Int}}(a))$$

Also, because A is not Harrop, $Sk(\exists x : s \bullet A)[\mathsf{extract_{Int}}(p)/f_P]$ is

$$Sk(A)[fst(\mathsf{extract_{Int}}(p))/x][snd(\mathsf{extract_{Int}}(p))/f_A]$$

Now, by the IH, there is a proof

$$\begin{array}{c} \Gamma' \\ \vdots \\ Sk(A)[v/x][\mathsf{extract_{Int}}(a)/f_A] \end{array} \tag{2.43}$$

Using the (red=) schema, we prove

$$\frac{fst(\mathsf{extract_{Int}}(p)) \ \rhd^*_{SML} \ v}{fst(\mathsf{extract_{Int}}(p)) = v} R_1 \tag{2.44}$$

where R_1 is red $= [[fst(\mathsf{extract_{Int}}(p)); v]; [s]]$. Similarly, we have

$$\frac{snd(\mathsf{extract_{Int}}(p)) \ \rhd^*_{SML} \ snd(\mathsf{extract_{Int}}(a))}{snd(\mathsf{extract_{Int}}(p)) = \mathsf{extract_{Int}}(a)} R_2 \tag{2.45}$$

where R_1 is red $= [[snd(\mathsf{extract_{Int}}(p)); \mathsf{extract_{Int}}(a)]; [\mathsf{etype}(A)]]$.

We apply the (subst) schema, setting P to $Sk(A)[y/x][\mathsf{extract_{Int}}(a)/f_A]$, setting u to be $fst(\mathsf{extract_{Int}}(p))$ and r to be v, obtaining

$$\frac{\dfrac{\begin{array}{c}\Gamma' \\ \vdots\ (2.43) \\ Sk(A)[v/x][\mathsf{extract_{Int}}(a)/f_A]\end{array} \quad \begin{array}{c} \vdots\ (2.44) \\ fst(\mathsf{extract_{Int}}(p)) = v\end{array}}{(Sk(A)[v/x][\mathsf{extract_{Int}}(a)/f_A] \wedge fst(\mathsf{extract_{Int}}(p)) = v)} (\wedge\text{-I})}{Sk(A)[fst(\mathsf{extract_{Int}}(p))/x][\mathsf{extract_{Int}}(a)/f_A]} R_2 \tag{2.46}$$

where R_2 is (subst)$[[Sk(A)[y/x][\mathsf{extract_{Int}}(a)/f_A]]; [fst(\mathsf{extract_{Int}}(p)); v]; [s]]$.

Then, we apply the (*subst*) schema a second time, letting P stand for the statement $Sk(A)[fst(\mathsf{extract_{Int}}(p))/x][y/f_A]$, setting u to $snd(\mathsf{extract_{Int}}(p))$ and r to $\mathsf{extract_{Int}}(a)$, obtaining the required proof

$$\cfrac{\cfrac{\begin{array}{c}\Gamma' \\ \vdots\ (2.46) \\ P[r/y]\end{array} \quad \begin{array}{c}\vdots\ (2.45) \\ u \doteq r\end{array}}{(P[r/y] \wedge u = r)}\ (\wedge\text{-I})}{Sk(A)[fst(\mathsf{extract_{Int}}(p))/x][\mathsf{extract_{Int}}(a)/f_A]}\ R_3$$

where R_3 is

$$(\text{subst})[[Sk(A)[fst(\mathsf{extract_{Int}}(p)/x][\mathsf{extract_{Int}}(a)/f_A]]]; \\ [snd(\mathsf{extract_{Int}}(p)); \mathsf{extract_{Int}}(a)]; [\mathsf{etype}(A)]]$$

This concludes case 2 and the case of ($\exists$-I).

Case: ($\exists$-E). Assume that p^P is of the form C, obtained by an application of ($\exists$-E)

$$\frac{\Gamma_1 \vdash_{\mathsf{Int}} a^{\exists y:s \bullet A} \quad \Gamma_2, u^{A[v/y]} \vdash_{\mathsf{Int}} b^C}{\Gamma_1, \Gamma_2 \vdash_{\mathsf{Int}} \mathsf{select}\ (a)\ \mathsf{in}\ v : s.u.b^C}\ (\exists\text{-E}) \tag{2.47}$$

So, $\Gamma' = \Gamma_1' \cup \Gamma_2'$.

There are two cases, dependent on whether A is Harrop or not.

1. If A is Harrop, then $\mathsf{extract_{Int}}(p)$ is

$$(\texttt{fn v : s =>}\ \mathsf{extract_{Int}}(b))\ \mathsf{extract_{Int}}(a)$$

Because A is Harrop, $Sk(A[v/y]) = A[v/y]$. So, by the IH, there is a proof such that

$$\begin{array}{c}\Gamma_2', A[v/y] \\ \vdots \\ Sk(C)[\mathsf{extract_{Int}}(b)/f_C]\end{array} \tag{2.48}$$

Taking this proof, we first apply ($\Rightarrow$-I) followed by ($\forall$-I) to give

$$\cfrac{\cfrac{\begin{array}{c}\Gamma_2', [A[v/y]] \\ \vdots \\ Sk(C)[\mathsf{extract_{Int}}(b)/f_C]\end{array}}{A[v/y] \Rightarrow Sk(C)[\mathsf{extract_{Int}}(b)/f_C]}\ (\Rightarrow\text{-I})}{\forall v : s \bullet A[v/y] \Rightarrow Sk(C)[\mathsf{extract_{Int}}(b)/f_C]}\ (\forall\text{-I}) \tag{2.49}$$

Because A is Harrop,

$$\begin{aligned} & Sk(\exists y : s \bullet A)[\mathsf{extract_{Int}}(a)/f_{\exists y:s \bullet A}] \\ = {} & A[f_{\exists y:s \bullet A}/y][\mathsf{extract_{Int}}(a)/f_{\exists y:s \bullet A}] \\ = {} & A[\mathsf{extract_{Int}}(a)/y] \end{aligned}$$

So, by the IH, there is a proof

$$\begin{array}{c} \Gamma_1' \\ \vdots \\ A[\mathsf{extract}_{\mathsf{Int}}(a)/y] \end{array} \tag{2.50}$$

We apply ($\forall$-E) on (2.49), setting v to $\mathsf{extract}_{\mathsf{Int}}(a)$, and then apply ($\Rightarrow$-E) on the result, instantiating with (2.50):

$$\dfrac{\dfrac{\begin{array}{c}\Gamma_2' \\ \vdots\ (2.49) \\ \forall v : s \bullet A[v/y] \Rightarrow Sk(C)[\mathsf{extract}_{\mathsf{Int}}(b)/f_A]\end{array}}{AC}(\forall\text{-E}) \quad \begin{array}{c}\Gamma_1' \\ \vdots\ (2.50) \\ A[\mathsf{extract}_{\mathsf{Int}}(a)/y]\end{array}}{Sk(C)[\mathsf{extract}_{\mathsf{Int}}(b)/f_C][\mathsf{extract}_{\mathsf{Int}}(a)/v]}(\Rightarrow\text{-I}) \tag{2.51}$$

where AC denotes the formula

$$A[\mathsf{extract}_{\mathsf{Int}}(a)/y] \Rightarrow Sk(C)[\mathsf{extract}_{\mathsf{Int}}(b)/f_C][\mathsf{extract}_{\mathsf{Int}}(a)/v]$$

Because, by definition of the ($\exists$-E) rule, v cannot occur in C, (2.51) is the same proof as

$$\begin{array}{c} \Gamma_1', \Gamma_2' \\ \vdots \\ Sk(C)[(\mathsf{extract}_{\mathsf{Int}}(b)[\mathsf{extract}_{\mathsf{Int}}(a)/v])/f_C] \end{array} \tag{2.52}$$

By application of the (red=) schema, we can obtain

$$\vdash_{\mathsf{Int}} (\mathit{fn}\ v => \mathsf{extract}_{\mathsf{Int}}(b))\ \mathsf{extract}_{\mathsf{Int}}(a) = \mathsf{extract}_{\mathsf{Int}}(b)[\mathsf{extract}_{\mathsf{Int}}(a)/v] \tag{2.53}$$

We then apply the (subst) schema, setting u to be

$$(\mathit{fn}\ v => \mathsf{extract}_{\mathsf{Int}}(b))\ \mathsf{extract}_{\mathsf{Int}}(a)$$

and r to be $\mathsf{extract}_{\mathsf{Int}}(b)[\mathsf{extract}_{\mathsf{Int}}(a)/v]$ and Q to be $Sk(C)[y/f_C]$, thereby obtaining:

$$\dfrac{\dfrac{\begin{array}{c}\Gamma_1', \Gamma_2' \\ \vdots\ (2.52) \\ Sk(C)[(\mathsf{extract}_{\mathsf{Int}}(b)[\mathsf{extract}_{\mathsf{Int}}(a)/v])/f_A]\end{array} \quad \begin{array}{c}\vdots\ (2.53) \\ u = r\end{array}}{(Q[r/y] \wedge u = r)}(\wedge\text{-I})}{Q[u/y]}R$$

where R is

$$\mathrm{subst}[[(\mathit{fn}\ v => \mathsf{extract}_{\mathsf{Int}}(b))\ \mathsf{extract}_{\mathsf{Int}}(a); \mathsf{extract}_{\mathsf{Int}}(b)[\mathsf{extract}_{\mathsf{Int}}(a)/v]]; \\ [\mathsf{etype}(C)]; [Sk(C)[y/f_C]]]$$

This is the required conclusion because

$$\begin{aligned} Q[u/y] &= Sk(C)[((\texttt{fn v =>}\ \mathsf{extract}_{\mathsf{Int}}(b)\ \mathsf{extract}_{\mathsf{Int}}(a)))/f_C] \\ &= Sk(C)[\mathsf{extract}_{\mathsf{Int}}(p)/f_C] \end{aligned}$$

2. If A is not Harrop, then $\mathsf{extract}_{\mathsf{Int}}(p)$ is

$$(\texttt{fn v : s => fn } \mathtt{x_u} : \mathsf{etype}(\mathtt{A}) \texttt{ => } \mathsf{extract}_{\mathsf{Int}}(b)) \quad \texttt{fst}(\mathsf{extract}_{\mathsf{Int}}(a))\ \texttt{snd}(\mathsf{extract}_{\mathsf{Int}}(a))$$

By the IH, there is a proof

$$\begin{array}{c} \Gamma_2', Sk(A)[v/y][x_u/f_A] \\ \vdots \\ Sk(C)[\mathsf{extract}_{\mathsf{Int}}(b)/f_C] \end{array} \tag{2.54}$$

To this proof we can apply ($\Rightarrow$-I), followed by two applications of ($\forall$-I), giving

$$\dfrac{\dfrac{\dfrac{\begin{array}{c} \Gamma_2', [Sk(A)[v/y][x_u/f_A]] \\ \vdots \\ Sk(C)[\mathsf{extract}_{\mathsf{Int}}(b)/f_C] \end{array}}{Sk(A)[v/y][x_u/f_A] \Rightarrow Sk(C)[\mathsf{extract}_{\mathsf{Int}}(b)/f_C]}\,(\Rightarrow\text{-I})}{\forall x_u : \mathsf{etype}(A) \bullet Sk(A)[v/y][x_u/f_A] \Rightarrow Sk(C)[\mathsf{extract}_{\mathsf{Int}}(b)/f_C]}\,(\forall\text{-I})}{\forall v : s \bullet \forall x_u : \mathsf{etype}(A) \bullet Sk(A)[v/y][x_u/f_A] \Rightarrow Sk(C)[\mathsf{extract}_{\mathsf{Int}}(b)/f_C]}\,(\forall\text{-I}) \tag{2.55}$$

Because A is not Harrop,

$$\begin{aligned} & Sk(\exists y : s \bullet A)[\mathsf{extract}_{\mathsf{Int}}(a)/f_{\exists y:s\bullet A}] \\ &= (Sk(A)[\mathit{fst}(f_{\exists y:s\bullet A})/y][\mathit{snd}(f_{\exists y:s\bullet A})/f_A])[\mathsf{extract}_{\mathsf{Int}}(a)/f_{\exists y:s\bullet A}] \\ &= Sk(A)[\mathit{fst}(\mathsf{extract}_{\mathsf{Int}}(a))/y][\mathit{snd}(\mathsf{extract}_{\mathsf{Int}}(a))/f_A] \end{aligned}$$

So, by the IH, there is a proof

$$\begin{array}{c} \Gamma_1' \\ \vdots \\ Sk(A)[\mathit{fst}(\mathsf{extract}_{\mathsf{Int}}(a))/y][\mathit{snd}(\mathsf{extract}_{\mathsf{Int}}(a))/f_A] \end{array} \tag{2.56}$$

We apply ($\forall$-E) on (2.55), setting v to $\mathit{fst}(\mathsf{extract}_{\mathsf{Int}}(a))$, then apply ($\forall$-E), setting x_u to $\mathit{snd}(\mathsf{extract}_{\mathsf{Int}}(a))$, and then apply ($\Rightarrow$-E) on the result, instantiating with (2.56):

$$\dfrac{\dfrac{\dfrac{\begin{array}{c} \Gamma_2' \\ \vdots\ (2.55) \\ AA \end{array}}{AC}\,(\forall\text{-E})}{A_2 \Rightarrow C_2}\,(\forall\text{-E}) \qquad \begin{array}{c} \Gamma_1' \\ \vdots\ (2.56) \\ A_2 \end{array}}{C_2}\,(\Rightarrow\text{-I}) \tag{2.57}$$

where AA denotes the formula

$$\forall v : s \bullet \forall x_u : \mathsf{etype}(A) \bullet Sk(A)[v/y][x_u/f_A] \Rightarrow Sk(C)[\mathsf{extract}_{\mathsf{Int}}(b)/f_A]$$

AC denotes the formula

$$A[\mathit{fst}(\mathsf{extract}_{\mathsf{Int}}(a))/y] \Rightarrow Sk(C)[\mathsf{extract}_{\mathsf{Int}}(b)/f_C][\mathit{fst}(\mathsf{extract}_{\mathsf{Int}}(a))/v],$$

where A_2 denotes the formula

$$Sk(A)[\mathit{fst}(\mathsf{extract}_{\mathsf{Int}}(a))/y][\mathit{snd}(\mathsf{extract}_{\mathsf{Int}}(a))/f_A]$$

and where C_2 denotes the formula

$$Sk(C)[\mathsf{extract}_{\mathsf{Int}}(b)/f_C][\mathit{fst}(\mathsf{extract}_{\mathsf{Int}}(a))/v][\mathit{snd}(\mathsf{extract}_{\mathsf{Int}}(a))/x_u].$$

Because, by the definition of the ($\exists$-E) rule, v cannot occur in C, and also we can assume x_u cannot occur in C,

$$Sk(C)[\mathsf{extract}_{\mathsf{Int}}(b)/f_C][\mathit{fst}(\mathsf{extract}_{\mathsf{Int}}(a))/v][\mathit{snd}(\mathsf{extract}_{\mathsf{Int}}(a))/x_u]$$

is the same formula as

$$Sk(C)[((\mathsf{extract}_{\mathsf{Int}}(b)[\mathit{fst}(\mathsf{extract}_{\mathsf{Int}}(a))/v][\mathit{snd}(\mathsf{extract}_{\mathsf{Int}}(a))/x_u]))/f_C]$$

and so (2.57) is the same proof as

$$\begin{array}{c} \Gamma_1', \Gamma_2' \\ \vdots \\ Sk(C)[((\mathsf{extract}_{\mathsf{Int}}(b)[\mathit{fst}(\mathsf{extract}_{\mathsf{Int}}(a))/v][\mathit{snd}(\mathsf{extract}_{\mathsf{Int}}(a))/x_u]))/f_C] \end{array} \tag{2.58}$$

We can apply the (red=) schema to obtain

$$\begin{array}{r} (\mathit{fn}\ v : s => \mathit{fn}\ x_u : \mathsf{etype}(A) => \mathsf{extract}_{\mathsf{Int}}(b)) \\ \mathit{fst}(\mathsf{extract}_{\mathsf{Int}}(a))\ \mathit{snd}(\mathsf{extract}_{\mathsf{Int}}(a)) = \\ \mathsf{extract}_{\mathsf{Int}}(b)[\mathit{fst}(\mathsf{extract}_{\mathsf{Int}}(a))/v][\mathit{snd}(\mathsf{extract}_{\mathsf{Int}}(a))/x_u] \end{array} \tag{2.59}$$

Recall that $\mathsf{extract}_{\mathsf{Int}}(p)$ is

$$\begin{array}{r} \mathit{fn}\ v : s => \mathit{fn}\ x_u : \mathsf{etype}(A) => \mathsf{extract}_{\mathsf{Int}}(b) \\ \mathit{fst}(\mathsf{extract}_{\mathsf{Int}}(a))\ \mathit{snd}(\mathsf{extract}_{\mathsf{Int}}(a)) \end{array}$$

So, another way of writing the conclusion of the proof (2.59) is

$$\vdash_{\mathsf{Int}} \mathsf{extract}_{\mathsf{Int}}(p) = \mathsf{extract}_{\mathsf{Int}}(b)[\mathit{fst}(\mathsf{extract}_{\mathsf{Int}}(a))/v][\mathit{snd}(\mathsf{extract}_{\mathsf{Int}}(a))/x_u] \tag{2.60}$$

So, we can apply the (subst) schema, using (2.59), setting u to be $\mathsf{extract}_{\mathsf{Int}}(p)$, setting r to be

$$\mathsf{extract_{Int}}(b)[fst(\mathsf{extract_{Int}}(a))/v][snd(\mathsf{extract_{Int}}(a))/x_u]$$

and P to be $Sk(C)[y/f_C]$, obtaining:

$$\dfrac{\dfrac{\begin{matrix}\Gamma'_1, \Gamma'_2 \\ \vdots\ (2.58) \\ Sk(C)[y/f_C][\mathsf{extract_{Int}}(b)[fst(\mathsf{extract_{Int}}(a))/v] \\ [snd(\mathsf{extract_{Int}}(a))/x_u]/y]\end{matrix} \qquad \begin{matrix}\vdots\ (2.60) \\ u = r\end{matrix}}{(Sk(C)[y/f_C][r/y] \wedge u = r)}\ (\wedge\text{-I})}{P[u/y]}\ R_2$$

where R_2 is

$$\mathrm{subst}[[\mathsf{extract_{Int}}(p), \mathsf{extract_{Int}}(b)[fst(\mathsf{extract_{Int}}(a))/v] \\ [snd(\mathsf{extract_{Int}}(a))/x_u]]; [\mathsf{etype}(C)]; [Sk(C)[y/f_C]]]$$

This proves the required conclusion, because

$$\begin{aligned} P[u/y] &= Sk(C)[y/f_C][\mathsf{extract_{Int}}(p)/y] \\ &= Sk(C)[\mathsf{extract_{Int}}(p)/f_C] \end{aligned}$$

Case: ($\Rightarrow$-I). Assume that p^P is of the form $(A \Rightarrow B)$ obtained by an application of ($\Rightarrow$-I)

$$\dfrac{\Gamma, u^A \vdash_{\mathsf{Int}} b^B}{\Gamma \vdash_{\mathsf{Int}} \mathsf{abstract}\ u.\ b^{(A \Rightarrow B)}}\ (\Rightarrow\text{-I})$$

There are two cases, dependent on whether A is Harrop or not.

1. Assume that A is Harrop. Then $\mathsf{extract_{Int}}(p)$ is $\mathsf{extract_{Int}}(b)$. By the IH, we know that there is a proof of the form

$$\begin{matrix}\Gamma', A \\ \vdots \\ Sk(B)[\mathsf{extract_{Int}}(b)/f_B]\end{matrix}$$

because $Sk(A)$ is A. Applying ($\Rightarrow$-I) to this proof, we obtain

$$\dfrac{\begin{matrix}\Gamma', [A] \\ \vdots \\ Sk(B)[\mathsf{extract_{Int}}(b)/f_B]\end{matrix}}{A \Rightarrow Sk(B)[\mathsf{extract_{Int}}(b)/f_B]}\ (\Rightarrow\text{-I})$$

This is the required proof because $Sk(A \Rightarrow B)[\mathsf{extract_{Int}}(p)/f_{A \Rightarrow B}]$ is the same formula as the conclusion $A \Rightarrow Sk(B)[\mathsf{extract_{Int}}(p)/f_B]$.

2. Assume that A is not Harrop. Then $\mathsf{extract}_{\mathsf{Int}}(p)$ is `fn xu => ` $\mathsf{extract}_{\mathsf{Int}}(b)$. By the IH, we know that there is a proof of the form

$$\begin{array}{c} \Gamma', Sk(A)[x_u/f_A] \\ \vdots \\ Sk(B)[\mathsf{extract}_{\mathsf{Int}}(b)/f_B] \end{array} \qquad (2.61)$$

Using the (red=) schema, it is easy to derive

$$\vdash_{\mathsf{Int}} (fn\ x_u => \mathsf{extract}_{\mathsf{Int}}(b))\ x_u = \mathsf{extract}_{\mathsf{Int}}(b) \qquad (2.62)$$

We apply the (subst) schema, using (2.62), setting u to be $(fn\ x_u => \mathsf{extract}_{\mathsf{Int}}(b))\ x_u$, r to be $\mathsf{extract}_{\mathsf{Int}}(b)$ and P to be $Sk(B)[y/f_B]$, obtaining:

$$\dfrac{\dfrac{\begin{array}{c}\Gamma', Sk(A)[x_u/f_A] \\ \vdots\ 2.61 \\ Sk(B)[\mathsf{extract}_{\mathsf{Int}}(b)/f_B]\end{array} \quad \begin{array}{c}\vdots\ (2.62) \\ (fn\ x_u => \mathsf{extract}_{\mathsf{Int}}(b))\ x_u = \mathsf{extract}_{\mathsf{Int}}(b)\end{array}}{(Sk(B)[\mathsf{extract}_{\mathsf{Int}}(b)/f_B] \wedge u = r)}\ (\wedge\text{-I})}{Sk(B)[(fn\ x_u => \mathsf{extract}_{\mathsf{Int}}(b)\ x_u)/f_B]}\ R_1 \qquad (2.63)$$

where R_1 is

$$(\text{subst})[[(fn\ x_u => \mathsf{extract}_{\mathsf{Int}}(b))\ x_u; \mathsf{extract}_{\mathsf{Int}}(b)]; [\mathsf{etype}(B)]; [Sk(B)[y/f_B]]]$$

By definition of $\mathsf{extract}_{\mathsf{Int}}(p)$, we know that the conclusion of (2.63) can be written $Sk(B)[\mathsf{extract}_{\mathsf{Int}}(p)\ x_u/f_B]$. We apply ($\Rightarrow$-I) on (2.63):

$$\dfrac{\begin{array}{c}[Sk(A)[x_u/f_B]] \\ \vdots\ (2.63) \\ Sk(B)[(\mathsf{extract}_{\mathsf{Int}}(p)\ x_u)/f_B]\end{array}}{Sk(A)[x_u/f_A] \Rightarrow Sk(B)[(\mathsf{extract}_{\mathsf{Int}}(p)\ x_u)/f_B]}\ (\Rightarrow\text{-I}) \qquad (2.64)$$

Finally, we apply ($\forall$-I) to (2.64), abstracting over x_u, to give

$$\forall x_u : \mathsf{etype}(A) \bullet Sk(A)[x_u/f_A] \Rightarrow Sk(B)[\mathsf{extract}_{\mathsf{Int}}(p)\ x_u/f_B]$$

This, by the definition of Skolem form, is the required proof.

Case: ($\Rightarrow$-E). Assume that p^P is of the form $\mathsf{app}(a,b)^C$, obtained by an application of ($\Rightarrow$-E)

$$\dfrac{\Gamma_1 \vdash_{\mathsf{Int}} a^{B \Rightarrow C} \quad \Gamma_2 \vdash_{\mathsf{Int}} b^B}{\Gamma_1, \Gamma_2 \vdash_{\mathsf{Int}} \mathsf{app}(a,b)^C}\ (\Rightarrow\text{-E})$$

so that $\Gamma' = \Gamma_1' \cup \Gamma_2'$.

There are two cases, dependent on whether B is Harrop or not.

1. Assume that B is Harrop. Then

$$\mathsf{extract}_{\mathsf{Int}}(p) = \mathsf{extract}_{\mathsf{Int}}(a) \tag{2.65}$$

Also, by the IH, we know that there are proofs

$$\begin{array}{c} \Gamma_1' \\ \vdots \\ B \Rightarrow Sk(C)[\mathsf{extract}_{\mathsf{Int}}(a)/f_C] \end{array} \tag{2.66}$$

and

$$\begin{array}{c} \Gamma_2' \\ \vdots \\ B \end{array} \tag{2.67}$$

We apply ($\Rightarrow$-E), instantiating (2.66) with (2.67) to give

$$\dfrac{\begin{array}{c} \Gamma_1' \\ \vdots\ (2.66) \\ B \Rightarrow Sk(C)[\mathsf{extract}_{\mathsf{Int}}(a)/f_C] \end{array} \quad \begin{array}{c} \Gamma_2' \\ \vdots\ (2.67) \\ B \end{array}}{Sk(C)[\mathsf{extract}_{\mathsf{Int}}(a)/f_C]}\ (\Rightarrow\text{-E})$$

Because of (2.65), the conclusion of this proof is the same as stating $Sk(C)[\mathsf{extract}_{\mathsf{Int}}(p)/f_C]$, as required.

2. Assume that B is not Harrop. Then $\mathsf{extract}_{\mathsf{Int}}(p)$ is $(\mathsf{extract}_{\mathsf{Int}}(a)\ \mathsf{extract}_{\mathsf{Int}}(b))$ Also, by the IH, we know that there are proofs

$$\begin{array}{c} \Gamma_1' \\ \vdots \\ \forall x : \mathsf{etype}(B) \bullet Sk(B)[x/f_B] \Rightarrow Sk(C)[\mathsf{extract}_{\mathsf{Int}}(a)\ x/f_C] \end{array} \tag{2.68}$$

and

$$\begin{array}{c} \Gamma_2' \\ \vdots \\ Sk(B)[\mathsf{extract}_{\mathsf{Int}}(b)/f_B] \end{array} \tag{2.69}$$

We apply ($\forall$-E) on (2.66), letting x be $\mathsf{extract}_{\mathsf{Int}}(b)$, and then apply ($\Rightarrow$-E), instantiating with (2.67) to give

$$\dfrac{\dfrac{\begin{array}{c} \Gamma_1' \\ \vdots\ (2.66) \\ \forall x : \mathsf{etype}(B) \bullet Sk(B)[x/f_B] \Rightarrow \\ Sk(C)[\mathsf{extract}_{\mathsf{Int}}(a)/f_C] \end{array}}{\begin{array}{c} Sk(B)[\mathsf{extract}_{\mathsf{Int}}(b)/f_B] \Rightarrow \\ Sk(C)[\mathsf{extract}_{\mathsf{Int}}(a)/f_C] \end{array}}\ (\forall\text{-E}) \quad \begin{array}{c} \Gamma_2' \\ \vdots\ (2.67) \\ Sk(B)[\mathsf{extract}_{\mathsf{Int}}(b)/f_B] \end{array}}{Sk(C)[\mathsf{extract}_{\mathsf{Int}}(a)\ \mathsf{extract}_{\mathsf{Int}}(b)/f_C]}\ (\Rightarrow\text{-E})$$

The conclusion of this proof is the same as stating $Sk(C)[\mathsf{extract}_{\mathsf{Int}}(p)/f_C]$ as required.

Case: ($\forall$-I). Assume that p^P is of the form $\mathsf{use}\ x : s.\ a^{\forall x:s \bullet A}$, obtained by an application of ($\forall$-I)

$$\frac{\Gamma \vdash_{\mathsf{Int}} a^A}{\Gamma \vdash_{\mathsf{Int}} \mathsf{use}\ x : s.\ a^{\forall x:s \bullet A}}\ (\forall\text{-I})$$

Because we have assumed that P is not Harrop (so $\forall x : s \bullet A$ is not Harrop), A must not be Harrop, and $\mathsf{extract_{Int}}(p)$ is `fn x =>` $\mathsf{extract_{Int}}(a)$.

By the IH, there is a proof

$$\begin{array}{c} \Gamma' \\ \vdots \\ Sk(A)[\mathsf{extract_{Int}}(a)/f_A] \end{array} \tag{2.70}$$

First, we use (red=) to derive

$$\vdash_{\mathsf{Int}} (\mathit{fn}\ x => \mathsf{extract_{Int}}(a)\ x) = \mathsf{extract_{Int}}(a) \tag{2.71}$$

Then, we apply the (*subst*) schema, using (2.71), setting u to be $(\mathsf{extract_{Int}}(p)\ x)$, r to be $\mathsf{extract_{Int}}(a)$ and P to be $Sk(A)[y/f_A]$, obtaining:

$$\dfrac{\dfrac{\begin{array}{c}\Gamma' \\ \vdots\ (2.70) \\ Sk(A)[\mathsf{extract_{Int}}(a)/f_A]\end{array} \quad \begin{array}{c}\vdots\ (2.71) \\ \mathsf{extract_{Int}}(p)\ x = \mathsf{extract_{Int}}(a)\end{array}}{(Sk(A)[\mathsf{extract_{Int}}(a)/f_A] \wedge u = r)}\ (\wedge\text{-I})}{Sk(A)[(\mathsf{extract_{Int}}(p)\ x)/f_A]}\ R \tag{2.72}$$

where R is $(\mathrm{subst})[[Sk(A)[y/f_A]]; [(\mathsf{extract_{Int}}(p)\ x); \mathsf{extract_{Int}}(a)]; [\mathsf{etype}(A)]]$.

Applying ($\forall$-I) on (2.72), abstracting over x, gives us the required conclusion

$$\vdash_{\mathsf{Int}} \forall x : \mathsf{etype}(A) \bullet Sk(A)[\mathsf{extract_{Int}}(p)\ x/f_A]$$

Case: ($\forall$-E). Assume that p^P is of the form $\mathsf{specific}(a,t)^{A[t/x]}$ obtained by an application of ($\forall$-E)

$$\frac{\Gamma \vdash_{\mathsf{Int}} a^{\forall x:s \bullet A}}{\Gamma \vdash_{\mathsf{Int}} \mathsf{specific}(a,t)^{A[t/x]}}\ (\forall\text{-E})$$

Because we have assumed that P (and so $\forall x : s \bullet A$) is not Harrop, this means that A must not be Harrop, and $\mathsf{extract_{Int}}(p)$ is $(\mathsf{extract_{Int}}(a)\ t)$

By the IH, there is a proof

$$\begin{array}{c} \Gamma' \\ \vdots \\ \forall x : s \bullet Sk(A)[\mathsf{extract_{Int}}(a)\ x/f_A] \end{array} \tag{2.73}$$

To obtain the required proof, we need only apply ($\forall$-E) over (2.73), instantiating with t, to give

$$\frac{\begin{array}{c}\Gamma' \\ \vdots\ (2.73) \\ \forall x : s \bullet Sk(A)[\mathsf{extract_{Int}}(a)\ x/f_A]\end{array}}{Sk(A)[\mathsf{extract_{Int}}(a)\ t/f_A]}\ (\forall\text{-E}) \tag{2.74}$$

This is the required proof, because

$$Sk(A)[\mathsf{extract_{Int}}(a)\ t/f_A]$$

is the same formula as

$$Sk(A)[\mathsf{extract_{Int}}(p)/f_A]$$

This last case concludes the proof. □

2.5.3 Extraction from proofs with axioms and schemata

The proof can be extended to include the use of axioms and application of schemata, if we make the following assumptions.

Assumption 2.3. We assume that, for each proof-term corresponding to an axiom rule:

$$\mathsf{Axiom}(A)^A$$

there is a function in Σ and a corresponding program in the *SML* preamble

$$PK_A : \mathsf{etype}(A)$$

such that PK_A is a modified realizer of A:

$$PK_A\ \mathsf{mr}\ A$$

Then, defining

$$\mathsf{extract_{Int}}(\mathsf{Axiom}(A)) = PK_A$$

the proof is extended trivially to include axioms.

Assumption 2.4. We assume that, for each proof-term corresponding to a rule generated from a schema,

$$\vdash \mathsf{Schema}(N, [\bar{e}; \bar{F}; \bar{t}; \bar{S}])^A$$

there is a function in Σ and a corresponding program in the *SML* preamble

$$\vdash_{\mathsf{Int}} PK_{N[\bar{e};\bar{F};\bar{t};\bar{S}]} : \mathsf{etype}(A)$$

such that

$$PK_{N[\bar{e};\bar{F};\bar{t};\bar{S}]}\ \mathsf{kr}\ A.$$

Defining

$$\mathsf{extract_{Int}}(\mathsf{Schema}(N, [\bar{e}; \bar{F}; \bar{t}; \bar{S}])) = PK_{N[\bar{e};\bar{F};\bar{t};\bar{S}]}$$

the proof extends trivially to include schemata application.

Example 2.6. We take the modified realizers for instances of the substitution schema (subst)

$$\frac{\vdash_{\mathsf{Int}} q_1^{P[r/y]} \quad \vdash_{\mathsf{Int}} q_2^{u=_s r}}{\vdash_{\mathsf{Int}} \mathsf{Schema}(subst, [[q_1; q_2]; P; \bar{y}; \bar{Z}])^{P[u/y]}} \text{(subst)}[[P]; [u; r]; [s]]$$

to be

$$\mathsf{PK}_{subst,[[q_1;q_2];P;\bar{y};\bar{Z}]} = \mathsf{extract}_{\mathsf{Int}}(q_1)[u/y] \tag{2.75}$$

This is permitted, by extending the proof by induction of Theorem 2.5.3 to include instances of (subst), with the following additional case. Assuming a proof ends in (subst)

$$\frac{\vdash_{\mathsf{Int}} q_1[r/y]^{P[r/y]} \quad \vdash_{\mathsf{Int}} q_2^{u=_s r}}{\vdash_{\mathsf{Int}} \mathsf{Schema}(subst, [[q_1; q_2]; P; \bar{y}; \bar{Z}])^{P[u/y]}} \text{(subst)}[[P]; [u; r]; [s]]$$

then, by the IH, there is a proof that

$$\vdash_{\mathsf{Int}} Sk(P[r/y])[\mathsf{extract}_{\mathsf{Int}}(q_1[r/y])/f_{P[r/y]}]$$

which means

$$\vdash_{\mathsf{Int}} Sk(P)[\mathsf{extract}_{\mathsf{Int}}(q_1)/f_P][r/y]$$

So, by applying (subst) again we have

$$\frac{\vdash_{\mathsf{Int}} Sk(P)[\mathsf{extract}_{\mathsf{Int}}(q_1)/f_P][r/y] \quad u =_s r}{\vdash_{\mathsf{Int}} Sk(P)[\mathsf{extract}_{\mathsf{Int}}(q_1)/f_P][u/y]} \text{(subst)}[[P]; [u; r]; [s]]$$

This maybe be rewritten to be the required conclusion, by (2.75),

$$\vdash_{\mathsf{Int}} Sk(P)[PK_{subst,[[q_1;q_2];P;\bar{y};\bar{Z}]}/f_P]$$

Remark 2.19. It is possible to define modified realizers for proof-terms corresponding to induction schema over recursive data types in a systematic way. However, we do this in Part IV (Chapter 9), for a constructive logic about structured specifications. That work is a more sophisticated treatment of the ideas here, and can be brought down to the simple, single signature treatment of $\mathcal{ADT}$ given here. For the moment, we simply give an example for integer induction.

Example 2.7. A definition of the sort int (integers), constructed from a constant $0 : int$ and successor function $suc : (int \to int)$ is associated with the induction schema

$$\frac{\vdash_{\mathsf{Int}} a^{X[0/z] \wedge (\forall x:int \bullet X[x/z] \Rightarrow X[suc(x)/z])}}{\vdash_{\mathsf{Int}} \mathsf{Schema}(recInt, [a; [X]])^{\forall y:int \bullet X[y/z]}} \text{(recInt}[[X]])$$

an application of this schema to a formula P is of the form

$$\frac{\vdash_{\mathsf{Int}} a^{P[0/z]\wedge(\forall x:int\bullet P[x/z]\Rightarrow P[suc(x)/z])}}{\vdash_{\mathsf{Int}} \mathsf{Schema}(recInt, [a; [P]])^{\forall y:int\bullet P[y/z]}} \text{ (recInt[[}P\text{]])}$$

We define the function $PK_{recInt[[P]]}$ to be shorthand for the SML program

```
let rec rho n y =
  begin match y with
    0 => fst(n)
  | x => snd(n)(x − 1)(rho n (x − 1))
```

Eval can be defined appropriately to correctly model the evaluation of this program in *SML*, so that $PK_{recInt[[P]]}$ is a modified realizer of $\forall y : int \bullet P[y/z]$. Rather than proving this, we shall simply assume this is true. See Part IV (Chapter 9) for a systematic treatment of modified realizers for induction schemata.

It immediately follows from this Theorem 2.5.3 and Assumptions 2.3 and 2.4 that logical proofs yield modified realizers of the proved formulae.

Corollary 2.5.1 (Logical proofs yield modified realizers). *If there is a proof* $\emptyset \vdash_{\mathsf{Int}} p^F$ *then*

$$\mathsf{extract}_{\mathsf{Int}}(p) \text{ mr } F$$

2.5.4 Relation between proofs and programs

It can be shown that the extraction results of this section, together with the strong normalization result of the previous section, lead to the following situation:

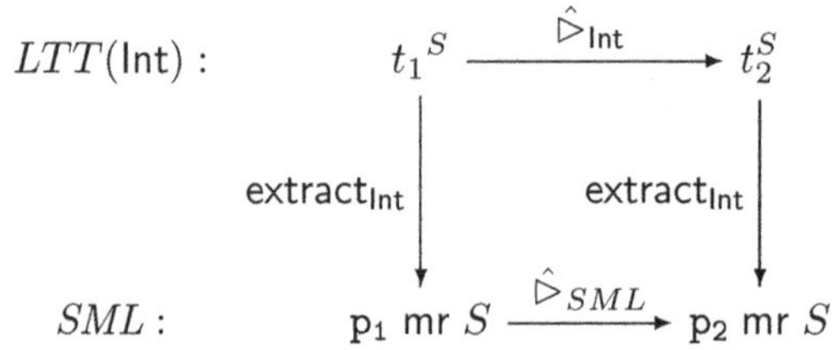

$$\begin{array}{lccc}
LTT(\mathsf{Int}): & t_1{}^S & \xrightarrow{\hat{\rhd}_{\mathsf{Int}}} & t_2^S \\
 & \Big\downarrow \mathsf{extract}_{\mathsf{Int}} & & \Big\downarrow \mathsf{extract}_{\mathsf{Int}} \\
SML: & \mathsf{p}_1 \text{ mr } S & \xrightarrow{\hat{\rhd}_{SML}} & \mathsf{p}_2 \text{ mr } S
\end{array}$$

Thus, in our state-of-the-art approach, proofs and programs are separate entities, related by extraction and a notion of realizability.

2.6 Example: Password checking system

We illustrate our concepts with the following example about an email service.

We consider a service that hosts email accounts for a number of users. When a user joins the service, he/she is required to define a new numerical password. We make the following assumptions concerning the password correctness functions for a new user joining, or logging onto, the system:

- Password numbers must be 4 digits long.
- If the number chosen is not of the right length, the system should output a response message, asking the user to select a new number of the correct length.
- If the number is of the correct length, then the system should output a response message to this effect.

We shall model the system by specifying these assumptions, defining notions of acceptable lengths of passwords and the correct responses for given passwords.

We model aspects of the password checking system by assuming the following functions and predicates to be in $\mathcal{ADT}$:

- We assume appropriate sorts and function symbols for the booleans, *bool*, natural numbers, *nat*, and strings *string*.
- We define a new boolean function $okLength(x)$ that will output *true* if the password number (x) is of the required length
- A new predicate $OkPwd(x)$ holds over a number, if the number is an acceptable password (that is, if $okLength(x) = true$).
- A new predicate $ValidMsg(x, y)$ that holds if a string y is a correct response message for the input of a password number x.

We take the following axioms in $\mathcal{ADT}$ to model our domain assumptions:

$$okLength(x) = true \Rightarrow OkPwd(x) \tag{2.76}$$
$$okLength(x) = false \Rightarrow \neg OkPwd(x) \tag{2.77}$$
$$OkPwd(x) \Rightarrow ValidMsg(x, \text{`Password acceptable'}) \tag{2.78}$$
$$\neg OkPwd(x) \Rightarrow \begin{array}{l} ValidMsg(x, \text{`Please choose a} \\ \text{password of correct length'}) \end{array} \tag{2.79}$$

The first two axioms tell us that a password number is acceptable if, and only if, it is of an acceptable length. The second two axioms define the appropriate response message strings for an acceptable password and for an unacceptable password.

Our goal formula will be that, given any password as input, the system will always output an appropriate response message. This is specified as follows:

$$\vdash \forall x : nat \bullet \exists y : string \bullet ValidMsg(x, y) \tag{2.80}$$

The Skolem form of the formula

$$A = \forall x : nat \bullet \exists y : string \bullet ValidMsg(x, y)$$

is

$$Sk(A) = \forall x : nat \bullet Valid(x, f_A(x))$$

Thus, by the definition of modified realizability, the theorem can be viewed as a specification of a function f_A that outputs an appropriate response message for a given password.

We assume the following axiom about the function $okLength(x)$:

$$\forall x : nat \bullet okLength(x) = true \vee okLength(x) = false \tag{2.81}$$

By our Assumptions 2.2 and 2.3, we assume there is a realizing term

$$PK_A : (nat \to Unit|Unit)$$

in the signature of $\mathcal{ADT}$ such that

$$PK_A \text{ mr } A$$

where A stands for the axiom $\forall x : nat \bullet okLength(x) = true \vee okLength(x) = false$. That is to say, we assume PK_A is present as a function symbol in the *SML* preamble and is defined with the following operational semantics

$$Eval(PK_A(x)) = \begin{cases} Inl(()) \text{ when } okLength(x) = true \\ Inr(()) \text{ when } okLength(x) = false \end{cases}$$

It is easy to see that a term with this semantics will produce the required modified realizer of (2.81): when the length of the password is acceptable, PK_A will be equal to $Inl(())$, and $Inr(())$ otherwise.

We derive (2.80) by reasoning over the possible cases that either $okLength(x) = true$ or $okLength(x) = false$.

Assuming the first case corresponds to the use of a proof-term variable

$$u^{okLength(x)=true}$$

in a derivation of the form

$$\dfrac{\dfrac{\begin{matrix} \vdots \\ p_7^{A_2} \end{matrix} \quad \dfrac{\begin{matrix} \vdots \\ p_6^{A_1} \end{matrix} \quad \dfrac{}{u^{okLength(x)=true}}(\text{Ass-I})}{\mathsf{app}(p_6, u)^{OkPwd(x)}}(\Rightarrow\text{-E})}{\mathsf{app}(p_7, \mathsf{app}(p_6, p_5))^{ValidMsg(x, \text{‘Password acceptable’})}}(\Rightarrow\text{-E})}{\dfrac{\mathsf{show}(\text{‘Password acceptable’}, \mathsf{app}(p_7, \mathsf{app}(p_6, p_5)))^{\exists y:string \bullet ValidMsg(x,y)}}{\mathsf{use}\ x : nat.\ \mathsf{show}(\text{‘Password acceptable’}, \mathsf{app}(p_7, \mathsf{app}(p_6, p_5)))^{\forall x:nat \bullet \exists y:string \bullet ValidMsg(x,y)}}(\forall\text{-I})}(\exists\text{-I})$$

(2.82)

where the proof-term $p_6^{A_1}$ corresponds to an instantiated axiom of $\mathcal{ADT}$, which when written

$$\mathsf{specific}(\mathsf{ax}(\forall x : nat \bullet okLength(x) = true \Rightarrow OkPwd(x)), x)$$

with type $okLength(x) = true \Rightarrow OkPwd(x)$. The proof-term $p_7^{A_2}$ also corresponds to an instantiated axiom of $\mathcal{ADT}$ written in full as

$$\mathsf{specific}(\mathsf{ax}(\forall x : nat \bullet OkPwd(x) \Rightarrow ValidMsg(x, \text{`Password acceptable'})), x)$$

with type $OkPwd(x) \Rightarrow ValidMsg(x, \text{`Password acceptable'})$

Similar reasoning over an assumption variable $v^{okLength(x)=false}$ will give a proof-term of the form

$$\dfrac{\dfrac{\dfrac{}{v^{okLength(x)=false}}\ (\text{Ass-I}) \quad \vdots \quad p_8^{ValidMsg(x,\text{`Please choose a password of correct length'})}}{\mathsf{show}\left(\begin{matrix}\text{`Please choose a} \\ \text{password of correct length'}\end{matrix}, p_8\right)^{\exists y:string\bullet\, ValidMsg(x,y)}}\ (\exists\text{-I})}{\mathsf{use}\ x : nat.\ \mathsf{show}\left(\begin{matrix}\text{`Please choose a} \\ \text{password of correct length'}\end{matrix}, p_8\right)^{\forall x:nat\bullet\exists y:string\bullet\, ValidMsg(x,y)}}\ (\forall\text{-I}) \quad (2.83)$$

where p_8 is a proof-term involving manipulation of our axioms.

Finally, by applying ($\vee$-E) over (2.81), (2.82) and (2.83), and then applying (hide) will give (2.80), as required

$$\vdash p^{\forall x:nat\bullet\exists y:string\bullet\, ValidMsg(x,y)}$$

with proof-term

$$\begin{aligned} p = {} & \mathsf{case}\ \mathsf{Axiom}(A)\ \mathsf{of} \\ & \quad \mathsf{inl}(u).\mathsf{use}\ x : nat.\ \mathsf{show}(\text{`Password acceptable'}, \mathsf{app}(p_7, \mathsf{app}(p_6, p_5))), \\ & \quad \mathsf{inr}(v).\mathsf{use}\ x : nat.\ \mathsf{show}(\text{`Please choose a password of correct length'}, p_8) \end{aligned}$$

The proof-term encodes constructive information obtained from the ($\exists$-I) steps used in the proof — in particular, the witness string y for a valid message such that $ValidMsg(x, y)$ given a password number, x, depending on the length of the password number.

We can apply Theorem 2.5.3 to obtain the function $\mathsf{extract}_{\mathsf{Int}}(q)$:

```
fn x : nat =>
match PK_A(x) with
        Inl(x_u) => 'Password acceptable',
        Inr(x_v) => 'Please choose a password in correct range'
```

such that

$$\vdash_{\mathsf{Int}} \forall x : nat \bullet Valid(x, (\mathsf{extract}_{\mathsf{Int}}(q)x)) \qquad (2.84)$$

as required.

Remark 2.20. We can see that (2.84) holds for any password input a, using Int and the axioms of $\mathcal{ADT}$. For example, if a is such that $okLength(a) = true$, then, by the operational semantics for PK_A,

$$\mathsf{extract}(q)(\mathsf{a}) \hat{\triangleright}_{SML} \text{ 'Password acceptable'}$$

Consequently, by schema (red=)

$$\vdash_{\mathsf{Int}} \mathsf{extract}(q) = \text{'Password acceptable'}$$

By the axioms (2.76) and (2.78) and the schemata and (subst), we know that

$$\vdash_{\mathsf{Int}} ValidMsg(a, \text{'Password acceptable'})$$

But then, by applying (subst) again, we arrive at

$$\vdash_{\mathsf{Int}} Valid(a, (\mathsf{extract}_{\mathsf{Int}}(q)a))$$

This is (2.84), instantiated with a, as required.

The advantage of our extraction methodology is that functions such as $\mathsf{extract}_{\mathsf{mod}}(q)$ are synthesized automatically from proofs, and their correctness as realizers is guaranteed by Theorem 2.5.3. There is no need for a further verification proof of function correctness.

Remark 2.21. Effective reasoning and program synthesis is dependent upon an adequate representation of the target domain. In the approach of this chapter, we axiomatize our domain using a single, unstructured specification $\mathcal{ADT}$. This approach is adequate for working with domains of small scale.

However, at the medium and large scales, it can be difficult to define, comprehend and maintain domain specifications without some notion of compositionality and hierarchy, instead of working within a single, unstructured theory $\mathcal{ADT}$, In Part IV, Chapters 7–9, we shall return to this domain. We shall specify and reason about the domain using structured specifications. That work has the advantage over the approach of this chapter that we can understand our domain using a compositional hierarchy of related theories. Consquently, we can utilize divide-and-conquer approaches to specifying and reasoning about a domain.

2.7 Discussion

There always remains the question of whether the proofs that we use are correct and whether the software we use preserves correctness. We are almost as vulnerable as any mathematician concerning the correctness of a proof. We say "almost" because our proofs are formalized and therefore checking the steps is a mechanical process. However, the extraction process uses software that may be unreliable. We minimise the effects of this because our procedures are simple and essentially syntactic. Ideally we would run the process on our own software to check it, but this is, at present, a daunting task.

In comparison to naïve functional program synthesis approaches, the advantages of the Curry–Howard protocol in the context of functional program synthesis are practical. We have chosen to use a *LTT* based on a first-order, many-sorted logic, because this is an easy type of deductive system to use. Additionally, adopting a loose coupling between terms and sorts of the *LTT* and programs and types of *SML* promotes a natural conceptual distinction: the logic is used for reasoning about programs, while *SML* is used for programming.

In some cases the protocol, or something similar, appears to be necessary in order to be able to define a simple Curry–Howard style extraction mechanism for more complicated logical systems. In the next two parts of this monograph, we will examine this assertion, for imperative program synthesis and Curry–Howard style synthesis over the proof system for reasoning about algebraic specifications. Without the protocol, these results would have been difficult to achieve.

This chapter serves as a reference for the kind of extraction that the Curry–Howard protocol should achieve within the familiar domain of functional programs synthesized from constructive proofs. The reader will therefore find it useful to refer back to this chapter for comparisons when we formally define the protocol and investigate less familiar domains.

3

The Curry–Howard Protocol

In this chapter, we define the Curry–Howard protocol, a framework for generalizing state-of-the-art (SOA) proofs-as-programs. The protocol specifies a minimal set of properties to be satisfied by a logic and programming language. If these properties are satisfied for a new logic and programming language, then we claim that the Curry–Howard isomorphism and SOA proofs-as-programs have been generalized.

The protocol is defined by identifying aspects of SOA proofs-as-programs that are relevant to program extraction. These relevant aspects are taken from examination of the previous chapter's functional synthesis technique. This is acceptable as that work is a simple example of SOA proofs-as-programs.

The protocol, of course, is useful insofar as it is applicable. In the next two parts of this monograph we shall apply the protocol to the synthesis of imperative and structured program synthesis. That work forms a basis for justifying the utility of the results of this chapter.

The chapter proceeds as follows:

- Section 3.1 provides a discussion on how SOA approaches should be generalized, including some arguments of what constitutes a good generalization.
- Section 3.2 uses that work to formally define the Curry–Howard protocol.
- Section 3.3 defines a process for application of the protocol to a given logic and programming language.
- Section 3.4 relates our results to the rest of the book and discusses some related work.

3.1 From ontology to protocol

We wish to define what makes a generalization of SOA proofs-as-programs. Moreover, we want this generalization to be "good."

Rôle	Purpose	Relationships with other rôles
Logic	Represent and make assertions about a problem space.	Logical statements are determined to be true or false by proofs; Logic should represent aspects of program behavior, by specifying required realizers, programs whose behavior satisfies the specification in some fashion.
Proofs	Show truth and falsity of assertions.	Proofs are made using a logic to derive true assertions; Logical proofs are transformed into programs.
Programming language	Write programs that perform computations; program behavior forms the basis of the problem space.	Aspects of programs are represented as mathematical objects; programs are extracted from proofs.

Fig. 3.1. Rôle-based ontology of SOA proofs-as-programs.

To generalize a concept is to define what is essential to its being, and then give a means of preserving this essence to form different concepts. In both philosophy and knowledge representation research, this essence is sometimes referred to as an *ontology*: the entities that are presumed to exist, and their interrelationships [Gru93].[1]

In any problem space it is possible to give a bad generalization. A good generalization of a concept can be aided by providing an abstract, rôle-based ontology. This kind of ontology identifies important rôles and relationships, but, as an abstraction, does not include the players of rôles. A good generalization of the concept preserves the rôles and relations, but permits possibly different players.

Example 3.1. For example, the notion of "marriage" involves an ontology consisting of husband and wife rôles. An anthropologist might compare a monogamous society with a polygamous society. He would take the familiar concept of monogamy and generalize it to polygamy. This act of generalization preserves the rôle-based ontology. It is understood and unquestioned by fellow anthropologists by virtue of the fact that the rôles of husband and wife are preserved (albeit now adapted for more than two players).

We believe that SOA approaches can be characterized by a rôle-based ontology, which is preserved by good generalizations. Examining the SOA method of the previous chapter, we identify several important rôles and relationships, which are displayed in Fig. 3.1. The three important rôles in program extraction are those of logic, proofs, and programming language.

As described in the review of the introductory chapter, SOA proofs-as-programs consists of several competing systems and methods. Due to its min-

[1] Ontologies have also been known as a notational, syntactic expression of so-called *conceptualizations* [GN87].

imal nature, our rôle-based ontology is common to all such approaches. This can be easily seen by examining each approach in turn — however, for reasons of brevity, we omit such a review.

The rôles of logic, proofs and programming language are central to all SOA approaches: *logical proofs* of specifications can be transformed into *programs* that satisfy specifications. The main relations between these rôles arise from realizability and extraction:

- logical statements form specifications of program behavior, in a sense defined by a realizability relation between programs and statements. Realizability formalizes how a statement is true of program behavior.
- extraction of realizers relates the rôles of proofs and programs by obtaining realizers from proofs of statements.

We place no restriction on the choice of the logic or programming language. This is because we want to generalize proofs-as-programs to logics and languages other than variants of intuitionistic logic and functional programming languages.

Remark 3.1. Alternative rôle-based ontologies can be formed, but we assert that, at least, they would involve the rôles we have identified. This is true by inspection of SOA approaches reviewed in Chapter 1 and those given in Chapter 2. Other rôles and relations could be added, but we believe that this would strengthen the ontology to an undesirable extent, making it difficult to use for a good generalization.

Remark 3.2. The ontology is more complicated than that of a naïve proofs-as-programs method. naïve proofs-as-programs involve a single type theory for defining algebraic theories, specifying programs, proving theorems and writing and running programs. Consequently, naïve approaches effectively combine the three rôles of Fig. 3.1 into a single rôle.

Note that the rôles are not usually made explicit in SOA approaches to functional program synthesis. Some proofs-as-programs involve powerful higher-order calculi in which the *players* of these rôles are represented in the same language.

Example 3.2. For example, the *Nuprl* type theory [BC85, CMH86] is often used for the rôles of logical reasoning and representing mathematical objects. The subset of *Nuprl* that corresponds to the simply typed lambda calculus is then used to play the rôle of a programming language [Sas86]. Programs, formulae, and proofs are all mathematical objects that may be predicated over in formulae. This approach involves a coincidence between mathematical objects, logical statements, and formal proofs.

However, we claim that, given any SOA method, even if some *players* are written in the same language, the three *rôles* are distinct for the purposes of program extraction, and may be treated as distinct to form a good generalization.

Rôle	Domain in the Curry–Howard protocol	Properties of domain
Logic	Natural deduction system	A formal system that defines a logical calculus.
Proofs	Logical type theory (LTT)	A type theory that enables encoding of proofs in the logical calculus, according to the Curry–Howard isomorphism: types denote statements, terms denotes proofs, and type inference corresponds to logical inference.
Programming language	Computational type theory (CTT)	A type theory for the programming language, equipped with an operational semantics.

Fig. 3.2. The rôles of the ontology and their corresponding domains in the Curry–Howard protocol.

Example 3.3. In the *Nuprl* approach, when a term is treated as a proof, its rôle is to derive logic statements and be transformed into programs, while, when a term is treated as a program, its rôle changes to performing computations and being extracted from proofs. The players of the rôles are identified in the epistemology of the theory because they are taken from the same domain. Nevertheless, the rôles themselves remain separate.

So, while we accept that a SOA approach may involve an epistemic identification of some rôles, we believe that an ontological demarcation is fundamental.

3.2 Formalizing the ontology

The ontology of the previous section was given in English. To use it as a guide for generalizing proofs-as-programs it is useful to explain the rôles and relations formally. In this way we can specify when the ontology is preserved by a logic and programming language. Hence, it will show precisely when SOA proofs-as-programs may be generalized. Our formalization will be given by a framework that represents the rôles and relationships of the ontology. We call this framework the *Curry–Howard protocol*.

To define the protocol, we need to specify the formal counterparts of the entities of Fig. 3.1: domains for representing the rôles and domain relationships that correspond to rôle relationships.

The rôles and the names of their corresponding domains are listed in Fig. 3.2. Proofs and programs both find natural representation in two distinct type theories:

- Type theory is ideal for denoting a logic with explicit representation of proofs. Proof-theoretic presentations of logics are often given in a type theory, where dependent products and sums are used to represent universal quantification and constructive existential quantification.

- All commonly used programming languages can be presented as type theories. Type inference rules determine how types of programs are deduced by a compiler. An accompanying operational semantics is given over the terms that defines how compiled programs are to be executed.

We now formally define the rôles and their inter-relationships.

3.2.1 Type theories

We will use the following broad definition of a type theory.

Definition 3.2.1 (Type theory). A type theory is of the form

$$TT = \langle Terms(TT), Types(TT), \vdash_{TT}, (:), TIR \rangle$$

consisting of terms, $Terms(TT)$, and types, $Types(TT)$, where

- $Terms(TT)$ and $Types(TT)$ are sets of terms and types with recursive grammars.
- We assume $Terms(TT)$ contains a distinguished subset of variables, $Var_{Terms(TT)}$.
- A binary *type judgement* relation (:) holds between terms and types.
- A type inference relation $\vdash_{TT}$ holds between a set of type judgements for variables $\Gamma = \{x_i : S_i\}_{i=1,\dots,n}$ ($x_i \in Var_{Terms(TT)}$ and $S_i \in Types(TT)$) (called a type context) and a single judgement for a term $t : S$ ($t \in Terms$, $S \in Types$). We call a relation of the form

$$\Gamma \vdash_{TT} t : S$$

  a type inference.
- The relation $\vdash$ is defined by a set of type inference rules, TIR, consisting of rules from several (premise) type inferences to a single (conclusion) inference, of the form

$$\frac{\Gamma_1 \vdash_{TT} t_1 : S_1 \quad \dots \quad \Gamma_n \vdash_{TT} t_n : S_n}{\Gamma \vdash_{TT} t : S} \text{ (R)}$$

  where (R) is a unique name of the rule.
  We permit SIR to contain an infinite number of rules.
  Given $t_1, t_2 \in Terms$ such that, for any context Γ,

$$\Gamma \vdash t_1 : S \Leftrightarrow \Gamma \vdash t_2 : S$$

  we say that t_1 and t_2 *have the same type* S.
  We will write $t^\Gamma : S$ to denote the term t and the fact that $\Gamma \vdash t : S$ can be derived, and call t *well-typed in context* Γ. We will simply write $t : S$ for $t^\Gamma : S$ if the context Γ can be unambiguously inferred.
- We require that $\vdash_{TT}$ is defined so that every term $t \in Terms(TT)$ is well-typed.

3.2.2 Logic

We provide a very general definition of a logic, consisting of formulae and rules for proving true formulae, based on Gabbay's notion of a deductive system in [Gab96].

Definition 3.2.2 (Natural deduction system). A natural deduction system

$$D = \langle Formulae(D), \vdash_D, DR \rangle$$

where

- $Formulae(D)$ is a set of well-formed statement formulae, generated by a recursive grammar, to be reasoned with in D.
- $\vdash_D$ is a relation that is defined between lists of *assumption* formulae $\Gamma = \{G_1, \ldots, G_n\}$ ($G_i \in Formulae(D)$) and a single *conclusion* formula $C \in Formulae(D)$,

  $$\Gamma \vdash_L C$$

  Such a relation is called an *inference*. This relation is defined by a set of deduction rules DR, consisting of rules from several (premise) inferences to a single (conclusion) inference, of the form

  $$\frac{\Gamma_1 \vdash_D F_1 \quad \ldots \quad \Gamma_n \vdash_D F_n}{\Gamma \vdash_D F} \text{ (R)}$$

  where (R) is the unique name of the rule.
  We permit DR to contain an infinite number of rules.

Example 3.4. Propositional logic, *Prop*, is one of the simplest logics that qualify as a deduction system, according to this definition. Let $ESig$ be the empty signature (consisting of an empty set of terms, sorts and rules). Then we define $Formulae(Prop)$ to be well-formed formulae constructed by:

- A set of basic propositions, $Pred(Prop)$, that are treated here as predicates that take no arguments. We set $Var_{Pred(Prop)}$ to be the empty set.
- The connectives, $Conn_{Prop} = \{\wedge, \neg, \vee, \Rightarrow\}$, with the obvious arities.
- There are no quantifiers: $Quant_{Prop} = \emptyset$.

There are a finite number of rules. Some of the rules are as follows.

$$\frac{\Gamma \vdash A}{\Gamma \vdash A \vee B} \qquad \frac{\Gamma \vdash B}{\Gamma \vdash A \vee B} \qquad \frac{}{\vdash A \vee \neg A} \qquad \frac{\Gamma \vdash \neg\neg A}{\Gamma \vdash A}$$

The soundness and completeness of this presentation may be proved with respect to a truth table semantics.

Example 3.5. The intuitionistic logic of the previous chapter is an example of a deduction system.

Well-formed formulae are built from the usual connectives and quantifiers, with predicates applying over elements of some signature.

The rules of the logic are the usual introduction and elimination rules of intuitionistic logic, together with axioms and schema for defining an algebraic theory over the logic's signature.

Example 3.6. Schemata are a common way of generating infinite rules for first-order logics, to simulate higher-order quantification over predicates.

For instance, to axiomatize the natural numbers in a first-order classical or intuitionistic logic, we require an induction schema. In a deduction system, it may be written as a parametrized rule with predicate variable X

$$\frac{\Gamma \vdash X(0) \wedge \forall n : nat \bullet X(n) \Rightarrow X(succ(n))}{\Gamma \vdash \forall x : nat \bullet X(x)} \ (ind\_Nat[X])$$

The parametrized rule is to be regarded as shorthand for an infinite set of rules

$$\frac{\Gamma \vdash P(0) \wedge \forall n : nat \bullet P(n) \Rightarrow P(succ(n))}{\Gamma \vdash \forall x : nat \bullet P(x)} \ (ind\_Nat[P])$$

for each predicate P available to the logic.

Example 3.7. Girard's Intuitionistic Linear Logic [Gir87] is an example of a natural deduction system. We describe the multiplicative fragment, MLL, as presented in [BBHdP93]. Well-formed formulae, $Formulae(MLL)$, are built from predicates that take no arguments. So, similar to $Prop$, we can take the signature of LL to be the empty signature, $ESig$. Formulae $Formulae(MLL)$ are defined from

- A set of basic propositions $Pred(MLL)$ that are treated here as predicates that take no arguments. There is a designated proposition $I \in Pred(MLL)$. We set $Var_{Pred(MLL)}$ to be the empty set.
- The connectives, $Conn_{MLL}$, consist of
  - $\otimes, \multimap, !$, where the first two connectives are binary, the last is unary, and
  - an infinite set of superscript indices, $()^n$ (n a natural number), that apply to the formulae, $Formulae(MLL)$, used to uniquely identify assumptions in rules.
- There are no quantifiers: $Quant_{Prop} = \emptyset$.

Some of the rules are as follows.

$$\frac{}{A^x \vdash A} \ (\text{Ass-I})_x$$

$$\frac{\Gamma_1 \vdash A \quad \Gamma_2 \vdash B}{\Gamma \vdash A \otimes B} \ (\otimes\text{-I}) \qquad \frac{\Gamma_1, A^x \vdash C \quad \Gamma_2, B^y \vdash C}{\Gamma \vdash C} \ (\otimes\text{-E})_{(x,y)}$$

$$\frac{\Delta_1 \vdash !A_1 \quad \ldots \quad \Delta_n \vdash !A_n \quad !A_1^{v_1}, \ldots, !A_n^{v_n} \vdash B}{\Delta_1, \ldots, \Delta_n \vdash !B} \ (\text{Promotion})_n$$

$$\frac{\Gamma \vdash !A \quad \Delta \vdash B}{\Gamma, \Delta \vdash B} \ (\text{Weakening}) \qquad \frac{\Gamma, A^x \vdash B}{\Gamma \vdash A \multimap B} \ (\multimap\text{-I})_x$$

where x, y range over natural numbers. Observe that $(\text{Ass-I})_x$, $(\otimes - E)_{(x,y)}$, $(Promotion)_n$ and $(\multimap\text{-I})_x$, denote an infinite set of rules, for each value of x and y. By associating indices with assumption formulae, the last rule tells us that we may only discharge one assumption from a deduction to form a linear implication.

3.2.3 Logical type theory

A logical type theory is a type theoretic presentation of a particular deduction system. We provide a very flexible definition, preserving relations between logic and proof rôles identified in the ontology of Fig. 3.1. These constraints form a generalization of the Curry–Howard isomorphism: the types of the *LTT* are formulae of the logic; terms encode the steps to derive their types, and a reduction relation can be assumed over terms, defining a normalization strategy over proofs. When a theorem can be proved in the logic, there is a term in the theory that has the theorem as a valid type, and vice versa.

Definition 3.2.3 (Logical type theory, *LTT*). A *logical type theory* (*LTT*) for a deduction system L is of the form

$$LTT(\mathsf{L}) = \langle PT(\mathsf{L}), Formulae(\mathsf{L}), (.)^{(.)}, \vdash_{\mathsf{L}}, PTR, \triangleright\rangle$$

with

$$\langle PT(\mathsf{L}), Formulae(\mathsf{L}), (.)^{(.)}, \vdash_{\mathsf{L}}, PTR\rangle$$

forming a type theory (see Definition 3.2.1) where $PT(\mathsf{L})$ are terms (called *proof-terms*), $Formulae(\mathsf{L})$ are taken as types, $(.)^{(.)}$ is a type judgement relation, and $\vdash_{\mathsf{L}}$ is a type inference relation defined by rules from PTR. We assume

- the set of proof-terms $PT(\mathsf{L})$ has a distinguished set of variables $Var_{PT(\mathsf{L})}$,
- $PT(\mathsf{L})$ includes a lambda calculus,
- the type judgement relation, written as a superscript, $(p)^{(T)}$, defined between proof-terms of $p \in PT(\mathsf{L})$, and the formulae $T \in Formulae(\mathsf{L})$ of L.
- We permit PTR to contain an infinite number of rules.
- There is a normalization relation $\hat{\triangleright}$, defined over proof-terms of $PT(\mathsf{L})$, generated as the transitive closure of a one-step reduction relation $\triangleright$. The relation $\triangleright$ is defined by a set of rules over proof-terms

  $$p_1 \triangleright p_2$$

  We require that:
  — Normalization preserves the typing of proof-terms, so that

  $$\Gamma \vdash_{PT(\mathsf{L})} p^A \text{ and } p_1 \triangleright p_2 \ \text{ entails } \ \Gamma \vdash_{PT(\mathsf{L})} p_2^A.$$

  — The normalization relation is strongly normalizing: that is, for each proof-term p_1, every sequence of one-step reductions is finite, terminating in a term p_n: $p_1 \triangleright p_2 \triangleright \ldots \triangleright p_n$.

— The relation satisfies the Church–Rosser property: for any proof-terms p, p_1 and p_2, such that $p \rhd p_1$ and $p \rhd p_2$, there must be a common term p_3 such that $p_1 \rhd p_3$ and $p_2 \rhd p_3$.

Example 3.8. Intuitionistic many-sorted predicate logic has several logical type theories, one of which was presented in the previous chapter. Martin-Löf's constructive type theory [ML84, ML85], Coquand's Calculus of Constructions [MLM90], or Luo's Extended Calculus of Constructions [Luo94] are other possibilities.

In all these theories, the normalization relation corresponds to removal of redundant introduction/elimination rule pairs in intuitionistic proofs.

If a type theory is a *LTT* for a deduction system, then we say that the Curry–Howard isomorphism has been *adapted* for the deduction system and type theory.

Example 3.9. There are several type theories proposed to adapt the Curry–Howard isomorphism to classical predicate logic: see, for instance, [Gri90, Mur91, Par93, BS93, BS95b, BS95a, Sch99b]. Inspection of these systems shows that each forms a *LTT* in our sense for classical logic.

Example 3.10. A new logic was defined by Nakano [Nak94], with connectives that enable explicit reasoning about catch-and-throw mechanisms. A logical type theory was also given that represents proofs according to the style of the Curry–Howard isomorphism.

Example 3.11. Many authors have showed how Girard's Intuitionistic Linear Logic [Gir87] can be associated with a logical type theory.

We briefly sketch the presentation of [BBHdP93] as a logical type theory for the multiplicative fragment. The deduction system is the *MLL* described in Example 3.7. The proof-terms consist of an extension of the lambda calculus. Some of the type inference rules are as follows.

$$\frac{}{X^{A^x} \vdash X^A}\ (\text{Ass-I})_x$$

$$\frac{\Gamma_1 \vdash e^A \quad \Gamma_2 \vdash f^B}{\Gamma \vdash e \otimes f^{A \otimes B}}\ (\otimes\text{-I}) \qquad \frac{\Gamma_1 \vdash e^{A \otimes B} \quad \Gamma_2, X^{A^x}, Y^{B^y} \vdash f^C}{\Gamma_1, \Gamma_2 \vdash \mathsf{let}\ e\ \mathsf{be}\ x \otimes y\ \mathsf{in}\ f^C}\ (\otimes\text{-E})_{(x,y)}$$

$$\frac{\Delta_1 \vdash e_1^{!A_1} \quad \ldots \quad \Delta_n \vdash e_n^{!A_n} \quad x_1^{!A_1^{v_1}}, \ldots, x_n^{!A_n^{v_n}} \vdash f^B}{\Delta_1, \ldots, \Delta_n, \Gamma \vdash \mathsf{promote}\ e_1, \ldots, e_n\ \mathsf{for} x_1, \ldots, x_n\ \mathsf{in}\ f^{!B}}\ (\text{Promotion})_n$$

$$\frac{\Gamma \vdash e^{!A} \quad \Delta \vdash f^B}{\Gamma, \Delta \vdash \mathsf{discard}\ e\ \mathsf{in}\ f^B}\ (\text{Weakening})$$

$$\frac{\Gamma, X^{A^x} \vdash e^B}{\Gamma \vdash \lambda X^{A^x}.e^{A \multimap B}}\ (\multimap\text{-I})_x$$

where x, y range over natural numbers. Observe that $(\text{Ass-I})_x$, $(\otimes\text{-E})_{(x,y)}$ and $(\multimap\text{-I})_x$ denote an infinite set of type rules, for each value of x and y. The normalization relation $\rhd$ is defined by the usual lambda reduction rules augmented with new rules for the new proof-terms. For instance, one of the new rules is

$$\mathsf{discard}\ (\mathsf{promote}\ e_1, \ldots, e_n\ \mathsf{for}\ x_1, \ldots, x_n\ \mathsf{in}\ t)\ \mathsf{in}\ u \ \rhd \\ \mathsf{discard}\ e_1\ \mathsf{in}(\ldots(\mathsf{discard}\ e_n\ \mathsf{in}\ u)\ldots)$$

This rule corresponds to normalizing logical proofs that involve (Promotion)$_n$ rules followed by the (Weakening) rule.

3.2.4 Computational type theory

The computational type theory (CTT) should have a simple definition because we wish to accommodate as wide a range of programming languages as possible.

Definition 3.2.4 (Computational type theory, CTT). A *computational type theory* (CTT) is a type theory (see Definition 3.2.1) of the form

$$\mathsf{C} = \langle Term(\mathsf{C}), Type(\mathsf{C}), :, \vdash_{\mathsf{C}}, TIR \rangle$$

where

- the set of terms, $Term(\mathsf{C})$, defines a set of programs,
- the set of types, $Type(\mathsf{C})$, defines a set of types for programs,
- typing judgements, written $t : T$, hold between programs and types according to the type inference relation $\vdash_{\mathsf{C}}$ and the rules TIR.

Remark 3.3. Observe that imperative languages can be included as computational type theories according to this definition.

3.2.5 The Curry–Howard protocol

The three rôles now associated with formal domains, and the relation between logical calculus and type theory having been defined, it remains to formalize the relationships pertaining to extraction. This will complete the formalization of the rôle-based ontology of Fig. 3.1, and provide us with the Curry–Howard protocol.

Recall that the three rôles have the following (circular) inter-relationships: aspects of program behavior are represented in the logic, logical statements are represented as formal proofs, and formal proofs may be transformed into programs.

We have already formalized the relation between logic and proofs.

From the perspective of program extraction, the most important relationship is that between the logic and programming language. This requires that:

- A formula of the logic can specify the behavior of a program as a *realizer* of the formula.
- There is a transformation of proofs of specifications into realizing programs.

Formalization requires

- a realizability relation to be defined between formulae of programs of the *CTT* and the *LTT*, and
- a program extraction and optimization map from proofs of the *LTT* to programs of the *CTT* such that, given a proof of a specification, the extracted program should realize the specification.

The protocol follows from these observations.

Definition 3.2.5 (The Curry–Howard protocol). The *Curry–Howard protocol* holds between a logical type theory, L, and computational type theory, C, when

1. There are extraction maps, etype, from formulae of L to types of C and $\mathsf{extract}$ from proof-terms of L to programs of C,
$$\begin{aligned} \mathsf{extract} &: PT(\mathsf{L}) \to Term(\mathsf{C}) \\ \mathsf{etype} &: Formulae(\mathsf{L}) \to Type(\mathsf{C}) \end{aligned}$$
such that, given a proof $d \in PT(\mathsf{L})$, such that
$$\Gamma \vdash d^A$$
then $\mathsf{extract}(d)$ is in C and is of type $\mathsf{etype}(A)$.
2. There is a realizability relation r between programs and formulae, such that, for any proof
$$\emptyset \vdash_{\mathsf{L}} p^A \in PT$$
it is true that
$$\mathsf{extract}(p) \mathsf{r} A$$

Remark 3.4. The concepts of the protocol are related according to the following diagram (with t^S denoting a proof-term t of S in the logical type theory):

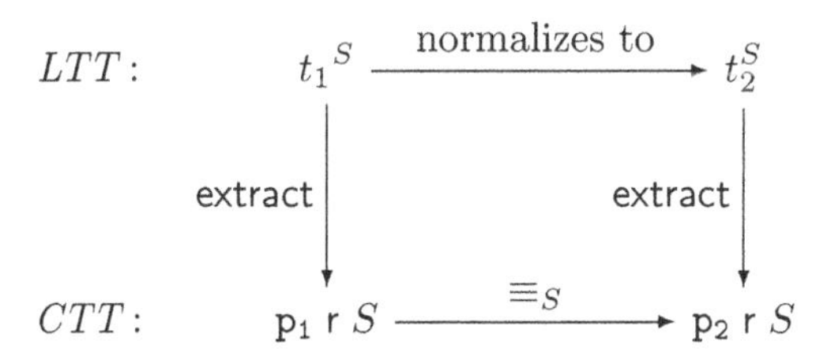

where $\mathsf{p}_1 \equiv_S \mathsf{p}_2$ holds between two programs when both $\mathsf{p}_1 \mathsf{~r~} S$ and $\mathsf{p}_2 \mathsf{~r~} S$.

3.3 Using the protocol

We now give a process for using the protocol to adapt proofs-as-programs to new contexts. The process involves the following steps:

1. Define a signature and a logical calculus that involves the signature. This might involve issues that are orthogonal to the protocol — e.g., finding a semantics for the calculus and proving soundness and completeness.
2. Define a logical type theory for the logical calculus. Again, other important properties that are not part of the protocol may need to be proved, such as the Church–Rosser theorem and strong normalization for the *LTT*'s proof-terms.
3. Identify a programming language and describe it by means of a computational type theory.
4. Prove the Curry–Howard protocol to hold over the domains.

We will refer to this process as *protocol application.*

3.4 Discussion

In the next chapters, we will use the protocol to adapt the Curry–Howard isomorphism and SOA proofs-as-programs to extracting imperative and structured programs from proofs.

This work will serve as an example of why the protocol is a useful and natural generalization of proofs-as-programs to new contexts.

In particular we shall see how the ontology of the protocol facilitates a more natural approach to adaptation than might otherwise be the case.

For example, the advantage of demarcating proofs from programs is more apparent in the imperative case of the next part than in the familiar constructive case. In Part II, we give a *LTT* for reasoning about side-effects and return values of imperative *SML* programs. The underlying calculus is a constructive version of the Hoare Logic with a natural deduction presentation. We then define an extract map from the *LTT* to an imperative *CTT* that preserves the protocol, and allows us to extract imperative programs from proofs of specifications.

Identification of domains would mean designing a unifying language in which the signature, *CTT* and *LTT* type theories could all be written. Any resulting *LTT* would have to be equipped with the *CTT*'s imperative operational semantics, with a call-by-value evaluation strategy over function applications. This semantics would coincide in some way with the *LTT*'s proof-normalization rules. The calculus of the *LTT* would be nonstandard, and would be difficult to learn and use. In contrast, distinguishing between the *LTT* and the *CTT* enables a logical calculus for reasoning about imperative programs that is closer to established logics.

Similar remarks can be made for the adaptation of Part III for synthesis of functions for structured specifications and structured programs.

To the best of our knowledge, no previous work has been done in identifying a framework for generalizing SOA proofs-as-programs.

In the area of type theory, Zhaohui Luo's *Extended Calculus of Constructions* (*ECC*, [Luo94]) has a similar philosophy to our framework. The *ECC* provides a predicative universe, *Prop*, to represent logical propositions and a Martin-Löf-style impredicative universe hierarchy to represent programs. As in Martin-Löf, the impredicative universes are open, so the same comparison holds. Like our protocol, the *ECC* has a similar division of labor between proving properties of programs (in *Prop*) and creating new programs and types (in the universe hierarchy). The *ECC* was designed to provide a unified framework for the two (recognized) separate tasks of logical reasoning and program development but not with program synthesis in mind. This means that in the *ECC* there is no notion of a simplifying extraction map between terms that represent proofs and program terms — they are identified.

We have presented the Curry–Howard protocol in an informal metalogic. Anderson [And93] used the Edinburgh Logical Framework (*ELF*) to provide a similar relationship between proofs in a logical type theory and programs in a computational type theory. That work was only concerned with constructive synthesis of functional programs, and in particular with defining relationships to obtain optimized programs within *ELF*. However, representations of optimized programs are not added to the logical type theory. Our metalogical results might benefit from a similar formal representation.

Part III

Imperative Proofs-as-Programs

III

Overview

Concurrent to the development of the proofs-as-programs paradigm has been research on the Hoare logic and related systems for the specification and synthesis of imperative and object-oriented programs. In this part of the monograph we combine the two approaches to synthesize imperative programs with return values.

Imperative programs involve both side-effects and side-effect-free return values. For instance, the *SML* program

$$\mathsf{s} :=!\mathsf{s} * 3; !\mathsf{s} * 2$$

involves a side-effect producing assignment statement, $\mathsf{s} :=!\mathsf{s} * 3$, followed by the return value $!\mathsf{s} * 2$. In many popular imperative languages (e.g., C# with delegates, C++ with *STL*, Eiffel with Agents, *SML* or *LISP*) such return values are potentially complex, involving higher-order functional aspects that are difficult to program correctly.

The goal of this part is to specify, reason about, and synthesize both aspects of imperative programs — side-effects and functional return values.

Our approach is as follows.

We use a version of Hoare logic to synthesize the side-effect producing aspect of a program, specified in terms of pre- and post-conditions. For instance, the formula

$$s_f > s_i$$

specifies a side-effect where the final value of state s, denoted by s_f, is greater than the initial value, denoted by s_i. We can use Hoare logic to synthesize a *SML* program that satisfies this specification, by producing, for example, a theorem of the form

$$\vdash \mathsf{s} :=!\mathsf{s} * 3 \diamond s_f > s_i$$

where the left-hand-side of $\diamond$ symbol is the required *SML* program, and the right-hand-side is a true statement about the program.

To specify and synthesize return values of a program we adapt realizability and proofs-as-programs. For instance, given the theorem

$$\mathtt{s} := \mathtt{s} * 3 \diamond s_f > s_i \wedge (\exists x : int.Even(x) \wedge x > s_i)$$

we can synthesize a program of the form

$$\mathtt{s} := \mathtt{s} * 3; \mathtt{f}$$

where the function f is a side-effect-free function (such as $!\mathtt{s}*2$) that realizes the existential part of the post-condition $(\exists x : int.Even(x) \wedge x > s_i)$, by providing a witness for the x. Our adaptation is done according to the Curry–Howard protocol framework outlined in the previous chapter (Chapter 3 of Part II).

Hoare logic is usually defined with respect to an internal logic. We take this to be constructive (intuitionistic) logic. This enables us to use some of the results of intuitionistic proofs-as-programs to synthesize correct return values from proofs. However, the adaptation is not trivial, as, unlike in functional program synthesis, our specifications involve initial and final values of state, and our extracted side-effect-free functions can sometimes involve state references.

The advantage of our approach to return value synthesis is that the user need not code the return value manually, but instead works within a logical theory to prove a theorem from which the return value is extracted. Constructive program synthesis has a successful track record in deriving side-effect-free functional programs. These methods are equipped to reason directly about and synthesize functions.

By combining Hoare-like approaches with constructive program synthesis, we obtain a best-of-both-worlds system for specifying and synthesizing the two aspects of imperative programs.

This part proceeds as follows:

- Chapter 4 presents a constructive version of Hoare logic for reasoning about and constructing imperative *SML* programs.
- The semantics of this logic is investigated in Chapter 5. That chapter also shows how the logic may be given a type-theoretic presentation in the style of the Curry–Howard isomorphism, where proofs and theorems are taken as terms and types of a logical type theory.
- The last chapter of this part, Chapter 6, shows how to use the type theory for imperative program synthesis by applying the Curry–Howard protocol. We discuss the uses of our approach to imperative program synthesis in a practical setting, and, in particular, show how the so-called design-by-contract approach to system design (see, e.g., [Mey97]) can be aided by our results.

4

Intuitionistic Hoare Logic

Hoare logic is a formal system for simultaneously reasoning about and constructing imperative programs. The system was first described by Hoare in [Hoa69]. In this chapter we define a constructive version of Hoare logic, called Intuitionistic Hoare logic, IHL, for reasoning about side-effects of imperative programs in *SML*. In later chapters we will adapt proofs-as-programs methods to this logic, to specify and synthesize imperative programs with side-effect-free return values.

The Hoare logic specifies side-effects through pre- and post-condition formulae [Flo67, Hoa69]. Pre- and post-conditions are assertions about the behavior of a program before and after execution, respectively. In the original presentation of the Hoare logic, pre- and post-condition formulae were treated as distinct [Hoa69]. We shall utilize the fact that pre- and post-conditions can be combined into a single formula description (common in, e.g., *OCL* [WK98], the B-method [Abr96], or Abadi [AL97]). For our purposes pre- and post-condition formulae are statements of many-sorted predicate logic, with special symbols used to denote values of state.

The Hoare logic manipulates theorems that, in our presentation, consist of pairs of programs and pre- and post-condition formulae, of the form

$$\mathtt{p} \diamond F$$

where the left hand side of the diamond is a program, and the right hand side of the diamond is a pre- and post-condition description of the program's behavior. The program's side-effect is described by the formula in terms of initial and final state reference values, prior to, and after, execution. We denote initial and final state reference values by the name of the state reference with a $()_i$ or a $()_f$ subscript respectively.

Example 4.1. An example of a theorem is

$$\vdash \mathtt{r} := !\mathtt{r} + \mathtt{1} \diamond r_f = r_i + 1$$

The example theorem is correct, because the pre- and post-condition statement asserts that the final value of state reference `r` is equal to the increment of the initial value, after executing the program `r :=!r + 1`.

The rules of the logic derive new theorems, enabling the user to develop programs and assertions about the program in tandem.

Hoare logic is always defined with respect to a separate logical subsystem. Usually, this is classical logic. This chapter introduces a version of Hoare logic that uses intuitionistic logic as its subsystem. In Hoare's 1969 paper [Hoa69], the logic was given with respect to a simple, toy, imperative language. Here, we will consider an imperative subset of *SML*.

The usefulness of our version of Hoare logic will be seen in later chapters where we use constructive aspects of IHL for adapting proofs-as-programs to synthesize *SML* programs with side-effects and side-effect-free return values.

Because Hoare logic involves programs and pre- and post-condition formulae, the first part of this chapter is dedicated to defining the syntax and semantics of programs and formulae. Then we define Hoare logic proper.

We proceed as follows:

- Section 4.1 defines the form of the signatures we use in this part of the book.
- In Section 4.2 we define an imperative subset of *SML*.
- Section 4.3 provides a simple operational and relational semantics for our subset.
- Section 4.4 defines well-formed formulae for our calculus and defines when a formula is true about an imperative *SML* program execution.
- We present Intuitionistic Hoare logic in Section 4.5. We explain why its rules allow us to infer true statements about programs.
- Section 4.6 compares our presentation to the more standard, Hoare-triple-based, presentations of Hoare logic.
- Finally we provide a summary and discussion of our results in Section 4.7.

Chapter 5 will discuss soundness and completeness of IHL and will show how IHL can be given a type theoretic presentation. Chapter 6 will achieve the goal of this part, identifying how correct imperative *SML* programs with return values can be synthesized from its proofs using proofs-as-programs.

The results of this chapter are standard, ultimately deriving from Hoare's original paper [Hoa69]. However, our use of *SML*, the choice of intuitionistic logic as a subsystem, and the natural deduction presentation, are points of difference from the usual literature.

4.1 Signatures

Our work will be parametrized over many-sorted signatures. We shall use signatures for two purposes:

- to define an assumed *SML* preamble of datatype and function declarations, which will be used in constructing larger *SML* programs, and
- to define terms for use in formulae of our Hoare logic.

Notation 4.1. First we take the following convention. As in the previous part, we will often let sets of things (terms, values, side-effect-free programs) be common to both programs and formulae. Depending on the context, we will interchange the names "sorts" and "types" to denote names of collections of things. We will use "types" when referring to collections of things used in programming (programs, values of programs, etc). We will use "sorts" when referring to collections of things used in formulae (terms used in predicates). We use a `typewriter` font to denote types and roman font to denote sorts.

We employ the usual definition of signature (specifically, we take that given in [CoF01, p. 3]), but extended with sorts corresponding to the relevant types of imperative *SML*: functional, product, disjoint union sorts, and the unit sort.

Definition 4.1.1 (Many-sorted signature with total functions). A *many-sorted signature* $\Sigma = \langle S, TF, Pred \rangle$ consists of:

- A set, S, of sorts. Sorts are generated from a set of *basic sorts*, $B(S)$ as according to the following inductive definition. First, $B(S) \subset S$. Also, if s_1 and s_2 are in S, then so are the
  — function sort $(s_1 \rightarrow s_2)$
  — product sort $(s_1 * s_2)$
  — disjoint union $(s_1|s_2)$.
- We assume that $B(S)$ includes a special sort, called $Unit$.
- Sets $TF_{w,s}$, of total function symbols, for each *function profile* (w, s). A function profile, (w, s), consists of a sequence word of argument sorts $w \in S^*$ and a result sort $s \in S$ (constants are treated as functions with no arguments). The length of w is called the *arity* of function symbols in $TF_{w,s}$.
- We assume that there is only one element in $TF_{\emptyset,Unit}$, a unit symbol, written $()$.
- Sets P_w of predicate symbols, for each *predicate profile* w. A predicate profile consists of a sequence of argument sorts $w \in S^*$. The length of w is called the *arity* of predicate symbols in P_w. For each basic sort $s \in B(S)$, there is a distinguished equality predicate $=_s \in P_{ss}$.

Our signature definition is the same as Definition 2.1.1, Chapter 2 of Part II. As in that part of this book, the symbols that identify operations and predicates may be overloaded, occurring in more than one of the above sets. Whenever there is ambiguity in sentences, function symbols f, and predicate symbols P should be qualified by profiles, written $f_{w,s}$ and P_w respectively. When no ambiguity is present, these profiles can be omitted.

4.1.1 Lambda terms

We define a lambda calculus for a signature as in Chapter 2 of Part II. The set of lambda terms $Terms(\Sigma)$, for a signature $\Sigma = \langle S, TF, Pred \rangle$, is given with respect to a denumerable set of term variables, Var, disjoint from the constants in TF. For reference, we repeat the grammar here in Fig. 4.1.

$a, b, c ::=$	elements of $Terms(\Sigma)$
e	any function or constant from TF
x	a variable $x \in Var$
$Inl(a)$	in left
$Inr(a)$	in right
$match\ a\ with\ Inl(x) \Rightarrow b \mid Inr(y) \Rightarrow c$	match case, $x, y \in Var$
$fun\ x : s\ \Rightarrow b$	lambda abstraction, $s \in Sorts(\Sigma)$
$(a\ b)$	application
(a, b)	pair
$fst(a)$	first projection
$snd(a)$	second projection

Fig. 4.1. Syntax terms of $Terms(\Sigma)$.

Terms are associated with sorts according to sort inference rules. We follow the same rules given in Fig. 2.2, p. 29, Chapter 2 of Part II. For reference, the rules are repeated in full in Fig. 4.2.

Free and bound variables are defined in the usual way.

Definition 4.1.2 (Free and bound variables of $Terms(\Sigma)$). Let x be any variable of Var, and t be a term of $Terms(\Sigma)$.

Then, x is *bound* in t if there is a subterm of t of the form

$$fn\ x : s \Rightarrow b$$

or

$$match\ a\ with\ Inl(x) \Rightarrow b \mid Inr(y) \Rightarrow c$$

or

$$match\ a\ with\ Inl(y) \Rightarrow b \mid Inr(x) \Rightarrow c$$

If x is not bound in t, then x is free in t. We write $BV(t)$ for the set of all bound variables of t, and $FV(t)$ for the set of all free variables of t. A term with no free variables is called *closed.*

We write $Closed(Terms(\Sigma))$ for the set of closed *PML* programs.

$$\frac{}{\Gamma, x : s \vdash_\Sigma x : s} \text{(Ass)}$$

$$\frac{f \in TF_{(s_1 \ldots s_n),s} \quad \Gamma_1 \vdash a_1 : s_1 \ldots \Gamma_n \vdash a_n : s_n}{\Gamma, \Gamma_1, \ldots, \Gamma_n \vdash_\Sigma f(a_1, \ldots, a_n) : s} \text{(Fn)}$$

$$\frac{\Gamma \vdash_\Sigma a : s_1}{\Gamma \vdash_\Sigma Inl(a) : (s_1|s_2)} \text{(Union}_1\text{)} \qquad \frac{\Gamma \vdash_\Sigma a : s_2}{\Gamma \vdash_\Sigma Inr(a) : (s_1|s_2)} \text{(Union}_2\text{)}$$

$$\frac{\Gamma_1 \vdash_\Sigma a : s_1 \quad \Gamma_2 \vdash_\Sigma b : s_2}{\Gamma_1, \Gamma_2 \vdash_\Sigma (a, b) : (s_1 * s_2)} \text{(Prod)}$$

$$\frac{\Gamma \vdash_\Sigma a : (s_1 * s_2)}{\Gamma \vdash_\Sigma fst(a) : s_1} \text{(Proj}_1\text{)} \qquad \frac{\Gamma \vdash_\Sigma a : (s_1 * s_2)}{\Gamma \vdash_\Sigma snd(a) : s_2} \text{(Proj}_2\text{)}$$

$$\frac{\Gamma, x : s_1 \vdash_\Sigma a : s_2}{\Gamma \vdash_\Sigma fn\, x : s_1 => a : s_1 \to s_2} \text{(Abs)} \qquad \frac{\Gamma_1 \vdash_\Sigma a : s_1 \quad \Gamma_2 \vdash_\Sigma b : (s_1 \to s_2)}{\Gamma_1, \Gamma_2 \vdash_\Sigma (b\, a) : s_2} \text{(App)}$$

$$\frac{\Gamma_1 \vdash_\Sigma a : (s_1|s_2) \quad \Gamma_2, x : s_1 \vdash_\Sigma b : s \quad \Gamma_3, y : s_2 \vdash_\Sigma c : s}{\Gamma_1, \Gamma_2, \Gamma_3 \vdash_\Sigma match\, a\, with\, Inl(x) => b \mid Inr(y) => c : s} \text{(Case)}$$

Fig. 4.2. Sort inference rules for terms of *PML*.

4.1.2 Evaluation

Because our terms constitute a lambda calculus we have the usual reduction rules. The rules are similar to those given in Section 2.4, Chapter 2 of Part II. The rules define a one-step reduction relation $\rhd_\Sigma$, given in Fig. 4.3. We write $\hat{\rhd}_\Sigma$ for the transitive closure of $\rhd_\Sigma$, and say that a *evaluates to* b if

$$a \;\hat{\rhd}_\Sigma\; b$$

We write

$$a \;\rhd^*_\Sigma\; b$$

if b can be obtained from a by one or more applications of the rules.

Definition 4.1.3 (Irreducible, normal terms). The set of *irreducible, normal terms*, $Normal(\Sigma)$, consists of elements a from $Terms(\Sigma)$ such that

$$a \;\hat{\rhd}_\Sigma\; a$$

Assumption 4.1. The rules are parametrized with respect to possible evaluation of function symbol application. When a function is applied to arguments of appropriate arity and types, we assume that there is always a resulting answer. This assumption is formalized by assuming a mapping, $Eval_\Sigma$, that gives the return value for function applications: given a function symbol $f \in TF_{s_1 \ldots s_n, s}$ and arguments $(a_1, \ldots, a_n)$ of sorts $s_1, \ldots, s_n$ respectively, $Eval_\Sigma(f(a_1, \ldots, a_n))$ returns a term from $Terms(\Sigma)$ of sort s.

$$
\begin{array}{r}
(fn\ x : s \texttt{=>} p)\ a \rhd_\Sigma p[a/x] \\
match\ Inl(a)\ with\ Inl(x) \texttt{=>} b \mid Inr(y) \texttt{=>} c \rhd_\Sigma b[a/x] \\
match\ Inr(a)\ with\ Inl(x) \texttt{=>} b \mid Inr(y) \texttt{=>} c \rhd_\Sigma c[a/y] \\
fst(a,b) \rhd_\Sigma a \\
snd(a,b) \rhd_\Sigma b \\
f(a_1,\ldots,a_n) \rhd_\Sigma Eval_\Sigma(f(a_1,\ldots,a_n))
\end{array}
$$

Fig. 4.3. Rules that define $\rhd_\Sigma$, giving the operational semantics of the data values.

This concludes our discussion of many-sorted signatures and their associated terms. In the next section, we will use these concepts in our definition of *SML*. In section 4.4, we will again use signatures to define the terms used in our formulae.

4.2 A subset of *SML*

We shall be reasoning about an imperative subset of *SML*, called *IML*, described in this section. We base our description on the *SML* standard given in [MTH90]. However, our definition is self-contained and it is not necessary for the reader to be well-versed in the standard. The language comprises basic imperative constructs (assignments, sequencing, conditionals, and loops) and has functional, product and disjoint union types. It includes side-effect-free programs that can be used to define return values and as boolean checks in loops and conditional statements. *SML* programs are usually written using precoded functions from libraries. In our work, we will assume a preamble library of datatypes, side-effect-free functions and black-box programs.

4.2.1 Preamble

Our preamble is formally defined now. The preamble will consist of two kinds of programs: pure *SML* functions, whose execution does not affect memory, and black-box programs with side-effect-effects, whose execution will result in changes to memory.

We shall use a signature to denote the pure *SML* functions of the preamble:

$$\Sigma_p = \langle S, TF, Pred \rangle$$

where the set of sorts, S, denotes the types available to the preamble, the set of functions TF denotes the set of side-effect-free functions defined in the preamble, and *Pred* is a set of predicate symbols used to make logical statements about the preamble. We defer discussion of the use of predicates to the next section. We will consider the lambda calculus of terms formed from our functions, $Terms(\Sigma_p)$, to define a set of pure *SML* programs, a subset of the programs of interest *IML*.

As in the previous part, we use `typewriter` font to distinguish terms and sorts of Σ when used as programs and types of *SML*, as opposed to terms and sorts in formulae.

Assumption 4.2. We assume that Σ_p includes the usual definition of the boolean data type *bool*, so that, for any closed boolean term $b : bool$, either

$$b \hat{\rhd}^*_{\Sigma_p} true$$

or

$$b \hat{\rhd}^*_{\Sigma_p} false$$

This assumption will be important when we use Σ_p to define boolean checks in conditional and while-loop constructs of *IML*.

Also, our results will assume a set of *side-effect-producing*, black-box programs. These black-box programs will be represented by sets of function symbols, BB. Each set BB_t of BB is sorted according to the sorts of Σ_p. A function symbol $f \in BB_t$ is intended to denote a side-effect-producing program named `f` of type `t`.

Assumption 4.3. We make the following assumptions about the precoded Σ_p and BB:

- All the functions and constants from the signature Σ_p and BB are defined in the preamble.
- We assume that the evaluation mapping $Eval_{\Sigma_p}$ models the behavior of the pure function symbols TF in the *SML* preamble. Thus the term $Eval_{\Sigma_p}(f(a_1, \ldots, a_n))$ should denote the return value obtained by evaluating $\mathtt{f(a_1, \ldots, a_n)}$ in *SML*.
- Basic sorts correspond to assumed *SML* types that have been defined in the preamble. Functional and product sorts $(t \rightarrow u)$ and $(t * u)$ are taken as functional and product *SML* types (`t-> u`) and (`t * u`) respectively. The disjoint union $t|u$ is taken as shorthand for

  ```
  (t,u) disjointUnion
  ```

  an instantiation of the parametrized *SML* datatype, given in the preamble as

  ```
  datatype ('a,'b) disjointUnion=Inl of 'a | Inr of 'b;;
  ```

These assumptions are comparable to those made in our treatment of pure functional *SML* in Section 2.4, Chapter 2 of Part II, p. 42.

Imperative *SML* programs involve state references — memory addresses — to store values. References are similar to pointers in C or C++, and must be *dereferenced* to obtain the value stored at the memory address. So we assume a set of state references, *StateRef*, sorted according to Σ_p. An element `r` $\in$

$StateRef_s$ is intended to represent a state reference that is available for use in a *SML* program. To obtain the value denoted by the state reference, *SML* uses the *dereferenced* expression `!r`, of *SML* type `s`.

We use the following sorted sets of dereferenced states:

$$!StateRef_s = \{!\mathtt{r} \mid r \in StateRef_s\}$$

Assumption 4.4. To use the state references of *StateRef* in actual *SML* programs, we assume the preamble provides an initialization of the form

```
val s1 = ref ... ;;
val s2 = ref ... ;;
val s3 = ref ... ;;
...
val sn = ref ... ;;
```

where $\mathtt{s1}, \ldots, \mathtt{sn}$ are all the state references of *StateRef*. Each initialization is of the form

$$\mathtt{val\ r} = \mathtt{ref\ a;;}$$

and must set the state identifier `r` to a reference `ref a` of the *SML* type `ref t` where $\mathtt{r} \in StateRef_t$. In this way state references, *StateRef*, are to be considered as global state references, available for any *SML* program we write.

4.2.2 Pure *SML* programs

We identify a pure, side-effect-free subset of *SML* constructed from a signature built from Σ_p with additional dereferenced state identifiers *!StateRef*.

$$\Sigma_{PML} = \langle S, TF \cup !StateRef, Pred \rangle$$

The pure programs consist of the lambda terms

$$PML = Terms(\Sigma_{PML})$$

Example 4.2 *(Terms of PML)*. Assuming we have the usual representations of integer arithmetic and lists of strings in *BB*, the following are example terms of *PML*, if `name1`, `name2`, `s1` and `s2` are state references in *StateRef*.

$$!\mathtt{s1}+!\mathtt{s2}$$

$$\mathtt{fn\ x : string} => [\mathrm{x}]@[!\mathtt{name1}, !\mathtt{name2}, \text{'Iman Poernomo'}]$$

$$\mathtt{fst}(!\mathtt{s1}, 22)$$

Remark 4.1. Observe that *PML* forms a lambda calculus. It is the same kind of lambda calculus subset of *SML* we reasoned with in Chapter 2 of Part II, but extended to include dereferenced states. So we have the usual notion of free and bound variables.

Remark 4.2. *PML* does not use state references in its lambda terms. That is to say, the names of *SML* memory addresses cannot be used. We can only use the *values* stored in these addresses by use of dereferenced states. It is possible to define a larger, pure subset of *SML* that includes state references, for use in pointer arithmetic expressions. However, *PML* will suffice for the purposes of this monograph.

4.2.3 Terms of *IML*

The terms of *PML* define a side-effect-free subset of *SML*. We now define an imperative extension of *PML*, called *IML*. This is the language we will use in our Hoare logic. The grammar of *IML* is given in Fig. 4.4, and involves the black-box programs of *BB*, together with the basic imperative constructs of assignments, sequencing, conditionals, and loops. Terms of *PML* are used to define Boolean terms, value expressions for state assignments and side-effect-free return value expressions.

Remark 4.3. Significant to the main result of this part of the book, given in Chapter 6, is the fact that *IML* includes a side-effect-free subset, *PML*. This permits the definition of complex side-effect-free return value expressions. In Chapter 6, we will show how an adaptation of proofs-as-programs permits us to synthesize correct *IML* programs with *PML* return value expressions, from proofs in our Hoare logic.

`a, b` ::=	*IML* terms
`t`	any side-effect producing function symbol from *BB*
`p`	any closed side-effect-free term from *Closed*(*PML*)
`a; b`	sequencing
`s:=v`	assignment of state reference `s` $\in$ *StateRef* to side-effect-free term `v` $\in$ *Closed*(*PML*)
`if c then a else b`	conditional, `c : bool` a Boolean term in *PML*
`while c do b`	while-loop, `c : bool` a Boolean term in *PML*
`()`	unit

Closed(*PML*) are terms of *PML* that do not contain free variables — see Definition 4.1.2.

Fig. 4.4. Syntax of *IML* constructed over black-box programs from *BB* and *PML*.

Remark 4.4. Note that the syntax of *IML* permits state references to be assigned to *closed* terms of *PML*. This reflects the situation in *SML* where assignment to open terms is illegal. For instance,

$$\texttt{s} := !\texttt{s} + \texttt{x}$$

is an illegal assignment because x is free in $!\texttt{s} + \texttt{x}$, while

$$\texttt{s} := \texttt{fn x : int => } !\texttt{s} + \texttt{x}$$

is a legal assignment, because $\texttt{fn x : int => } !\texttt{s} + \texttt{x}$ is a closed term.

4.2.4 Types of *IML*

We take the set of types of our programs, *Types*(*IML*), to consist of the sorts S of Σ_p (the same sorts used by *PML* and *BB*) — (we refer to these sorts as *types* when using them in *SML* programs).

Terms are associated with types according to the inference rules $TIR(IML)$. These rules involve a sorting relation (:) between terms and types. An inference takes the form

$$\Gamma \vdash_{IML} \texttt{a} : \texttt{s} \tag{4.1}$$

where Γ is a context of variables with types of the form $\{\texttt{x}_1 : \texttt{s}_1, \ldots, \texttt{x}_n : \texttt{s}_n\}$. The inference's intended meaning is that the term a has the type t, given when its free variables $\texttt{x}_1, \ldots, \texttt{x}_n$ denote possible terms of types $\texttt{s}_1, \ldots, \texttt{s}_n$. If Γ can be uniquely inferred from a, we say a is *well-typed* with type s and simply write $\texttt{a} : \texttt{s}$.

The *type* inference rules for the *PML* subset are simply the *sort* inference rules for

$$\Sigma_{PML} = \langle S, TF \cup \mathit{!StateRef}, \mathit{Pred} \rangle$$

that is to say,

$$\Gamma \vdash_{\Sigma_{PML}} a : s \Leftrightarrow \Gamma \vdash_{IML} \texttt{a} : \texttt{s}$$

Remark 4.5. The rules include typing for dereferenced states, treated as function symbols of the signature. This is achieved by the rule

$$\frac{\texttt{s} \in \mathit{!StateRef}_t}{\Gamma \vdash_{IML} !\texttt{s} : \texttt{t}} \text{(Fn)}$$

The typing rules for the *imperative part* of *IML* are given in Fig. 4.5.

4.2.5 *IML* is a computational type theory

We observe that *IML* defines a computational type theory (*CTT*) in the sense used by the Curry–Howard protocol, of the form

$$C(IML) = \langle \mathit{Terms}(IML), \mathit{Type}(IML), :, \vdash_{IML}, TIR(IML) \rangle$$

(See Chapter 3, Definition 3.2.4, p. 86 for the general definition of a *CTT*.)

Terms, *Terms*(*IML*), and types, *Type*(*IML*), are the terms and types of *IML*, with the typing rules for $\vdash_{IML}$ defining a set $TIR(IML)$.

$$\frac{b \in BB_t}{\Gamma \vdash_{IML} \mathtt{b} : \mathtt{t}}\ (BB)$$

$$\frac{\Gamma_1 \vdash_{IML} \mathtt{s} : \mathtt{t} \quad \Gamma_2 \vdash_{IML} \mathtt{v} : \mathtt{t}}{\Gamma_1, \Gamma_2 \vdash_{IML} \mathtt{s} := \mathtt{v} : \mathtt{t}}\ (\text{assign})$$

$$\frac{\Gamma_1 \vdash_{IML} \mathtt{a} : \mathtt{U} \quad \Gamma_2 \vdash_{IML} \mathtt{b} : \mathtt{t}}{\Gamma_1, \Gamma_2 \vdash_{IML} \mathtt{a}; \mathtt{b} : \mathtt{T}}\ (\text{seq})$$

$$\frac{\Gamma_1 \vdash_{IML} \mathtt{a} : \mathtt{Bool} \quad \Gamma_2 \vdash_{IML} \mathtt{b} : \mathtt{t} \quad \Gamma_3 \vdash_{IML} \mathtt{c} : \mathtt{t}}{\Gamma_1, \Gamma_2, \Gamma_3 \vdash_{IML} \mathtt{if\ a\ then\ b\ else\ c} : \mathtt{t}}\ (\text{conditional})$$

$$\frac{\Gamma_1 \vdash_{IML} \mathtt{w} : \mathtt{t} \quad \Gamma_2 \vdash_{IML} \mathtt{c} : \mathtt{Bool}}{\Gamma_1, \Gamma_2 \vdash_{IML} \mathtt{while\ c\ do\ w} : \mathtt{unit}}\ (\text{loop})$$

Type inference rules for the subset of side-effect-free terms *PML* are given in Fig. 4.2.

Fig. 4.5. Type inference rules for the imperative constructs of *IML*.

4.3 Semantics of *SML*

The semantics of programs is given as usual, according to a model of an abstract environment with mutable memory *state*. The machine's state consists of state identifiers that store data values. A program's evaluation changes the state of the machine by affecting the contents of these identifiers.

We provide two related forms of semantics — an operational and a relational semantics — that tell us how programs evaluate. The former semantics tells us how we expect programs to execute, in terms of reduction sequences that produce (possibly non-terminating) sequences of states of the abstract environment, possibly resulting in a return value. The latter semantics provides a more abstract understanding of programs, as side-effect relations between final and initial states. A side-effect is formally understood as a relation between two states, representing the result of executing a terminating imperative program, to make a transition from an initial state to a final state.

The operational semantics will be useful for two reasons:

1. Operational semantics models how a *SML* program evaluates. It is therefore at a lower level of abstraction than relational semantics, and we can define the relational semantics in terms of the operational.
2. The operational semantics tells us how programs evaluate, and what their *return value* is. In contrast, the relational semantics ignores return values. This is important for Chapter 6, where we will be concerned with synthesis of correct return values.

The relational semantics of programs is a higher-level, less detailed, description of how programs can evaluate, and is useful in understanding the semantics of Hoare logic given in this chapter and for deriving soundness results in Chapter 5.

It is possible to define a third, denotational, semantics for our programs, but this is not necessary for our results.

4.3.1 Data values

Our programs manipulate data. Data can be thought of as mathematical *values.* Unlike programs, values are static, and are not open to further reduction but are independent of changes to the memory of the computer.

In *SML* these data terms are always not amenable to further reduction and are closed — they cannot have free variables.

For the purposes of this book we shall define our values to be the closed, irreducible, side-effect-free terms of $Terms(\Sigma_p)$. Because repeated application of the reduction rules for these always terminates, we shall consider data values to be equivalent modulo the $\hat{\triangleright}_{\Sigma_p}$ relation.

Definition 4.3.1 (*Values).* We define the set of *data values* to be the closed, irreducible terms of $Terms(\Sigma_p)$,

$$Values = Closed(Normal(\Sigma_p))$$

Remark 4.6. We do not permit state references in our data. In full *SML* it is possible to have state references themselves as values. This is useful when pointer manipulation is required. However, we do not consider such values here. See remark 4.2 for a similar note about *PML.*

4.3.2 States

Intuitively we can imagine a state of a *SML* environment to be a configuration of the memory of a computer. Recall that *SML state references* are named locations of memory (see p. 101 in the previous section). In our semantics we will be interested in understanding program execution in terms of state "snapshots" of the memory — all the values stored at state references for a particular instant in the evaluation of a program.

The formal definition of a *state* of is as follows.

Definition 4.3.2 (State). A *state* is a function σ from the state references *StateRef* to values from *Values.* We let *MLStates* consist of the set of all states. We assume that any $\sigma \in MLStates$ maps state references to $Closed(PML)$ terms of the same type. That is to say, σ must satisfy the constraint

$$s \in StateRef \quad StateType(s) = t \Leftrightarrow \vdash_{IML} \sigma(s) : t$$

This requirement is standard for any modern imperative language: state references are typed, and may only store values of their types.

Each function $\sigma \in MLStates$ represents a possible memory configuration, where $\sigma(s)$ is the value of the state reference s.

We will use the following notation.

Definition 4.3.3. Given a term $\mathtt{n} \in PML$ and a state σ we write $\sigma(\mathtt{n})$ for the term

$$\mathtt{n}[\sigma(!\bar{\mathtt{s}})/!\bar{\mathtt{s}}]$$

(which is in *Values*) where $!\bar{\mathtt{s}}$ is a list of every dereferenced state reference in $\mathtt{n}$.

Remark 4.7. If $\mathtt{n}$ is a side-effect-free term, only the values of its dereferenced state references need to be computed in order to transform $\mathtt{n}$ into a value (that is to say, into an element of *Values*). Thus, the term $\sigma(\mathtt{n})$ should be thought of as an evaluation of $\mathtt{n}$ in state σ.

4.3.3 Operational semantics

An operational semantics will now be given in terms of possibly infinite reduction sequences of programs and memory states, each representing a "snapshot" of how the program execution affects the state of the *SML* interpreter.

We use a typical call-by-value operational semantics (see, e.g., [Gun93]), given in Fig. 4.6. The semantics defines an evaluation relation $\rhd_{IML}$ over pairs of *IML* programs and states,

$$\langle \mathtt{p}, \sigma \rangle$$

called *configurations*.

Notation 4.2 *(Variants of maps).* We will use the following notation throughout this monograph. Given any mapping $\theta : A \to B$, we will write $\theta[a \mapsto b] : A \to B$ for the mapping that is identical to θ over all elements of the domain, except (possibly) a, which is mapped to b.

Remark 4.8. The intended meaning of

$$\langle \mathtt{p}, \sigma \rangle \rhd_{IML} \langle \mathtt{p}', \sigma' \rangle$$

is that $\mathtt{p}$ can evaluate to $\mathtt{p}'$ with a change in state from σ to σ'. So the value of the state location $\mathtt{s}$ is $\sigma(\mathtt{s})$ prior to running $\mathtt{p}$. Similarly, the value stored in $\mathtt{s}$ is $\sigma'(\mathtt{s})$ after $\mathtt{p}$ evaluates to $\mathtt{p}'$.

Remark 4.9. Observe that the (*PML*) rule shows how to evaluate *PML* terms, by

1. evaluating all dereferenced state identifiers, to give a data value, and then
2. applying lambda reduction rules, to obtain a final value that cannot be reduced further.

As a consequence of this, all function symbols used in *PML* (that is, those symbols from the signature Σ_p) are evaluated according to the $Eval_p$ mapping.

The semantics assumes we know how to evaluate side-effect-producing black-box programs of the preamble. Recall that such functions are given in *BB*. We assume that evaluation is correctly described by a mapping $Eval_i$, from

$$\frac{\mathtt{n} \in \mathit{Closed(PML)} \quad \sigma(\mathtt{n}) \rhd^*_{\Sigma_p} \mathtt{v}}{\langle \mathtt{n}, \sigma \rangle \rhd_{IML} \langle \mathtt{v}, \sigma \rangle} \; (PML)$$

$$\frac{\mathit{Eval}_i(\mathtt{f(a_1, \ldots, a_n)}, \sigma) = (\mathtt{b}, \sigma')}{\langle \mathtt{f(a_1, \ldots, a_n)}, \sigma \rangle \rhd_{IML} \langle \mathtt{b}, \sigma' \rangle} \; (\mathrm{se})$$

$$\frac{\langle \mathtt{p_1}, \sigma \rangle \rhd_{IML} \langle \mathtt{p_1'}, \sigma' \rangle \quad \langle \mathtt{p_2}, \sigma' \rangle \rhd_{IML} \langle \mathtt{p_3}, \sigma'' \rangle}{\langle \mathtt{p_1}; \mathtt{p_2}, \sigma \rangle \rhd_{IML} \langle \mathtt{p_3}, \sigma'' \rangle} \; (\mathrm{seq})$$

$$\frac{\sigma' = \sigma[\mathtt{s} \mapsto \sigma(\mathtt{v})]}{\langle \mathtt{s} := \mathtt{v}, \sigma \rangle \rhd_{IML} \langle !\mathtt{s}, \sigma' \rangle} \; (\mathrm{assign})$$

$$\frac{\sigma(\mathtt{b}) \rhd_{\Sigma_p} \mathtt{true}}{\langle \mathtt{if\ b\ then\ p\ else\ q}, \sigma \rangle \rhd_{IML} \langle \mathtt{p}, \sigma \rangle} \; (\mathrm{conditional})_1$$

$$\frac{\sigma(\mathtt{b}) \rhd_{\Sigma_p} \mathtt{false}}{\langle \mathtt{if\ b\ then\ p\ else\ q}, \sigma \rangle \rhd_{IML} \langle \mathtt{q}, \sigma \rangle} \; (\mathrm{conditional})_2$$

$$\frac{\sigma(\mathtt{b}) \rhd_{\Sigma_p} \mathtt{false}}{\langle \mathtt{while\ b\ do\ c}, \sigma \rangle \rhd_{IML} \langle \mathtt{()}, \sigma \rangle} \; (\mathrm{while})_1$$

$$\frac{\sigma(\mathtt{b}) \rhd_{\Sigma_p} \mathtt{true} \quad \langle \mathtt{c}, \sigma \rangle \rhd_{IML} \langle \mathtt{r}, \sigma' \rangle}{\langle \mathtt{while\ b\ do\ r}, \sigma \rangle \rhd_{IML} \langle \mathtt{while\ b\ do\ c}, \sigma' \rangle} \; (\mathrm{while})_2$$

Fig. 4.6. Operational Semantics for *IML*

function applications and states to values and states. Given a function symbol $f \in BB_{s_1 \ldots s_n, s}$, arguments $(\mathtt{a_1}, \ldots, \mathtt{a_n})$ of sort $(\mathtt{s_1} * \ldots * \mathtt{s_n})$, and initial state σ,

$$\mathit{Eval}_i(\mathtt{f(a_1, \ldots, a_n)}, \sigma)$$

gives a value from *Values* of sort s and a new state σ'. We assume that $\mathit{Eval}(\mathtt{f(a_1, \ldots, a_n)}, \sigma)$ is exactly the return value and final state obtained by evaluating $\mathtt{f(a_1, \ldots, a_n)}$ in *SML*.

Assumption 4.5. Recall that Assumption 4.2 (p. 101) took Σ_p and consequently the closed irreducible terms of that signature, *Values*, to include booleans. So, for any $\mathtt{b} \in \mathit{Values}$ and any state σ, either

$$\sigma(\mathtt{b}) \rhd^*_{IML} \mathtt{true} \quad \text{or} \quad \sigma(\mathtt{b}) \rhd^*_{IML} \mathtt{false}$$

This ensures that the boolean checks of conditional and while-loop statements will always reduce to true or false, just as they should do in *SML*.

4.3.4 Evaluation and return values

We define evaluation of a program to be the repeated application the reduction rules of Fig. 4.6. It is important to note that these rules will sometimes never terminate — we can have *SML* programs that continue in infinite loop.

For instance, to evaluate a program p for a given initial state σ, we consider the configuration

$$\langle \mathtt{p}, \sigma \rangle$$

and apply a rule to obtain a new configuration

$$\langle \mathtt{p}_1, \sigma_1 \rangle$$

such that

$$\langle \mathtt{p}, \sigma \rangle \rhd_{IML} \langle \mathtt{p}_1, \sigma_1 \rangle$$

We then repeat the process, obtaining a *possibly non-terminating* sequence of configurations:

$$\begin{aligned} \langle \mathtt{p}, \sigma \rangle &\rhd_{IML} \langle \mathtt{p}_1, \sigma_1 \rangle \\ &\rhd_{IML} \cdots \\ &\rhd_{IML} \langle \mathtt{p}_n, \sigma_n \rangle \\ &\rhd_{IML} \cdots \end{aligned}$$

If the sequence terminates at a final configuration, say, $\langle \mathtt{p}_j, \sigma_j \rangle$ some j, then we say that p in initial state σ evaluates to return value $\mathtt{p}_j$ with final state σ'.

Example 4.3. Consider the loop

```
while !conEstab == false do
  tryConnect();
'Connection established'
```

This program tries to establish a connection with a database, by repeatedly calling a function called `tryConnect()`. The state `conEstab` determines if a connection has been established or not. The program continues until this state has a `true` value.

The evaluation sequence of the program will either continue indefinitely or else terminate with a string return value,

```
'Connection established'
```

Example 4.4. Assume *StateRef* includes the state `i` with $StateType(\mathtt{i}) = \mathtt{int}$.

Let σ be a state such that $\sigma(\mathtt{i}) = 0$. The program

$$\mathtt{while\ !i < 3\ do\ i := !i + 1}$$

results in a terminating execution sequence

$$\begin{aligned} &\langle \mathtt{while\ !i < 3\ do\ i := !i + 1}, \sigma \rangle \\ &\qquad \rhd_{IML} \langle \mathtt{while\ !i < 3\ do\ i := !i + 1}, \sigma[\mathtt{i} \mapsto 1] \rangle \\ &\qquad \rhd_{IML} \langle \mathtt{while\ !i < 3\ do\ i := !i + 1}, \sigma[\mathtt{i} \mapsto 2] \rangle \\ &\qquad \rhd_{IML} \langle \mathtt{while\ !i < 3\ do\ i := !i + 1}, \sigma[\mathtt{i} \mapsto 3] \rangle \\ &\qquad \rhd_{IML} \langle (), \sigma[\mathtt{i} \mapsto 3] \rangle \end{aligned}$$

by repeated application of the (*loop*) reduction rule, because, for any τ,

$$\langle \mathtt{i} :=\,!\mathtt{i} + \mathtt{1}, \tau\rangle \rhd_{IML} \langle (), \tau[\mathtt{i} \mapsto \tau(\mathtt{i}) + 1]\rangle$$

Thus this program in initial state σ evaluates to return value () with final state $\sigma[\mathtt{i} \mapsto 3]$.

Remark 4.10. Inspection of our operational semantics easily reveals that return values are always elements in *Values.*

4.3.5 Relational semantics

We now turn our attention to a more abstract, *relational semantics* of programs. This semantics considers programs solely in terms of a range of possible side-effects, that is, changes in state that result from execution. The relational view, in contrast to the more detailed view afforded by operational semantics, is not concerned with sequences of state and program configurations, nor in the return values given by some executions.

We write *MLRel* for the set of side-effect relations given over *MLStates*,

$$MLRel = \mathcal{P}(MLStates \times MLStates)$$

(where $\mathcal{P}(A)$ denotes the power set of a set A).

Given a relation $R \in MLRel$ and $\sigma, \sigma' \in MLStates$, we will often write

$$\sigma\ R\ \sigma'$$

for

$$(\sigma, \sigma') \in R$$

A side-effect relation provides an abstract view of program behavior in terms of input and output states.

Example 4.5. Using our notation, the increment function $\mathtt{g} := \mathtt{g} + \mathtt{1}$ is represented by the side-effect relation from *MLRel*

$$R = \{(\sigma, \sigma') \mid \sigma' = \sigma[\mathtt{g} \mapsto \sigma(\mathtt{g}) + 1]\}$$

A relational semantics for *IML* programs is given by associating a side-effect relation with each *IML* program, using a semantic valuation map

$$[\![.]\!] : IML \to MLRel \tag{4.2}$$

The definition of $[\![.]\!]$ simply uses the operational semantics of *terminating programs*, ignoring intermediate states.

Definition 4.3.4 (Relational semantics of programs). Given any program $\mathtt{p} \in IML$, if there are states $\sigma, \sigma' \in MLStates$ such that

$$\langle \mathtt{p}, \sigma\rangle \hat{\rhd}_{IML} \langle \mathtt{r}, \sigma'\rangle$$

for some return value $\mathtt{r} \in IML$ then

$$(\sigma, \sigma') \in [\![\mathtt{p}]\!]$$

If there are no such states for p (that is to say, if the program does not terminate), then we take $[\![\mathtt{p}]\!]$ to be the emptyset $\emptyset$.

Remark 4.11. It is possible to define a relational semantics for our programs that is independent of the operational semantics, by defining operators over the side-effect relations that correspond to imperative constructs. Such a semantics can be built using, for instance, Kleene Algebras with tests (see, e.g., [Koz97]).

Remark 4.12. *Values* constitutes a term model for data stored in states. It is possible to define a wider range of models for data, and, as a consequence, a more general relational semantics for *IML*. This type of general relational semantics is commonly used for the semantics of Hoare logic.

However, for our purposes, we are only interested in states *MLStates* that involve *Values* as data. This is because, in *SML*, it is the terms of *Values* that are stored in *SML* state references, and we are primarily concerned with a model of execution that is as close to *SML* implementation as possible. When we introduce formulae of intuitionistic Hoare logic and the calculus itself, we will provide soundness results using *MLRel*. The interested reader is referred to [Cou90] for a more general semantic treatment of Hoare logic, given over a range possible relational semantics. It would be easy to adapt that treatment to our case.

4.4 Formulae

We define a set of formulae that describe side-effect relations and, as a consequence, program behavior. A range of possible side-effects is specified by pre- and post-conditions, in the single formula style of, for instance, OCL [WK98], the B-method [Abr96], or Abadi's object-oriented extension of Hoare logic [AL97].

A side-effect is described in terms of initial and final state reference values, prior to, and after, execution. Such initial and final state reference values are respectively denoted by the name of the state reference with a $()_i$ or with a $()_f$ subscript. For instance, the formula $r_f > r_i$ specifies side-effects where the initial value of r, denoted by r_f, is greater than the initial value, denoted by r_i. A program that satisfies this specification is $\mathtt{r} := !\mathtt{r} + 1$.

The formulae of our logic are first-order, many-sorted, and use the quantifiers and connectives of classical/intuitionistic logic. Sorts, terms, and predicates are obtained from the signature Σ_p.

To enable the specification of side-effects our predicates range over the usual set of terms of Σ_p extended by the subscripted *StateRef* symbols that denote initial and final state reference values. For instance, if $r \in \mathit{StateRef}$ then $r_i * 20 + r_f$ is a well-formed term that may be used in our logic.

We can define when a formula is true of a side-effect relation from *MLRel*. Recall that a program's execution is defined by a particular side-effect relation.

We therefore can define when a formula is true of a program's execution: this is exactly when it is true of the program's side-effect relation. For instance,

$$s_f > s_i$$

is true of the side-effect relation

$$R = \{(\sigma, \sigma') \mid \sigma' = \sigma[\mathtt{s} \mapsto \sigma(\mathtt{s}) + 1]\}$$

because the new value of $\mathtt{s}$ is greater than the old value of $\mathtt{s}$ for R (that is, $\sigma'(\mathtt{s})$ is greater than $\sigma(\mathtt{s})$ for every $(\sigma, \sigma') \in R$). The side-effect relation R denotes the range of possible side-effects for $\mathtt{s} := \mathtt{s} + 1$, and so the formula is true of this program's execution.

We proceed as follows. We first describe a set of terms used by our formulae. Then we define the set of well-formed formulae. Finally, we define when a formula is true of a side-effect relation, and, hence, true of a program's execution.

4.4.1 Terms

Our terms consist of the lambda calculus built from Σ_p, extended by extra symbols to refer to the initial and final values of state identifers in a given side-effect relation.

Definition 4.4.1 (Initial and final values). We reserve a sorted set of symbols *IStateVal* to represent initial values of state references, and a sorted set *FStateVal* to represent final values, i.e.,

$$\begin{aligned} IStateVal_t &= \{s_i \mid \mathtt{s} \in StateRef_t\} \\ FStateVal_t &= \{s_f \mid \mathtt{s} \in StateRef_t\} \end{aligned}$$

We write *StateVal* for *IStateVal* $\cup$ *FStateVal*. Given a term $s_i \in IStateVal$, we write $state\!-\!id(s_i)$ for the corresponding state reference, $\mathtt{s}$. We overload this function, so that, given a term $s_f \in FStateVal$, we write $state\!-\!id(s_f)$ for the corresponding state reference, $\mathtt{s}$.

Our terms are defined as follows.

Definition 4.4.2 (Elements of $Terms(\Sigma_t)$). We define the signature of our terms to be

$$\Sigma_t = \langle S, TF \cup StateVal, Pred \rangle$$

formed from Σ_p extended by *StateVal*, treated as function symbols.

Then we define $Terms(\Sigma_t)$ according to the grammar of Fig. 4.1, p. 98.

Remark 4.13. $Terms(\Sigma_t)$ is a lambda calculus equipped with the usual notion of free and bound variables.

Remark 4.14. Note that we have now defined three distinct lambda calculi — *PML*, *Values*, and now $Terms(\Sigma_t)$. The calculi all share the sorts S and function symbols TF. However, they each serve different purposes:

- *PML* defines the side-effect-free expressions of *IML*,
- *Values* defines the data values of states and evaluated return values of programs.
- $Terms(\Sigma_t)$ is used in Hoare logic as terms of predicates.

Example 4.6. Assuming Σ_t contains symbols for representing integer arithmetic, strings and lists of strings, the following are elements of $Terms(\Sigma_t)$:

$$\mathit{fn}\ x : \mathit{int} => x + s_f + s_i$$

$$s1_i + s2_i$$

$$\mathit{fn}\ x : \mathit{string} => [x]@[name1_f, name2_f, 'Iman\ Poernomo']$$

$$\mathit{fst}(f1_i, 22)$$

Because $Terms(\Sigma_t)$ is a lambda calculus we have the reduction relation $\triangleright_{\Sigma_t}$, with rules as defined in Fig. 4.3, p. 100.

Remark 4.15. The reduction rules treat the state identifiers as irreducible constant symbols. State identifiers are only associated with values when interpretated with respect to a given side-effect relation, described later in this section.

We will require the following definition.

Definition 4.4.3 (State reference values for a term). Let t be a term of $Terms(\Sigma_t)$. We define $initial(t)$ to be the list of *IStateVal* symbols that occur in t. We define $final(t)$ similarly, to be the list of *FStateVal* symbols that occur in t.

The sort inference rules for our terms are simply those of Σ_t, as given in Fig. 4.2, p. 99.

Remark 4.16. Note that the sort inference rules treat the initial and final state references as function symbols for the purpose of associating sorts:

$$\frac{s_i \in \mathit{IStateVal}_t}{\Gamma \vdash_{\Sigma_t} s_i : t}\ (\mathrm{Fn})_i \qquad \frac{s_f \in \mathit{FStateVal}_t}{\Gamma \vdash_{\Sigma_t} s_f : t}\ (\mathrm{Fn})_f$$

4.4.2 Well-formed formulae

We use first-order, many-sorted formulae to make assertions about programs. All predicates and terms are taken from Σ_t.

The definition well-formed formulae over the signature Σ_t is as in Definition 2.1.3 of Chapter 2, Part II. We repeat the definition for reference, made specific for our signature Σ_t.

Definition 4.4.4 (Well-formed formulae for Σ_t). Let $\Sigma_t = \langle S, TF \cup \mathit{StateVal}, P\rangle$. The set of well-formed formulae for a signature, $WFF(\Sigma_t)$ is the least set containing

- every $P(s_1, \ldots, t_n)$ where $P \in P_{\{s_1,\ldots,s_n\}}$ is a predicate symbol in P and every t_j $(j = 1, \ldots, n)$ is a well-sorted term of sort s_j,
- every formula $(A \wedge B)$, $A, B \in WFF(\Sigma_t)$,
- every formula $(A \vee B)$, $A, B \in WFF(\Sigma_t)$,
- every formula $(A \Rightarrow B)$, $A, B \in WFF(\Sigma_t)$,
- every formula $\forall x : s \bullet F$ where $x \in Var$ and $F \in WFF(\Sigma_t)$,
- every formula $\exists x : s \bullet F$ where $x \in Var$ and $F \in WFF(\Sigma_t)$,
- the formula $\perp$.

We often write $\neg A$ for $A \Rightarrow \perp$.

The usual definitions of free and bound variables apply to our formulae.

Definition 4.4.5 (Bound and free variables of formulae). A formula G binds a variable of Var if, and only if, either

- G contains a subformula of form $\forall x : s \bullet A$ or $\exists x : s \bullet A$, or
- G contains an atomic subformula of the form $P(a_1, \ldots, a_n)$ where $x : s$ occurs in $BV(a_j)$ for some a_j $(j \in \{1, \ldots, n\})$.

In this case, we say x is bound in G. If a variable x is not bound by a formula G, but occurs in a $Terms(\Sigma_t)$ term used in an atomic subformula of G, then we say that it *occurs free* in G.

We write $FV(G)$ for the set of free variables of G, and $BV(G)$ for the set of bound variables.

4.4.3 Interpreting terms of $Terms(\Sigma_t)$

To understand how our formulae specify *SML* program behavior, we need to interpret them over the data that is stored in state references used by *IML* programs. Recall that, for the purposes of our work, we take this data to be *Values*.

Definition 4.4.6 (Interpreting $Terms(\Sigma_t)$ terms over *Values*). Let σ and σ' be states of *MLStates*.

Let $\hat{\iota}$ be an interpretation of Var by *Values*. We define ι to be the inductive extension of $\hat{\iota}$ to all of $Terms(\Sigma_p)$, so that, for any variable $x \in Var$ and any function $f : \bar{T} \to T \in Terms(\Sigma_p)$,

$$\iota(x) = \hat{\iota}(x)$$

$$\iota(f(a_1, \ldots, a_n)) = f(\iota(a_1), \ldots, \iota(a_n))$$

We write $\iota(t)_{\sigma}^{\sigma'} \in Values$ for the interpretation of t formed by extending the interpretation $\hat{\iota}$ to all of $Terms(\Sigma_t)$,

$$\iota(t)_{\sigma}^{\sigma'} = \iota'(t)$$

where ι' interprets every symbol $s_i \in initial(t)$ ($\mathbf{s} \in StateRef$) by the corresponding initial state reference value $\sigma(\mathbf{s})$, and similarly for final state references:

$$\iota' = \iota[\bar{i} \mapsto \sigma(state\!-\!id(\bar{i}))][\bar{f} \mapsto \sigma'(state\!-\!id(\bar{i}))]$$

with $\bar{i}$ and $\bar{f}$ the lists of initial and final state reference identifiers, $initial(t)$ and $final(t)$, respectively.

Lemma 4.4.1. *Let a and b be elements of $Terms(\Sigma_t)$, σ, σ' be states and ι some interpretation. Then*

$$\iota(a)_{\sigma}^{\sigma'} \rhd_{\Sigma_p} \iota(b)_{\sigma}^{\sigma'} \Leftrightarrow a \rhd_{\Sigma_t} b$$

Proof. By induction on the possible forms of a, using the fact that the operational semantics of Σ_t has the same reduction rules as Σ_p, and treats the state identifiers as irreducible symbols. □

4.4.4 Truths about side-effect relations

We can define when a formula P is true of a particular side-effect transition pair (σ, σ'), i.e., $\sigma, \sigma' \in MLStates$.

The terms of *IStateVal* and *FStateVal* denote the initial and final values of corresponding state identitifers for a side-effect relation. This leads to a formal definition of when a formula is true of a side-effect transition pair.

We define a formula to be *true of a side-effect relation* when it is true of every side-effect transition of the relation.

We required the following definition.

Definition 4.4.7 (Truth valuation function). Given the signature $\Sigma = \langle S, TF, Pred \rangle$ a truth valuation function is a map

$$h : Pred \to \mathcal{P}(Values)$$

where

$$(a_1, \ldots, a_n) \in h(P)$$

only when, for every predicate P in $Pred_{s_1 \ldots s_n}$, we have that $a_1 : s_1, \ldots, a_n : s_n$.

Assumption 4.6. Our results will assume a fixed truth valuation function h for

$$\Sigma_p = \langle S, TF, Pred \rangle$$

Intuitively, this function tells us what elements of $Values$ the predicates from $Pred$ hold over.

Definition 4.4.8 (Truth about side-effects). Let G be a formula. Let ι be an interpretation of terms of $Terms(\Sigma_p)$ by elements of *Values* (formed by an interpretation $\hat{\iota}$ of elements of Var by elements of *Values*).

Let $\sigma, \sigma' \in MLStates$.

We say that G *is true of the side-effect* (σ, σ') *under the interpretation* ι, written $(\sigma, \sigma') \Vdash_{\iota} G$, when

- if G is atomic, of the form $P(a_1, \ldots, a_n)$, then

$$h(P) \ni (\iota(a_1)_{\sigma}^{\sigma'}, \ldots, \iota(a_n)_{\sigma}^{\sigma'})$$

- if G is $(A \vee B)$, then

$$(\sigma, \sigma') \Vdash_{\iota} A \text{ or } (\sigma, \sigma') \Vdash_{\iota} B$$

- if G is $(A \wedge B)$, then

$$(\sigma, \sigma') \Vdash_{\iota} A \text{ and } (\sigma, \sigma') \Vdash_{\iota} B$$

- if G is $(A \Rightarrow B)$, then

$$(\sigma, \sigma') \Vdash_{\iota} A \text{ entails } (\sigma, \sigma') \Vdash_{\iota} B$$

- if G is $\forall x : t \bullet P$, then
$$(\sigma, \sigma') \Vdash_{\iota} G[a/x]$$
for every $a : t \in Values$.
- if G is $\exists x : s \bullet P$, then
$$(\sigma, \sigma') \Vdash_{\iota} G[a/x]$$
for some $a : t \in Values$.
- it is never the case that $(\sigma, \sigma') \Vdash_{\iota} \bot$.
- it is always the case that $(\sigma, \sigma') \Vdash_{\iota} true$.

Let R be an element of *MLRel*.
We say that G is *true of* R *under interpretation* ι, written $R \Vdash_{\iota} G$, when,

$$\text{all } (\sigma, \sigma') \in R.\ (\sigma, \sigma') \Vdash_{\iota} G$$

We say that G is *true of* R *with initial state* σ *under* ι, written $R \Vdash_{\iota}^{\sigma} G$ when, if $\sigma\ R\ \sigma'$, then
$$(\sigma, \sigma') \Vdash_{\iota} G$$

We say that G is *true of* R *with initial state* σ, written $R \Vdash^{\sigma} G$, when, for every interpretation ι, $R \Vdash_{\iota}^{\sigma} G$.

We say that G is *true of* R, writing $R \Vdash G$, when, for every interpretation ι, $R \Vdash_{\iota} G$.

Remark 4.17. The notion of truth given uses models which interpret $Terms(\Sigma_p)$ and $Terms(\Sigma_t)$ by elements of *Values*. This is because *Values* terms are the data used in *SML*, and we are primarily concerned with models that are as close to *SML* implementation as possible.

However, there is a wider range of models of $Terms(\Sigma_p)$. It is possible to interpret $Terms(\Sigma_p)$ and $Terms(\Sigma_t)$ over a range of models to arrive at a more general definition of truth of formulae. This is the usual case when providing a semantics for Hoare logic. See Remark 4.12 (p. 111) for a similar observation about side-effect relations of *IML*.

The interested reader is referred to [Cou90] for a more general semantic treatment of the formulae of Hoare logic given over a range of models of data. It would be easy to adapt that treatment to our case.

4.5 Calculus

We now define a constructive version of Hoare logic, which we call Intuitionistic Hoare logic (IHL), for reasoning about side-effects of imperative programs in *SML*.

Hoare logic is traditionally given with respect to a simple imperative programming language and a language for expressing pre- and post-conditions [Flo67, Hoa69]. In our case, IHL uses *IML* as the imperative programming language, and the formulae $WFF(\Sigma_t)$ for specifying pre- and post-conditions. The programs were described in Sections 4.2 and 4.3 and the formulae were defined in Section 4.4.

Hoare logic is often parametrized over a deductive system, usually classical logic. For the purposes of adapting proofs-as-programs in later chapters, we will instead use an intuitionistic deduction system, based on that given in Chapter 2 of Part II.

4.5.1 Program/formula pairs

The theorems of our calculus involve program/formula pairs of the form

$$\mathtt{p} \diamond A$$

consisting of a program $\mathtt{p} \in IML$ and a formula A from $WFF(\Sigma_t)$. We shall refer to the set of such pairs as $Pairs(\mathsf{IHL})$.

Formally, the meaning of program/formula pair in our calculus is given by an interpretation of $Terms(\Sigma_t)$.

Definition 4.5.1 (Truth of program/formula pair). We say that a formula $\mathtt{p} \diamond A$ is *true* for an interpretation τ when

$$[\![\mathtt{p}]\!] \Vdash_\tau A$$

(Recall the semantic valuation map $[\![.]\!]$ was defined on p. 110 above.) We say that $\mathtt{p} \diamond A$ is true, if, and only if, $\mathtt{p} \diamond A$ is true for every possible interpretation.

The formula is a true statement about the side-effect relation, $[\![\mathtt{p}]\!]$, associated with $\mathtt{p}$ *provided that* $\mathtt{p}$ *terminates.*

4.5.2 Rules

Our calculus provides a set of rules for constructing new true program/formula pairs from known true program/formula pairs.

The basic calculus is presented in Figs. 4.7 and 4.8. The former rules are the basic rules of Hoare logic, while the latter are rules for intuitionistic logic.

The basic rules are used to construct new programs and new truths about the programs from old.

Remark 4.18. The rule (seq) of the logic enables us to build a new program (p; q) and a new truth $A[\bar{s}_i/\bar{v}] \Rightarrow C[\bar{s}_f/\bar{v}]$ about the program from known truths $A[\bar{s}_i/\bar{v}] \Rightarrow B[\bar{s}_f/\bar{v}]$ and $B[\bar{s}_i/\bar{v}] \Rightarrow C[\bar{s}_f/\bar{v}]$ about the subprograms p and q respectively.

The (loop) and (ite) rules are similar.

Remark 4.19. The (assign) permits us to make a simple assertion about assignment statements: that is, given a program s := v we know that the final value of s must be equal to v.

Remark 4.20. Our version of Hoare logic is concerned with so-called *partial correctness* of statements about programs [Cou90]. Given a theorem, the formula is true about the program, assuming that the program terminates. However, the logic is not equipped with a means to prove program termination. In particular the (loop) rule does not prove termination of the while-loop of the conclusion. This can be rectified by considering a version of Hoare logic for total correctness. That version of Hoare logic is well understood, but is complicated, for our current purpose of adapting proofs-as-programs. We limit ourselves to partial correctness, and leave the extension to total correctness as future work.

Remark 4.21. It will be shown that all theorems are true, by the *soundness* property, which is proved in the next chapter.

The (ite) and (loop) rules require the map $\mathsf{tologic_i}$, which transforms a *SML* boolean function b into a $Terms(\Sigma_t)$ boolean term, for use in formulae. The map replaces all dereferenced state references of the form !s with initial state identifiers of the form s_i.

Definition 4.5.2. Given any term b, we define

$$\mathsf{tologic_i}(\mathsf{b}) = b[\bar{s}_i/!\bar{s}]$$

where $!\bar{s}$ is every dereferenced state reference in b, and $\bar{s}_i$ the corresponding list of initial state identifiers.

We also define

$$\mathsf{tologic_f}(\mathsf{b}) = b[\bar{s}_f/!\bar{s}]$$

where $!\bar{s}$ is every dereferenced state reference in b, and $\bar{s}_f$ the corresponding list of final state identifiers.

$$\frac{}{\vdash_{\mathsf{IHL}} \mathtt{s} := \mathtt{v} \diamond s_f = \mathsf{tologic_i}(\mathtt{v})} \text{ (assign)}$$

where $s \in \mathit{StateRef}$.

$$\frac{\vdash_{\mathsf{IHL}} \mathtt{p} \diamond (\mathsf{tologic_i}(\mathtt{b}) = \mathit{true} \Rightarrow C) \quad \vdash_{\mathsf{IHL}} \mathtt{q} \diamond (\mathsf{tologic_i}(\mathtt{b}) = \mathit{false} \Rightarrow C)}{\vdash_{\mathsf{IHL}} \mathtt{if\ b\ then\ p\ else\ q} \diamond C} \text{ (ite)}$$

$$\frac{\vdash_{\mathsf{IHL}} \mathtt{p} \diamond (A[\bar{s}_i/\bar{v}] \Rightarrow B[\bar{s}_f/\bar{v}]) \quad \vdash_{\mathsf{IHL}} \mathtt{q} \diamond (B[\bar{s}_i/\bar{v}] \Rightarrow C[\bar{s}_f/\bar{v}])}{\vdash_{\mathsf{IHL}} \mathtt{p;q} \diamond (A[\bar{s}_i/\bar{v}] \Rightarrow C[\bar{s}_f/\bar{v}])} \text{ (seq)}$$

where A and B are free of state identifiers.

$$\frac{\vdash_{\mathsf{IHL}} \mathtt{q} \diamond (\mathsf{tologic_i}(\mathtt{b}) = \mathit{true} \wedge A[\bar{s}_i/\bar{v}]) \Rightarrow A[\bar{s}_f/\bar{v}]}{\vdash_{\mathsf{IHL}} \mathtt{while\ b\ do\ q} \diamond A[\bar{s}_i/\bar{v}] \Rightarrow (A[\bar{s}_f/\bar{v}] \wedge \mathsf{tologic_f}(\mathtt{b}) = \mathit{false})} \text{ (loop)}$$

where A is free of state identifiers.

$$\frac{\vdash_{\mathsf{IHL}} \mathtt{p} \diamond P \quad \vdash_{\mathsf{Int}} (P \Rightarrow A)}{\vdash_{\mathsf{IHL}} \mathtt{p} \diamond A} \text{ (cons)}$$

Intuitionistic deduction $\vdash_{\mathsf{Int}}$ is given in Fig. 4.8.

Fig. 4.7. The basic rules of IHL.

The rule (cons) of Fig. 4.7 is given with respect to intuitionistic deduction $\vdash_{\mathsf{Int}}$, as described in Chapter 2, Section 2.2. We use the same rules but given over formulae from $WFF(\Sigma_t)$. The core, basic rules for $\vdash_{\mathsf{Int}}$ are repeated for reference in Fig. 4.8.

The intutionistic rules are concerned with truths that are universal to all programs. that is to say, they can be used to infer properties that hold over any side-effect.

Example 4.7. For instance, an application of the logical ($\wedge$-I) rule

$$\frac{s_f = s_i * 2 \vdash_{\mathsf{Int}} s_f \geq s_i \quad s_f = s_i * 2 \vdash_{\mathsf{Int}} Even(s_f)}{s_f = s_i * 2 \vdash_{\mathsf{Int}} s_f \geq s_i \wedge Even(s_f)} \text{ ($\wedge$-I)}$$

tells us that, for any program that makes $s_f = s_i * 2$ true, because it follows that $s_f \geq s_i$ and $Even(s_f)$ must also be true, $s_f \geq s_i \wedge Even(s_f)$ must be true of the program.

Remark 4.22. Our calculus is a form of natural deduction, using sequents to represent proofs. The sequent format presentation of proofs is equivalent to a "tree" format presentation (the former preferred when space needs to be conserved, the latter preferred when the steps of a deduction need to be displayed clearly). A sequent $\vdash_{\mathsf{IHL}} \mathtt{p} \diamond F$ is equivalent to the following tree format presentation:

$$\begin{array}{c} \vdots \\ \mathtt{p} \diamond F \end{array}$$

Assume that x, y are arbitrary variables of sort s from signature Σ, and that a and c are well-sorted terms of sort s.

$$\frac{}{A \vdash_{\mathsf{Int}} A} \text{(Ass-I)}$$

$$\frac{\Delta, A \vdash_{\mathsf{Int}} B}{\Delta \vdash_{\mathsf{Int}} (A \Rightarrow B)} (\Rightarrow\text{-I}) \qquad \frac{\Delta \vdash_{\mathsf{Int}} A \quad \Delta' \vdash_{\mathsf{Int}} (A \Rightarrow B)}{\Delta, \Delta' \vdash_{\mathsf{Int}} B} (\Rightarrow\text{-E})$$

$$\frac{\Delta \vdash_{\mathsf{Int}} A}{\Delta \vdash_{\mathsf{Int}} \forall x : s \bullet A} (\forall\text{-I}) \qquad \frac{\Delta \vdash_{\mathsf{Int}} \forall x : s \bullet A}{\Delta \vdash_{\mathsf{Int}} A[c/x]} (\forall\text{-E})$$

x is free in A, not free in Δ

$$\frac{\Delta \vdash_{\mathsf{Int}} P[a/y]}{\Delta \vdash_{\mathsf{Int}} \exists y : s \bullet P} (\exists\text{-I}) \qquad \frac{\Delta_1 \vdash_{\mathsf{Int}} \exists y : s \bullet P \quad \Delta_2, P[x/y] \vdash_{\mathsf{Int}} C}{\Delta_1, \Delta_2 \vdash_{\mathsf{Int}} C} (\exists\text{-E})$$

where x is not free in C

$$\frac{\Delta \vdash_{\mathsf{Int}} A \quad \Delta' \vdash_{\mathsf{Int}} B}{\Delta, \Delta' \vdash_{\mathsf{Int}} (A \wedge B)} (\wedge\text{-I})$$

$$\frac{\Delta \vdash_{\mathsf{Int}} (A_1 \wedge A_2)}{\Delta \vdash_{\mathsf{Int}} A_1} (\wedge\text{-E}_1) \qquad \frac{\Delta \vdash_{\mathsf{Int}} (A_1 \wedge A_2)}{\Delta \vdash_{\mathsf{Int}} A_2} (\wedge\text{-E}_2)$$

$$\frac{\Delta \vdash_{\mathsf{Int}} A_1}{\Delta \vdash_{\mathsf{Int}} (A_1 \vee A_2)} (\vee\text{-I}_1) \qquad \frac{\Delta \vdash_{\mathsf{Int}} A_2}{\Delta \vdash_{\mathsf{Int}} (A_1 \vee A_2)} (\vee\text{-I}_2)$$

$$\frac{\Delta \vdash_{\mathsf{Int}} A \vee B \quad \Delta_1, A \vdash_{\mathsf{Int}} C \quad \Delta_2, B \vdash_{\mathsf{Int}} C}{\Delta_1, \Delta_2, \Delta \vdash_{\mathsf{Int}} C} (\vee\text{-E})$$

$$\frac{\Delta \vdash_{\mathsf{Int}} \bot}{\Delta \vdash_{\mathsf{Int}} A} (\bot\text{-E})$$

provided A is Harrop

Fig. 4.8. The basic rules of many-sorted intuitionistic logic, Int, ranging over $WFF(\Sigma_t)$.

4.5.3 Axioms and schemata

In addition to the rules described, our deduction system permits for use in the intuitionistic subsystem and Hoare logic proper. We use axioms and schemata to define knowledge about a problem domain by extra-logical constraints about the behavior of signature terms.

We assume a set of axioms $\mathcal{AX}$, which consists of two sets: a set of $WFF(\Sigma_t)$ formulae, $\mathcal{AX}_{\mathsf{Int}}$ and a set of pairs of *IML* programs and $WFF(\Sigma_t)$ formulae, $\mathcal{AX}_{BB}$. The former forms axioms used in the intuitionistic subsystem, while the latter forms axioms about programs, used in the Hoare logic proper.

To use axioms from $\mathcal{AX}_{\mathsf{Int}}$, we use the introduction rule

$$\frac{A \in \mathcal{AX}_{\mathsf{Int}}}{\vdash_{\mathsf{Int}} A} \; (\text{Ax-I})_{\mathsf{Int}}$$

Similarly, to use axioms from $\mathcal{AX}_{BB}$, we have the rule

$$\frac{(\mathtt{p} \diamond A) \in \mathcal{AX}_{BB}}{\vdash_{\mathsf{IHL}} \mathtt{p} \diamond A} \; (\text{Ax-I})_{BB}$$

Assumption 4.7. We leave the set of axioms to be a parameter of our system, to be specified for use in a particular domain. However, we shall assume that, at least, axioms for (Heyting) integer arithmetic are included for reasoning in the intuitionistic subset.

We use schemata rules as a metalogical device for generating (a potentially infinite number of) axioms in $\mathcal{AX}$.

Schemata for the intuitionistic subsystem are as defined in Chapter 2, Part II, Definition 2.2.2, p. 34. We will require several standard intuitionistic schemata for reasoning about equality and disjoint unions in lambda terms, given in Fig. 4.9. These are the same as those schemata required for the intuitionistic logic of Chapter 2. These schemata provide a consistent theory about notions of equality and properties of the lambda calculus. The schemata require that we have an equality predicate for every basic sort.

Assumption 4.8. We require that, for every basic sort s from Σ_t (and consequently from Σ_p), there is a binary equality predicate $=_s \in Pred_s$. The subscript is omitted when there is no confusion regarding the sort.

Remark 4.23. A result of the schemata for equality is that, in the semantics of IHL, the valuation function h is constrained as follows. For each $=_s$ and terms $a_1 : s$ and $a_2 : s$ of $Terms(\Sigma_t)$, any interpretation function ι and states σ and σ':

$$h(=_s) \ni (\iota(a_1)_{\sigma}^{\sigma'}, \iota(a_2)_{\sigma}^{\sigma'})$$

if, and only if,

$$a_1 \rhd_{\Sigma_t} a_2$$

By Lemma 4.4.1, this can be shown to be equivalent to the condition that

$$\iota(a_1)_{\sigma}^{\sigma'} \rhd_{\Sigma_p} \iota(a_2)_{\sigma}^{\sigma'}$$

We also use schemata to generate program/formula pair axioms in $\mathcal{AX}_{BB}$. These are particularly useful when we wish to define how a black-box program from BB will react to input. The general form of a schema is as follows.

Definition 4.5.3 (Schemata for parametrized black-box programs). Let $\mathtt{f}$ be a black-box program of $BB_{s_1 \ldots s_n, s}$, parametrized over arguments of types $(\mathtt{s_1}, \ldots, \mathtt{s_n})$. Let $\bar{x}$ be a list of n term variables $x_1, \ldots, x_n$.

A black-box schema is of the form

$$\frac{}{\vdash_{\mathsf{IHL}} \mathtt{f(x_1,\ldots,x_n)} \diamond F} R[\bar{x}]$$

where R is the name of the schema. An application of the schema is

$$\frac{}{\vdash_{\mathsf{IHL}} \mathtt{f(a_1,\ldots,a_n)} \diamond F} R[\bar{a}]$$

where $\bar{a}$ is a list of n terms $a_1, \ldots, a_n$ of sorts $s_1, \ldots, s_n$.

$$\frac{}{\vdash_{\mathsf{Int}} u =_s r \Rightarrow r =_s u} \text{(ref)}$$

where s is a basic sort

$$\frac{P[r/y] \wedge u =_s r}{P[u/y]} \text{(subst)}[[P];[u;r];[s]]$$

where u and r are well-sorted of *basic sort* s and y is the only free variable in P

$$\frac{\vdash_{\mathsf{Int}} \forall y_1 : s_1 \bullet P[Inl(y_1)/x] \wedge \forall y_2 : s_2 \bullet P[Inr(y_2)/x]}{\vdash_{\mathsf{Int}} \forall x : s_1|s_2 \bullet P} \text{(disj-ind)}[P;[s_1;s_2]]$$

$$\frac{\vdash_{\mathsf{Int}} Inl(u) = Inl(r)}{\vdash_{\mathsf{Int}} u =_{s_1} r} \text{(union=}_1\text{)}[[u;r];[s_1;s_2]]$$

where $Inl(u)$ and $Inl(r)$ are well-sorted terms of sort $(s_1|s_2)$

$$\frac{\vdash_{\mathsf{Int}} Inr(u) = Inr(r)}{\vdash_{\mathsf{Int}} u =_{s_1} r} \text{(union=}_2\text{)}[[u;r];[s_1;s_2]]$$

where $Inr(u)$ and $Inr(r)$ are well-sorted terms of sort $(s_1|s_2)$

$$\frac{}{\vdash_{\mathsf{Int}} Inl(u) = Inr(r) \Rightarrow \bot} \text{(union}\neq\text{)}[[u;r];[s_1;s_2]]$$

where u and r are well-sorted terms of sorts s_1 and s_2 respectively

Fig. 4.9. Equality schemata and schemata for reasoning about disjoint unions.

4.5.4 Intuitionistic Hoare logic as a natural deduction system

Our calculus is a natural deduction system, in the sense identified by the Curry–Howard protocol of Definition 3.2.2, Chapter 3, Part II, p. 82.

$$\mathsf{IHL} = \langle Pairs(\mathsf{IHL}), \vdash_{\mathsf{IHL}}, DR(\mathsf{IHL}) \rangle$$

where $DR(\mathsf{IHL})$ consists of the rules that define $\vdash_{\mathsf{IHL}}$. Observe, however, that $\vdash_{\mathsf{IHL}}$ and $DR(\mathsf{IHL})$ are dependent on the deduction system for Int:

$$\mathsf{IHL} = \langle WFF(\Sigma_t), \vdash_{\mathsf{Int}}, DR(\mathsf{Int}) \rangle$$

where $DR(\mathsf{Int})$ consists of the rules that define $\vdash_{\mathsf{Int}}$.

Recall that Definition 3.2.2 takes a deduction system to be a means of proving "statements." In the case of IHL, these statements are program/formula pairs. This will be significant in the next chapter, when we define a type theory for IHL, in the style of constructive type theories satisfying the Curry–Howard isomorphism for intuitionistic logic. In that work, we shall represent program/formula pairs as types.

4.5.5 Example: electronic banking system

We illustrate Hoare logic with a medium sized example, involving code for part of an electronic banking system. We shall return to this example in the next two chapters, to show how proofs-as-programs methodology can be applied.

Consider an Automatic Bank Teller machine (ATM) example with the following domain conditions:

- The ATM permits the user to enter a Personal Identification Number (PIN) and to withdraw money. In order to withdraw money, the user must enter their PIN and a database connection to the bank's server must be made. The machine has a screen on which it displays messages to the user.
- The integer state reference `pin` stores the PIN entered by the user, the boolean state reference `canWithdraw` stores a flag to determine whether or not the user may withdraw money from the machine, and the boolean state reference `isConnected` stores a flag to determine whether or not there is a connection to the bank's server.
- We use the predicate $AppMessage(m)$ to assert that a string m is an appropriate message to display on the screen for the user, given that the ATM is in some particular state.
- There is a program p satisfying the following property. Given the user has entered their PIN correctly, the program allows the user to withdraw money. This property is formally given by an axiom

$$\vdash_{\mathsf{IHL}} \mathsf{p} \diamond PINCorrect(pin_i) \Rightarrow canWithdraw_f = true$$

- There is a program q such that, if the user is permitted to withdraw money, then a database connection is established, and also it is the case that there is an appropriate message that can be displayed. These properties are formally given by the axiom

$$\vdash_{\mathsf{IHL}} \mathsf{q} \diamond canWithdraw_i = true \Rightarrow \\ (isConnected_f = true \wedge \exists x : string \bullet AppMessage(x))$$

For the sake of argument, we simplify our domain with the following assumptions:

- We assume two *SML* record datatypes have been defined, `user` and `account`. Instances of the former contain information to represent a user in the system, while instances of the latter represent bank accounts. We do not detail the full definition of these types.
  However, we assume that an `account` record type that contains a `user` element in the `owner` field to represent the owner of the account. So the owner of the account element `myAccount` : `account` is accessed by `myAccount.owner`.
  We also assume that `user` is an equivalence type in *SML*, so that its elements may be compared using the boolean valued comparison function =.
  We assume a constant `currentUser` : `user` that represents the current user who is the subject of the account search.
- The database is represented in *SML* as an array of accounts,

  $$\texttt{db} : \texttt{account array}$$

  Following the *SML* API, the array is 0-indexed, with the ith element accessed as

  $$\texttt{sub(db, i)}$$

  and the size of the array given as

  $$\texttt{length db}$$

  Assume we have an array of size `Size`, called `accounts`. Although *SML* arrays are mutable, for the purposes of this example, we will consider `db` to be an immutable value. Consequently, it will be represented in our logic as a constant.
- We assume a state reference `counter` : `int ref`, to be used as a counter in searches through the database.

  We take a predicate

  $$allAccountsAt(u : user, x : account\ list, y : int)$$

whose meaning is that x is a list of all accounts found to be owned by the user u, up to the point y in the database `db`. The predicate defined by the following axioms in $\mathcal{AX}$

$$\forall u : user \bullet \forall x : (account\ list) \bullet \forall y : int \bullet (allAccountsAt(u, x, y) \Rightarrow \\ (\forall z : int \bullet z \leq y \Rightarrow sub(db, z).owner = u)) \quad (4.3)$$

$$\forall u : user \bullet \forall x : (account\ list) \bullet \forall y : int \bullet \\ ((y < (length\ db) - 1) \wedge sub(db, y+1).user = u \wedge allAccountsAt(u, x, y)) \Rightarrow \\ allAccountsAt(u, sub(db, y+1) :: x, y+1) \quad (4.4)$$

$$\forall u : user \bullet \forall x : (account\ list) \bullet \forall y : int \bullet$$
$$(y < (length\ db) - 1 \wedge \neg sub(l, y+1).user = u \wedge allAccountsAt(u, x, y)) \Rightarrow$$
$$allAccountsAt(u, x, y+1) \quad (4.5)$$

$$\forall u : user \bullet \forall y : int \bullet y = 0 \Rightarrow allAccountsAt(u, [], y) \quad (4.6)$$

Observe that these are intuitionistic axioms, for use in intuitionistic deduction, and so they do not involve programs. However, deductions that use these axioms can be used within the Hoare logic by means of the (cons) rule.

We will develop a program that satisfies the following property: given a user's details, it is *possible* to obtain a list of all accounts held at the bank by the user, by searching through the database. This is formally stated as the following requirement

$$\exists y : (account\ list) \bullet listAllAccounts(currentUser, y, counter_f) \wedge$$
$$(counter_f < (length\ db) - 1) = false \quad (4.7)$$

The post-condition requirement of $counter_f$ signifies that a complete search of the database should be completed by the program.

The previous axioms are Harrop. We also have a non-Harrop axiom

$$y < (length\ db) - 1 \Rightarrow sub(l, y+1).owner = u \vee \neg sub(l, y+1).owner = u \quad (4.8)$$

From (4.4), (4.5) and (4.8), we can derive an intuitionistic proof of the form

$$y < (length\ db) - 1, allAccountsAt(u, x, y)$$
$$\vdash_{\mathsf{Int}} \exists l : (account\ list) \bullet allAccountsAt(u, l, y+1) \quad (4.9)$$

By assuming $\exists l : (account\ list) \bullet allAccountsAt(u, l, y)$, we can apply ($\exists$-E) on (4.9) and then obtain

$$\vdash_{\mathsf{Int}} \forall y : int \bullet \forall u : user \bullet$$
$$(y < (length\ db) - 1) \wedge \exists l : (account\ list) \bullet allAccountsAt(u, l, y) \Rightarrow$$
$$\exists l : (account\ list) \bullet allAccountsAt(u, l, y+1) \quad (4.10)$$

We can transform (4.10) into

$$\vdash_{\mathsf{Int}} \forall y : int \bullet \forall v : int \bullet v = y + 1 \Rightarrow \forall u : user \bullet$$
$$(y < (length\ db) - 1) \wedge \exists l : (account\ list) \bullet allAccountsAt(u, l, y) \Rightarrow$$
$$\exists l : (account\ list) \bullet allAccountsAt(u, l, v) \quad (4.11)$$

We then instantiate (4.12) with $counter_i$ and $counter_f$ and $currentUser$ for y, v and u, to give

$$\vdash_{\mathsf{Int}} counter_f = counter_i + 1 \Rightarrow (counter_i < (length\ db) - 1) \wedge \\ \exists l : (account\ list) \bullet allAccountsAt(currentUser, l, counter_i) \Rightarrow \\ \exists l : (account\ list) \bullet allAccountsAt(currentUser, l, counter_f) \quad (4.12)$$

We also have the following, by the (assign) rule of Hoare logic:

$$\vdash \mathtt{counter} := !\mathtt{counter} + 1 \diamond counter_f = counter_i + 1 \quad (4.13)$$

And so, by applying (cons) to (4.13) and (4.12),

$$\vdash \mathtt{counter} := !\mathtt{counter} + 1 \diamond (counter_i < (length\ db) - 1) \wedge \\ \exists l : (account\ list) \bullet allAccountsAt(currentUser, l, counter_i) \Rightarrow \\ \exists l : (account\ list) \bullet allAccountsAt(currentUser, l, counter_f) \quad (4.14)$$

Then we apply (loop) to (4.14)

$$\vdash \mathtt{while}\ !\mathtt{counter} < (\mathtt{length\ db}) - 1\ \mathtt{do\ counter} := !\mathtt{counter} + 1 \diamond \\ \exists l : (account\ list) \bullet allAccountsAt(currentUser, l, counter_i) \Rightarrow \\ \exists l : (account\ list) \bullet allAccountsAt(currentUser, l, counter_f) \wedge \\ (counter_f < (length\ db) - 1) = false \quad (4.15)$$

From the axiom (4.6) we can derive

$$\vdash_{\mathsf{Int}} counter_f = 0 \Rightarrow \\ \exists y : (account\ list) \bullet allAccountsAt(currentUser, y, counter_f) \quad (4.16)$$

By application of (assign)

$$\vdash \mathtt{counter} := 0 \diamond counter_f = 0 \quad (4.17)$$

Then, applying (cons) to (4.17) and (4.16) gives

$$\mathtt{counter} := 0 \diamond \\ \exists y : (account\ list) \bullet allAccountsAt(currentUser, y, counter_f) \quad (4.18)$$

This can be weakened to include a true hypothesis *true*:

$$\mathtt{counter} := 0 \diamond true \Rightarrow \\ \exists y : (account\ list) \bullet allAccountsAt(currentUser, y, counter_f) \quad (4.19)$$

So, using (seq) on (4.19) and (4.15), we can obtain

$$\vdash \texttt{counter} := !\texttt{counter} + \texttt{1};$$
$$\texttt{while}\ !\texttt{counter} < (\texttt{length db}) - \texttt{1}\ \texttt{do}\ \texttt{counter} := !\texttt{counter} + \texttt{1} \diamond$$
$$true \Rightarrow \exists y : (account\ list) \bullet allAccountsAt(currentUser, y, counter_f) \wedge$$
$$(counter_f < (length\ db) - 1) = false \quad (4.20)$$

which can be simplified to the required form

$$\vdash \texttt{counter} := !\texttt{counter} + \texttt{1};$$
$$\texttt{while}\ !\texttt{counter} < (\texttt{length db}) - \texttt{1}\ \texttt{do}\ \texttt{counter} := !\texttt{counter} + \texttt{1} \diamond$$
$$\exists y : (account\ list) \bullet allAccountsAt(currentUser, y, counter_f) \wedge$$
$$(counter_f < (length\ db) - 1) = false \quad (4.21)$$

The program of this pair satisfies the specification (4.7).

Remark 4.24. We shall return to this example in the following chapters. By adapting proofs-as-programs to IHL, it is possible to augment the imperative program of (4.21) with a side-effect-free return value function. The return value satisfies the specification (4.7), according to a notion of realizability adapted from the functional program synthesis of Chapter 2, Part II. Essentially, when viewed as a specification of a return value, (4.7) requires a program that, given a user's details, will search through a database to obtain all accounts held at the bank by the user, and then return this list.

In Chapter 5, we will define a logical type theory that can encode proofs of IHL in the style of the Curry–Howard isomorphism. Section 5.3 of that chapter will show how our example proof is represented in this way. Then, in Chapter 6, we will formally define return value realizability of IHL and provide a method of program extraction, returning to our example (in Section 6.4) to synthesize an imperative program (with correct side-effects and return values) from this proof.

4.6 Comparison to Hoare logic with triples and its extensions

We make some brief observations about IHL and its relation to the original Hoare logic and its variants. This section may be skipped by readers interested only in using our calculus for program extraction. It is a necessary preliminary for the discussion of completeness of SSL given in the next chapter.

4.6.1 Single formula versus Hoare triples

We have chosen to present our logic using program/formula pairs rather than the original Hoare triples. We choose to do this, because, in the next chapter, we will present a logical type theory of IHL, where program/formula pairs are treated as types. Then, in Chapter 6, we will use the Skolem form of a formula in a pair to define how the formula specifies a required return value for the program. It is more convenient to use a single formula in these cases.

Single formulae with initial and final state identifiers are common in program specification methodologies, such as OCL [WK98] or the B-method [Abr96]. Hoare logics in which post-conditions of triples are predicates involving initial and final state identifiers are used in, for instance, [HH86, HHH+87, HHS87]. Also, program/formula pairs are sometimes a more convenient notation compared to triples [AL97].

4.6.2 Hoare triples

However, it is possible to show that our calculus is equivalent to a Hoare logic that uses triples, TIHL. We briefly sketch TIHL and outline how the equivalence holds.

4.6.3 Terms

Our terms consist of the lambda calculus built from Σ_p, extended by extra symbols to refer to values of state identifers for a particular point in a program's execution (not initial and final points, as was the case for terms of $Terms(\Sigma_t)$ above).

The set of terms $Terms(\mathsf{TIHL})$ consists of $Terms(\Sigma_p)$ inductively extended by *StateRef*, treated as special constant symbols.

Our terms are defined as follows.

Definition 4.6.1 (Elements of $Terms(\Sigma_{t*})$). We define the signature of our terms to be

$$\Sigma_t = \langle S, TF \cup StateVal, Pred \rangle$$

formed from Σ_p extended by *StateRef* treated as constant function symbols.

Then we define $Terms(\Sigma_{t*})$ according to the grammar of Fig. 4.1, p. 98.

Remark 4.25. $Terms(\Sigma_t)$ is a lambda calculus equipped with the usual notion of free and bound variables.

Sort inference rules of terms $Terms(\Sigma_{t*})$ are given by the the inference rules for $Terms(\Sigma_p)$, augmented with a rule for dealing with initial and final state references:

$$\frac{s \in StateRef \text{ and } StateType(s) = T}{\emptyset \vdash s}$$

We define tologic∗ to be a map from *Values* to $Terms(\Sigma_{t*})$ so that, tologic∗(t) consists of t with every dereferenced state reference !s replaced by the corresponding state reference s.

The formulae of TIHL are $WFF(\Sigma_t)$, the first-order many-sorted formulae that use terms from $Terms(\Sigma_{t*})$. Intuitively, these formulae make statements about the state of a program at a particular point in the program's execution. In contrast to the formulae of $WFF(\Sigma_t)$, we cannot make assertions about initial and final states within the same formula. Instead, such statements are made using the formulae in a Hoare triple.

Hoare triples are defined to be of the form

$$\{A\}\mathtt{p}\{B\}$$

where A, B are well-formed formulae of $Formula(\mathsf{TIHL})$, and $\mathtt{p}$ is a program of *IML*. Intuitively, A and B are statements about the state of the abstract machine. The triple is *correct* for a terminating $\mathtt{p}$, if whenever A is true prior to executing $\mathtt{p}$, then B will be true afterwards.

4.6.4 Calculus

The core rules of Hoare logic TIHL are given in Fig. 4.10. As in the case of IHL, these rules are defined with respect to the intuitionistic deduction of Fig. 4.8, but now with formulae taken from $WFF(\Sigma_{t*})$. We use the same rules that define $\vdash_{\mathsf{Int}}$, now given over formulae from $WFF(\Sigma_{t*})$, to define a deduction relation $\vdash_{\mathsf{Int(TIHL)}}$.

$$\frac{}{\vdash_{\mathsf{TIHL}} \{true\}\mathtt{s} := \mathtt{v}\{s = \mathsf{tologic}*(\mathtt{v})\}}\ (\text{assign\_}t)$$

where $s \in StateRef$.

$$\frac{\vdash_{\mathsf{TIHL}} \{B \wedge \mathsf{tologic}*(\mathtt{b}) = true\}\mathtt{p}\{C\} \quad \vdash_{\mathsf{TIHL}} \{B \wedge \mathsf{tologic}*(\mathtt{b}) = false\}\mathtt{q}\{C\}}{\vdash_{\mathsf{TIHL}} \{B\}\mathtt{if\ b\ then\ p\ else\ q}\{C\}}\ (\text{ite\_}t)$$

$$\frac{\vdash_{\mathsf{TIHL}} \{A\}\mathtt{p}\{B\} \quad \vdash_{\mathsf{TIHL}} \{B\}\mathtt{q}\{C\}}{\vdash_{\mathsf{TIHL}} \{A\}\mathtt{p};\mathtt{q}\{C\}}\ (\text{seq\_}t)$$

$$\frac{\vdash_{\mathsf{TIHL}} \{\mathsf{tologic}*(\mathtt{b}) = true \wedge A\}\mathtt{w}\{A\}}{\vdash_{\mathsf{TIHL}} \{A\}\mathtt{while\ b\ do\ q}\{A \wedge \mathsf{tologic}*(\mathtt{b}) = false\}}\ (\text{loop\_}t)$$

$$\frac{\vdash_{\mathsf{TIHL}} \{A'\}\mathtt{p}\{B'\} \quad A' \vdash_{\mathsf{Int(TIHL)}} A \quad B' \vdash_{\mathsf{Int(TIHL)}} B}{\vdash_{\mathsf{TIHL}} \{A\}\mathtt{p}\{B\}}\ (\text{cons\_}t)$$

Intuitionistic deduction $\vdash_{\mathsf{Int(TIHL)}}$ is given by the same rules that define $\vdash_{\mathsf{Int}}$ in Fig. 4.8, but now ranging over formulae from $WFF(\Sigma_{t*})$.

Fig. 4.10. Rules of the Hoare calculus with the triple notation.

4.6.5 Equivalence of Hoare triples to program/formula pairs

Any Hoare triple can be mapped to a program/formula pair and vice versa, in such a way that provability is preserved. That is to say, we can give a bijection ϕ from triples $\{A\}\mathtt{p}\{B\}$ to pairs $\mathtt{p} \diamond F$ such that provability of triples is preserved as provability of pairs, and vice versa.

Assume that $\bar{s}$ is a list that, at least, includes all elements of *StateRef* that occur in A and B, with $\bar{s}_i$ and $\bar{s}_f$ being the corresponding lists of initial and final state identifiers, then we define ϕ by

$$\phi(\{A\}\mathtt{p}\{B\}) = \mathtt{p} \diamond (A[\bar{s}_i/\bar{s}] \Rightarrow B[\bar{s}_f/\bar{s}])$$
$$\phi^{-1}(\mathtt{p} \diamond F) = \{\bar{x} = \bar{s}\}\mathtt{p}\{F[\bar{s}/\bar{s}_f][\bar{x}/\bar{s}_i]\}$$

The following theorem tells us that ϕ preserves triple provability in TIHL as program/formula provability in IHL.

Theorem 4.6.2 (ϕ preserves provability). *Take any program* $\mathtt{p}$ *and any formulae of* P *and* Q *of* $WFF(\Sigma_{t*})$.

$$\vdash_{\mathsf{TIHL}} \{P\}\mathtt{p}\{Q\} \textit{ entails } \vdash_{\mathsf{IHL}} \mathtt{p} \diamond \phi(\{P\}\mathtt{p}\{Q\})$$

Proof. By induction on the length of the proof $\Gamma \vdash_{\mathsf{TIHL}} \{P\}\mathtt{p}\{Q\}$.

Case: proof ends in (assign). Assume $\{P\}\mathtt{p}\{Q\}$ is of the form $\{true\}\mathsf{s} := \mathtt{v}\{s = \mathsf{tologic*}(\mathtt{v})\}$, deduced by (assign_$t$):

$$\frac{}{\vdash_{\mathsf{TIHL}} \{true\}\mathsf{s} := \mathtt{v}\{s = \mathsf{tologic*}(\mathtt{v})\}} \text{(assign\_}t\text{)}$$

We are required to show

$$\frac{}{\vdash_{\mathsf{IHL}} \mathsf{s} := \mathtt{v} \diamond true \Rightarrow s_f = \mathsf{tologic_f}(\mathtt{v})} \tag{4.22}$$

By applying the (assign) rule of IHL, we have

$$\frac{}{\vdash_{\mathsf{IHL}} \mathsf{s} := \mathtt{v} \diamond s_f = \mathsf{tologic_f}(\mathtt{v})} \text{(assign)} \tag{4.23}$$

We have the following intuitionistic proof:

$$\dfrac{\dfrac{\dfrac{\dfrac{}{s_f = \mathsf{tologic_f}(\mathtt{v}) \vdash_{\mathsf{Int}} s_f = \mathsf{tologic_f}(\mathtt{v})}\text{(Ass-I)} \quad \dfrac{}{true \vdash_{\mathsf{Int}} true}\text{(Ass-I)}}{s_f = \mathsf{tologic_f}(\mathtt{v}), true \vdash_{\mathsf{Int}} s_f = \mathsf{tologic_f}(\mathtt{v}) \wedge true}\text{(}\wedge\text{-I)}}{s_f = \mathsf{tologic_f}(\mathtt{v}), true \vdash_{\mathsf{Int}} s_f = \mathsf{tologic_f}(\mathtt{v})}\text{(}\wedge\text{-E}_2\text{)}}{s_f = \mathsf{tologic_f}(\mathtt{v}) \vdash_{\mathsf{Int}} true \Rightarrow s_f = \mathsf{tologic_f}(\mathtt{v})}\text{(}\Rightarrow\text{-I)} \tag{4.24}$$

We apply (*cons*) to 4.23 and 4.24 to obtain 4.22, as required.

Case: proof ends in (ite_t). Assume $\{P\}\mathtt{p}\{Q\}$ is of the form

$$\{B\}\texttt{if b then p else q}\{C\}$$

deduced by (ite_t):

$$\frac{\begin{array}{c}\vdash_{\mathsf{TIHL}} \{B \wedge \mathsf{tologic*}(\mathtt{b}) = true\}\mathtt{p}\{C\} \\ \vdash_{\mathsf{TIHL}} \{B \wedge \mathsf{tologic*}(\mathtt{b}) = false\}\mathtt{q}\{C\}\end{array}}{\vdash_{\mathsf{TIHL}} \{B\}\mathtt{if\ b\ then\ p\ else\ q}\{C\}}\ (\text{ite\_}t)$$

Let $\bar{s}$ be a list that includes all elements of *StateRef* that occur in B and C, with $\bar{s}_i$ and $\bar{s}_f$ being the corresponding lists of initial and final state identifiers.

We are required to show

$$\vdash_{\mathsf{IHL}} \mathtt{if\ b\ then\ p\ else\ q} \diamond B[\bar{s}_i/\bar{s}] \Rightarrow C[\bar{s}_f/\bar{s}] \tag{4.25}$$

By the IH, we have

$$\vdash_{\mathsf{IHL}} \mathtt{p} \diamond B[\bar{s}_i/\bar{s}] \wedge \mathsf{tologic_i}(\mathtt{b}) = true \Rightarrow C[\bar{s}_f/\bar{s}] \tag{4.26}$$

$$\vdash_{\mathsf{IHL}} \mathtt{q} \diamond B[\bar{s}_i/\bar{s}] \wedge \mathsf{tologic_i}(\mathtt{b}) = false \Rightarrow C[\bar{s}_f/\bar{s}] \tag{4.27}$$

It is easy to prove the following using the intuitionistic calculus:

$$B[\bar{s}_i/\bar{s}] \wedge \mathsf{tologic_i}(\mathtt{b}) = true \Rightarrow C[\bar{s}_f/\bar{s}], \mathsf{tologic_i}(\mathtt{b}) = true \\ \vdash_{\mathsf{Int}} B[\bar{s}_i/\bar{s}] \Rightarrow C[\bar{s}_f/\bar{s}] \tag{4.28}$$

$$B[\bar{s}_i/\bar{s}] \wedge \mathsf{tologic_i}(\mathtt{b}) = false \Rightarrow C[\bar{s}_f/\bar{s}], \mathsf{tologic_i}(\mathtt{b}) = false \\ \vdash_{\mathsf{Int}} B[\bar{s}_i/\bar{s}] \Rightarrow C[\bar{s}_f/\bar{s}] \tag{4.29}$$

By applying the (cons) rule to (4.26) and (4.28), and to (4.27) and (4.29), we obtain

$$\vdash_{\mathsf{IHL}} \mathsf{tologic_i}(\mathtt{b}) = true \Rightarrow (\mathtt{p} \diamond B[\bar{s}_i/\bar{s}] \Rightarrow C[\bar{s}_f/\bar{s}]) \tag{4.30}$$

$$\vdash_{\mathsf{IHL}} \mathsf{tologic_i}(\mathtt{b}) = false \Rightarrow (\mathtt{q} \diamond B[\bar{s}_i/\bar{s}] \Rightarrow C[\bar{s}_f/\bar{s}]) \tag{4.31}$$

Application of (ite) to (4.30) and (4.31) gives us (4.25), as required.

Case: proof ends in (seq_t). Assume $\{P\}\mathtt{p}\{Q\}$ is of the form $\{A\}\mathtt{p};\mathtt{q}\{C\}$, deduced by (seq_$t$):

$$\frac{\vdash_{\mathsf{TIHL}} \{A\}\mathtt{p}\{B\} \quad \vdash_{\mathsf{TIHL}} \{B\}\mathtt{q}\{C\}}{\vdash_{\mathsf{TIHL}} \{A\}\mathtt{p};\mathtt{q}\{C\}}\ (\text{seq\_}t)$$

Let $\bar{s}$ be a list that, at least, includes all elements of *StateRef* that occur in A, B and C, with $\bar{s}_i$ and $\bar{s}_f$ being the corresponding lists of initial and final state identifiers

We are required to show

$$\vdash_{\mathsf{IHL}} \mathtt{p};\mathtt{q} \diamond (A[\bar{s}_i/\bar{s}] \Rightarrow C[\bar{s}_f/\bar{s}]) \tag{4.32}$$

By the IH, we know that

$$\vdash_{\mathsf{IHL}} \mathtt{p} \diamond (A[\bar{s}_i/\bar{s}] \Rightarrow B[\bar{s}_f/\bar{s}]) \tag{4.33}$$
$$\vdash_{\mathsf{IHL}} \mathtt{q} \diamond (B[\bar{s}_i/\bar{s}] \Rightarrow C[\bar{s}_f/\bar{s}]) \tag{4.34}$$

We can then apply (seq) to (4.33) and (4.34) to obtain (4.32), as required.

Case: proof ends in (loop_t). Assume $\{P\}\mathtt{p}\{Q\}$ is of the form $\{A\}\mathtt{while}\ \mathsf{b}\ \mathsf{do}\ \mathsf{q}\{A \wedge \mathsf{tologic}{*}(\mathsf{b}) = \mathit{false}\}$, deduced by $(loop_t)$:

$$\frac{\vdash_{\mathsf{TIHL}} \{\mathsf{tologic}{*}(\mathsf{b}) = \mathit{true} \wedge A\}\mathtt{w}\{A\}}{\vdash_{\mathsf{TIHL}} \{A\}\mathtt{while}\ \mathsf{b}\ \mathsf{do}\ \mathsf{q}\{A \wedge \mathsf{tologic}{*}(\mathsf{b}) = \mathit{false}\}}\ (\text{loop\_}t)$$

We are required to show

$$\vdash_{\mathsf{IHL}} \mathtt{while}\ \mathsf{b}\ \mathsf{do}\ \mathsf{q} \diamond A[\bar{s}_i/\bar{v}] \Rightarrow A[\bar{s}_f/\bar{v}] \wedge \mathsf{tologic}_\mathsf{f}(\mathsf{b}) = \mathit{false} \tag{4.35}$$

Let $\bar{s}$ be a list that includes all elements of *StateRef* that occur in A, with $\bar{s}_i$ and $\bar{s}_f$ being the corresponding lists of initial and final state identifiers

By the IH, we know

$$\vdash_{\mathsf{IHL}} \mathsf{q} \diamond (\mathsf{tologic}_\mathsf{i}(\mathsf{b}) = \mathit{true} \wedge A[\bar{s}_i/\bar{v}] \Rightarrow A[\bar{s}_f/\bar{v}]) \tag{4.36}$$

We can easily prove the following:

$$\mathsf{tologic}_\mathsf{i}(\mathsf{b}) = \mathit{true} \wedge A[\bar{s}_i/\bar{v}] \Rightarrow A[\bar{s}_f/\bar{v}], \mathsf{tologic}_\mathsf{i}(\mathsf{b}) = \mathit{true} \wedge A[\bar{s}_i/\bar{v}] \vdash_{\mathsf{Int}} A[\bar{s}_f/\bar{v}] \tag{4.37}$$

Then, applying (cons) to (4.36) and (4.37) gives

$$\mathsf{tologic}_\mathsf{i}(\mathsf{b}) = \mathit{true} \wedge A[\bar{s}_i/\bar{v}] \vdash_{\mathsf{Int}} \mathsf{q} \diamond A[\bar{s}_f/\bar{v}] \tag{4.38}$$

Then, applying (loop) to (4.38) will give the required conclusion (4.35).

Case: proof ends in (cons_t). Assume $\{P\}\mathtt{p}\{Q\}$ is of the form $\{A\}\mathtt{p}\{B\}$ obtained from an application of (cons_t):

$$\frac{\vdash_{\mathsf{TIHL}} \{A'\}\mathtt{p}\{B'\} \quad A' \vdash_{\mathsf{Int(TIHL)}} A \quad B' \vdash_{\mathsf{Int(TIHL)}} B}{\vdash_{\mathsf{TIHL}} \{A\}\mathtt{p}\{B\}}\ (\text{cons\_}t)$$

Let $\bar{s}$ be a list that includes all elements of *StateRef* that occur in A, B, A' and B' with $\bar{s}_i$ and $\bar{s}_f$ being the corresponding lists of initial and final state identifiers

We are required to show

$$\vdash_{\mathsf{IHL}} \mathtt{p} \diamond (A[\bar{s}_i/\bar{s}] \Rightarrow B[\bar{s}_f/\bar{s}]) \tag{4.39}$$

By the IH, we know

$$\vdash_{\mathsf{IHL}} \mathtt{p} \diamond (A'[\bar{s}_i/\bar{s}] \Rightarrow B'[\bar{s}_f/\bar{s}]) \tag{4.40}$$

Now, we can transform any proof $M \vdash_{\mathsf{Int(TIHL)}} N$ into Int proofs: $M[\bar{s}_i/\bar{s}] \vdash_{\mathsf{Int}} N[\bar{s}_i/\bar{s}]$ and $M[\bar{s}_f/\bar{s}] \vdash_{\mathsf{Int}} N[\bar{s}_f/\bar{s}]$, where $\bar{s}$ is the list of all *StateRef* elements that occur in M and N.

So we have the following Int proofs

$$A'[\bar{s}_i/\bar{s}] \vdash_{\mathsf{Int})} A[\bar{s}_i/\bar{s}] \tag{4.41}$$

$$B'[\bar{s}_f/\bar{s}] \vdash_{\mathsf{Int})} B[\bar{s}_f/\bar{s}] \tag{4.42}$$

Using (4.41) and (4.42), it is easy to derive

$$A'[\bar{s}_i/\bar{s}], B'[\bar{s}_f/\bar{s}] \vdash_{\mathsf{Int})} A[\bar{s}_i/\bar{s}] \wedge B[\bar{s}_f/\bar{s}] \tag{4.43}$$

We arrive at the required conclusion (4.39) by applying (cons) to (4.40) and (4.43).

This last case concludes the proof. □

The following theorem tells us that ϕ^{-1} preserves triple provability in TIHL as program/formula provability in IHL.

Theorem 4.6.3 (ϕ^{-1} preserves provability). *Take any program* p *and any formulae of* P *of* $WFF(\Sigma_t)$.

$$\vdash_{\mathsf{IHL}} \mathsf{p} \diamond P \text{ entails } \vdash_{\mathsf{TIHL}} \phi^{-1}(\mathsf{p} \diamond P)$$

Proof. By induction on the form of the length of the proof of

$$\vdash_{\mathsf{IHL}} \mathsf{p} \diamond P$$

The proof is symmetric to that of Theorem 4.6.2. □

4.6.6 Axioms and schemata

The axioms and schemata of $\mathcal{ADT}$ can still be used in TIHL to reason about $\mathcal{ADT}$ in TIHL. It is a straightforward task to transform the axioms and schemata rules of $BB(\mathcal{ADT})$ into equivalent forms for use in TIHL using ϕ.

4.6.7 Nondeterministic assignment

Our calculus deals with the same imperative constructs as the original version of Hoare Logic (that is, while-loops, conditional statements, sequencing and assignment) [Hoa69]. However, unlike the original Hoare logic, we do not deal with nondeterministic assignment. This construct could be added to our calculus with minimal changes to our semantics. But this is a future research topic that would extend the results of the next two chapters in order to deal with this construct.

4.6.8 Total correctness

Because this semantics of program/formula pairs is given over terminating programs, our version of Hoare logic is concerned with so-called *partial correctness* of statements about programs [Cou90]. The original presentation of Hoare logic dealt with partial correctness [Hoa69]. However, extensions have been made to deal with total correctness, yielding a variant of Hoare logic that derives truths about programs that always terminate [Hoa81]. This is a future research topic for extending our calculus, and the results that follow in order to deal with total correctness.

4.7 Discussion

This chapter presented a version of Hoare logic, called IHL, that

- uses constructive deduction for its internal logic, and
- reasons and constructs programs taken from an imperative subset of *SML*.

Our presentation employed program/formula pairs instead of the more traditional Hoare triples. We showed that the two presentations are equivalent, which shall be the reason for this presentation is that it aids representation of IHL as a type theory, in the style of the Curry–Howard isomorphism for intuitionistic logic.

On its own the logic can be used to develop correct imperative programs that satisfy pre- and post-condition specifications of side-effects. This is the usual benefit of Hoare logic.

We extend this result, adapting the proofs-as-programs paradigm to IHL, to develop correct *SML* programs that satisfy side-effect *and* return value specifications. This will be shown in the last chapter of this part (Chapter 6).

In order to arrive at that result, we must:

- Understand the semantics of IHL.
- Provide a type theoretic representation of IHL.

These topics are dealt with in the next chapter (Chapter 5).

5

Properties of Intuitionistic Hoare Logic

In this chapter, we discuss important model- and proof-theoretic properties of Intuitionistic Hoare Logic (IHL).

We outline a soundness proof for IHL, using the relational semantics and the definition of the truth of a formula given in the previous chapter. We also examine completness issues for IHL, which, like ordinary Hoare logic, is incomplete. As noted in the previous chapter, this semantics and notion of truth is given over a specific model — a relational semantics that is close to actual *SML* execution. This result is important for us: it tells us that theorems of IHL produce true statements about *SML* program execution.

We next define a logical type theory for representing proofs in our logic. The type theory is analogous to the constructive type theory for intuitionistic logic described in Chapter 3. That is:

- A type represents a specification.
- A term for a type represents a proof of a specification.
- Our terms form a lambda calculus.
- Term reduction rules defines how redundancies in proofs may be removed.

Our logical type theory's distinguishing feature is that its types are the program/formula pairs of IHL. This logical type theory is important for the main result of this part of the monograph, given in the next chapter, where we apply the Curry–Howard protocol to extract imperative programs with return values from IHL proofs.

We proceed as follows:

- Section 5.1 examines soundness and completeness issues for IHL.
- The logical type theory for IHL is defined in Section 5.2.
- A discussion and summary is provided in Section 5.4.

5.1 Model theoretic properties

In this section we outline a soundness proof for IHL using the relational semantics and the definition of the truth of a formula given in Chapter 4. As noted in that chapter, this semantics and notion of truth is given over a specific model — a relational semantics that is close to actual *SML* execution.

This result is important for us: it tells us that theorems of IHL produce true statements about *SML* program execution.

It is possible to give a more general semantics of program execution, over a wider range of models. Such a general semantics has been well understood by other authors, and is outside the scope of this monograph. However, we briefly discuss the form of the semantics. We then outline and discuss soundness and completeness issues and results for the general semantics.

5.1.1 Soundness with respect to *SML* semantics

Recall the relational semantics for *IML* programs given in Chapter 4. It was given over a set of states, $MLStates$, and maps from state references, $StateRef$, to *Values* (the closed, irreducible terms of $Terms(\Sigma_p)$). This semantics is close to the actual execution of *SML* programs because it reflects the fact that *SML* states store closed, side-effect-free terms. The relational semantics associates terminating programs of *IML* with relations between pairs of states representing the initial and final states of a program.

First, we show that intuitionistic proofs that use formulae from $WFF(\Sigma_t)$ allow us to infer truths about arbitrary side-effects. This means that we can use the core rules of Int to reason about properties that are universal to all programs.

Lemma 5.1.1. *Let Γ be a set of assumption formulae $G_1, \ldots, G_n$ and G a formula such that there is a proof constructed from the core rules of* Int

$$\Gamma \vdash_{\mathsf{Int}} G$$

Take any interpretation ι and any two states $\sigma, \sigma' \in MLStates$. Assume, for every $i = 1, \ldots, n$,

$$(\sigma, \sigma') \Vdash_\iota G_i$$

Then

$$(\sigma, \sigma') \Vdash_\iota G$$

(In this case, we say that the sequent $\Gamma \vdash_{\mathsf{Int}} G$ is valid.*)*

Proof. Given a set of formulae $H = \{H_1, \ldots, H_m\}$, we write

$$Valid(H, \iota, (\sigma, \sigma'))$$

for the assumption that, for every $i = 1, \ldots, m$, for any relation $R \in MLRel$,

$$(\sigma, \sigma') \Vdash_\iota H_i$$

So, the assumption of the lemma is

$$Valid(\Gamma, \iota, (\sigma, \sigma')) \tag{5.1}$$

We proceed by induction on the length of the proof $\Gamma \vdash_{\mathsf{Int}} G$.

(Ass-I). If G is derived by (Ass-I)

$$\frac{}{G \vdash_{\mathsf{Int}} G} \text{(Ass-I)}$$

the IH tells us that

$$Valid(\{G\}, \iota, (\sigma, \sigma')) \text{ entails } (\sigma, \sigma') \Vdash_\iota G \tag{5.2}$$

But the requirement of (5.2) is satisfied by (5.1), and so we are done.

($\wedge$-I). Assume G is of the form $(A \wedge B)$, and is derived by ($\wedge$-I)

$$\frac{\Gamma_1 \vdash A \quad \Gamma_2 \vdash B}{\Gamma_1, \Gamma_2 \vdash (A \wedge B)} (\wedge\text{-I})$$

The IH and (5.1) make the following true: $(\sigma, \sigma') \Vdash_\iota A$ and $(\sigma, \sigma') \Vdash_\iota B$. So, by definition of $\Vdash_\iota$, $(\sigma, \sigma') \Vdash_\iota (A \wedge B)$, as required.

($\wedge$-E$_1$). Assume G is of the form A_1, and is derived by ($\wedge$-E$_1$)

$$\frac{\Gamma \vdash_{\mathsf{Int}} (A_1 \wedge A_2)}{\Gamma \vdash_{\mathsf{Int}} A_1} (\wedge\text{-E}_1)$$

The IH and (5.1) make the following true: $(\sigma, \sigma') \Vdash_\iota (A_1 \wedge A_2)$. But then, by definition of $\Vdash_\iota$, $(\sigma, \sigma') \Vdash_\iota A_1$, as required.

($\wedge$-E$_2$). This case is similar to the case of a proof ending in ($\wedge$-E$_1$).

($\Rightarrow$-I). Assume G is of the form $(A \Rightarrow B)$, and is derived by ($\Rightarrow$-I)

$$\frac{\Gamma, A \vdash B}{\Gamma \vdash (A \Rightarrow B)} (\Rightarrow\text{-I})$$

The IH dictates that $(\sigma, \sigma') \Vdash_\iota A$ entails $(\sigma, \sigma') \Vdash_\iota B$. So, in particular, $(\sigma, \sigma') \Vdash_\iota (A \Rightarrow B)$, as required.

($\Rightarrow$-E). Assume $\Gamma = \Gamma_1 \cup \Gamma_2$, and that our sequent is derived via ($\Rightarrow$-E):

$$\frac{\Gamma_1 \vdash (A \Rightarrow B) \quad \Gamma_2 \vdash A}{\Gamma_1, \Gamma_2 \vdash B} (\Rightarrow\text{-E})$$

The IH over the second premise of the rule means that

$$\text{if } Valid(\Gamma_2, \iota, (\sigma, \sigma')) \text{ holds, then } (\sigma, \sigma') \Vdash_\iota A \tag{5.3}$$

and applying IH over the first premise gives

$$\text{if } Valid(\Gamma_1, \iota, (\sigma, \sigma')) \text{ holds, then } (\sigma, \sigma') \Vdash_\iota (A \Rightarrow B) \tag{5.4}$$

Assumption (5.1) entails $Valid(\Gamma_1 \cup \Gamma_2, \iota, (\sigma, \sigma'))$. This means that $Valid(\Gamma_1, \iota, (\sigma, \sigma'))$, and, as a consequence of (5.4), $(\sigma, \sigma') \Vdash_\iota (A \Rightarrow B)$. By definition, this means

$$(\sigma, \sigma') \Vdash_\iota A \text{ entails } (\sigma, \sigma') \Vdash_\iota B \tag{5.5}$$

We are required to show that $(\sigma, \sigma') \Vdash_\iota B$ holds. The IH entails that $Valid(\Gamma_2, \iota, (\sigma, \sigma'))$. We instantiate (5.3) with this, to give $(\sigma, \sigma') \Vdash_\iota A$ and hence

$$(\sigma, \sigma') \Vdash_\iota A \tag{5.6}$$

We then instantiate (5.5) with (5.6), to give $(\sigma, \sigma') \Vdash_\iota B$, as required.

($\forall$-I). Assume G is of the form $\forall x : t \bullet A$, and that the sequent is derived by

$$\frac{\Gamma \vdash A}{\Gamma \vdash \forall x : t \bullet A} \ (\forall\text{-I})$$

This rule requires that $x : t \notin FV(\Gamma)$.

By the IH, for every interpretation η,

$$Valid(\Gamma, \eta, (\sigma, \sigma')) \text{ entails } (\sigma, \sigma') \Vdash_\eta A \tag{5.7}$$

Take any $x : t$-variant ι' (that is, any interpretation ι' that differs from ι only over $x : t$). By assumption (5.1), the fact that $x : t \notin FV(\Gamma)$, we can prove

$$Valid(\Gamma, \iota', (\sigma, \sigma')) \tag{5.8}$$

Setting η to ι' in (5.7), and then instantiating with (5.8),

$$(\sigma, \sigma') \Vdash_{\iota'} A$$

for every $x : t$-variant ι' of ι, as required.

($\forall$-E). Assume G is $A[a/x]$, derived by

$$\frac{\Gamma \vdash \forall x : t \bullet A}{\Gamma \vdash A[a/x]} \ (\forall\text{-E})$$

By the IH,

$$Valid(\Gamma, \iota, (\sigma, \sigma')) \text{ entails } (\sigma, \sigma') \Vdash_\iota \forall x : t \bullet A \tag{5.9}$$

Instantiating (5.9) with the assumption (5.1) yields $(\sigma, \sigma') \Vdash_\iota \forall x : t \bullet A$ which means that, for every $x : t$-variant ι' of ι, $(\sigma, \sigma') \Vdash_{\iota'} A$. Take the x-variant $\iota' = \iota[x \mapsto \iota(a)_\sigma^{\sigma'}]$. By the IH, $(\sigma, \sigma') \Vdash_{\iota'} A$. From this it is easy to show that $(\sigma, \sigma') \Vdash_\iota A[a/x]$ for any ι, as required.

($\exists$-I). Assume G is of the form $\exists x : t \bullet A$, and that the sequent is derived by

$$\frac{\Gamma \vdash A[a/x]}{\Gamma \vdash \exists x : t \bullet A} \text{ } (\exists\text{-I})$$

By the IH and (5.1),

$$(\sigma, \sigma') \Vdash_\iota A[a/x] \tag{5.10}$$

Let ι' be a $x : t$-variant of ι (that is, any interpretation ι' that differs from ι only over $x : t$), defined

$$\iota' = \iota[x \mapsto \iota(a)_\sigma^{\sigma'}]$$

From this and (5.10) we can derive $(\sigma, \sigma') \Vdash_{\iota'} A$. So, by definition of $\Vdash_\iota$, $(\sigma, \sigma') \Vdash_\iota \exists x : t \bullet A$, as required.

($\exists$-E). Assume G is Q derived by

$$\frac{\Gamma_1 \vdash_{\mathsf{Int}} \exists y : t \bullet P \quad \Gamma_2, P[x/y] \vdash_{\mathsf{Int}} Q}{\Gamma_1, \Gamma_2 \vdash_{\mathsf{Int}} Q} \text{ } (\exists\text{-E})$$

The rule requires that $x : t$ does not occur free in Q or Γ_2. By the IH and (5.1) $(\sigma, \sigma') \Vdash_\iota \exists y : t \bullet P$. This means that there is a $y : t$-variant ι' of ι such that

$$(\sigma, \sigma') \Vdash_{\iota'} P \tag{5.11}$$

In turn, (5.11) entails that there is a $x : t$-variant ι'' of ι such that

$$(\sigma, \sigma') \Vdash_{\iota'} P[x/y] \tag{5.12}$$

Also, by the IH, for every interpretation η

$$Valid(\Gamma_2, \eta, (\sigma, \sigma')) \text{ and } (\sigma, \sigma') \Vdash_\eta P[x/y] \text{ entails } (\sigma, \sigma') \Vdash_\eta Q \tag{5.13}$$

Because $x : t$ does not occur free in Γ_2, it is possible to derive from (5.1) that

$$Valid(\Gamma_2, \iota'', (\sigma, \sigma')) \tag{5.14}$$

So we can instantiate (5.13) with (5.14) and (5.12) to obtain

$$(\sigma, \sigma') \Vdash_{\iota''} Q$$

but because $x : t$ does not occur in Q, we have that

$$(\sigma, \sigma') \Vdash_\iota Q$$

as required.

($\vee$-I$_1$). Assume G is of the form $(A_1 \vee A_2)$, and is derived by

$$\frac{\Gamma \vdash_{\mathsf{Int}} A_1}{\Gamma \vdash_{\mathsf{Int}} (A_1 \vee A_2)} \text{ } (\vee\text{-I}_1)$$

The IH and (5.1) make the following true: $(\sigma, \sigma') \Vdash_\iota A_1$. But then, by definition of $\Vdash_\iota$, $(\sigma, \sigma') \Vdash_\iota (A_1 \vee A_2)$, as required.

($\vee$-I$_2$). This case is similar to the case of a proof ending in ($\vee$-I$_1$).

($\vee$-E). Assume that $\Gamma = \Gamma_1 \cup \Gamma_2 \cup \Gamma_3$ and G is of the form C, derived via ($\vee$-E)

$$\frac{\Gamma_1 \vdash (A \vee B) \quad \Gamma_2, A \vdash C \quad \Gamma_3, B \vdash C}{\Gamma_1, \Gamma_2, \Gamma_3 \vdash C} \ (\vee\text{-E})$$

The IH entails that

$$Valid(\Gamma_1, \iota, (\sigma, \sigma')) \text{ entails } (\sigma, \sigma') \Vdash_\iota (A \vee B) \tag{5.15}$$

$$Valid(\Gamma_2 \cup \{A\}, \iota, (\sigma, \sigma'), w) \text{ entails } (\sigma, \sigma') \Vdash_\iota C \tag{5.16}$$

$$Valid(\Gamma_3 \cup \{B\}, \iota, (\sigma, \sigma')) \text{ entails } (\sigma, \sigma') \Vdash_\iota C \tag{5.17}$$

Assumption (5.1) and the fact that $\Gamma = \Gamma_1 \cup \Gamma_2 \cup \Gamma_3$ entail

$$\begin{array}{c} Valid(\Gamma_1, \iota, (\sigma, \sigma')) \\ Valid(\Gamma_2, \iota, (\sigma, \sigma')) \\ Valid(\Gamma_3, \iota, (\sigma, \sigma')) \end{array}$$

These statements satisfy the premises of (5.15), (5.16) and (5.17). So,

$$(\sigma, \sigma') \Vdash_\iota (A \vee B) \tag{5.18}$$

$$(\sigma, \sigma') \Vdash_\iota A \text{ entails } (\sigma, \sigma') \Vdash_\iota C \tag{5.19}$$

$$(\sigma, \sigma') \Vdash_\iota B \text{ entails } (\sigma, \sigma') \Vdash_\iota C \tag{5.20}$$

must hold.

By (5.18) and the definition of $\Vdash$, either $(\sigma, \sigma') \Vdash_\iota A$ or $(\sigma, \sigma') \Vdash_\iota B$. We argue over these two cases. Assume $(\sigma, \sigma') \Vdash_\iota A$. This assumption, together with (5.19), gives $(\sigma, \sigma') \Vdash_\iota C$. For the other case, assume $(\sigma, \sigma') \Vdash_\iota B$. This assumption, together with (5.20), gives $(\sigma, \sigma') \Vdash_\iota C$. Thus, we may conclude that $(\sigma, \sigma') \Vdash_\iota C$, as required.

($\bot$-E). Assume that G is obtained via an application of $(\bot - E)$

$$\frac{\Gamma \vdash \bot}{\Gamma \vdash G} \ (\bot\text{-E})$$

By the IH and (5.1), we know that $(\sigma, \sigma') \Vdash_\iota \bot$. By definition, this is never the case. So, we can conclude anything (by a metalogical application of the classical absurdity rule) $(\sigma, \sigma') \Vdash_\iota G$, as required. □

We are now able to give a soundness theorem for the core rules of IHL proofs over our relational semantics.

Theorem 5.1.1 (Soundness). *Let* t *be a terminating program and G be a formula. Take any proof constructed from the core rules of* IHL

$$\vdash_{\mathsf{IHL}} \mathtt{t} \diamond G$$

Take any interpretation ι. It is the case that

$$[\![t]\!] \Vdash_\iota G$$

Proof. We proceed by induction on the derivation of $\Gamma \vdash \mathtt{t} \diamond G$.

For a set of formulae Δ and interpretation η, let $Valid(\Delta, \eta)$ denote the assumption that, for every $D \in \Delta$,

$$\text{for every } R \in MLRel,\ R \Vdash_\iota D$$

Assume

$$Valid(\Gamma, \iota) \tag{5.21}$$

Let σ, σ' be any states such that

$$\sigma\ [\![t]\!] \sigma'$$

We need only show that

$$(\sigma, \sigma') \Vdash_\iota G$$

Our proof uses the definition of relational semantics (Chapter 4, Definition 4.3.4, Section 4.2, p. 110), which is given in terms of the operational semantics (Chapter 4, Fig. 4.6, p. 108). Given any program $\mathtt{p} \in IML$, if there are states $\sigma, \sigma' \in MLStates$ such that

$$\langle \mathtt{p}, \sigma \rangle \hat{\rhd} \langle \mathtt{r}, \sigma' \rangle$$

for some return value $\mathtt{r} \in IML$, then

$$(\sigma, \sigma') \in [\![\mathtt{p}]\!]$$

(ite). Assume that $t \diamond G$ is of the form $w \diamond C$, $\Gamma = \Gamma_1 \cup \Gamma_2 \cup \Gamma_3$, and that the sequent is derived via (ite) as

$$\frac{\vdash_{\mathsf{IHL}} \mathsf{tologic_i}(\mathsf{b}) = true \Rightarrow C \quad \vdash_{\mathsf{IHL}} \mathsf{tologic_i}(\mathsf{b}) = false \Rightarrow C}{\vdash_{\mathsf{IHL}} \mathtt{if\ b\ then\ p\ else\ q} \diamond C}\ (\text{ite})$$

The IH and (5.21) entail that

$$[\![\mathtt{p}]\!] \Vdash_\iota \mathsf{tologic_i}(\mathsf{b}) = true \Rightarrow C \tag{5.22}$$
$$[\![\mathtt{q}]\!] \Vdash_\iota \mathsf{tologic_i}(\mathsf{b}) = false \Rightarrow C \tag{5.23}$$

The boolean function $\mathsf{b} \in Closed(Terms(IML))$ is such that

$$\sigma(\mathsf{b}) \rhd_{\Sigma_p} \mathtt{true}$$

or

$$\sigma(\mathsf{b}) \rhd_{\Sigma_p} \mathtt{false}$$

We reason over these two possibilities.

Case: $\sigma(\mathsf{b}) \rhd_{\Sigma_p} \mathtt{true}$. By the operational semantics of `if b then p else q`, we see that, because $\mathsf{b} \rhd_{\Sigma_p} \mathtt{true}$ holds,

$$[\![\mathtt{if\ b\ then\ p\ else\ q}]\!] = [\![\mathtt{p}]\!]$$

So (5.22) entails that

$$[\![\mathtt{p}]\!] \Vdash_\iota \mathsf{tologic_i}(\mathsf{b}) = \mathit{true} \Rightarrow C \tag{5.24}$$

Then, by definition of $\Vdash_\iota$, it must be the case that

$$(\sigma, \sigma') \Vdash_\iota \mathsf{tologic_i}(\mathsf{b}) = \mathit{true}$$

This fact, (5.24) and the definition of $\Vdash_\iota$ yield

$$(\sigma, \sigma') \Vdash_\iota C$$

as required.

Case: $\sigma(\mathsf{b}) \rhd_{\Sigma_p} \mathtt{false}$. Similar to the previous case.

(loop). Assume $t \diamond G$ is of the form $\mathtt{while\ b\ do\ q} \diamond A[\bar{s}_i/\bar{v}] \Rightarrow (A[\bar{s}_f/\bar{v}] \wedge \mathsf{tologic_f}(\mathsf{b}) = \mathit{false})$, whose derivation ends in (*loop*):

$$\frac{\vdash_{\mathsf{IHL}} \mathtt{q} \diamond (\mathsf{tologic_i}(\mathsf{b}) = \mathit{true} \wedge A[\bar{s}_i/\bar{v}]) \Rightarrow A[\bar{s}_f/\bar{v}]}{\vdash_{\mathsf{IHL}} \mathtt{while\ b\ do\ q} \diamond A[\bar{s}_i/\bar{v}] \Rightarrow A[\bar{s}_f/\bar{v}] \wedge \mathsf{tologic_f}(\mathsf{b}) = \mathit{false}}\ (\mathrm{loop})$$

The rule requires the following constraints that no state identifiers occur in A.

Assume that

$$(\sigma, \sigma') \Vdash_\iota A[\bar{s}_i/\bar{v}] \tag{5.25}$$

By the IH, we know that,

$$[\![\mathtt{q}]\!] \Vdash_\iota (\mathsf{tologic_i}(\mathsf{b}) = \mathit{true} \wedge A[\bar{s}_i/\bar{v}]) \Rightarrow A[\bar{s}_f/\bar{v}]$$

and so, for any τ, τ' such that $\tau[\![\mathtt{q}]\!]\tau'$,

$$(\tau, \tau') \Vdash_\iota (\mathsf{tologic_i}(\mathsf{b}) = \mathit{true} \wedge A[\bar{s}_i/\bar{v}]) \Rightarrow A[\bar{s}_f/\bar{v}] \tag{5.26}$$

By assumption, `while b do q` terminates, so by the operational semantics, either

1. $\sigma = \sigma'$ with $\sigma(\mathsf{b}) \rhd \mathtt{false}$ holds, or
2. there is an $n > 0$ such that

$$\sigma = \sigma_0 [\![\mathtt{q}]\!] \sigma_1 [\![\mathtt{q}]\!] \sigma_2 \ldots \sigma_{n-1} [\![\mathtt{q}]\!] \sigma_n$$

and

$$\text{for every } j,\ 0 \le j < n \text{ entails } \sigma_j(\mathsf{b}) \rhd_{\Sigma_p} \mathtt{true} \text{ and } \sigma_n(\mathsf{b}) \rhd_{\Sigma_p} \mathtt{false} \tag{5.27}$$

In case 1), by definition of $\Vdash_\iota$,

$$(\sigma, \sigma') \Vdash_\iota \mathsf{tologic_i}(\mathsf{b}) = false \tag{5.28}$$

Also, because $\sigma = \sigma'$, (5.25) and the fact that there are no state identifiers in A, we have that

$$(\sigma, \sigma') \Vdash_\iota A[\bar{s}_f/\bar{v}] \tag{5.29}$$

So, (5.28) and (5.29) give

$$(\sigma, \sigma') \Vdash_\iota A[\bar{s}_f/\bar{v}] \wedge \mathsf{tologic_f}(\mathsf{b}) = false \tag{5.30}$$

Because we have assumed (5.25), the definition of $\Vdash_\iota$ and (5.30) give us

$$(\sigma, \sigma') \Vdash_\iota A[\bar{s}_i/\bar{v}] \Rightarrow (A[\bar{s}_f/\bar{v}] \wedge \mathsf{tologic_f}(\mathsf{b}) = false)$$

In case 2), the first clause of (5.27) entails that

$$(\sigma_i, \sigma_{i+1}) \Vdash_\iota (\mathsf{tologic_i}(\mathsf{b}) = true) \tag{5.31}$$

for $(0 < i < n - 1)$. We show

$$(\sigma_i, \sigma_{i+1}) \Vdash_\iota A[\bar{s}_f/\bar{v}] \tag{5.32}$$

by induction for $i = 0, \ldots, n - 1$.

Base case. By (5.25), and the fact that state identifiers do not occur in A, we have

$$(\sigma = \sigma_0, \sigma_1) \Vdash_\iota A[\bar{s}_f/\bar{v}] \tag{5.33}$$

This and (5.31) give us

$$(\sigma_0, \sigma_1) \Vdash_\iota A[\bar{s}_f/\bar{v}]$$

as required.

Inductive step. Assume (5.32) holds for some $k < n - 2$:

$$(\sigma_{k-1}, \sigma_k) \Vdash_\iota A[\bar{s}_f/\bar{v}] \tag{5.34}$$

Because A does not contain any state identifiers, this is equivalent to writing

$$(\sigma_k, \sigma_{k+1}) \Vdash_\iota A[\bar{s}_i/\bar{v}] \tag{5.35}$$

This and (5.31) give us

$$(\sigma_k, \sigma_{k+1}) \Vdash_\iota (\mathsf{tologic_i}(\mathsf{b}) = true \wedge A[\bar{s}_i/\bar{v}]) \tag{5.36}$$

Instantiating the IH (5.26) with (5.36) gives

$$(\sigma_k, \sigma_{k+1}) \Vdash_\iota A[\bar{s}_f/\bar{v}]$$

as required, concluding the proof of (5.32).

Now, (5.32) means that, in particular,

$$(\sigma_{n-1}, \sigma_n = \sigma') \Vdash_\iota A[\bar{s}_f/\bar{v}] \tag{5.37}$$

Because A does not contain any state identifiers, this means

$$(\sigma, \sigma') \Vdash_\iota A[\bar{s}_f/\bar{v}] \tag{5.38}$$

The second clause of (5.27) entails that

$$(\sigma, \sigma') \Vdash_\iota \mathsf{tologic_f b} = false \tag{5.39}$$

So, (5.38) and (5.39), together with the assumption (5.25) give

$$(\sigma, \sigma') \Vdash_\iota A[\bar{s}_i/\bar{v}] \Rightarrow (A[\bar{s}_f/\bar{v}] \wedge \mathsf{tologic_f(b)} = false)$$

as required.

(seq). Assume $\mathsf{t} \diamond G$ is of the form $\mathsf{p;q} \diamond A[\bar{s}_i/\bar{v}] \Rightarrow C[\bar{s}_f/\bar{v}]$ derived from a proof ending in

$$\frac{\vdash_{\mathsf{IHL}} \mathsf{p} \diamond (A[\bar{s}_i/\bar{v}] \Rightarrow B[\bar{s}_f/\bar{v}]) \quad \vdash_{\mathsf{IHL}} \mathsf{q} \diamond (B[\bar{s}_i/\bar{v}] \Rightarrow C[\bar{s}_f/\bar{v}])}{\vdash_{\mathsf{IHL}} \mathsf{p;q} \diamond (A[\bar{s}_i/\bar{v}] \Rightarrow C[\bar{s}_f/\bar{v}])} \text{(seq)}$$

where A and B are free of state identifiers.

The IH tells us that

$$[\![\mathsf{p}]\!] \Vdash_\iota (A[\bar{s}_i/\bar{v}] \Rightarrow B[\bar{s}_f/\bar{v}]) \tag{5.40}$$

and

$$[\![\mathsf{q}]\!] \Vdash_\iota (B[\bar{s}_i/\bar{v}] \Rightarrow C[\bar{s}_f/\bar{v}]) \tag{5.41}$$

By the operational and relational semantics of $\mathsf{p;q}$, we know that, because $\sigma[\![\mathsf{p;q}]\!]\sigma'$, there is an intermediate state σ'' such that

$$\sigma[\![\mathsf{p}]\!]\sigma''[\![\mathsf{q}]\!]\sigma'$$

So, (5.40) entails

$$(\sigma, \sigma'') \Vdash_\iota (A[\bar{s}_i/\bar{v}] \Rightarrow B[\bar{s}_f/\bar{v}]) \tag{5.42}$$

and (5.41) entails

$$(\sigma'', \sigma') \Vdash_\iota (B[\bar{s}_i/\bar{v}] \Rightarrow C[\bar{s}_f/\bar{v}]) \tag{5.43}$$

Assume

$$(\sigma, \sigma') \Vdash_\iota A[\bar{s}_i/\bar{v}] \tag{5.44}$$

Because A does not contain any state identifiers, this means

$$(\sigma, \sigma'') \Vdash_\iota A[\bar{s}_i/\bar{v}] \tag{5.45}$$

with which we may instantiate (5.42), to give

$$(\sigma, \sigma'') \Vdash_\iota B[\bar{s}_f/\bar{v}] \tag{5.46}$$

Because B does not contain any state identifiers, (5.46) means

$$(\sigma, \sigma'') \Vdash_\iota B[\bar{s}_i/\bar{v}] \tag{5.47}$$

We instantiate (5.43) with (5.47), to give

$$(\sigma'', \sigma') \Vdash_\iota C[\bar{s}_f/\bar{v}] \tag{5.48}$$

Now, because C does not contain state identifiers, (5.48) entails

$$(\sigma, \sigma') \Vdash_\iota C[\bar{s}_f/\bar{v}] \tag{5.49}$$

Because (5.49) follows from the assumption (5.44), we have

$$(\sigma, \sigma') \Vdash_\iota A[\bar{s}_i/\bar{v}] \Rightarrow C[\bar{s}_f/\bar{v}]$$

as required.

(cons). Assume $\mathtt{t} \diamond G$ is of the form $\mathtt{p} \diamond A$, derived from a proof ending in (cons):

$$\frac{\vdash_{\mathsf{IHL}} \mathtt{p} \diamond P \quad P \vdash_{\mathsf{Int}} A}{\vdash_{\mathsf{IHL}} \mathtt{p} \diamond A} \text{ (cons)}$$

By the IH, we know

$$(\sigma, \sigma') \Vdash_\iota P \tag{5.50}$$

Also, because $P \vdash_{\mathsf{Int}} A$, by Lemma 5.1.1,

$$(\sigma, \sigma') \Vdash_\iota P \text{ entails } (\sigma, \sigma') \Vdash_\iota A \tag{5.51}$$

We instantiate (5.51) with (5.50), to give

$$(\sigma, \sigma') \Vdash A$$

as required.

This last case concludes our proof. □

5.1.2 Axioms and schemata

The proof of soundness is given over the core rules of IHL and Int. As discussed in the previous chapter, these rules are to be used in conjunction with non-logical axioms and schemata to reason about a domain problem and given black-box *IML* programs.

Our soundness result can be easily extended to include proofs that involve axioms and schemata, assuming that these are true. For instance, given an intuitionistic axiom $A \in \mathcal{AX}_{\mathsf{Int}}$, we must assume that

$$(\sigma, \sigma') \Vdash_\iota A$$

for any interpretation ι and initial and final states σ and σ'. Similarly, given an axiom about programs, $(\mathtt{p} \diamond A) \in \mathcal{AX}_{BB}$, we must assume that, for any interpretation ι,

$$[\![\mathtt{p}]\!] \Vdash_\iota A$$

The axioms generated by schemata must also be subject to the same assumptions for soundness to be extended.

5.1.3 Soundness and completeness over general models

The semantics of Hoare logic is usually given over a range of possible models for imperative programs. Such semantics are well understood, and need not be discussed in detail here. However, we make some salient remarks regarding soundness and completeness issues for such a general semantics.

In our semantics, *Values* forms a term model of the data that we want to store in the states. For our current purpose this semantics is sufficient, because we wish to use a model of execution that is as close to *SML* implementation as possible.

To achieve such a wider range of models, we could take different models of *Values*, which, in turn, achieve different states for different models. In this way, it is possible to give a more general relational semantics for *IML*, parametrized over a range of possible models for *Values*.

Such a general semantics is given in [Cou90, pp. 897–898] for a Hoare logic that uses triples. This semantics can easily be adapted to the logic TIHL of the previous chapter. It defines a notion of when a triple is true for a model M of *Values*, which we shall write as

$$M \Vdash_{Gen\_t} \{A\}\mathsf{p}\{B\}$$

and also when a single formula is true for a model, written as

$$M \Vdash_{GenInt\_t} A$$

We can use this notion of truth to define truth for program/formulae pairs:

$$M \Vdash_{Gen} p \diamond P$$

holds whenever

$$M \Vdash_{Gen\_t} \phi^{-1}(p \diamond P)$$

Soundness

Soundness for Hoare logic was originally given in [Hoa69] for a general semantics of this form. See [Cou90, pp. 901–902] for a proof of soundness over a general relational semantics, based upon a proof by [Coo78]. This proof can be adapted to TIHL, and therefore to a proof that, for all models of *Values* in the semantics, obeys

$$\vdash_{triple} \{A\}\mathsf{p}\{B\} \text{ entails } M \Vdash_{Gen\_t} \{A\}\mathsf{p}\{B\} \tag{5.52}$$

Then, by Theorem 4.6.3, p. 133, Section 4.6 of Chapter 4,

$$\vdash_{\mathsf{IHL}} \mathsf{p} \diamond P \text{ entails } \vdash_{\mathsf{TIHL}} \phi^{-1}(\mathsf{p} \diamond P)$$

So we can use this to extend the proof of soundness over the general relational semantics to proofs in IHL, using (5.52),

$$\begin{aligned} \vdash_{\mathsf{IHL}} \mathtt{p} \diamond P \text{ entails } & \vdash_{\mathsf{TIHL}} \phi^{-1}(\mathtt{p} \diamond P) \\ \text{entails } & M \Vdash_{Gen\_t} \phi^{-1}(\mathtt{p} \diamond P) \\ \text{entails } & M \Vdash_{Gen} \mathtt{p} \diamond P \end{aligned}$$

for any model M.

Completeness

Completeness means that every true statement is provable. The property for completeness for TIHL is stated as follows:

$$\Vdash_{Gen\_t} \{A\}\mathtt{p}\{B\} \text{ entails } \vdash_{\mathsf{TIHL}} \{A\}\mathtt{p}\{B\}$$

Completeness is not possible for Hoare logic.

The incompleteness for Hoare logic is deep rooted. This fact is not necessarily a consequence of unprovability problems with the formulae used in Hoare logic that might be inherited through the consequence rule. We can define versions of TIHL in which all formulae (all pre- and post-conditions) are decidable, but for which, for all models M, it is not the case that

$$M \Vdash_{Gen\_t} \{A\}\mathtt{p}\{B\} \text{ entails } \vdash_{\mathsf{TIHL}} \{A\}\mathtt{p}\{B\}$$

See [Cou90, pp. 909–910] for a proof of this, based upon proofs of [GS78] and [Wan78].

In our intuitionistic version of Hoare logic, we assume at least that the axioms equivalent to those of Heyting arithmetic are included for reasoning using the consequence rule (cons) (Assumption 4.7 of the previous chapter, p. 121). So we reason using statements whose truth values are undecidable. It was shown in [Coo78] that the completeness of Hoare logic with Peano arithmetic is equivalent to the semidecidability of the nonhalting problem. This can adapted to Heyting arithmetic. So our version of Hoare logic is incomplete for this situation.

Cook defined a notion of *relative completeness* [Coo78], which holds for Hoare logic, given some restrictions.

Relative completeness requires a notion of *expressiveness* of an interpretation of commands and formulae in a model M. Intuitively, expressiveness entails that,

- given a true triple involving sequential programs

  $$M \Vdash_{Gen\_t} \{A\}\mathtt{p};\mathtt{q}\{C\}$$

  we can always obtain an intermediate invariant formula B such that

  $$M \Vdash_{Gen\_t} \{A\}\mathtt{p}\{B\}$$

- given a true triple involving a loop

$$M \Vdash_{Gen\_t} \{A\}\texttt{while b do q}\{B\}$$

we know that there is an intermediate invariant formula I such that

$$M \Vdash_{GenInt\_t} (I \wedge \mathsf{tologic*}(\mathtt{b}) = false) \Rightarrow A$$

$$M \Vdash_{GenInt\_t} B \Rightarrow (I \wedge \mathsf{tologic*}(\mathtt{b}) = false)$$

and

$$M \Vdash_{Gen\_t} \{\mathsf{tologic*}(\mathtt{b}) = true \wedge I\}\mathtt{w}\{I\}$$

Relative completeness states that, if

1. a model is expressive for TIHL, and
2. we restrict proofs within Int(TIHL) to known truths in the model, that is, P for which

$$\vdash_{\mathsf{Int(TIHL)}} P \Leftrightarrow M \Vdash_{\mathsf{Int(TIHL)}} P$$

then

$$M \Vdash_{Gen\_t} \{A\}\mathtt{p}\{B\} \text{ entails } \vdash_{\mathsf{TIHL}} \{A\}\mathtt{p}\{B\}$$

See [Cou90, 914–918] for a proof of this result.

As in the case of soundness for general models, we can easily adapt this result to show a form of relative completeness for IHL, using the ϕ translation between the triple notion for TIHL and the program/formula pair notion of IHL.

5.2 Proof-theory of Intuitionistic Hoare Logic

We shall define a logical type theory LTT(IHL) for representing proofs in IHL. The main result of this part of the book, presented in the next chapter, is the application of the Curry–Howard protocol to extract imperative programs, with return values, from proofs in IHL. This section is therefore important, because a logical type theory is an essential requirement for the protocol to be used.

Our type theory has analogous properties to the type theory for intuitionistic logic described in Chapter 3. A type inference is of the form

$$\vdash_{LTT(\mathsf{IHL})} t^{\mathtt{w}\diamond P}$$

where

- the type $(\mathtt{w} \diamond P)$ corresponds to a theorem of IHL, and
- the term t (called a "proof-term") represents a proof of the theorem.

Similar to intuitionistic type theories, our type theory is a kind of lambda calculus. In this sense our logic and logical type theory satisfy a form of the Curry–Howard isomorphism. The distinguishing feature of this form of the isomorphism is that types are the program/formula pairs of IHL, whereas in the intuitionistic case, types were simply formulae.

Because IHL uses an intuitionistic deductive system, Int, the theory $LTT(\mathsf{IHL})$ involves a corresponding separate type theory, $LTT(\mathsf{Int})$. This theory is the logical type theory for Int described in Chapter 3, but with types now ranging over $WFF(\Sigma_t)$ (first-order, many-sorted formulae with state identifiers) instead of the formula of ADT.

As with the Curry–Howard isomorphism for intuitionistic logic, we define proof-term reduction rules that correspond to a proof normalization process. The normalization strategy over IHL only involves normalization of the intuitionistic proofs used in the (cons) rule. Normalization is not done over the basic, program building rules of IHL. As a result, strong normalization and the Church–Rosser property follow trivially from the corresponding theorems for Int.

5.2.1 Full form of the type theory for IHL

Written in full, our logical type theory $LTT(\mathsf{IHL})$ is

$$LTT(\mathsf{IHL}) = \langle PT(\mathsf{IHL}), Pairs(\mathsf{IHL}), (.)^{(.)}, \vdash_{LTT(\mathsf{IHL})}, PTR(LTT(\mathsf{IHL})), \rhd_{\mathsf{IHL}}\rangle$$

(see Definition 3.2.3, Chapter 3, p. 84 for the general form of a logical type theory). The set of proof-terms are denoted by $PT(\mathsf{IHL})$. Types are taken to be pairs of *IML* programs and $WFF(\Sigma_t)$ formulae, $Pairs(\mathsf{IHL})$. Type inference given by $\vdash_{LTT(\mathsf{IHL})}$ and rules $PTR(LTT(\mathsf{IHL}))$, explained in Fig. 5.2. The normalization relation $\rhd_{\mathsf{IHL}}$ is described in 5.2.7.

Following Definition 3.2.3 of Chapter 3, to be a logical type theory for IHL, $LTT(\mathsf{IHL})$ must be such that type inference in $LTT(\mathsf{IHL})$ and deduction in IHL are isomorphic, in the sense that

$$(\text{there is a } p \in PT(\mathsf{IHL}) \text{ where } \vdash_{\mathsf{IHL}} p^A) \Leftrightarrow \vdash_{\mathsf{IHL}} A$$

This isomorphism is proved in Theorem 5.2.5.

The elements of our type theory are now discussed in detail.

5.2.2 A logical type theory for Int

The definition of $LTT(\mathsf{IHL})$ involves a logical type theory for Int, intuitionistic proofs about $WFF(\Sigma_t)$:

$$LTT(\mathsf{Int}) = \langle PT(\mathsf{Int}), WFF(\Sigma_t), (.)^{(.)}, \vdash_{LTT(\mathsf{Int})}, PTR(LTT(\mathsf{Int})), \rhd_{\mathsf{Int}}\rangle$$

Because $WFF(\Sigma_t)$ is the same as $Formula(\mathsf{Int})$ with terms expanded to include state identifiers, and the inference rules are unchanged:

- We can take proof-terms $PT(\mathsf{Int})$ to be identical to those of $PT(\mathsf{Int})$ (given in Chapter 2), with the extension that existential witness terms v of $\mathsf{show}(v, a)$ now range over elements of $Terms(\mathsf{IHL})$, instead of $Terms(\mathcal{ADT})$.
- We can retain the proof-term inference rules of $LTT(\mathsf{Int})$, $PTR(LTT(\mathsf{Int}))$.
- We can keep the normalization relation $\rhd_{\mathsf{Int}}$ defined by the rules used in $LTT(\mathsf{Int})$.

5.2.3 Proof-terms

The proof-terms of our logical type theory, $PT(\mathsf{IHL})$, are intended to represent proofs in IHL. Their grammar is given in Fig. 5.1. The grammar is given with respect to a denumerable set of proof-term variables, $Var_{PT(\mathsf{IHL})}$.

The elements of $PT(\mathsf{IHL})$ that represent proofs of the subsystem Int are also displayed in Fig. 5.1.

5.2.4 Typing relation

Proof-terms represent proofs of program/formula pairs. This relationship is defined by a typing relation between proof-terms and types. A typed proof-term of our theory is written

$$p^{\mathtt{l} \diamond F}$$

where p is a proof-term, $\mathtt{l}$ is a program of IML and F is a formula of $WFF(\Sigma_t)$.

From a proof-theoretic perspective the rules of IHL define how we can use proofs in the construction of larger proofs. Making this fact explicit, the rules of IHL lead to the typing rules for LTT(IHL) given in Fig. 5.2.

Example 5.1. Consider the a proof in IHL that involves an application of the (loop) rule

$$\frac{\vdash_{\mathsf{IHL}} w \diamond (\mathsf{tologic_i}(\mathtt{b}) = true \wedge A[\bar{s}_i/\bar{v}]) \Rightarrow A[\bar{s}_f/\bar{v}]}{\vdash_{\mathsf{IHL}} \mathtt{while\ b\ do\ w} \diamond (A[\bar{s}_i/\bar{v}] \Rightarrow (A[\bar{s}_f/\bar{v}] \wedge \mathsf{tologic_f}(\mathtt{b}) = false))} \text{(loop)}$$

This application requires that a proof of the premise is given in order to construct a proof of the conclusion. This construction is made explicit as the proof-term $\mathsf{wd}(q)$ in the type inference

$$\frac{\vdash_{LTT(\mathsf{IHL})} q^{\mathtt{w} \diamond (\mathsf{tologic_i}(\mathtt{b}) = true \wedge A[\bar{s}_i/\bar{v}]) \Rightarrow A[\bar{s}_f/\bar{v}]}}{\vdash_{LTT(\mathsf{IHL})} \mathsf{wd}(q)^{\mathtt{while\ b\ do\ w} \diamond (A[\bar{s}_i/\bar{v}] \Rightarrow (A[\bar{s}_f/\bar{v}] \wedge \mathsf{tologic_f}(\mathtt{b}) = false))}} \text{(loop)}$$

with q the proof-term denoting the proof of the premise.

The (cons) rule is important. In IHL, this rule uses deduction in the intuitionistic subsystem Int to derive new truths about programs. In the logical type theory, the corresponding type inference rule permits proof-terms of the subsystem $LTT(\mathsf{Int})$ to be used to construct new proof-terms, via the cons proof-term operator.

$a, b, c ::=$	$PT(\mathsf{Int})$, proof-terms of Int
x	proof-term variable, $x \in Var_{PT(\mathsf{IHL})}$
$\mathsf{Axiom}(A)$	intuitionistic axiom, A is a formula
$\mathsf{Schema}(N, [\bar{e}; \bar{F}; \bar{t}; \bar{S}])$	intuitionistic schema application, N is a schema name, $\bar{e}$, $\bar{F}$, $\bar{t}$ and $\bar{S}$ are lists of proof-terms, formulae, terms and sorts, respectively
$\mathsf{abstract}\ x.\ a$	abstraction
$\mathsf{app}(a, b)$	application
$\mathsf{use}\ i.\ a$	$\mathcal{ADT}$-abstraction, $i \in Var(Term)$
$\mathsf{specific}(a, v)$	$\mathcal{ADT}$-application, $v \in Terms(\mathsf{IHL})$
$\langle a, b \rangle$	pair
$\mathsf{fst}(a)$	first projection
$\mathsf{snd}(b)$	second projection
$\mathsf{inl}(a)$	in left
$\mathsf{inr}(b)$	in right
$\mathsf{case}\ a\ \mathsf{of}\ \mathsf{inl}(x).b,\ \mathsf{inr}(y).c$	case
$\mathsf{abort}(a, F)$	abort, $F \in WFF(\Sigma_t)$
$\mathsf{show}(v, a)$	witness, $v \in Terms(\mathsf{IHL})$
$\mathsf{select}\ (a)\ \mathsf{in}\ x.y.b$	select, $x, y \in Var_{\mathcal{ADT}}$
$a, b, c\ ::=$	$PT(\mathsf{IHL})$, proof-terms of IHL
$\mathsf{IHLAxiom}(\mathtt{w} \diamond A)$	black-box axiom, $\mathtt{w} \diamond A$ is a program/formula pair
$\mathsf{IHLSchema}(N[\bar{e}])$	black-box schema application, $\bar{e}$ is a list of terms and N is a schema name
$\mathsf{assign}(r, i)$	r a state reference and i a term
$\mathsf{seq}(d, e)$	sequence
$\mathsf{ite}(d, e)$	if-then-else
$\mathsf{wd}(d)$	loop
$\mathsf{cons}(d, a)$	consequence, $a \in PT(\mathsf{Int})$

Fig. 5.1. Syntax of proof-terms $PT(\mathsf{IHL})$ for the calculus IHL.

Proof-terms for the subsystem $LTT(\mathsf{Int})$ are written

$$p^F$$

where p is a proof-term and F is a formula of $WFF(\Sigma_t)$. The typing rules are identical to those given in Chapter 2, Fig. 2.6, p. 38, but using the signature Σ_t. For completeness, we repeat the typing rule in Fig. 5.3. See Section 2.3 of that chapter for a deeper discussion of how these rules correspond to proofs in the subsystem.

$$\frac{}{\vdash_{LTT(\mathsf{IHL})} \mathsf{assign}(s, v)^{\mathtt{s:=v} \diamond s_f = \mathsf{tologic_i}(\mathtt{v})}} \text{ (assign)}$$

where $s \in StateRef$

$$\frac{\vdash_{LTT(\mathsf{IHL})} q_1^{\mathtt{l_1} \diamond (\mathsf{tologic_i}(\mathtt{b}) = true \Rightarrow C)} \quad \vdash_{LTT(\mathsf{IHL})} q_2^{\mathtt{l_2} \diamond (\mathsf{tologic_i}(\mathtt{b}) = false \Rightarrow C)}}{\vdash_{LTT(\mathsf{IHL})} \mathsf{ite}(q_1, q_2)^{\mathtt{if\ b\ then\ l_1\ else\ l_2} \diamond C}} \text{ (ite)}$$

$$\frac{\vdash_{LTT(\mathsf{IHL})} p^{\mathtt{w_1} \diamond (A[\bar{s}_i/\bar{v}] \Rightarrow B[\bar{s}_f/\bar{v}])} \quad \vdash_{LTT(\mathsf{IHL})} q^{\mathtt{w_2} \diamond (B[\bar{s}_i/\bar{v}] \Rightarrow C[\bar{s}_f/\bar{v}])}}{\vdash_{LTT(\mathsf{IHL})} \mathsf{seq}(p^{\mathtt{w_1} \diamond (A[\bar{s}_i/\bar{v}] \Rightarrow B[\bar{s}_f/\bar{v}])}, q^{\mathtt{w_2} \diamond (B[\bar{s}_i/\bar{v}] \Rightarrow C[\bar{s}_f/\bar{v}])})^{\mathtt{w_1;w_2} \diamond (A[\bar{s}_i/\bar{v}] \Rightarrow C[\bar{s}_f/\bar{v}])}} \text{ (seq)}$$

where A and B do not contain any state identifiers

$$\frac{\vdash_{LTT(\mathsf{IHL})} q^{\mathtt{w} \diamond (\mathsf{tologic_i}(\mathtt{b}) = true \wedge A[\bar{s}_i/\bar{v}]) \Rightarrow A[\bar{s}_f/\bar{v}]}}{\vdash_{LTT(\mathsf{IHL})} \mathsf{wd}(q)^{\mathtt{while\ b\ do\ w} \diamond (A[\bar{s}_i/\bar{v}] \Rightarrow (A[\bar{s}_f/\bar{v}] \wedge \mathsf{tologic_f}(\mathtt{b}) = false))}} \text{ (loop)}$$

where A does not contain any state identifiers

$$\frac{\vdash_{LTT(\mathsf{IHL})} q_1^{\mathtt{p} \diamond P} \quad \vdash_{LTT(\mathsf{Int})} q_2^{(P \Rightarrow A)}}{\vdash_{LTT(\mathsf{IHL})} \mathsf{cons}(q_1, q_2)^{\mathtt{p} \diamond A}} \text{ (cons)}$$

Fig. 5.2. Type inference rules of $LTT(\mathsf{IHL})$ corresponding to the structural rules of Hoare logic IHL.

5.2.5 Axioms and schemata

We require that there are type inference rules for all axioms and schemata of IHL and Int. These typing rules of $LTT(\mathsf{IHL})$ and $LTT(\mathsf{Int})$ are generated from the corresponding rules of IHL and Int is the following way.

Definition 5.2.1 (Type inference for black-box axioms). Given an axiom introduction rule (Ax-I)$_{BB}$ from IHL

$$\frac{(\mathtt{p} \diamond A) \in \mathcal{AX}_{BB}}{\vdash_{\mathsf{IHL}} \mathtt{p} \diamond A} \text{ (Ax-I)}_{BB}$$

we use the proof-term $\mathsf{IHLAxiom}(\mathtt{p} \diamond A)$ to denote an application of this rule, with the corresponding type formation rule

$$\frac{(\mathtt{p} \diamond A) \in \mathcal{AX}_{BB}}{\vdash_{LTT(\mathsf{IHL})} \mathsf{IHLAxiom}(\mathtt{p} \diamond A)^{\mathtt{p} \diamond A}} \text{ (Ax-I)}_{BB}$$

The type inference rule corresponding to an intuitionistic axiom rule (Ax-I)$_{\mathsf{Int}}$ is the same as that given for Int in Chapter 2, Definition 2.3.1, p. 37, repeated here for completeness.

Definition 5.2.2 (Type inference for intuitionistic axioms). Given an inference

Assume that x, y are arbitrary variables of sort s, and that a and c are well-sorted terms of sort s. We abbreviate the relation $\vdash_{LTT(\mathsf{Int})}$ by $\vdash$.

$$\frac{}{x^A \vdash x^A} \text{(Ass-I)}$$

$$\frac{\Delta, x^A \vdash b^B}{\Delta \vdash \mathsf{abstract}\ x.\ b^{(A \Rightarrow B)}} (\Rightarrow\text{-I}) \qquad \frac{\Delta \vdash a^A \quad \Delta' \vdash p^{(A \Rightarrow B)}}{\Delta, \Delta' \vdash \mathsf{app}(p, a)^B} (\Rightarrow\text{-E})$$

$$\frac{\Delta \vdash p^A}{\Delta \vdash \mathsf{use}\ x : s.\ p^{\forall x:s \bullet A}} (\forall\text{-I}) \qquad \frac{\Delta \vdash p^{\forall x:s \bullet A}}{\Delta \vdash \mathsf{specific}(p, c)^{A[c/x]}} (\forall\text{-E})$$

x is free in A, not free in Δ

$$\frac{\Delta \vdash p^{P[a/y]}}{\Delta \vdash \mathsf{show}(a, p)^{\exists y:s \bullet P}} (\exists\text{-I}) \qquad \frac{\Delta_1 \vdash p^{\exists y:s \bullet P} \quad \Delta_2, x^{P[x/y]} \vdash q^C}{\Delta_1, \Delta_2 \vdash C} (\exists\text{-E})$$

where x not occur free in C

$$\frac{\Delta \vdash a^A \quad \Delta' \vdash b^B}{\Delta, \Delta' \vdash \langle a, b \rangle^{(A \wedge B)}} (\wedge\text{-I})$$

$$\frac{\Delta \vdash p^{(A_1 \wedge A_2)}}{\Delta \vdash \mathsf{fst}(p)^{A_1}} (\wedge\text{-E}_1) \qquad \frac{\Delta \vdash p^{(A_1 \wedge A_2)}}{\Delta \vdash \mathsf{snd}(p)^{A_2}} (\wedge\text{-E}_2)$$

$$\frac{\Delta \vdash p^{A_1}}{\Delta \vdash \mathsf{inl}(p)^{(A_1 \vee A_2)}} (\vee\text{-I}_1) \qquad \frac{\Delta \vdash p^{A_2}}{\Delta \vdash \mathsf{inr}(p)^{(A_1 \vee A_2)}} (\vee\text{-I}_2)$$

$$\frac{\Delta \vdash p^{A \vee B} \quad \Delta_1, x^A \vdash a^C \quad \Delta_2, y^B \vdash b^C}{\Delta_1, \Delta_2, \Delta \vdash \mathsf{case}\ p\ \mathsf{of}\ \mathsf{inl}(x).a,\ \mathsf{inr}(y).b^C} (\vee\text{-E})$$

$$\frac{\Delta \vdash a^\perp}{\Delta \vdash \mathsf{abort}(a)^A} (\perp\text{-E})$$

provided A is Harrop

Fig. 5.3. Type inference rules for the subsystem $LTT(\mathsf{Int})$.

$$\frac{A \in \mathcal{AX}_{\mathsf{Int}}}{\vdash_{\mathsf{Int}} A} (\text{Ax-I})_{\mathsf{Int}}$$

we use the proof-term $\mathsf{Axiom}(A)$ to denote an application of this rule, with the corresponding type formation rule

$$\frac{A \in \mathcal{AX}_{\mathsf{Int}}}{\vdash_{LTT(\mathsf{Int})} \mathsf{Axiom}(A)^A} (\text{Ax-I})_{\mathsf{Int}}$$

Type inference rules corresponding to schemata of the intuitionistic subset are of the same form as those given in Chapter 2, Definition 2.3.1, p. 38. We repeat this definition for completeness. These are defined as follows.

Definition 5.2.3 (General form of type inference rules for schemata). Given a schema rule $R[\bar{X};\bar{y};\bar{Z}]$ from Int, where $\bar{X}$, $\bar{y}$ and $\bar{Z}$ are lists of variables ranging over formulae, terms and sorts, respectively:

$$\frac{\Gamma_1 \vdash_{\mathsf{Int}} F_1 \quad \dots \quad \Gamma_n \vdash_{\mathsf{Int}} F_n}{\vdash_{\mathsf{Int}} F} \; R[\bar{X};\bar{y};\bar{Z}]$$

we define corresponding type formation schemata for proof-terms of the form

$$\mathsf{Schema}(R, [[q_1; \dots; q_n]; \bar{X}; \bar{y}; \bar{Z}])$$

written

$$\frac{\vdash_{\mathsf{Int}} q_1^{F_1} \quad \dots \quad \vdash_{\mathsf{Int}} q_n^{F_n}}{\vdash_{\mathsf{Int}} \mathsf{Schema}(R, [[q_1; \dots; q_n]; \bar{X}; \bar{y}; \bar{Z}])^F} \; R[\bar{X};\bar{y};\bar{Z}]$$

Definition 5.2.4 (General form of type inference rules for black-box schemata). Let $\mathtt{f}$ be a black-box program of $BB_{s_1 \dots s_n, s}$, parametrized over arguments of types $(\mathtt{s_1}, \dots, \mathtt{s_n})$. Let $\bar{x}$ be a list of n term variables $x_1, \dots, x_n$. Take a black-box schema

$$\frac{}{\vdash_{\mathsf{IHL}} \mathtt{f}(\mathtt{x_1}, \dots, \mathtt{x_n}) \diamond F} \; R[\bar{x}]$$

The corresponding type inference schema is

$$\frac{}{\vdash_{LTT(\mathsf{IHL})} \mathsf{IHLSchema}(R[\bar{x}])^{\mathtt{f}(\mathtt{x_1},\dots,\mathtt{x_n}) \diamond F}} \; R[\bar{x}]$$

Take a IHL proof involving an application of the schema

$$\frac{}{\vdash_{\mathsf{IHL}} \mathtt{f}(\mathtt{x_1}, \dots, \mathtt{x_n}) \diamond F} \; R[\bar{a}]$$

where $\bar{a}$ is a list of n terms $a_1, \dots, a_n$ of sorts $s_1, \dots, s_n$. This corresponds to an application of the type inference schema

$$\frac{}{\vdash_{LTT(\mathsf{IHL})} \mathsf{IHLSchema}(R[\bar{a}])^{\mathtt{f}(\mathtt{a_1},\dots,\mathtt{a_n}) \diamond [a_1,\dots,a_n / x_1,\dots,x_n]}} \; R[\bar{a}]$$

5.2.6 The Curry–Howard correspondence

We now show that proof-terms of LTT(IHL) represent proofs of theorems in IHL.

Theorem 5.2.5 (Curry–Howard correspondence between LTT(IHL) and IHL). *The following properties hold*

1. *Given a natural deduction proof D of $\vdash_{\mathsf{IHL}} \mathtt{w} \diamond A$, we can construct a proof-term $f^{\mathtt{w} \diamond A}$ such that*

$$\vdash_{LTT(\mathsf{IHL})} f^{\mathtt{w} \diamond A}$$

2. *Given a proof-term* $f^{\mathtt{w}\diamond A}$ *such that*

$$\vdash_{LTT(\mathsf{IHL})} f^{\mathtt{w}\diamond A}$$

we can construct a natural deduction proof D *of* $\vdash \mathtt{w} \diamond A$.

Proof. Item 1) is derived by straightforward induction on the structure of the deduction D. Item 2) is given by induction on the structure of the inference $\vdash_{LTT(\mathsf{IHL})} f^{\mathtt{w}\diamond A}$. □

5.2.7 Proof normalization

We can define a normalization strategy for removing redundant parts of IHL proofs.

We do not define any reduction rules for applications of program/formula inference rules of IHL. That is to say, applications of these rules never result in redundant proof steps.

The only proof reductions that we apply are over the intuitionistic proofs, which occur within IHL proofs because of the (cons) rule. As in intuitionistic logic, normalization is done by matching applications of introduction and elimination rules (see, e.g., [Gen69] or [GLT89] and also Chapter 2).

For example, a proof of the form

$$\dfrac{\dfrac{\begin{matrix}[A] \\ \vdots\, b \\ B\end{matrix}}{(A \Rightarrow B)}\,(\Rightarrow\text{-I}) \qquad \begin{matrix}\vdots\, a \\ A\end{matrix}}{B}\,(\Rightarrow\text{-E})$$

involves the redundant use of a (⇒-I) followed by (⇒-E). The proof can be simplified to

$$\begin{matrix}\vdots\, a \\ A \\ \vdots\, b \\ B\end{matrix}$$

The normalization strategy for Int is given by defining a reduction relation $\rhd_{\mathsf{Int}}$ over typed terms of $LTT(\mathsf{Int})$. See Fig. 5.4 for the seven reduction rules that constitute an inductive definition of $\rhd_{\mathsf{Int}}$ (previously given in Chapter 3). The LHD and the RHD of a rule are called the *redex* and the *reduct* of the rule, respectively.

We define $\hat{\rhd}_{\mathsf{Int}}$ to be the transitive closure of the application of these rules. When $a \hat{\rhd}_{\mathsf{Int}} b$ holds, when b is obtainable from a by a sequence of replacements of subterms using the rules of Fig. 5.4. In this case, we say that a is *reducible* to b.

We then use this normalizing relation to define normalization of IHL proofs, given by the relation $\rhd_{\mathsf{IHL}}$ over $LTT(\mathsf{IHL})$ proof-terms. This relation is given by a single reduction rule

$$\boxed{a \hat{\rhd}_{\mathsf{Int}} a' \text{ entails } \mathsf{cons}(d, a) \rhd_{\mathsf{IHL}} \mathsf{cons}(d, a')} \tag{5.53}$$

We take $\hat{\rhd}_{\mathsf{IHL}}$ to be the transitive closure of this rule. Observe that,

$$d \hat{\rhd}_{\mathsf{IHL}} d'$$

if, and only if, d' is obtained from d by replacing occurrences of intuitionistic proof-terms in $\mathsf{cons}(t, b)$ terms by equivalent normalized forms by repeated application of the rules for $\rhd_{\mathsf{Int}}$.

We have the following result.

Lemma 5.2.1. *Take any $LTT(\mathsf{IHL})$ proof-terms $a^{\mathtt{l} \diamond A}$ and $b^{\mathtt{m} \diamond B}$.*

If $a^{\mathtt{l} \diamond A} \rhd_{\mathsf{IHL}} b^{\mathtt{m} \diamond B}$, then the type of a is the same as the type of b. That is to say, the program/formula pair $\mathtt{l} \diamond A$ is the same as $\mathtt{m} \diamond B$.

Proof. Using Lemma 2.3.1 of Chapter 2 for $LTT(\mathsf{Int})$, which entails that, if an intuitionistic proof-term $c \in PT(\mathsf{Int})$ reduces to $c' \in PT(\mathsf{Int})$ by one of the rules of Fig. 5.4, then the type of c is the same as the type of c'.

Consequently, any application of the rule (5.53)

$$c \rhd_{\mathsf{Int}} c' \text{ entails } \mathsf{cons}(d, c) \rhd_{\mathsf{IHL}} \mathsf{cons}(d, c')$$

will mean that $\mathsf{cons}(d, c)$ and $\mathsf{cons}(d, c')$ will have the same type. The lemma follows immediately from this fact. □

1.	$\mathsf{app}(\mathsf{abstract}\ x.\ a^{(A \Rightarrow B)}, b^A) \rhd_{\mathsf{Int}} a[b/x]^B$
2.	$\mathsf{specific}(\mathsf{use}\ x : s.\ a^{\forall x:s \bullet A}, v : s) \rhd_{\mathsf{Int}} a[v/i]^{A[v/x]}$
3.	$\mathsf{fst}(\langle a, b \rangle^{(A \wedge B)}) \rhd_{\mathsf{Int}} a^A$
4.	$\mathsf{snd}(\langle a, b \rangle^{(A \wedge B)}) \rhd_{\mathsf{Int}} b^B$
5.	$\mathsf{case}\ \mathsf{inl}(a)^{(A \vee B)}\ \mathsf{of}\ \mathsf{inl}(x^A).b^C,\ \mathsf{inr}(y^B).c^C \rhd_{\mathsf{Int}} b[a/x]^C$
6.	$\mathsf{case}\ \mathsf{inr}(a)^{(A \vee B)}\ \mathsf{of}\ \mathsf{inl}(x^A).b^C,\ \mathsf{inr}(y^B).c^C \rhd_{\mathsf{Int}} c[a/y]^C$
7.	$\mathsf{select}\ (\mathsf{show}(v, a)^{\exists y:s \bullet P})\ \mathsf{in}\ x^P.y.b^C \rhd_{\mathsf{Int}} b[a/x][v/y]^C$

Fig. 5.4. The reduction rules that inductively define $\rhd_{\mathsf{Int}}$.

5.2.8 Strong Normalization and the Church–Rosser property

The strong normalization property tells us that the normalization process will always terminate. To show that this property holds over our calculus, we need to show that the proof-terms of LTT(IHL) are strongly normalizable, according to the following definition.

Definition 5.2.6 (Strongly normalizable proof-terms — Definition 2.3.4 of Chapter 3). We say that a proof-term is *normal* if it contains no redex — that is, if it is irreducible.

Given a proof-term t, we let $\mathcal{N}(t)$ denote the least upper bound of lengths of reduction sequences for t. We say that t is *strongly normalizable* if all reduction sequences are finite.

By Theorem 2.3.5 of Chapter 3, p. 41, proof-terms of $PT(\mathsf{Int})$ are strongly normalizing with respect to the relation Int. We use exactly the same set of proof-terms, but extended with a larger set of witnesses within $\mathsf{show}(v, a)$ terms. However, because witness terms do not affect application of $\rhd_{\mathsf{Int}}$, it follows that our proof-terms are strongly normalizing with respect to $\rhd_{\mathsf{Int}}$.

Strong normalization of $LTT(\mathsf{IHL})$ follows easily from the fact that Int proofs are strongly normalizing.

Theorem 5.2.7 (Strong normalization for IHL proofs). *Each proof-term of $PT(\mathsf{IHL})$ is strongly normalizing.*

Proof. Any IHL proof has a finite number of applications of the (cons) rule. Consequently, for any $PT(\mathsf{IHL})$ term, there are only a finite number of subterms of the form $\mathsf{cons}(d, a)$.

Any reduction using $\rhd_{\mathsf{IHL}}$ can only involve application of the rule (5.53). That is to say, a reduction according to the rule for $\rhd_{\mathsf{IHL}}$ can occur if, and only if, a reduction according to the rules for $\rhd_{\mathsf{Int}}$ can occur over a subterm of the form $\mathsf{cons}(d, a)$. Because proof-terms of $LTT(\mathsf{Int})$ are strongly normalizing, there are only a finite number of times the intuitionistic proof-term a can be reduced in $\mathsf{cons}(d, a)$ subterms.

By these two facts, the theorem holds. □

The Church–Rosser property says that divergent proof normalization sequences always eventually converge to yield the same proof.

As in the previous part of this book, we formalize this notion using the Curry–Howard correspondence, proving the Church–Rosser property in terms of the relation $\rhd$ over a proof's corresponding proof-term. We say that proofs in IHL satisfy the Church–Rosser property when the normalizing relation $\rhd$ satisfies the diamond property.

Definition 5.2.8 (Diamond property). A relation # over a set S satisfies the *diamond property* when for every x, x_1 and x_2 in S

$$(x\#x_1 \text{ and } x\#x_2 \text{ entails there is a } x_3 \text{ such that } (x_1\#x_3 \text{ and } x_2\#x_3))$$

As a consequence of the Church–Rosser property for Int (Theorem 2.3.7, Chapter 2, p. 42) we know that $PT(\mathsf{Int})$ proof-terms satisfy the Church–Rosser property. From this, we have the Church–Rosser property for $PT(\mathsf{IHL})$.

Theorem 5.2.9 (Church–Rosser property for IHL proofs). *The relation* $\rhd_{\mathsf{IHL}}$ *over* $PT(\mathsf{IHL})$ *satisfies the diamond property. Consequently,* $PT(\mathsf{IHL})$ *satisfies the Church–Rosser property.*

Proof. A consequence of the Church–Rosser property for $PT(\mathsf{Int})$ and the definition of $\rhd_{\mathsf{IHL}}$. □

5.3 Example: Electronic Banking System

Recall the electronic banking system example of Chapter 4, Section 4.5.5. The system consists of a database of account details, indexed by user identification. We used IHL to develop a program and a description of the program. The program searches through the database, making it possible to obtain a list of all accounts held at the bank by the user, given a user's details. This described by the program/formula pair in the theorem

$$\begin{aligned}\vdash\ & \texttt{counter} := 0;\\ & \quad \texttt{while}\ !\texttt{counter} < (\texttt{length db}) - 1\ \texttt{do counter} :=!\texttt{counter} + 1 \diamond\\ & \quad \exists y : (account\ list) \bullet allAccountsAt(currentUser, y, counter_f) \wedge\\ & \qquad\qquad\qquad (counter_f < (length\ db) - 1) = false \qquad (5.54)\end{aligned}$$

5.3.1 Axioms

The domain was axiomatized as follows. We take a predicate

$$allAccountsAt(u : user, x : account\ list, y : int)$$

whose meaning is that x is a list of all accounts found to be owned by the user u, up to the point y in the database db. The predicate is defined by the following four axioms in $\mathcal{AX}$, denoted by $A_{5.55}$, $A_{5.56}$, $A_{5.57}$ and $A_{5.58}$ respectively (see Chapter 4, Section 4.5.5 for the explanations of the axioms).

$$\begin{aligned}&\forall u : user \bullet \forall x : (account\ list) \bullet \forall y : int \bullet allAccountsAt(u, x, y) \Rightarrow\\ &\qquad\qquad \forall z : int \bullet z \leq y \Rightarrow sub(db, z).owner = u \qquad (5.55)\end{aligned}$$

$$\begin{aligned}&\forall u : user \bullet \forall x : (account\ list) \bullet \forall y : int \bullet\\ &\qquad\qquad sub(db, y + 1).owner = u \wedge\\ &allAccountsAt(u, x, y) \Rightarrow allAccountsAt(u, sub(db, y + 1) :: x, y + 1) \qquad (5.56)\end{aligned}$$

$$\forall u : user \bullet \forall x : (account\ list) \bullet \forall y : int \bullet \neg sub(l, y+1).owner = u \wedge allAccountsAt(u, x, y) \Rightarrow allAccountsAt(u, x, y+1) \quad (5.57)$$

$$\forall u : user \bullet \forall y : int \bullet y = 0 \Rightarrow allAccountsAt(u, [], y) \quad (5.58)$$

(These axioms are available for introduction in the intuitionistic subsystem of IHL, so they do not involve programs.)

The previous axioms are Harrop. We also have a non-Harrop axiom, denoted by $A_{5.59}$:

$$y < (length\ db) - 1 \Rightarrow sub(l, y+1).owner = u \vee \neg sub(l, y+1).owner = u \quad (5.59)$$

Applications of these axioms are used in the LTT by writing $\mathsf{Axiom}(A)$, where A is the axiom. For instance, use of (5.55) is denoted by $\mathsf{Axiom}(A_{5.55})$.

5.3.2 Constructing the proof-term

We follow the original proof, building the corresponding proof-term. From (5.56), (5.57) and (5.59), we can derive an intuitionistic proof-term of the form

$$e^{y<(length\ db)-1}, f^{allAccountsAt(u,x,y)} \vdash_{\mathsf{Int}} p_{5.60}^{\exists l:(account\ list)\bullet allAccountsAt(u,l,y+1)} \quad (5.60)$$

where $p_{5.60}$ is a proof-term denoting proof by cases

$$\mathsf{case\ app}(A_{(5.59)}, e)\ \mathsf{of\ inl}(g^{sub(l,y+1).owner=u}).\mathsf{show}(sub(db, y+1) :: x, p_2), \mathsf{inr}(h^{\neg sub(l,y+1).owner=u}).\mathsf{show}(x, p_3)$$

p_2 uses (5.56) and the assumption $sub(l, y+1).owner = u$ to derive

$$\mathsf{app}(\mathsf{app}(\mathsf{app}(\mathsf{app}(A_{(5.56)}, u), x), y), \langle g, f\rangle)^{allAccountsAt(u,sub(db,y+1)::x,y+1)}$$

and p_3 uses (5.57) and the assumption $\neg sub(l, y+1).owner = u$ to derive

$$\mathsf{app}(\mathsf{app}(\mathsf{app}(\mathsf{app}(A_{(5.56)}, u), x), y), \langle g, f\rangle)^{allAccountsAt(u,x,y+1)}$$

By assuming $\exists l : (account\ list) \bullet allAccountsAt(u, l, y)$, we can apply ($\exists$-E) on (5.60) and then obtain

$$\vdash_{\mathsf{Int}} \forall y : int \bullet \forall u : user \bullet (y < (length\ db) - 1) \wedge \exists l : (account\ list) \bullet allAccountsAt(u, l, y) \Rightarrow \exists l : (account\ list) \bullet allAccountsAt(u, l, y+1) \quad (5.61)$$

by (⇒-I) over our assumptions, and successive (∀-I) over the free variables. The corresponding proof-term $p_{5.61}$ is of the form

$$\mathsf{use}\ y : int.\ \mathsf{use}\ u : user.\ \mathsf{abstract}\ m^{(y<(length\ db)-1)\wedge\exists l:(account\ list)}.$$
$$\mathsf{app}(\mathsf{app}(\mathsf{abstract}\ e^{y<(length\ db)-1}.\ \mathsf{abstract}\ i^{\exists l:(account\ list)\bullet allAccountsAt(u,l,y)}.$$
$$\mathsf{specific}(i, x.f^{allAccountsAt(u,x,y).p_{5.60}}), \mathsf{fst}(m)), \mathsf{snd}(m))$$

Using (subst) and taking a variable v such that $v = y + 1$, we can transform (5.61) into

$$\vdash_{\mathsf{Int}} \forall y : int \bullet \forall v : int \bullet v = y + 1 \Rightarrow \forall u : user \bullet$$
$$(y < (length\ db) - 1) \wedge \exists l : (account\ list) \bullet allAccountsAt(u, l, y) \Rightarrow$$
$$\exists l : (account\ list) \bullet allAccountsAt(u, l, v) \quad (5.62)$$

with proof-term $p_{5.62}$ of the form

$$\mathsf{use}\ y : int.\ \mathsf{use}\ v : int.\ \mathsf{abstract}\ r^{v=y+1}.$$
$$\mathsf{Schema}(subst, [[(\mathsf{specific}(p_{5.61}, y)); r^{v=y+1}];$$
$$[\forall u : user \bullet (y < (length\ db) - 1) \wedge \exists l : (account\ list) \bullet allAccountsAt(u, l, y) \Rightarrow$$
$$\exists l : (account\ list) \bullet allAccountsAt(u, l, m)][y + 1; v]; [int]])$$

We then instantiate (5.62) with $counter_i$ and $counter_f$ and $currentUser$ for y, v and u, respectively, to give

$$\vdash_{\mathsf{Int}} counter_f = counter_i + 1 \Rightarrow (counter_i < (length\ db) - 1) \wedge$$
$$\exists l : (account\ list) \bullet allAccountsAt(currentUser, l, counter_i) \Rightarrow$$
$$\exists l : (account\ list) \bullet allAccountsAt(currentUser, l, counter_f) \quad (5.63)$$

The proof-term corresponding to this proof, which we will denote by $p_{5.63}$, is of the form

$$\mathsf{specific}(\mathsf{specific}(\mathsf{specific}(p_{5.62}, counter_i), counter_f), currentUser)$$

We also have the following, by the (assign) rule of Hoare logic:

$$\vdash \mathtt{counter} := !\mathtt{counter} + 1 \diamond counter_f = counter_i + 1 \quad (5.64)$$

This has a corresponding proof-term $\mathsf{assign}(counter, counter + 1)$.

And so, by applying (cons) to (5.64) and (5.63)

$$\vdash \mathtt{counter} := !\mathtt{counter} + 1 \diamond (counter_i < (length\ db) - 1) \wedge$$
$$\exists l : (account\ list) \bullet allAccountsAt(currentUser, l, counter_i) \Rightarrow$$
$$\exists l : (account\ list) \bullet allAccountsAt(currentUser, l, counter_f) \quad (5.65)$$

The corresponding proof-term is

$$\mathsf{cons}(\mathsf{assign}(counter, counter + 1), p_{5.63})$$

Then we apply (*loop*) on (5.65)

$$\vdash \texttt{while !counter < (length db) - 1 do counter :=!counter + 1} \diamond \exists l : (account\ list) \bullet allAccountsAt(currentUser, l, counter_i) \Rightarrow \exists l : (account\ list) \bullet allAccountsAt(currentUser, l, counter_f) \wedge (counter_f < (length\ db) - 1) = false \quad (5.66)$$

with resulting proof-term

$$\mathsf{wd}(\mathsf{cons}(\mathsf{assign}(counter, counter + 1), p_{5.63}))$$

From the axiom (5.58) we can derive

$$\vdash_{\mathsf{Int}} counter_f = 0 \Rightarrow \exists y : (account\ list) \bullet allAccountsAt(currentUser, y, counter_f) \quad (5.67)$$

with a proof-term $p_{5.67}$. By application of (assign) we have

$$\vdash \texttt{counter} := 0 \diamond counter_f = 0 \quad (5.68)$$

with proof-term $\mathsf{assign}(counter, 0)$. Then, applying (cons) to (5.68) and (5.67) gives

$$\texttt{counter} := 0 \diamond \exists y : (account\ list) \bullet allAccountsAt(currentUser, y, counter_f) \quad (5.69)$$

with proof-term $\mathsf{cons}(\mathsf{assign}(counter, 0), p_{5.67})$. This can be weakened to include a true hypothesis *true*:

$$\texttt{counter} := 0 \diamond true \Rightarrow \exists y : (account\ list) \bullet allAccountsAt(currentUser, y, counter_f) \quad (5.70)$$

with a proof-term of the form

$$\mathsf{cons}(\mathsf{cons}(\mathsf{assign}(counter, 0), p_{5.67}), ptrue)$$

where $ptrue$ is a proof-term for an intuitionistic proof of $P \Rightarrow (true \Rightarrow P)$ where P is $\exists y : (account\ list) \bullet allAccountsAt(currentUser, y, counter_f)$.

So, using (seq) on (5.70) and (5.66), we can obtain

$$\vdash \texttt{counter := 0; while y < (length db) - 1 do counter :=!counter + 1} \diamond true \Rightarrow \exists y : (account\ list) \bullet allAccountsAt(currentUser, y, counter_f) \wedge (counter_f < (length\ db) - 1) = false \quad (5.71)$$

with proof-term

$$\mathsf{seq}(\mathsf{cons}(\mathsf{cons}(\mathsf{assign}(counter, 0), p_{5.67}), ptrue), \mathsf{wd}(\mathsf{cons}(\mathsf{assign}(counter, counter + 1), p_{5.63})))$$

which can be simplified to the required form of (5.54)

$$\vdash \texttt{counter} :=!\texttt{counter} + \texttt{1}; \; \texttt{while y} < (\texttt{length db}) - \texttt{1 do counter} :=!\texttt{counter} + \texttt{1} \diamond \; \exists y : (account\ list) \bullet allAccountsAt(currentUser, y, counter_f) \wedge (counter_f < (length\ db) - 1) = false$$

with proof-term $p_{5.72}$

$$\mathsf{cons}(\mathsf{seq}(\mathsf{cons}(\mathsf{cons}(\mathsf{assign}(counter, 0), p_{5.67}), ptrue), \mathsf{wd}(\mathsf{cons}(\mathsf{assign}(counter, counter + 1), p_{5.63}))), qtrue) \quad (5.72)$$

where $qtrue$ is a proof of $(true \Rightarrow A) \Rightarrow A$ with A standing for

$$\exists y : (account\ list) \bullet allAccountsAt(currentUser, y, counter_f) \wedge (counter_f < (length\ db) - 1) = false$$

Remark 5.1. We have shown how to encode (5.54) in our logical type theory. In the next chapter, we will define a notion of return value realizability of IHL and provide a method of program extraction, returning to our example in Section 6.4. There, we shall see how to transform our proof-term into a program that, given a user's details, will search through a database to obtain all accounts held at the bank by the user, and then returns this list.

5.3.3 Normalization

The proof can by normalized by application of the reduction rule (5.53), (and so using the intuitionistic proof-term reduction rules of Fig. 5.4). These rules only reduce the subterms corresponding to intuitionistic subproofs used within our IHL proof. The only applicable subterm of $p_{5.72}$ is $p_{5.63}$. The proof-term was of the form

$$\mathsf{specific}(\mathsf{specific}(\mathsf{specific}(p_{5.62}, counter_i), counter_f), currentUser)$$

Normalization involves repeated applications of rule 2 of Fig. 5.4, reducing pairs of specific and use proof-term constructors.

This results in a proof-term $p'_{5.63}$ of the form

$$\begin{array}{l}\mathsf{specific}(\mathsf{abstract}\ r^{counter_f = counter_i + 1}.\\ \quad \mathsf{Schema}(subst, [[p'_{5.61}; r^{counter_f = counter_i + 1}];\\ \quad [\forall u : user \bullet (counter_i < (length\ db) - 1) \wedge\\ \quad \exists l : (account\ list) \bullet allAccountsAt(u, l, counter_i) \Rightarrow\\ \exists l : (account\ list) \bullet allAccountsAt(u, l, m)][counter_i + 1; counter_f]; [int]]),\\ \quad currentUser)\end{array}$$

where the normalized form of $\mathsf{specific}(\mathsf{specific}(p_{5.61}, counter_i), counter_f)$ is $p'_{5.61}$:

$$\begin{array}{l}\mathsf{use}\ u : user.\ \mathsf{abstract}\ m^{(counter_i < (length\ db) - 1) \wedge \exists l : (account\ list)}.\\ \quad \mathsf{app}(\mathsf{app}(\mathsf{abstract}\ e^{counter_i < (length\ db) - 1}.\\ \quad \mathsf{abstract}\ i^{\exists l : (account\ list) \bullet allAccountsAt(u, l, counter_i)}.\\ \quad \mathsf{specific}(i, x.f^{allAccountsAt(u, x, counter_i)}.p'_{5.60}), \mathsf{fst}(m)), \mathsf{snd}(m))\end{array}$$

We define $p'_{5.60}$ by

$$\begin{array}{l}\mathsf{case}\ \mathsf{app}(A_{(5.59)}, e)\ \mathsf{of}\\ \quad \mathsf{inl}(g^{sub(l, counter_i + 1).owner = u}).\mathsf{show}(sub(db, counter_i + 1) :: x, p'_2),\\ \quad \mathsf{inr}(h^{\neg sub(l, counter_i + 1).owner = u}).\mathsf{show}(x, p'_3)\end{array}$$

where p'_2 is

$$\mathsf{app}(\mathsf{app}(\mathsf{app}(\mathsf{app}(A_{(5.56)}, u), x), counter_i), \langle g, f \rangle)$$

with type $allAccountsAt(u, sub(db, counter_i + 1) :: x, counter_i + 1)$ and p'_3 is

$$\mathsf{app}(\mathsf{app}(\mathsf{app}(\mathsf{app}(A_{(5.56)}, u), x), counter_i), \langle g, f \rangle)^{allAccountsAt(u, x, counter_i + 1)}$$

Because this is the only reducible subterm of $p_{5.72}$, the normalized proof $p'_{5.72}$ is given by taking $p_{5.72}$ and substituting $p'_{5.63}$ for $p_{5.63}$, to give

$$\begin{array}{l}\mathsf{cons}(\mathsf{seq}(\mathsf{cons}(\mathsf{cons}(\mathsf{assign}(counter, 0),\\ \quad p_{5.67}), ptrue), \mathsf{wd}(\mathsf{cons}(\mathsf{assign}(counter, counter + 1), p'_{5.63}))), qtrue)\end{array}$$

Remark 5.2. The outermost specific application argument $currentUser$ of $p'_{5.63}$ cannot be matched with the abstraction use $u : user$ variable of $p'_{5.61}$, because the proof-term constructor for the (subst) schema separates the former from the latter. In certain cases it is possible to add additional reduction rules to move schemata up and down a proof, to facilitate matchings for further reductions. These rules would be similar to those defined by the authors for their structured specification logic — see [CPW00, PCW02] and also Chapter 8 in Part IV of this book. However, for the purposes of our work, we will be satisfied with the current reduction rules.

Remark 5.3. When extracting a program from our example proof in the next chapter, we will show that this sub-proof can be transformed into a *SML* program that helps to build a list of accounts for a user. We will see that the extracted program's structure reflects parts of the structure of the proof-term. Consequently, normalization of proof-terms aids the extraction of simpler programs.

5.4 Discussion

This chapter presented some important semantic and proof-theoretic properties of IHL. First we examined soundness and completeness issues for IHL. Then we defined logical type theory for IHL, providing a property analogous to the Curry–Howard correspondence for intuitionistic logic.

The logical type theory is of particular importance for the results of this part of the monograph, as it is necessary for us to apply the Curry–Howard protocol to IHL.

Our proof-terms form an augmented lambda calculus. Our domain of reasoning is imperative programs. Observe that, in contrast to naïve functional proof-as-programs, proof-term normalization does not correspond to imperative program execution. We claim that it is not possible to provide a naïve proofs-as-programs correspondence for Hoare logic, because proof normalization and imperative operational semantics are too different to be identified. However, in the next chapter, we will show how to transform our proof-terms into provably correct imperative programs with side-effects and side-effect-free return values. In this way, following the analogy to state-of-the-art functional proofs-as-programs, program evaluation and proof-normalization are clearly distinguished from one another.

6

Proofs-as-Imperative-Programs

In this chapter we show how to synthesize correct imperative programs from proofs in IHL according to the Curry–Howard protocol of Chapter 3. We take a novel approach to specifying imperative program behavior, using the formulae of IHL to specify, not only pre- and post- conditions of imperative programs, but also the required return values. A required return value is taken as a Skolem function for the formula (using the definition of Skolemization for first-order many-sorted formulae given in Chapter 2 of Part II).

Hoare logic is good for reasoning about and developing side-effect producing imperative programs. However, Hoare logic on its own is arguably not as well suited as constructive methods for developing side-effect-free aspects of imperative programs, such as return values.

Return values can be complex — and difficult to reason about. Most imperative languages permit the definition of higher-order functions as return values. This can be achieved in C++, for instance, using the Standard Template Library (*STL*), or in Eiffel using the Agents library. In an impure functional language such as *SML*, complex return values take the form of side-effect-free lambda expressions. In general, such expressions are difficult to code from a given specification, due to their functional nature and the fact that they may use state values that have been manipulated by code preceding the return value.

Hoare logic usually specifies properties of return values, or a view of states, by associating these values with designated state symbols. The construction of a return value consists of designing an assignment statement for a designated state reference, and proving a required property holds about the resulting final value of the state. For instance, imagine we want to derive a program whose return value is an even number. In Hoare logic, if we let $\mathtt{r}$ be the state symbol associated with a return value $!\mathtt{s} * 2$, we can prove

$$\mathtt{s} := 3; \mathtt{r} := !\mathtt{s} * 2 \diamond Even(r) \tag{6.1}$$

The program on the left hand side of the tuple $\mathtt{s} := 3; \mathtt{r} := !\mathtt{s} * 2$ is *not* the required program — it is represents the required program with the state reference $\texttt{r}$ representing the required return value. The program needs to be transformed into a program where the required return value replaces the assignment $\texttt{r} := \texttt{s} * 2$:

$$\texttt{s} := 10; !\texttt{s} * 2 \tag{6.2}$$

might be synthesized. Thus, the second assignment $\texttt{r} :=!\texttt{s} * 2$ of (6.1) is shorthand for the fact that $!\texttt{s} * 2$ should be returned by the required program (6.2), and the assertion is shorthand for specifying that the return value is even.

The problem with this traditional approach is that return values are not synthesized — they are hand-written. The designer is required to explicitly define a return value before proving properties about it. For instance, in a proof of (6.1), the required return value $!\texttt{s} * 2$ must be identified and assigned to $\texttt{r}$ so that we can prove that $Even(r)$ holds for the program.

Such an explicit definition is an implementation detail — it involves writing the code for a return value. We would prefer to be able to specify, prove properties about, and synthesize required return values, while hiding such implementation details.

Constructive methods enable us to achieve this. In constructive program synthesis, a proof of a statement can be used to synthesize a realizer of the statement. The realizer is a functional program that satisfies the statement as a specification. For example, an existential statement

$$\exists x : t \bullet A(x)$$

can be used to synthesize a function that returns a value p such that $A(p)$ is provable. Because the realizer is synthesized from a proof, the details about its definition are hidden from the prover. The prover need only be concerned with using logic to reason about a problem, not with the definition of a program.

In this chapter, we will adapt this property of constructive functional synthesis to the imperative context.

Example 6.1. Given a constructive proof in IHL of

$$\texttt{s} := \texttt{s} * 3 \diamond s_f > s_i \wedge (\exists x : int \bullet Even(x) \wedge x > s_i)$$

our techniques will synthesize a program of the form

$$\texttt{s} := \texttt{s} * 3; \texttt{f}$$

where the function f is a side-effect-free function (such as $!\texttt{s} * 2$) that realizes the existential statement of the post-condition $(\exists x : int \bullet Even(x) \wedge x > s_i)$, acting as a witness for the x.

Example 6.2. Given the proof of the Example 4.5.5 of Chapter 4 (p. 123)

$$\frac{\begin{array}{l}\vdash \mathtt{p} \diamond PINCorrect(pin_i) \Rightarrow canWithdraw_f = true \\ \vdash \mathtt{q} \diamond canWithdraw_i = true \Rightarrow \\ \quad (isConnected_f = true \wedge \exists x : string \bullet AppMessage(x))\end{array}}{\begin{array}{l}\vdash \mathtt{p;q} \diamond PINCorrect(pin_i) \Rightarrow \\ (isConnected_f = true \wedge \exists x : string \bullet AppMessage(x))\end{array}} \text{ (seq)}$$

our adaptation will enable us to synthesize a program m that makes

$$\mathtt{p;q} \diamond PINCorrect(pin_i) \Rightarrow (isConnected_f = true \wedge \\ \exists x : string \bullet AppMessage(x))$$

true (just as p;q does), but also returns an appropriate message, given that the user has entered their PIN correctly. That is to say, we will synthesize a program m that, when executed, will return a string r such that

$$AppMessage(r)$$

is true, given that the user has entered their PIN correctly.

By virtue of the synthesis process the user should have no need to manually code the return value, but instead works within IHL to prove a theorem, from which the return value is then synthesized.

We proceed in the following way, ensuring that our approach adheres to the Curry–Howard protocol, where *IML* and *LTT*(IHL) are considered as the computational type theory and the logical type theory respectively. We define a notion of realizability between *IML* programs and formulae of IHL. Essentially, a program is a realizer of a formula when the formula correctly describes the side-effects of the program *and* the the possible return values of the program. After this, we define an extraction map from proof-terms of *LTT*(IHL) (given in the previous chapter) to *IML* programs. The map is used to transform a proof of a program/formula pair $\mathtt{p} \diamond F$ into a program m with return value that realizes the formula F. The program m will be similar to p in that it makes F true, but more complicated in that it will involve a return value realizer for F.

This chapter is organized as follows.

- Our notion of realizability is explained in section 6.1 and program extraction is explained in section 6.2.
- We explain how these results lead to a successful application of the Curry–Howard protocol in section 6.3.
- In Section 6.4, we illustrate our program synthesis methods for the electronic banking example used throughout this part of the monograph.
- Section 6.5 provides a second example application of our methods. We show how the synthesis of programs with return values can correspond to the synthesis of programs with complex contracts, to be used in systems built according to Bertrand Meyer's principle of design-by-contract [Mey97].
- A chapter summary and concluding discussion are provided in section 6.6.

6.1 Realizability

This section discusses a new notion of realizability between *IML* programs and the formulae of IHL. Intuitively, an imperative program p is a realizer of a program/formula pair $1 \diamond F$ when:

- The side-effects of p are correctly described by F. That is to say, F is true of p in the sense defined in Chapter 4.
- The side-effects of 1 involving state references used in F are the same as those of p involving those references.
- The return value of p is correctly described by F. That is to say, for every execution, the return value of p is a modified realizer of F, as defined for intuitionistic logic in Section 2.5 of Chapter 2, Part II, p. 2.5.

The first two requirements use definitions given in Chapter 4. The third requirement requires further elucidation.

We define how return values may be specified by formulae by adapting intuitionistic modified realizability. A formula of IHL can be true about the return value of a program, when the return value is a *modified realizer* for the formula. This is done by adapting the intuitionistic definition to our context. To make this adaptation, we require a notion of Skolem form for formulae of IHL, $WFF(\Sigma_t)$.

6.1.1 Skolemization

Our notion of Skolemization follows the intuitionistic case of Chapter 2, and so involves Harrop formulae.

Harrop formulae are defined as in Definition 2.2.1 of Chapter 2, Part II. We repeat the definition of completeness here.

Definition 6.1.1 (Harrop). A formula F of $WFF(\Sigma_t)$ is a *Harrop formula* if it is

1. an atomic formula,
2. of the form $(A \wedge B)$ where A and B are Harrop formulae,
3. of the form $(A \Rightarrow B)$ where B (but not necessarily A) is a Harrop formula, or
4. of the form $(\forall x : s \bullet A)$ where A is a Harrop formula.

We write $H(F)$ if F is a Harrop formula, and $\neg H(F)$ if F is not a Harrop formula.

We also need to define a sort extraction map etype from formulae to sorts of Σ_t (and so of Σ_p). This is given in Fig. 6.1.

We can now define the Skolem form of a $WFF(\Sigma_t)$ formula, in the same way as we did for formulae of Chapter 2.

F	etype(F)
any Harrop formula	$Unit$
$(A \wedge B)$	$\begin{cases} \text{etype}(A) & \text{if not } H(B) \\ \text{etype}(B) & \text{if not } H(A) \\ \text{etype}(A) * \text{etype}(B) & \text{otherwise} \end{cases}$
$(A \vee B)$	etype(A)|etype(B)
$(A \Rightarrow B)$	$\begin{cases} \text{etype}(B) & \text{if not } H(A) \\ \text{etype}(A) \rightarrow \text{etype}(B) & \text{otherwise} \end{cases}$
$(\forall x : s \bullet A)$	$s \rightarrow$ etype(A)
$(\exists x : s \bullet A)$	$\begin{cases} s & \text{if } H(A) \\ s * \text{etype}(A) & \text{otherwise} \end{cases}$
$\bot$	$Unit$

P is an atomic predicate.

Fig. 6.1. Definition of etype.

Definition 6.1.2 (Skolem form and Skolem functions). Given a closed formula A, we define the *Skolem form* of A to be the Harrop formula $Sk(A) = Sk'(A, \emptyset)$, where $Sk'(A, AV)$ is defined as follows.

A unique function letter f_A, called the *Skolem function*, is associated with each such formula A, of sort etype(A). AV represents a list of application variables for A (that is, the variables that will be arguments of f_A). If AV is $\{x_1 : s_1, \ldots, x_n : s_n\}$ then $f(AV)$ stands for the function application $app(f, (x_1, \ldots, x_n))$.

1. If A is Harrop, then $Sk'(A, AV) = A$.
2. If $A = (B \vee C)$, then

$$Sk'(A, AV) = \\ (\forall x : \text{etype}(B) \bullet f_A(AV) = Inl(x) \Rightarrow Sk'(B, AV)[x/f_B]) \wedge \\ (\forall y : \text{etype}(C) \bullet f_A(AV) = Inr(y) \Rightarrow Sk'(C, AV)[y/f_C])$$

3. If $A = (B \wedge C)$, then
   a) If B is Harrop and C is not Harrop,
   $$Sk'(A, AV) = B \wedge Sk'(C, AV)[snd(f_A)/f_C]$$
   b) If B is not Harrop and C is Harrop,
   $$Sk'(A, AV) = (Sk'(B, AV)[fst(f_A)/f_B] \wedge C)$$

c) If B and C are not Harrop,

$$Sk'(A, AV) = (Sk'(B, AV)[fst(f_A)/f_B] \wedge Sk'(C, AV)[snd(f_A)/f_C])$$

4. If $A = (B \Rightarrow C)$, then
   a) If B is Harrop,

$$Sk'(A, AV) = (B \Rightarrow Sk'(C, AV)[f_A/f_C])$$

   b) If B is not Harrop and C is not Harrop,

$$Sk'(A, AV) = \forall x : \mathsf{etype}(B) \bullet (Sk'(B, AV)[x/f_B] \Rightarrow Sk'(C, AV)[(f_A x)/f_C])$$

5. If $A = \exists y : s \bullet P$, then
   a) when P is Harrop, $Sk'(A, AV) = Sk'(P, AV)[f_A(AV)/y]$.
   b) when P is not Harrop,

$$Sk'(A, AV) = Sk'(P, AV)[fst(f_A(AV))/y][snd(f_A(AV))/f_P]$$

6. If $A = \forall x : s \bullet P$, then $Sk'(A, AV) = \forall x : s \bullet Sk'(P, AV)[(f_A x)/f_P]$.

For intuitionistic proofs, we retain the same notion of modified realizability used in Chapter 2.

Definition 6.1.3 (Modified realizability). A program p is a *modified realizer* of a formula F if, and only if, $\vdash_{\mathsf{Int}} Sk(F)[p/f_F]$ is provable. In this case, we write p mr F.

6.1.2 Adapting modified realizability for specifying return values

We say that a formula specifies a program's possible return values if, for every execution of the program, the return value can be used as a Skolem function in the Skolemized version of the formula, which is made true by the program's execution. In this case, we say that the return value is a modified realizer for the program's side-effects or a *return value realizer.*

Definition 6.1.4 (Return value realizability). Let p be an *IML* program. We say that the program is a is a *return value realizer* of F *for the initial state* σ and interpretation ι when, for any $\sigma' \in MLStates$, if

$$\langle \mathrm{p}, \sigma \rangle \hat{\triangleright} \langle \texttt{answer}, \sigma' \rangle$$

then

$$(\sigma, \sigma') \Vdash_\iota Sk(F)[answer/f_F]$$

In this case, we write $\mathsf{p}\ \mathsf{retr}^{\sigma}_{\iota}\ F$.

If $\mathsf{p}\ \mathsf{retr}^{\sigma}_{\iota}\ F$ for some σ, we say that the program is a is a *return value realizer* of F under ι and write $\mathsf{p}\ \mathsf{retr}_{\iota}\ F$.

If $\mathsf{p}\ \mathsf{retr}_{\iota}\ F$ for some ι, we say that the program is a is a *return value realizer* of F and write $\mathsf{p}\ \mathsf{retr}\ F$.

Our definition of return value realizability is analogous to realizability for intuitionistic logic, where a modified realizer is a term that provides the constructive content of a formula, making the Skolemized version of the formula true when substituted for the Skolem function. However, our definition is further complicated because of to the presence of state identifiers in the formula that require realizability to be given with respect to a program's side-effects.

6.1.3 Specifying side-effects and return values

Using the definition of return value realizability and the definition of truth about program side-effects, we can now define how program/formula pairs specify programs *and* required return values of programs.

We use the following definition.

Definition 6.1.5 (Visible side-effect equivalence). Take two programs, p and q of *IML* and take any formula A.

We say that p and q are *side-effect equivalent* over the states visible from A, and write

$$\mathsf{p} \equiv_A \mathsf{q}$$

when

$$(\sigma, \sigma') \in [\![\mathsf{p}]\!] \text{ entails } (\sigma, \sigma'') \in [\![\mathsf{q}]\!]$$

provided σ' and σ'' differ only over state references that are *not* used in A. That is to say, σ' and σ'' differ only over references not in $state\!-\!id(\bar{s}_i :: \bar{r}_f)$, where $\bar{s}_i$ and $\bar{r}_f$ are the initial and final state identifiers that occur in A.

The notion of a program/formula pair being true of both program's side-effects and return values, informally stated at the beginning of this section, is called IHL-realizability, and is defined formally as follows.

Definition 6.1.6 (IHL-realizability). Let p be an *IML* program, and let $\mathtt{w} \diamond P$ be a program/formula pair of IHL.

We say p is a IHL-*realizer* of $\mathtt{w} \diamond P$ if, and only if,

1. $[\![\mathsf{p}]\!] \Vdash P$, and
2. $\mathsf{p}\ \mathsf{retr}\ P$
3. $\mathsf{p} \equiv_P \mathtt{w}$

When these hold, we write $\mathsf{p}\ \mathsf{kr}\ w \diamond P$.

In the next section, we will show how to extract correct programs as IHL-realizers of specifications from proofs in Hoare logic.

6.2 Extraction

In this section we prove the main result of this part of the book: that there is an extraction map, $\mathsf{extract}_{\mathsf{IHL}}$, from proof-terms of $LTT(\mathsf{IHL})$ to IML programs that generates IHL-realizers from proofs of specifications. Due to the presence of intuitionistic proofs in Hoare logic (proofs via the (cons) rule), our map also involves an extraction map over intuitionistic proof-terms, of the form given in Chapter 2.

We first need to make some assumptions about the treatment of black-box programs, axioms, and schemata in extraction. Then we will revisit intuitionistic extract, provide our full extraction map, and finally derive our main result.

6.2.1 Assumptions about black-box programs and Σ_p

Our results assume that we can obtain IHL-realizers from axioms and applications of schemata.

Assumption 6.1 *(IHL-realizers for axioms and schemata).* We assume that, for each proof-term corresponding to an axiom,

$$\mathsf{IHLAxiom}(\mathtt{w} \diamond A)^{\mathtt{w} \diamond A}$$

there is a program $\mathsf{PK}_{\mathtt{w} \diamond A} : \mathsf{etype}(A)$ such that

$$\mathsf{PK}_{\mathtt{w} \diamond A} \ \mathsf{kr} \ \mathtt{w} \diamond A$$

Similarly, we assume that, for each proof-term corresponding to a rule generated from a black-box schema,

$$\mathsf{IHLSchema}(N[\bar{e}])^{\mathtt{w} \diamond A}$$

there is a program $\mathsf{PK}_{N[\bar{e}]} : \mathsf{etype}(A)$ such that

$$\mathsf{PK}_{N[\bar{e}]} \ \mathsf{kr} \ \mathtt{w} \diamond A$$

We also assume that all non-Harrop axioms used in intuitionistic proofs have associated modified realizers.

Assumption 6.2. We assume that, for each proof-term corresponding to an axiom:

$$\mathsf{Axiom}(A)^A$$

there is a function in Σ_p and a corresponding program in the SML preamble

$$PK_A : \mathsf{etype}(A)$$

such that

$$PK_A \ \mathsf{mr} \ A$$

Similarly, we assume that, for each proof-term corresponding to a rule generated from a schema,

$$\mathsf{Schema}(N, [\bar{e}; \bar{F}; \bar{t}; \bar{S}])^A$$

there is a function in Σ and a corresponding program in the *SML* preamble

$$\vdash_{\mathsf{Int}} PK_{N[\bar{e};\bar{F};\bar{t};\bar{S}]} : \mathsf{etype}(A)$$

such that

$$PK_{N[\bar{e};\bar{F};\bar{t};\bar{S}]} \ \mathsf{kr}\ A$$

This assumption is similar to the one made in the presentation of functional proofs-as-programs in Chapter 2 (see Assumption 2.3, p. 68).

Example 6.3. Following Chapter 2, we take the modified realizer for instances of the substitution schema (subst)

$$\frac{q_1^{P[\beta/y]} \quad q_2^{\alpha =_s \beta}}{\mathsf{Schema}(subst, [[q_1; q_2]; P; \bar{y}; \bar{Z}])^{P[\alpha/y]}} \ (\mathrm{subst})[[P]; [\alpha; \beta]; [s]]$$

to be

$$\mathsf{PK}_{subst,[[q_1;q_2];P;\bar{y};\bar{Z}]} = \mathsf{extract}_{\mathsf{Int}}(q_1[\alpha/y])$$

It is easy to see that this is the required realizer, using the same reasons given in Example 2.6, p. 69.

6.2.2 Extraction over intuitionistic proofs

Because of the (cons) rule, where intuitionistic proofs of $\mathsf{Int}(\mathsf{IHL})$ are used in IHL proofs, $LTT(\mathsf{IHL})$ proof-terms involve proof-terms taken from the intuitionistic subsystem of $LTT(\mathsf{IHL})$, and our definition of $\mathsf{extract}_{\mathsf{IHL}}$ must be built on an extraction map over intuitionistic proof-terms, of the form given in Chapter 2. For reference, we provide this map again in Fig. 6.2.

The map in Fig. 6.2 extracts $Terms(\Sigma_t)$ terms from $PT(\mathsf{Int}(\mathsf{IHL}))$ proofs. Note that we can treat the resulting terms of $Terms(\Sigma_t)$ as pure *IML* program terms with state identifiers taken to be free variables. The extracted terms are intuitionistic modified realizers in the sense defined in Section 2.5 of Chapter 2, Part II, p. 44. The map presumes a set of variables in Var, each corresponding to a proof-term variable from $Var_{PT(\mathsf{IHL})}$, $\{x_u \mid u \in Var_{PT(\mathsf{Int})}\}$.

Theorem 6.2.1. *Take any proof*

$$\vdash_{\mathsf{Int}} T$$

Then $\mathsf{extract}_{\mathsf{Int}}(t)$ *is an intuitionistic modified realizer of* T,

$$\mathsf{extract}_{\mathsf{Int}}(t) \ \mathsf{mr}\ T$$

That is to say,

$$\vdash_{\mathsf{Int}} Sk(T)[\mathsf{extract}_{\mathsf{Int}}(t)/f_T]$$

Proof. The proof follows from the proof of Theorem 2.5 of Chapter 2, Part II, p. 44. The extraction map $\mathsf{extract}_{\mathsf{Int}}(p)$ is the same. The only difference is that state identifiers are used in the terms, $Terms(\Sigma_t)$, of our version of Int. However, the presence of state identifiers does not affect the proof as these are treated as special constant symbols in Σ_t, and so the proof can be retained. □

6.2.3 Imperative program extraction

Using the intuitionistic extraction map of Fig. 6.2, we can define $\mathsf{extract}_{\mathsf{IHL}}$ over all the proof-terms of $LTT(\mathsf{IHL})$, as in Fig. 6.3.

Remark 6.1. The extraction map for producing *IML* programs (Fig. 6.3) treats terms $\mathsf{extract}_{\mathsf{Int}}(t) \in Terms(\Sigma_t)$ as *IML* programs. This is possible when we treat state identifiers as free variables in an *IML* term.

So, for instance, the term

$$s_i + v_f + 10$$

from $Terms(\Sigma_t)$ can be treated as an *IML* program with free variables $\mathtt{s_i}$ and $\mathtt{v_f}$

$$\mathtt{s_i} + \mathtt{v_f} + 10$$

The required results are provided in Theorems 6.2.4 and 6.2.5.

6.2.4 Preliminary results

To prove our results, we require the following lemmata.

Theorem 6.2.2. *Take any proof*

$$\vdash_{LTT(\mathsf{IHL})} d^{\mathtt{w}\diamond A}$$

Then

$$\vdash_{IML} \mathsf{extract}(d) : \mathsf{etype}(A)$$

is a correct type inference.

Proof. By induction on the possible forms of d.

In the case where d represents the application of an axiom or schema, we have the theorem from the assumptions in Section 6.2.1, p. 172.

The other cases follow easily using the type inference rules of Fig. 4.5, Chapter 4, p. 105. □

Lemma 6.2.1. *Take an arbitrary interpretation ι and states σ and σ'. If there is a term a such that $(\sigma, \sigma') \Vdash_\iota Sk(A)[a/f_A]$ then $(\sigma, \sigma') \Vdash_\iota A$.*

Proof. By induction on the form of A. □

p^P	$\mathsf{extract}_{\mathsf{Int}}(p^P)$
any proof-term where $H(P)$	`()`
u^A	x_u not $H(A)$ `()` $H(A)$
$\mathsf{Axiom}(N)^A$	$\mathtt{PK}_N$
$\mathsf{Schema}(N, [\bar{e}; \bar{F}; \bar{t}; \bar{S}])^A$	$\mathtt{PK}_{N[\bar{e};\bar{F};\bar{t};\bar{S}]}$
$\mathsf{abstract}\ u^A.\ a^B$	`fn` $\mathtt{x_u}$ `=>` $\mathsf{extract}_{\mathsf{Int}}(a)$ not $H(A)$ $\mathsf{extract}_{\mathsf{Int}}(a)$ $H(A)$
$\mathsf{app}(c^{A \Rightarrow B}, a^A)$	$\mathsf{extract}_{\mathsf{Int}}(c)$ $H(A)$ $(\mathsf{extract}_{\mathsf{Int}}(c)\ \mathsf{extract}_{\mathsf{Int}}(a))$ not $H(A)$
$\mathsf{use}\ x : s.\ a^A$	`fn x : s =>` $\mathsf{extract}_{\mathsf{Int}}(a)$
$\mathsf{specific}(a^{\forall x:s \bullet A}, v)$	$(\mathsf{extract}_{\mathsf{Int}}(a)\ \mathtt{v})$
$\langle a^A, b^B \rangle$	$(\mathsf{extract}_{\mathsf{Int}}(a), \mathsf{extract}_{\mathsf{Int}}(b))$
$\mathsf{case}\ a^{A \vee B}\ \mathsf{of}\ \mathsf{inl}(t^A).b^C,$ $\mathsf{inr}(u^B).c^C$	`match` $\mathsf{extract}_{\mathsf{Int}}(a)$ `with` $\mathtt{Inl(x_t)}$ `=>` $\mathsf{extract}_{\mathsf{Int}}(b)$, $\mathtt{Inr(x_u)}$ `=>` $\mathsf{extract}_{\mathsf{Int}}(c)$
$\mathsf{show}(v, a^A)$	`v` $H(A)$ $(\mathtt{v}, \mathsf{extract}_{\mathsf{Int}}(a))$ not $H(A)$
$\mathsf{select}\ (a^{\exists y \bullet A})\ \mathsf{in}\ x.u^{A[x/y]}.b^B$	(`fn x =>` $\mathsf{extract}_{\mathsf{Int}}(b)$) $\mathsf{extract}_{\mathsf{Int}}(a)$ } $H(A)$ (`fn x =>` `fn` $\mathtt{x_u}$ `=>` $\mathsf{extract}_{\mathsf{Int}}(b)$) `fst`$(\mathsf{extract}_{\mathsf{Int}}(a))$ `snd`$(\mathsf{extract}_{\mathsf{Int}}(a))$ } not $H(A)$
$\#(a)$ where $\#$ is inl, inr, fst or snd	$\#(\mathsf{extract}_{\mathsf{Int}}(a))$
$\mathsf{abort}(a^{\perp})$	`()`

Fig. 6.2. The extraction map $\mathsf{extract}_{\mathsf{Int}}$, defined from intuitionistic proof-terms $LTT(\mathsf{Int})$ to terms of $Terms(\Sigma_t)$.

Lemma 6.2.2. *Take an arbitrary interpretation ι. Let A be a formula that does not contain any initial or final state identifiers. Let $\sigma, \sigma', \sigma''$ be arbitrary states. Then*

$$(\sigma, \sigma') \Vdash_\iota Sk(A[\bar{s}_f/\bar{v}])[p/f_{A[\bar{s}_f/\bar{v}]}] \text{ entails}$$
$$(\sigma', \sigma'') \Vdash_\iota Sk(A[\bar{s}_i/\bar{v}])[p/f_{A[\bar{s}_i/\bar{v}]}]$$

Lemma 6.2.3. *Let A be an arbitrary formula and σ be a state. Let σ' and σ'' be states that differ from σ only over a state reference $\mathtt{r}$. Assume r_i and r_f do not occur in $Sk(A)[\bar{s}_i/\bar{v}][answer\ x/f_{A[\bar{s}_i/\bar{v}]}]$. Then*

$$(\sigma, \sigma') \Vdash_{\iota'} Sk(A)[\bar{s}_i/\bar{v}][answer\ x/f_{A[\bar{s}_f/\bar{v}]}] \text{entails}$$
$$(\sigma, \sigma'') \Vdash_{\iota'} Sk(A)[\bar{s}_f/\bar{v}][answer\ x/f_{A[\bar{s}_f/\bar{v}]}]$$

$t^{\mathtt{w}\diamond T}$	$\mathsf{extract}_{\mathsf{IHL}}(t)$	
any proof-term where $H(T)$	`w`	
$\mathsf{IHLAxiom}(\mathtt{w} \diamond A)^{\mathtt{w}\diamond A}$	$\mathsf{PK}_{\mathtt{w}\diamond A}$	
$\mathsf{IHLSchema}(N[\bar{e}])^{\mathtt{w}\diamond A}$	$\mathsf{PK}_{N[\bar{e}]}$	

$\mathsf{wd}(u)^{\mathtt{while\ b\ do\ l}\diamond T}$ where T is $A[\bar{s}_i/\bar{v}] \Rightarrow (A[\bar{s}_f/\bar{v}] \wedge \mathsf{tologic_f}(\mathsf{b}) = \mathit{false})$	$\mathtt{rv}_1$:= fn x : etype(A) => x;
	while b do
	  $\mathtt{rv}_2$:= $\mathsf{extract}_{\mathsf{IHL}}(q)$;
	  $\mathtt{rv}_1$:= (fn $\mathtt{x}_2$ :: $\mathtt{x}_1$ =>
	    fn x : etype(A) => $\mathtt{x}_2$ ($\mathtt{x}_1$ x))
	    !$\mathtt{rv}_2$!$\mathtt{rv}_1$;
	!$\mathtt{rv}_1$;	

$\mathsf{ite}(q_1, q_2)^{\mathtt{if\ b\ then\ l_1\ else\ l_2}\diamond C}$	if b then $\mathsf{extract}_{\mathsf{IHL}}(q_1)$ else $\mathsf{extract}_{\mathsf{IHL}}(q_2)$	

$\mathsf{seq}(p^{\mathtt{w}_1\diamond P}, q^{\mathtt{w}_2\diamond Q})^{\mathtt{w}_1;\mathtt{w}_2\diamond T}$ where T is $A[\bar{s}_i/\bar{v}] \Rightarrow C[\bar{s}_f/\bar{v}]$, P is $A[\bar{s}_i/\bar{v}] \Rightarrow B[\bar{s}_f/\bar{v}]$, Q is $B[\bar{s}_i/\bar{v}] \Rightarrow C[\bar{s}_f/\bar{v}]$	$\mathtt{rv}_p$:= $\mathsf{extract}_{\mathsf{IHL}}(p)$;
	$\mathtt{rv}_q$:= $\mathsf{extract}_{\mathsf{IHL}}(q)$;
	(fn $\mathtt{x}_p$ => fn $\mathtt{x}_q$ =>
	  fn x : etype(A) =>
	  $\mathtt{x}_q$ ($\mathtt{x}_p x$))!$\mathtt{rv}_p$!$\mathtt{rv}_q$
		not $H(A)$ not $H(B)$ and not $H(C)$

	$\mathtt{rv}_p$:= $\mathsf{extract}_{\mathsf{IHL}}(p)$; $\mathtt{rv}_q$:= $\mathsf{extract}_{\mathsf{IHL}}(q)$; $\mathtt{rv}_q$ $\mathtt{rv}_p$	$H(A)$ not $H(B)$ and not $H(C)$
	w; $\mathtt{rv}_q$:= $\mathsf{extract}_{\mathsf{IHL}}(q)$; !$\mathtt{rv}_q$	$H(A)$ $H(B)$ and not $H(C)$

	w;
	$\mathtt{rv}_q$:= $\mathsf{extract}_{\mathsf{IHL}}(q)$;
	(fn $\mathtt{x}_q$ =>
	  fn x : etype(A) =>
	  $\mathtt{x}_q$ x) !$\mathtt{rv}_q$
		not $H(A)$ and $H(B)$ and not $H(C)$

$\mathsf{cons}(p^{\mathtt{w}\diamond P}, q^{P\Rightarrow A})^{\mathtt{w}\diamond A}$	$\bar{\mathtt{i}}$:= $\bar{\mathtt{s}}$;
	$\mathtt{rv}_p$:= $\mathsf{extract}_{\mathsf{IHL}}(p)$;
	$\bar{\mathtt{f}}$:= $\bar{\mathtt{s}}$;
	(fn $\bar{\mathtt{s}}_\mathtt{i}$ :: $\bar{\mathtt{s}}_\mathtt{f}$ =>
	  $\mathsf{extract}_{\mathsf{Int}}(q)$!$\mathtt{rv}_p$)
	  !$\bar{\mathtt{i}}$ :: $\bar{\mathtt{f}}$
		not $H(P)$ and not $H(A)$

	$\bar{\mathtt{i}}$:= $\bar{\mathtt{s}}$; w; $\bar{\mathtt{f}}$:= $\bar{\mathtt{s}}$; (fn $\bar{\mathtt{s}}_\mathtt{i}$ :: $\bar{\mathtt{s}}_\mathtt{f}$ => $\mathsf{extract}_{\mathsf{Int}}(q)$) !$\bar{\mathtt{i}}$ :: $\bar{\mathtt{f}}$	$H(P)$ and not $H(A)$

Where we assume that $\bar{s}$ is the list of state references corresponding to the initial and final state identifiers occurring in formula T, and that $\mathtt{rv}_1$, $\mathtt{rv}_2$, $\mathtt{rv}_p$, $\mathtt{rv}_q$, and lists $\bar{\mathtt{i}}$ and $\bar{\mathtt{f}}$ are state references that do not occur in extract(p) and extract(q), and whose corresponding state identifiers *never* occur in any formula used in the proof of p or q.

Fig. 6.3. The extraction map $\mathsf{extract}_{\mathsf{IHL}} : \mathit{LTT}(\mathsf{IHL}) \to \mathit{IML}$.

Proof. The proof is straightforward, by induction over the possible forms of A. □

Lemma 6.2.4. *Take any interpretation ι. Take any set of state references*

$$R = \{\mathtt{r}^1, \ldots, \mathtt{r}^n\}$$

Let A be a formula in which the corresponding final and initial state identifiers,

$$R_{if} = \{r_i, r_f \mid \mathtt{r} \in R\}$$

do not occur. Let σ, σ' be arbitrary states.

If τ' is a state that differs from σ' only over the state references from R, then

$$(\sigma, \sigma') \Vdash_\iota A \text{ entails } (\sigma, \tau') \Vdash_\iota A$$

Also, if τ is a state that differs from σ only over the state references R, then

$$(\sigma, \sigma') \Vdash_\iota A \text{ entails } (\tau, \sigma') \Vdash_\iota A$$

Proof. By induction over the form of A, using the definition of $\Vdash$ (Chapter 4, Section 4.4.3, Definition 4.4.8, p. 115).

We only exhibit the base case where A is a predicate over terms. The other cases follow easily.

Assume A is of the form $P(a_1, \ldots, a_n)$. Then

$$(\sigma, \sigma') \Vdash_\iota A \Leftrightarrow h(P) \ni P(\iota(a_1)_\sigma^{\sigma'}, \ldots, \iota(a_n)_\sigma^{\sigma'})$$

Now, for each a_j $(i = 1, \ldots, n)$,

$$\begin{aligned} \iota(a_j)_\sigma^{\sigma'} &= \iota'(a_j) \\ &\quad \text{where } \iota' = \iota\,[initial(a_j) \mapsto \sigma(state\!-\!id(initial(a_j)))] \\ &\qquad\qquad [final(a_j) \mapsto \sigma'(state\!-\!id(final(a_j)))] \end{aligned}$$

But, because the elements of R_{if} do not occur in A, $initial(a_j)$ and $final(a_j)$ cannot contain any elements of R_{if}. Consequently, $state\!-\!id(initial(a_j))$ and $state\!-\!id(final(a_j))$ cannot contain any elements of R. This means that

$$\sigma(state\!-\!id(initial(a_j))) = \tau(state\!-\!id(initial(a_j)))$$

and

$$\sigma'(state\!-\!id(final(a_j))) = \tau'(state\!-\!id(final(a_j)))$$

It follows that

$$\iota(a_j)_\sigma^{\sigma'} = \iota(a_j)_\sigma^{\tau'}$$

and

$$\iota(a_j)_\tau^{\sigma'} = \iota(a_j)_\tau^{\sigma'}$$

By the definition of $\Vdash$, it is easy to see that

$$(\sigma, \sigma') \Vdash_\iota A \text{ entails } (\sigma, \tau') \Vdash_\iota A$$

and

$$(\sigma, \sigma') \Vdash_\iota A \text{ entails } (\tau, \sigma') \Vdash_\iota A$$

follow from these facts. □

Lemma 6.2.5. *Let A be a formula and let σ and σ' be arbitrary states. Let $p : t$ and $q : t$ be terms of programs of $Term(\Sigma_p)$ such that*

$$p \rhd_{\Sigma_p} q$$

Then, for any interpretation ι,

$$(\sigma, \sigma') \Vdash_\iota A[p/v] \text{ entails } (\sigma, \sigma') \Vdash_\iota A[q/v]$$

Proof. By induction over the form of A, using the definition of $\Vdash$ (Chapter 4, Section 4.4.3, Definition 4.4.8, p. 115).

Similarly, to the previous lemma, we only exhibit the base case, where A is a predicate over terms. The other cases follow easily.

Assume A is of the from $P(a_1, \dots, a_n)$. Then

$$\begin{aligned}(\sigma, \sigma') \Vdash_\iota A[p/v] &\Leftrightarrow (\sigma, \sigma') \Vdash_\iota P(a_1[p/v], \dots, a_n[p/v]) \\ &\Leftrightarrow h(P) \ni P(\iota(a_1[p/v])_\sigma^{\sigma'}, \dots, \iota(a_n[p/v])_\sigma^{\sigma'}) \\ &\Leftrightarrow h(P) \ni P(\iota'(a_1)_\sigma^{\sigma'}, \dots, \iota'(a_n)_\sigma^{\sigma'})\end{aligned}$$

where $\iota' = \iota[v \mapsto \iota(p)]$. Because $p \rhd_{\Sigma_p} q$, by definition of interpretations (Chapter 4, Section 4.4.3, Definition 4.4.6, p. 114), this means $\iota' = \iota[v \mapsto \iota(q)]$. So, we know

$$\begin{aligned}&h(P) \ni P(\iota'(a_1)_\sigma^{\sigma'}, \dots, \iota'(a_n)_\sigma^{\sigma'}) \\ \Leftrightarrow\ &h(P) \ni P(\iota(a_1[q/v])_\sigma^{\sigma'}, \dots, \iota(a_n[q/v])_\sigma^{\sigma'}) \\ \Leftrightarrow\ &(\sigma, \sigma') \Vdash_\iota P(a_1[q/v], \dots, a_n[q/v]) \\ \Leftrightarrow\ &(\sigma, \sigma') \Vdash_\iota A[q/v]\end{aligned}$$

as required. □

Lemma 6.2.6. *Let A be a formula that does not contain initial state identifers and let B be a formula that does not contain final state identifiers.*

Then, for any states $\sigma, \sigma', \sigma''$

$$(\sigma, \sigma') \Vdash A \text{ entails } (\sigma'', \sigma') \Vdash A$$

and

$$(\sigma, \sigma') \Vdash B \text{ entails } (\sigma, \sigma'') \Vdash B$$

Proof. The proof is straightforward, but tedious, by induction over the possible forms of A and B. □

Lemma 6.2.7. *Let* $\mathtt{p}$ *be an element of PML and* a *an element of Values (that is, of the closed, irreducible subset of* $Terms(\Sigma_p)$*, Values* $=$ $Closed(Normal(\Sigma_p))$*) such that*

$$\langle \mathtt{p}, \sigma \rangle \rhd_{IML} \langle \mathsf{a}, \sigma \rangle$$

for some state σ.

Then

$$(\sigma, \sigma) \Vdash \mathsf{tologic_i}(p) = a$$

Proof. The proof is straightforward, using the operational semantics of *IML*. □

Corollary 6.2.1. *Let* $\mathtt{p}$ *be an element of PML and* a *an element of Values such that*

$$\langle \mathtt{p}, \sigma \rangle \rhd \langle \mathsf{a}, \sigma' \rangle$$

Then, for any state σ',

$$(\sigma, \sigma') \Vdash \mathsf{tologic_i}(p) = a$$

and

$$(\sigma', \sigma) \Vdash \mathsf{tologic_f}(p) = a$$

Proof. The proof is straightforward, using interpretation of equality in our models and the definitions of $\mathsf{tologic_i}$ and $\mathsf{tologic_f}$. □

6.2.5 Extraction yields visible side-effect equivalence

We are now ready to derive one important part of our main result: that a program $\mathtt{m}$ extracted from a proof of $\mathtt{w} \diamond P$ is equivalent to $\mathtt{w}$ over the state references used by P. This is the third of the requirements given by Definition 6.1.6 for extracted programs to be IHL-realizers of program/formula pairs. We will soon derive the other two requirements.

Theorem 6.2.3 (Extraction yields visible side-effect equivalence). *Given a proof*

$$\vdash_{LTT(\mathsf{IHL})} p^{\mathtt{w} \diamond P}$$

we have that

$$[\![\mathsf{extract_{IHL}}(p)]\!] \equiv_P \mathtt{w}$$

Proof. When P is Harrop, by the definition of $\mathsf{extract_{IHL}}$,

$$\mathsf{extract}(p) = \mathtt{w}$$

and so clearly

$$[\![\mathsf{extract_{IHL}}(p)]\!] \equiv_P \mathtt{w}$$

Case: Axioms and schemata. By the assumptions given in Assumption 6.1 (p. 172), if p is of the form

$$\mathsf{IHLAxiom}(\mathtt{w} \diamond A)^{\mathtt{w} \diamond A}$$

then

$$\mathsf{extract}_{\mathsf{IHL}}(p) \equiv_A \mathtt{w}$$

as required.

Similarly, by Assumption 6.1 (p. 172), if p is of the form

$$\mathsf{IHLSchema}(N[\bar{e}])^{\mathtt{w} \diamond A}$$

then

$$\mathsf{extract}_{\mathsf{IHL}}(p) \equiv_A \mathtt{w}$$

as required.

Case: Proof ends in an application of (loop). Assume p is of the form

$$\mathsf{wd}(q)^{\mathtt{while\ b\ do\ l} \diamond A[\bar{s}_i/\bar{v}] \Rightarrow (A[\bar{s}_f/\bar{v}] \wedge \mathsf{tologic_f}(\mathtt{b}) = false)}$$

obtained by

$$\frac{q^{\mathtt{w} \diamond (\mathsf{tologic_i}(\mathtt{b}) = true \wedge A[\bar{s}_i/\bar{v}])} \vdash_{LTT(\mathsf{IHL})} q^{\mathtt{w} \diamond A[\bar{s}_f/\bar{v}]}}{\vdash_{LTT(\mathsf{IHL})} \mathsf{wd}(q)^{\mathtt{while\ b\ do\ q} \diamond A[\bar{s}_i/\bar{v}] \Rightarrow A[\bar{s}_f/\bar{v}] \wedge \mathsf{tologic_f}(\mathtt{b}) = false}}\ loop$$

Then

$$\mathsf{extract}_{\mathsf{IHL}}(p) = \begin{pmatrix} \mathtt{rv_1} := \mathtt{fn\ x} : \mathsf{etype}(\mathtt{A}) \mathtt{=>\ x;} \\ \mathtt{while\ b\ do} \\ \quad \mathtt{rv_2} := \mathsf{extract}_{\mathsf{IHL}}(q); \\ \quad \mathtt{rv_1} := \mathtt{(fn\ x_2\ =>\ fn\ x_1\ =>} \\ \quad\quad \mathtt{fn\ x} : \mathsf{etype}(\mathtt{A}) \mathtt{=>\ x_2\ (x_1\ x))} \\ \quad\quad \mathtt{!rv_2\ !rv_1;} \\ \mathtt{!rv_1;} \end{pmatrix}$$

By the IH,

$$\mathsf{extract}_{\mathsf{IHL}}(q) \equiv_{(\mathsf{tologic_i}(\mathtt{b}) = true \wedge A[\bar{s}_i/\bar{v}]) \Rightarrow A[\bar{s}_f/\bar{v}]} \mathtt{l} \tag{6.3}$$

Assume that $\bar{r}$ are all the state references used in $\mathtt{b}$, so that $\bar{r}_i$ are all the state identifiers used in $\mathsf{tologic_i}(\mathtt{b})$ and $\bar{r}_f$ are all the state identifiers used in $\mathsf{tologic_f}(\mathtt{b})$.

Observe that the state identifiers used in the formula

$$(\mathsf{tologic_i}(\mathtt{b}) = true \wedge A[\bar{s}_i/\bar{v}]) \Rightarrow A[\bar{s}_f/\bar{v}]$$

are exactly

$$\bar{r}_i :: \bar{s}_i :: \bar{s}_f$$

and the state identifiers used in the conclusion

$$A[\bar{s}_i/\bar{v}] \Rightarrow (A[\bar{s}_f/\bar{v}] \wedge \mathsf{tologic_f(b)} = false)$$

are

$$\bar{r}_f :: \bar{s}_i :: \bar{s}_f$$

Because

$$state\!-\!id(\bar{r}_i :: \bar{s}_i :: \bar{s}_f) = \bar{r} :: \bar{s} = state\!-\!id(\bar{r}_f :: \bar{s}_i :: \bar{s}_f)$$

we can use (6.3) and Definition 6.1.5 to obtain

$$\mathsf{extract_{IHL}}(q) \equiv_P \mathtt{1} \tag{6.4}$$

By the operational semantics of *IML* (Fig. 4.6 in Chapter 4, p. 108) it follows that, because $\mathtt{rv}_1$ and $\mathtt{rv}_2$ do not occur in $\mathsf{extract_{IHL}}(q)$,

$$\left(\begin{array}{l} \mathtt{rv}_2 := \mathsf{extract_{IHL}}(q); \\ \mathtt{rv}_1 := (\mathtt{fn}\ \mathtt{x}_2 \Rightarrow \mathtt{fn}\ \mathtt{x}_1 \Rightarrow \\ \quad \mathtt{fn}\ \mathtt{x} : \mathsf{etype}(\mathtt{A}) \Rightarrow \mathtt{x}_2\ (\mathtt{x}_1\ \mathtt{x})) \\ \quad !\mathtt{rv}_2\ !\mathtt{rv}_1; \end{array}\right) \equiv_P \mathtt{1}$$

Finally, again by the operational semantics of *IML*, because $\mathtt{rv}_1$ does not occur in $\mathsf{extract_{IHL}}(q)$,

$$\mathsf{extract_{IHL}}(p) \equiv_P \mathtt{while}\ \mathsf{b}\ \mathtt{do}\ \mathtt{1}$$

as required.

Case: Proof ends in an application of (seq).

Assume $p^{\mathtt{w}\diamond P}$ is of the form

$$\mathsf{seq}(q^{\mathtt{w}_1\diamond(A[\bar{s}_i/\bar{v}]\Rightarrow B[\bar{s}_f/\bar{v}])}, r^{\mathtt{w}_2\diamond(B[\bar{s}_i/\bar{v}]\Rightarrow C[\bar{s}_f/\bar{v}])})^{\mathtt{w}_1;\mathtt{w}_2\diamond(A[\bar{s}_i/\bar{v}]\Rightarrow C[\bar{s}_f/\bar{v}])}$$

obtained by application of (seq)

$$\frac{\vdash q^{\mathtt{w}_1\diamond(A[\bar{s}_i/\bar{v}]\Rightarrow B[\bar{s}_f/\bar{v}])} \qquad \vdash r^{\mathtt{w}_2\diamond(B[\bar{s}_i/\bar{v}]\Rightarrow C[\bar{s}_f/\bar{v}])}}{\vdash \mathsf{seq}(q^{\mathtt{w}_1\diamond(A[\bar{s}_i/\bar{v}]\Rightarrow B[\bar{s}_f/\bar{v}])}, r^{\mathtt{w}_2\diamond(B[\bar{s}_i/\bar{v}]\Rightarrow C[\bar{s}_f/\bar{v}])})^{\mathtt{w}_1;\mathtt{w}_2\diamond(A[\bar{s}_i/\bar{v}]\Rightarrow C[\bar{s}_f/\bar{v}])}}$$

By the IH,

$$\mathsf{extract_{IHL}}(q) \equiv_{A[\bar{s}_i/\bar{v}]\Rightarrow B[\bar{s}_f/\bar{v}]} \mathtt{w}_1 \tag{6.5}$$

Observe that, because A is free of state identifiers, the state identifiers used in the formula

$$(A[\bar{s}_i/\bar{v}] \Rightarrow B[\bar{s}_f/\bar{v}])$$

are the same as those used in the conclusion

$$(A[\bar{s}_i/\bar{v}] \Rightarrow C[\bar{s}_f/\bar{v}])$$

and so by (6.3)

$$\mathsf{extract_{IHL}}(q) \equiv_P \mathtt{w}_1 \tag{6.6}$$

By similar reasoning, we can conclude

$$\mathsf{extract}_{\mathsf{IHL}}(r) \equiv_P \mathtt{w}_2 \tag{6.7}$$

There are four cases: 1) A is not Harrop, B is not Harrop, and C is not Harrop, 2) A is Harrop, B is not Harrop, and C is not Harrop, 3) A is Harrop and B is Harrop and C is not Harrop, 4) A is Harrop and B is Harrop and C is not Harrop. (There are only four cases to deal with, because in other possibilities, the conclusion formula is Harrop and the extracted program is identical to the program of the program/formula pair and we are done).

Case 1: A is not Harrop, B is not Harrop, and C is not Harrop. Then

$$\mathsf{extract}_{\mathsf{IHL}}(p) = \begin{pmatrix} \mathtt{rv}_q := \mathsf{extract}_{\mathsf{IHL}}(q); \\ \mathtt{rv}_r := \mathsf{extract}_{\mathsf{IHL}}(r); \\ (\mathtt{fn\ x_q \Rightarrow fn\ x_r \Rightarrow} \\ \quad \mathtt{fn\ x} : \mathsf{etype}(\mathtt{A}) \mathtt{\Rightarrow x_r\ (x_q x)}) \\ \quad \mathtt{!rv}_q\ \mathtt{!rv}_r \end{pmatrix}$$

Because $\mathtt{rv}_p$ and $\mathtt{rv}_q$ do not occur in $\mathsf{extract}_{\mathsf{IHL}}(p)$ and $\mathsf{extract}_{\mathsf{IHL}}(q)$, by (6.6) and (6.7) and the operational semantics, we can conclude that

$$\mathsf{extract}_{\mathsf{IHL}}(p) \equiv_P \mathtt{w}_1; \mathtt{w}_2$$

as required.

Cases 2–4. These cases follow by similar reasoning.

Case: Proof ends in an application of (ite). Assume that $p^{\mathtt{w} \diamond P}$ is of the form

$$\mathsf{ite}(p, q)^{\mathtt{if\ b\ then\ w_1\ else\ w_2} \diamond C}$$

obtained

$$\frac{\vdash q^{\mathtt{w}_1 \diamond (\mathsf{tologic_i}(\mathtt{b}) = true \Rightarrow C)} \quad \vdash r^{\mathtt{w}_2 \diamond (\mathsf{tologic_i}(\mathtt{b}) = false \Rightarrow C)}}{\vdash \mathsf{ite}(q, r)^{\mathtt{if\ b\ then\ w_1\ else\ w_2} \diamond C}} \text{(ite)}$$

By the IH

$$\mathsf{extract}_{\mathsf{IHL}}(q) \equiv_{(\mathsf{tologic_i}(\mathtt{b}) = true \Rightarrow C)} \mathtt{w}_1 \tag{6.8}$$

and

$$\mathsf{extract}_{\mathsf{IHL}}(r) \equiv_{(\mathsf{tologic_i}(\mathtt{b}) = false \Rightarrow C)} \mathtt{w}_2 \tag{6.9}$$

Assume that $\bar{r}$ are all the state references used in $\mathtt{b}$, so that $\bar{r}_i$ are all the state identifiers used in $\mathsf{tologic_i}(\mathtt{b})$.

Assume that $\bar{s}_i :: \bar{t}_f$ are all the state identifiers used in C.

If $\bar{r} \cap state{-}id(\bar{s}_i :: \bar{t}_f) = \emptyset$, then by (6.8) and Definition 6.1.5,

$$(\sigma, \sigma') \in [\![\mathtt{w}_1]\!] \text{ entails } (\sigma, \sigma'') \in [\![\mathsf{extract}_{\mathsf{IHL}}(q)]\!]$$

where σ' and σ'' only differ over states *not* in

$$r :: state{-}id(\bar{s}_i :: \bar{r}_f)$$

We can weaken this to obtain

$$(\sigma, \sigma') \in [\![\mathtt{w}_1]\!] \text{ entails } (\sigma, \sigma'') \in [\![\mathsf{extract}_{\mathsf{IHL}}(q)]\!]$$

where σ' and σ'' only differ over states *not* in

$$state{-}id(\bar{s}_i :: \bar{r}_f)$$

And so

$$\mathsf{extract}_{\mathsf{IHL}}(q) \equiv_C \mathtt{w}_1 \tag{6.10}$$

Similar reasoning over (6.9) shows

$$\mathsf{extract}_{\mathsf{IHL}}(r) \equiv_C \mathtt{w}_2 \tag{6.11}$$

Then, by the operational semantics (Fig. 4.6 in Chapter 4, p. 108) and Definition 6.1.5, (6.10) and (6.11) yield

$$\mathtt{if\ b\ then}\ \mathsf{extract}(q)\ \mathtt{else}\ \mathsf{extract}(r)$$

as required.

Case: Proof ends in an application of (cons). Assume $p^{\mathtt{w}\diamond P}$ is of the form

$$\mathsf{cons}(p^{\mathtt{w}\diamond R}, q^{R \Rightarrow A})^{\mathtt{w}\diamond A}$$

derived by

$$\frac{\vdash r^{\mathtt{w}\diamond R} \quad \vdash_{\mathsf{Int}} q^{R \Rightarrow A}}{\vdash \mathsf{cons}(r^{\mathtt{w}\diamond R}, q^{R \Rightarrow A})^{\mathtt{w}\diamond A}}\ (\text{cons})$$

Assume

$$(\sigma, \sigma') \in [\![\mathtt{w}]\!]$$

If R and A are not Harrop, then

$$\mathsf{extract}_{\mathsf{IHL}}(r) = \begin{pmatrix} \bar{\mathtt{i}} := \bar{\mathtt{s}}; \\ \mathtt{rv}_p := \mathsf{extract}(p); \\ \bar{\mathtt{f}} := \bar{\mathtt{s}}; \\ (\mathtt{fn}\ \bar{\mathtt{s}}_{\mathtt{i}} :: \bar{\mathtt{s}}_{\mathtt{f}} \Rightarrow \\ \quad \mathtt{fn}\ \mathtt{x}_{\mathtt{v}} : \mathsf{etype}(\mathtt{R}) \Rightarrow \mathsf{extract}_{\mathsf{Int}}(q)) \\ \quad !\mathtt{rv}_p\ !\bar{\mathtt{i}} :: \bar{\mathtt{f}} \end{pmatrix}$$

By the operational semantics for $\mathsf{extract}_{\mathsf{IHL}}(r)$, if

$$(\sigma, \sigma'') \in [\![\mathsf{extract}_{\mathsf{IHL}}(r)]\!]$$

then σ'' and σ differ only over the references, $\bar{\mathtt{i}}$, $\mathtt{rv}_p$ and $\bar{\mathtt{f}}$. But, because we always assume that state identifiers corresponding to these references cannot occur in R or A, we then have

$$\mathsf{extract}_{\mathsf{IHL}}(r) \equiv_A \mathtt{w}$$

as required.

If P is Harrop and A is not Harrop, then

$$\mathsf{extract}_{\mathsf{IHL}}(r) = \begin{pmatrix} \bar{\mathtt{i}} := \bar{\mathtt{s}}; \\ \mathtt{w}; \\ \bar{\mathtt{f}} := \bar{\mathtt{s}}; \\ (\mathtt{fn}\ \bar{\mathtt{s}}_{\mathtt{i}} :: \bar{\mathtt{s}}_{\mathtt{f}}\ \mathtt{=>}\ \mathsf{extract}_{\mathsf{Int}}(q)) \\ \quad !\bar{\mathtt{i}} :: \bar{\mathtt{f}} \end{pmatrix}$$

By the operational semantics for $\mathsf{extract}_{\mathsf{IHL}}(r)$, if

$$(\sigma, \sigma'') \in [\![\mathsf{extract}_{\mathsf{IHL}}(r)]\!]$$

then σ'' and σ differ only over the references, $\bar{\mathtt{i}}$ and $\bar{\mathtt{f}}$. But, because we always assume that state identifiers corresponding to these references cannot occur in R or A, we then have

$$\mathsf{extract}_{\mathsf{IHL}}(r) \equiv_A \mathtt{w}$$

as required.

This last case concludes the proof. □

6.2.6 Extraction results

We are now ready to show that we can extract correct programs with return values from IHL proofs. Theorems 6.2.4 and 6.2.5, proved below, together with Theorem 6.2.3 above tell us that any proof

$$\vdash_{LTT(\mathsf{IHL})} p^{\mathtt{w} \diamond P}$$

can be transformed into an IHL-realizer,

$$\mathsf{extract}_{\mathsf{IHL}}(p)\ \mathsf{kr}\ P$$

Part of the proof of this theorem was presented in [PC03], but this is the first full presentation.

Theorem 6.2.4 (Extraction produces programs that satisfy proved formulae). *Given a proof*

$$\vdash_{LTT(\mathsf{IHL})} p^{\mathtt{w} \diamond P}$$

then

$$[\![\mathsf{extract}_{\mathsf{IHL}}(p)]\!] \Vdash P$$

Proof. When P is Harrop, by the definition of $\mathsf{extract}_{\mathsf{IHL}}$,

$$\mathsf{extract}(p) = \mathtt{w}$$

and so

$$[\![\mathsf{extract}_{\mathsf{IHL}}(p)]\!] \Vdash P$$

follows from soundness (Theorem 5.1.1, Chapter 5, p. 140).

When P is not Harrop, we proceed as follows.

Let $\bar{s}_i :: \bar{r}_f$ be all the state identifiers used in P.

By Theorem 6.2.3,

$$\mathsf{extract}_{\mathsf{IHL}}(p) \equiv_P \mathtt{w}$$

This means that

$$(\sigma, \sigma') \in [\![\mathsf{extract}_{\mathsf{IHL}}(p)]\!] \text{ entails } (\sigma, \sigma'') \in [\![\mathtt{w}]\!] \tag{6.12}$$

only when σ' and σ'' differ over states *not* in

$$state\!-\!id(\bar{s}_i :: \bar{r}_f)$$

By soundness (Theorem 5.1.1, Chapter 5, p. 140),

$$[\![\mathtt{w}]\!] \Vdash P \tag{6.13}$$

We apply Lemma 6.2.4 to (6.13) and (6.12) to obtain

$$[\![\mathsf{extract}_{\mathsf{IHL}}(p)]\!] \Vdash P$$

as required. □

This theorem proves another requirement of our programs in order to be IHL-realizers of the theorems they are extracted from.

The next theorem is the last requirement of our main result — it tells us that extracted programs result in correct return value realizers.

Theorem 6.2.5 (Program extraction produces return value realizers). *Take any proof*

$$\vdash_{LTT(\mathsf{IHL})} t^{\mathtt{w} \diamond T}$$

Let ι be any interpretation.

Then

$$\mathsf{extract}_{\mathsf{IHL}}(t) \;\; \mathsf{retr}_\iota \; T$$

Proof. To prove this, we proceed by induction over the form of T.

We use the following induction hypothesis:

Take any proof

$$\vdash_{LTT(\mathsf{IHL})} t^{\mathtt{w} \diamond T}$$

Take any pair of states (σ, σ') such that $\mathsf{extract}_{\mathsf{IHL}}(t)$ terminates with an execution sequence of the form

$$\langle \mathsf{extract}_{\mathsf{IHL}}(t), \sigma \rangle \;\hat{\triangleright}\; \langle \mathtt{answer}, \sigma' \rangle \tag{6.14}$$

yielding a return value `answer`. Then

$$(\sigma, \sigma') \Vdash_\iota Sk(T)[answer/f_T]$$

Observe that answer has a representation $answer = \mathsf{tologic}(\mathtt{answer})$ in *Values*.

Case 1: T is Harrop. In this case, by the definition of Skolem form, we are required to prove that, if

$$\langle \mathtt{w}, \sigma \rangle \hat{\triangleright} \langle \mathtt{answer}, \sigma' \rangle$$

then

$$(\sigma, \sigma') \Vdash_\iota T \tag{6.15}$$

But this is the case by soundness (Theorem 5.1.1).

Case 2: Proof ends in an application of (loop). Assume that $t^{\mathtt{w} \diamond T}$ is of the form

$$\mathsf{wd}(q)^{\mathtt{while\ b\ do\ l} \diamond A[\bar{s}_i/\bar{v}] \Rightarrow (A[\bar{s}_f/\bar{v}] \wedge \mathsf{tologic_f}(\mathtt{b}) = false)}$$

By the IH, we know that

$$\mathsf{extract_{IHL}}(q) \ \mathsf{retr}_\iota \ A[\bar{s}_i/\bar{v}] \wedge \mathsf{tologic_i}(b) = true \Rightarrow A[\bar{s}_f/\bar{v}] \tag{6.16}$$

This means that, for any τ, τ' and pure program value $\mathtt{answer}_\tau$, if

$$\langle \mathsf{extract_{IHL}}(q), \tau \rangle \hat{\triangleright} \langle \mathtt{answer}_\tau, \tau' \rangle$$

we know that, for $answer_\tau = \mathsf{tologic}(\mathtt{answer}_\tau)$,

$$(\tau, \tau') \Vdash Sk(A[\bar{s}_i/\bar{v}] \wedge \mathsf{tologic_i}(b) = true \Rightarrow A[\bar{s}_f/\bar{v}])$$
$$[answer_\tau / f_{A[\bar{s}_i/\bar{v}] \wedge \mathsf{tologic_i}(b) = true \Rightarrow A[\bar{s}_f/\bar{v}]}] \tag{6.17}$$

There are two cases, depending on whether A is Harrop or not. We only deal with the latter case, as the approach is similar for the former case.

We wish to show that answer is such that

$$(\sigma, \sigma') \Vdash_\iota Sk(A[\bar{s}_i/\bar{v}] \Rightarrow (A[\bar{s}_f/\bar{v}] \wedge \mathsf{tologic_f}(\mathtt{b}) = false))[answer/f_P] \tag{6.18}$$

As $A[\bar{s}_i/\bar{v}]$ is not Harrop, and $\mathsf{extract_{IHL}}(t)$ is

```
rv1 := fn x : etype(A) => x;
while b do
   rv2 := extractIHL(q);
   rv1 := (fn x2 => fn x1 => fn x : etype(A) => x2(x1x)) !rv2 !rv1;
!rv1
```

By the definition of Skolem form and the fact that $A[\bar{s}_i/\bar{v}]$ and $A[\bar{s}_f/\bar{v}]$ are not Harrop, the required statement (6.18) may be rewritten as

$$(\sigma, \sigma') \Vdash_\iota \forall x : \mathsf{etype}(A[\bar{s}_i/\bar{v}]) \bullet Sk(A)[\bar{s}_i/\bar{v}][x/f_{A[\bar{s}_i/\bar{v}]}] \Rightarrow$$
$$(Sk(A)[\bar{s}_f/\bar{v}][answer\ x/f_{A[\bar{s}_f/\bar{v}]}] \wedge \mathsf{tologic_f}(\mathtt{b}) = false) \tag{6.19}$$

First we make some observations about the execution of the extracted program.

Beginning of observations.

Because we know that $\mathsf{extract_{IHL}}(t)$ terminates, by the definition of $\rhd$, the program must have an execution sequence that results in states

$$\sigma = \sigma_0, \sigma_1, \ldots, \sigma_n = \sigma'$$

where

$$\begin{array}{l}
\left\langle \left(\begin{array}{l} \mathtt{rv}_1 := \mathtt{fn\ x} : \mathsf{etype}(\mathtt{A}) \mathtt{=>}\ \mathtt{x}; \\ \mathtt{while\ b\ do} \\ \quad \mathtt{rv}_2 := \mathsf{extract_{IHL}}(q); \\ \quad \mathtt{rv}_1 := (\mathtt{fn\ x_2 =>} \\ \quad \mathtt{fn\ x_1 => fn\ x} : \mathsf{etype}(\mathtt{A}) \mathtt{=>}\ \mathtt{x_2(x_1x)})\ \mathtt{!rv_2\ !rv_1}; \\ \mathtt{!rv_1} \end{array} \right), \sigma_0 \right\rangle \\
\rhd \left\langle \left(\begin{array}{l} \mathtt{while\ b\ do} \\ \quad \mathtt{rv}_2 := \mathsf{extract_{IHL}}(q); \\ \quad \mathtt{rv}_1 := (\mathtt{fn\ x_2 =>} \\ \quad \mathtt{fn\ x_1 => fn\ x} : \mathsf{etype}(\mathtt{A}) \mathtt{=>}\ \mathtt{x_2(x_1x)})\ \mathtt{!rv_2\ !rv_1}; \\ \mathtt{!rv_1} \end{array} \right), \sigma_1 \right\rangle \\
\hat{\rhd}\ \langle \mathtt{!rv_1}, \sigma_n \rangle
\end{array} \tag{6.20}$$

so $\mathtt{answer} = \sigma_n(\mathtt{rv}_1)$ and

$$\begin{array}{l}
\left\langle \left(\begin{array}{l} \mathtt{rv}_2 := \mathsf{extract_{IHL}}(q); \\ \mathtt{rv}_1 := (\mathtt{fn\ x_2 => fn\ x_1 =>} \\ \quad \mathtt{fn\ x} : \mathsf{etype}(\mathtt{A}) \mathtt{=>}\ x_2(\mathtt{x}_1 x))\ \mathtt{!rv_2\ !rv_1} \end{array} \right), \sigma_1 \right\rangle \rhd \langle \mathtt{!rv_1}, \sigma_2 \rangle \\
\left\langle \left(\begin{array}{l} \mathtt{rv}_2 := \mathsf{extract_{IHL}}(q); \\ \mathtt{rv}_1 := (\mathtt{fn\ x_2 => fn\ x_1 =>} \\ \quad \mathtt{fn\ x} : \mathsf{etype}(\mathtt{A}) \mathtt{=>}\ x_2(\mathtt{x}_1 x))\ \mathtt{!rv_2\ !rv_1} \end{array} \right), \sigma_2 \right\rangle \rhd \langle \mathtt{!rv_1}, \sigma_3 \rangle \\
\text{and} \\
\ldots \\
\text{and} \\
\left\langle \left(\begin{array}{l} \mathtt{rv}_2 := \mathsf{extract_{IHL}}(q); \\ \mathtt{rv}_1 := (\mathtt{fn\ x_2 => fn\ x_1 =>} \\ \quad \mathtt{fn\ x} : \mathsf{etype}(\mathtt{A}) \mathtt{=>}\ x_2(\mathtt{x}_1 x))\ \mathtt{!rv_2\ !rv_1} \end{array} \right), \sigma_{n-1} \right\rangle \rhd \langle \mathtt{!rv_1}, \sigma_n \rangle
\end{array} \tag{6.21}$$

with

$$\langle \mathsf{b}, \sigma_i \rangle \rhd \langle \mathtt{true}, \sigma_i \rangle \tag{6.22}$$

$(i = 1, \ldots, n-1)$ and

$$\langle \mathsf{b}, \sigma_n \rangle \rhd \langle \mathtt{false}, \sigma_n \rangle \tag{6.23}$$

Observe that (6.22) and Corollary 6.2.1 entail

$$(\sigma_i, \sigma_{i+1}) \Vdash \mathsf{tologic_i}(\mathsf{b}) = \mathit{true} \tag{6.24}$$

for $i = 1, \ldots, n-1$. Similarly, (6.23) and Corollary 6.2.1 tell us that, for any state τ

$$(\tau, \sigma_n) \Vdash \mathsf{tologic_f}(\mathsf{b}) = \mathit{false} \tag{6.25}$$

Let σ_i'' $(i = 1, \ldots, n-1)$ denote the state such that

$$\langle \mathsf{extract}_{\mathsf{IHL}}(q), \sigma_i\rangle \hat{\rhd} \langle \mathtt{answer}_{\sigma_i}, \sigma_i''\rangle \tag{6.26}$$

for some return value $\mathtt{answer}_{\sigma_i}$. Let σ_i' ($i = 1, \ldots, n-1$) denote the state such that

$$\langle \mathtt{rv}_1 := \mathsf{extract}_{\mathsf{IHL}}(q), \sigma_i\rangle \hat{\rhd} \langle \mathtt{answer}_{\sigma_i}, \sigma_i'\rangle \tag{6.27}$$

so that

$$\begin{aligned}\langle \mathtt{rv}_2 := \mathsf{extract}_{\mathsf{IHL}}(q), \sigma_i\rangle &\rhd \langle !\mathtt{rv}_2, \sigma_i'\rangle \\ &\hat{\rhd} \langle \mathtt{answer}_{\sigma_i}, \sigma_i'\rangle\end{aligned}$$

The last line holds by definition of the operational semantics for assignment (rule (assign) of Fig. 4.6 in Chapter 4, p. 108) and by (6.26), because it must be the case that

$$\sigma_i'(\mathtt{rv}_2) = \mathtt{answer}_{\sigma_i} \tag{6.28}$$

Observe that σ_i'' and σ_i' differ only over $\mathtt{rv}_2$. Also, by the operational semantics for assignment (rule (assign) of Fig. 4.6 in Chapter 4, p. 108), σ_i' and σ_{i+1} must differ only over $\mathtt{rv}_1$ in

$$\begin{aligned}&\langle \mathtt{rv}_1 := (\mathtt{fn\ x}_2 \Rightarrow \mathtt{fn\ x}_1 \Rightarrow \mathtt{fn\ x} : \mathsf{etype}(\mathtt{A}) \Rightarrow \mathtt{x}_2(\mathtt{rv}_1\mathtt{x})) \\ &\qquad\qquad !\mathtt{rv}_2\ !\mathtt{rv}_1, \sigma_i'\rangle \rhd \langle !\mathtt{rv}_1, \sigma_{i+1}\rangle\end{aligned}$$

Thus, it must be the case that

σ_i'' and σ_{i+1} differ only over $\mathtt{rv}_1$ and $\mathtt{rv}_2$ in

$$\left\langle \begin{pmatrix} \mathtt{rv}_2 := \mathsf{extract}_{\mathsf{IHL}}(q); \\ \mathtt{rv}_1 := (\mathtt{fn\ x}_2 \Rightarrow \\ \mathtt{fn\ x}_1 \Rightarrow \mathtt{fn\ x} : \mathsf{etype}(\mathtt{A}) \Rightarrow \mathtt{x}_2(\mathtt{x}_1\mathtt{x}))\ !\mathtt{rv}_2\ !\mathtt{rv}_1 \end{pmatrix}, \sigma_i \right\rangle$$
$$\rhd \langle \mathtt{rv}_1, \sigma_{i+1}\rangle \tag{6.29}$$

for $i = 1, \ldots, n-1$.

By inspection of the evaluation sequence (6.20),

$$\sigma_1(\mathtt{rv}_1) = \mathtt{fn\ x} : \mathsf{etype}(\mathtt{A}) \Rightarrow \mathtt{x} \tag{6.30}$$

Also, because $\mathtt{rv}_1$ does not occur in $\mathsf{extract}_{\mathsf{IHL}}(q)$, the execution of $\mathsf{extract}_{\mathsf{IHL}}(q)$ from σ_i to σ_i' will not affect the value of $\mathtt{rv}_1$: that is, $\sigma_i'(\mathtt{rv}_1) = \sigma_i(\mathtt{rv}_1)$. So, by inspection of the evaluation sequence (6.21):

$\sigma_{i+1}(\mathtt{rv}_1)$ is the normal form of $\mathtt{fn\ x} : \mathsf{etype}(\mathtt{A}) \Rightarrow \sigma_{\mathtt{i}}'(\mathtt{rv}_2)(\sigma_{\mathtt{i}}(\mathtt{rv}_1)\mathtt{x})$

($i = 1, \ldots, n-1$). (see Definition 4.1.3 of Chapter 4, p. 99 for the definition of normal form). That is to say,

$\sigma_{i+1}(\mathtt{rv}_1)$ is the normal form of $\mathtt{fn\ x} : \mathsf{etype}(\mathtt{A}) \Rightarrow \mathtt{answer}_{\sigma_{\mathtt{i}}}(\sigma_{\mathtt{i}}(\mathtt{rv}_1)\mathtt{x})$ (6.31)

($i = 1, \ldots, n-1$).

So, because $\texttt{answer} = \sigma_n(\texttt{rv}_1)$, when $n > 1$, $\texttt{answer}$ must be the normal form of

$$\texttt{fn x} : \mathsf{etype}(\texttt{A}) \texttt{=>} (\texttt{answer}_{\sigma_{n-1}}(\texttt{fn x} : \mathsf{etype}(\texttt{A}) \texttt{=>} \\ \texttt{answer}_{\sigma_{n-1}}(\ldots \texttt{answer}_{\sigma_1}(\texttt{fn x} : \mathsf{etype}(\texttt{A}) \texttt{=>} x\texttt{x})\ldots)\texttt{x})\texttt{x})$$

That is, if $n > 1$

$$\texttt{answer} = \texttt{fn x} : \mathsf{etype}(\texttt{A}) \texttt{=>} \texttt{answer}_{\sigma_{n-1}}(\texttt{answer}_{\sigma_{n-2}} \ldots (\texttt{answer}_{\sigma_1}\texttt{x})\ldots) \tag{6.32}$$

Also, by (6.30), when $n = 1$ (that is, when $\sigma' = \sigma_1$),

$$\texttt{answer} = \texttt{fn x} : \mathsf{etype}(\texttt{A}) \texttt{=>} \texttt{x} \tag{6.33}$$

Take arbitrary τ, τ'' such that

$$\langle \mathsf{extract_{IHL}}(q), \tau\rangle \mathrel{\hat{\triangleright}} \langle \texttt{answer}_\tau, \tau''\rangle$$

By (6.16) and (6.17), the definition of Skolem form and the fact that $A[\bar{s}_i/\bar{v}]$ and $A[\bar{s}_f/\bar{v}]$ are not Harrop,

$$(\tau, \tau'') \Vdash_\iota \forall x : \mathsf{etype}(A[\bar{s}_i/\bar{v}]) \bullet Sk(A[\bar{s}_i/\bar{v}])[x/f_{A[\bar{s}_i/\bar{v}]}] \\ \wedge \mathsf{tologic_i}(b) = \mathit{true} \Rightarrow A[\bar{s}_f/\bar{v}][\mathit{answer}_\tau\ x/f_{A[\bar{s}_f/\bar{v}]}] \tag{6.34}$$

Recall that $\texttt{rv}_1$ and $\texttt{rv}_2$ do not occur in

$$\forall x : \mathsf{etype}(A[\bar{s}_i/\bar{v}]) \bullet Sk(A[\bar{s}_i/\bar{v}])[x/f_{A[\bar{s}_i/\bar{v}]}] \wedge \mathsf{tologic_i}(b) = \mathit{true} \Rightarrow \\ A[\bar{s}_f/\bar{v}][\mathit{answer}_\tau\ x/f_{A[\bar{s}_f/\bar{v}]}]$$

So, by Lemma 6.2.4, for any τ' that differs from τ'' only over state variables $\texttt{rv}_1$ and $\texttt{rv}_2$, it must be the case that

$$(\tau, \tau') \Vdash_\iota \forall x : \mathsf{etype}(A[\bar{s}_i/\bar{v}]) \bullet Sk(A[\bar{s}_i/\bar{v}])[x/f_{A[\bar{s}_i/\bar{v}]}] \\ \wedge \mathsf{tologic_i}(b) = \mathit{true} \Rightarrow A[\bar{s}_f/\bar{v}][\mathit{answer}_\tau\ x/f_{A[\bar{s}_f/\bar{v}]}] \tag{6.35}$$

Also, recall (6.29): that σ_i'' and σ_{i+1} differ only over $\texttt{rv}_1$ and $\texttt{rv}_2$ in

$$\left\langle \begin{pmatrix} \texttt{rv}_2 := \mathsf{extract_{IHL}}(q); \\ \texttt{rv}_1 := (\texttt{fn x}_2 \texttt{=>} \\ \texttt{fn x}_1 \texttt{=>} \texttt{fn x} : \mathsf{etype}(\texttt{A}) \texttt{=>} \texttt{x}_2(\texttt{x}_1\texttt{x}))\ !\texttt{rv}_2\ !\texttt{rv}_1 \end{pmatrix}, \sigma_i \right\rangle \\ \triangleright \langle \texttt{rv}_1, \sigma_{i+1}\rangle$$

for $i = 1, \ldots, n-1$.

This fact and (6.35) mean that

$$(\sigma_i, \sigma_{i+1}) \Vdash_\iota \forall x : \mathsf{etype}(A[\bar{s}_i/\bar{v}]) \bullet Sk(A[\bar{s}_i/\bar{v}])[x/f_{A[\bar{s}_i/\bar{v}]}] \wedge \mathsf{tologic_i}(b) = true \Rightarrow A[\bar{s}_f/\bar{v}][answer_{\sigma_i}\ x/f_{A[\bar{s}_f/\bar{v}]}] \quad (6.36)$$

for $i = 1, \ldots, n-1$.

End of observations.

We wish to show (6.19):

$$(\sigma, \sigma') \Vdash_\iota \begin{array}{l} \forall x : \mathsf{etype}(A[\bar{s}_i/\bar{v}]) \bullet Sk(A)[\bar{s}_i/\bar{v}][x/f_{A[\bar{s}_i/\bar{v}]}] \Rightarrow \\ (Sk(A)[\bar{s}_f/\bar{v}][answer\ x/f_{A[\bar{s}_f/\bar{v}]}] \wedge \mathsf{tologic_f(b)} = false) \end{array}$$

To do this, we take an arbitrary x : $\mathsf{etype}(A[\bar{s}_i/\bar{v}])$-variant ι' of ι with the assumption

$$(\sigma, \sigma') \Vdash_{\iota'} Sk(A)[\bar{s}_i/\bar{v}][x/f_{A[\bar{s}_i/\bar{v}]}] \quad (6.37)$$

and we prove

$$(\sigma, \sigma') \Vdash_{\iota'} Sk(A)[\bar{s}_f/\bar{v}][answer\ x/f_{A[\bar{s}_f/\bar{v}]}] \quad (6.38)$$

and

$$(\sigma, \sigma') \Vdash_{\iota'} \mathsf{tologic_f(b)} = false \quad (6.39)$$

Proof of (6.39).

By (6.25),

$$(\tau, \sigma_n) \Vdash_\iota \mathsf{tologic_f(b)} = false$$

for any τ. So, in particular,

$$(\sigma_0, \sigma_n) \Vdash_{\iota'} \mathsf{tologic_f(b)} = false$$

which is the same as writing (6.38), as required.

End of proof of (6.39).

Proof of (6.38)

There are two subcases:

1. $\sigma = \sigma_0$ and $\sigma' = \sigma_1$ $(n = 1)$.
2. $\sigma = \sigma_0$ and $\sigma' = \sigma_n$ $(n > 1)$.

Subcase (1). In this case, by (6.33),

$$\texttt{answer} = \texttt{fn x} : \mathsf{etype}(\texttt{A}) \texttt{ => x}$$

and so,

$$\texttt{answer x} \rhd_{\Sigma_p} \texttt{x}$$

Then, by Lemma 6.2.5, (6.37) may be rewritten as

$$(\sigma_0, \sigma_1) \Vdash_{\iota'} Sk(A)[\bar{s}_i/\bar{v}][answer\ x/f_{A[\bar{s}_i/\bar{v}]}] \quad (6.40)$$

Now, $\sigma_0 = \sigma$ and $\sigma_1 = \sigma'$ only differ over $\texttt{rv}_1$, which does not occur in

$$Sk(A)[\bar{s}_i/\bar{v}][answer\ x/f_{A[\bar{s}_i/\bar{v}]}]$$

we may use Lemma 6.2.3 on (6.40) to give

$$(\sigma, \sigma') \Vdash_{\iota'} Sk(A)[\bar{s}_f/\bar{v}][answer\ x/f_{A[\bar{s}_f/\bar{v}]}] \tag{6.41}$$

Subcase (2). If $\sigma' = \sigma_n$ for $n > 1$, we proceed as follows.
Define

$$\begin{aligned} \mathtt{a}_1 &= x \\ \mathtt{a}_k &= \mathtt{answer}_{\sigma_{k-1}}(\mathtt{a}_k) \end{aligned}$$

for $k = 2, \ldots, n$. As usual, we take a_i to be defined as $\mathsf{tologic}(\mathtt{a}_i)$.

It will be important to note that, as $\mathtt{answer}_{\sigma_i}$ is state-free, it is the case that each $\mathtt{a}_k$ is also state-free. Consequently, the only state references in

$$Sk(A)[\bar{s}_f/\bar{v}][a_j/f_{A[\bar{s}_f/\bar{v}]}]$$

are $\bar{s}_f$.

By expanding the definition of $\mathtt{a}_n$, we obtain

$$\mathtt{a}_n = \mathtt{answer}_{\sigma_{n-1}}(\mathtt{answer}_{\sigma_{n-2}} \ldots (\mathtt{answer}_{\sigma_1}\mathtt{x}) \ldots)$$

We will next show, for any $j = 2, \ldots, n-1$

$$(\sigma_j, \sigma_{j+1}) \Vdash_{\iota'} Sk(A)[\bar{s}_f/\bar{v}][a_{j+1}/f_{A[\bar{s}_f/\bar{v}]}] \tag{6.42}$$

We proceed by induction.

Base case. First, note that (6.37) can be written as

$$(\sigma_0, \sigma_n) \Vdash_{\iota'} Sk(A)[\bar{s}_i/\bar{v}][x/f_{A[\bar{s}_i/\bar{v}]}]$$

But, because σ_0 and σ_1 differ only over $\mathtt{rv}_1$, which does not occur in $Sk(A)[\bar{s}_i/\bar{v}][x/f_{A[\bar{s}_i/\bar{v}]}]$, by Lemma 6.2.4, we know this means

$$(\sigma_1, \sigma_n) \Vdash_{\iota'} Sk(A)[\bar{s}_i/\bar{v}][x/f_{A[\bar{s}_i/\bar{v}]}]$$

Also, because final states are not used in $Sk(A)[\bar{s}_i/\bar{v}][x/f_{A[\bar{s}_i/\bar{v}]}]$, by Lemma 6.2.6, we know that

$$(\sigma_1, \sigma_2) \Vdash_{\iota'} Sk(A)[\bar{s}_i/\bar{v}][x/f_{A[\bar{s}_i/\bar{v}]}] \tag{6.43}$$

So, we can instantiate (6.35) with (6.43) and (6.24 with $i = 1$), to give

$$(\sigma_1, \sigma_2) \Vdash_{\iota'} A[\bar{s}_f/\bar{v}][answer_{\sigma_1}\ x/f_{A[\bar{s}_f/\bar{v}]}]$$

and we are done.

Inductive step. Assume that

$$(\sigma_k, \sigma_{k+1}) \Vdash_{\iota'} Sk(A)[\bar{s}_f/\bar{v}][a_{k+1}/f_{A[\bar{s}_f/\bar{v}]}]$$

holds for some $k < n-2$.

Because no initial state references occur in $Sk(A)[\bar{s}_f/\bar{v}]$, by Lemma 6.2.2, this means

$$(\sigma_{k+1}, \sigma_{k+2}) \Vdash_{\iota'} Sk(A[\bar{s}_i/\bar{v}])[\bar{s}_i/\bar{v}][a_{k+1}/f_{A[\bar{s}_i/\bar{v}]}] \quad (6.44)$$

We can instantiate (6.35) setting $\tau = \sigma_{k+1}$ and $\tau' = \sigma_{k+2}$ and with (6.44) and (6.24) setting $i = k+1$ we obtain

$$(\sigma_{k+1}, \sigma_{k+2}) \Vdash_{\iota'} A[\bar{s}_f/\bar{v}][answer_{\sigma_{k+1}}\, a_k/f_{A[\bar{s}_f/\bar{v}]}]$$

which means

$$(\sigma_{k+1}, \sigma_{k+2}) \Vdash_{\iota'} A[\bar{s}_f/\bar{v}][a_{k+2}/f_{A[\bar{s}_f/\bar{v}]}]$$

as required and (6.42) is proven.

So, by (6.42), we know, in particular, that

$$(\sigma_{n-1}, \sigma_n) \Vdash_{\iota'} Sk(A)[\bar{s}_f/\bar{v}][a_n/f_{A[\bar{s}_f/\bar{v}]}]$$

Now, because initial state references do not occur in $Sk(A[\bar{s}_f/\bar{v}])$, by Lemma 6.2.6 this means that

$$(\sigma_0, \sigma_n) \Vdash_{\iota'} Sk(A)[\bar{s}_f/\bar{v}][a_n/f_{A[\bar{s}_f/\bar{v}]}]$$

Also, because $n > 1$, (6.32) must hold, that is,

$$\texttt{answer} = \texttt{fn x : etype(A) =>}$$
$$\texttt{answer}_{(\sigma_{n-1}, \sigma n-1')}(\texttt{answer}_{(\sigma_{n-2}, \sigma n-2')} \ldots (\texttt{answer}_{(\sigma_1, \sigma 1')} x) \ldots)$$

we know that

$$\texttt{answer}\ x \rhd_{\Sigma_p} a_n$$

and by Lemma 6.2.5,

$$(\sigma_0, \sigma_n) \Vdash_{\iota'} Sk(A)[\bar{s}_f/\bar{v}][answer\ x/f_{A[\bar{s}_f/\bar{v}]}]$$

End of proof of (6.38). Finally, by the definition of $\Vdash$, because we took an arbitrary ι', we have

$$(\sigma, \sigma') \Vdash_{\iota} \forall x : \mathsf{etype}(A[\bar{s}_i/\bar{v}]) \bullet Sk(A)[\bar{s}_i/\bar{v}][x/f_{A[\bar{s}_i/\bar{v}]}] \Rightarrow$$
$$(Sk(A)[\bar{s}_f/\bar{v}][answerx/f_{A[\bar{s}_f/\bar{v}]}] \wedge \mathsf{tologic_f(b)} = false)$$

Case 3: Proof ends in an application of (seq). Assume $t^{\mathtt{w} \diamond T}$ is of the form

$$\mathsf{seq}(p^{\mathtt{w}_1 \diamond A[\bar{s}_i/\bar{v}] \Rightarrow B[\bar{s}_f/\bar{v}]}, q^{\mathtt{w}_2 \diamond B[\bar{s}_i/\bar{v}] \Rightarrow C[\bar{s}_f/\bar{v}]})^{\mathtt{w}_1;\mathtt{w}_2 \diamond A[\bar{s}_i/\bar{v}] \Rightarrow C[\bar{s}_f/\bar{v}]}$$

derived by

$$\frac{\vdash p^{\mathtt{w}_1 \diamond A[\bar{s}_i/\bar{v}] \Rightarrow B[\bar{s}_f/\bar{v}]} \qquad \vdash q^{\mathtt{w}_2 \diamond B[\bar{s}_i/\bar{v}] \Rightarrow C[\bar{s}_f/\bar{v}]}}{\vdash \mathsf{seq}(p^{\mathtt{w}_1 \diamond A[\bar{s}_i/\bar{v}] \Rightarrow B[\bar{s}_f/\bar{v}]}, q^{\mathtt{w}_2 \diamond B[\bar{s}_i/\bar{v}] \Rightarrow C[\bar{s}_f/\bar{v}]})^{\mathtt{w}_1;\mathtt{w}_2 \diamond A[\bar{s}_i/\bar{v}] \Rightarrow C[\bar{s}_f/\bar{v}]}} \ (\text{seq})$$

By the IH, we know that

$$\mathsf{extract}_{\mathsf{IHL}}(p)\ \mathsf{retr}_\iota\ (A[\bar{s}_i/\bar{v}] \Rightarrow B[\bar{s}_f/\bar{v}]) \tag{6.45}$$

and

$$\mathsf{extract}_{\mathsf{IHL}}(p)\ \mathsf{retr}_\iota\ (B[\bar{s}_i/\bar{v}] \Rightarrow C[\bar{s}_f/\bar{v}]) \tag{6.46}$$

There are four cases: 3a) A is not Harrop, B is not Harrop, and C is not Harrop, 3b) A is Harrop, B is not Harrop, and C is not Harrop, 3c) A is Harrop and B is Harrop and C is not Harrop, 3d) A is Harrop and B is Harrop and C is not Harrop. (All other possibilities involve the conclusion formula being Harrop, and so are dealt with already.)

Case 3a: A is not Harrop, B is not Harrop, and C is not Harrop. In this case, by the definition of Skolem form, and because $\mathsf{etype}(A) = \mathsf{etype}(A[\bar{s}_i/\bar{v}])$, we are required to prove

$$(\sigma, \sigma') \Vdash_\iota \forall x : \mathsf{etype}(A) \bullet Sk(A[\bar{s}_i/\bar{v}])[x/f_{A[\bar{s}_i/\bar{v}]}] \Rightarrow Sk(C[\bar{s}_f/\bar{v}])[(answer\ x)/f_{C[\bar{s}_f/\bar{v}]}] \tag{6.47}$$

To show this, we pick an arbitrary $x : \mathsf{etype}(A)$-variant of ι, ι', with the following assumption

$$(\sigma, \sigma') \Vdash_{\iota'} Sk(A[\bar{s}_i/\bar{v}])[x/f_{A[\bar{s}_i/\bar{v}]}] \tag{6.48}$$

and wish to derive

$$(\sigma, \sigma') \Vdash_{\iota'} Sk(C[\bar{s}_f/\bar{v}])[(answer\ x)/f_{C[\bar{s}_f/\bar{v}]}] \tag{6.49}$$

We require the following observations.

Beginning of observations for Case 3a.

By (6.45), given any τ, τ' and side-effect-free program value a_1, if

$$\langle \mathsf{extract}_{\mathsf{IHL}}(p), \tau\rangle \mathrel{\hat{\triangleright}} \langle \mathsf{a}_1, \tau'\rangle$$

then we know that, for $a_1 = \mathsf{tologic}(\mathsf{a}_1)$,

$$(\tau, \tau') \Vdash \forall x : \mathsf{etype}(A[\bar{s}_i/\bar{v}]) \bullet Sk(A[\bar{s}_i/\bar{v}])[x/f_{A[\bar{s}_i/\bar{v}]}] \Rightarrow Sk(B[\bar{s}_f/\bar{v}])[(a_1\ x)/f_{B[\bar{s}_f/\bar{v}]}] \tag{6.50}$$

Similarly, (6.46) means that, given any τ, τ' and side-effect-free program value a_2, if

$$\langle \mathsf{extract}_{\mathsf{IHL}}(q), \tau\rangle \mathrel{\hat{\triangleright}} \langle \mathtt{answer}_\tau, \tau'\rangle$$

then we know that, for $a_2 = \mathsf{tologic}(\mathsf{a}_2)$,

$$(\tau, \tau') \Vdash \forall x : \mathsf{etype}(B[\bar{s}_i/\bar{v}]) \bullet Sk(B[\bar{s}_i/\bar{v}])[x/f_{A[\bar{s}_i/\bar{v}]}] \Rightarrow Sk(C[\bar{s}_f/\bar{v}])[(a_2\ x)/f_{B[\bar{s}_f/\bar{v}]}] \tag{6.51}$$

In this case,

$$\mathsf{extract_{IHL}}(t) \text{ is } \begin{array}{l} \mathtt{rv}_p := \mathsf{extract_{IHL}}(p); \\ \mathtt{rv}_q := \mathsf{extract_{IHL}}(q); \\ (\mathtt{fn}\ \mathtt{x}_p \ \texttt{=>}\ \mathtt{fn}\ \mathtt{x}_q \ \texttt{=>} \\ \quad \mathtt{fn}\ \mathtt{x} : \mathsf{etype}(\mathtt{A}) \ \texttt{=>}\ \mathtt{x_q}\ (\mathtt{x_p x})) \\ \quad !\mathtt{rv}_p\ !\mathtt{rv_q} \end{array}$$

Because we assume $\mathsf{extract_{IHL}}(t)$ terminates, by the operational semantics the exection of $\mathsf{extract_{IHL}}(t)$ must result in a series of five states,

$$\sigma = \sigma_0, \sigma_1, \sigma_2, \sigma_3, \sigma_4 = \sigma'$$

such that

$$\begin{array}{l} \left\langle \left(\begin{array}{l} \mathtt{rv}_p := \mathsf{extract_{IHL}}(p); \\ \mathtt{rv}_q := \mathsf{extract_{IHL}}(q); \\ (\mathtt{fn}\ \mathtt{x}_p \ \texttt{=>}\ \mathtt{fn}\ \mathtt{x}_q \ \texttt{=>} \\ \quad \mathtt{fn}\ \mathtt{x} : \mathsf{etype}(\mathtt{A}) \ \texttt{=>}\ \mathtt{x_q}\ (\mathtt{x_p x})) \\ \quad !\mathtt{rv}_p\ !\mathtt{rv_q} \end{array} \right), \sigma_0 \right\rangle \\ \hat{\rhd} \left\langle \left(\begin{array}{l} \mathtt{fn}\ \mathtt{x}_p \ \texttt{=>}\ \mathtt{fn}\ \mathtt{x}_q \ \texttt{=>} \\ \quad \mathtt{fn}\ \mathtt{x} : \mathsf{etype}(\mathtt{A}) \ \texttt{=>}\ \mathtt{x_q}\ (\mathtt{x_p x})) \\ \quad !\mathtt{rv}_p\ !\mathtt{rv_q} \end{array} \right), \sigma_4 \right\rangle \\ \hat{\rhd} \langle \mathtt{answer}, \sigma_4 \rangle \end{array}$$

with

$$\begin{array}{r} \langle \mathsf{extract_{IHL}}(p), \sigma_0 \rangle \ \hat{\rhd}\ \langle \mathtt{a}_1, \sigma_1 \rangle \\ \langle \mathtt{rv}_p := \mathsf{extract_{IHL}}(p), \sigma_1 \rangle \rhd \langle !\mathtt{rv}_p, \sigma_2 \rangle \\ \langle \mathsf{extract_{IHL}}(q), \sigma_2 \rangle \ \hat{\rhd}\ \langle \mathtt{a}_2, \sigma_3 \rangle \\ \langle \mathtt{rv}_q := \mathsf{extract_{IHL}}(q), \sigma_3 \rangle \rhd \langle !\mathtt{rv}_q, \sigma_4 \rangle \end{array} \tag{6.52}$$

where

$$\sigma_2 = \sigma_1[\mathtt{rv}_p \mapsto \mathtt{a}_1], \tag{6.53}$$

$$\sigma_4 = \sigma_3[\mathtt{rv}_q \mapsto \mathtt{a}_2] \tag{6.54}$$

and

$$\mathtt{answer} = \sigma_4 \left(\begin{array}{l} \mathtt{fn}\ \mathtt{x}_p \ \texttt{=>}\mathtt{fn}\ \mathtt{x}_q \ \texttt{=>} \\ \quad (\mathtt{fn}\ \mathtt{x} : \mathsf{etype}(\mathtt{A}) \ \texttt{=>}\ \mathtt{x_q}\ (\mathtt{x_p x})) \\ \quad !\mathtt{rv}_p\ !\mathtt{rv_q} \end{array} \right) \tag{6.55}$$

Also, because both $\mathtt{rv}_p$ does not occur in $\mathsf{extract_{IHL}}(q)$, it must be the case that

$$\sigma_4(\mathtt{rv}_p) = \sigma_2(\mathtt{rv}_p) = \mathtt{a}_1 \sigma_4(\mathtt{rv}_q) = \mathtt{a}_2 \tag{6.56}$$

So, (6.55) and (6.56) together entail

$$\mathtt{answer} = \mathtt{fn}\ \mathtt{x} : \mathsf{etype}(\mathtt{A}) \ \texttt{=>}\ \mathtt{a}_2\ (\mathtt{a}_1 \mathtt{x}) \tag{6.57}$$

a pure, state-free term.

End of observations for Case 3a.

Now, as $Sk(A[\bar{s}_i/\bar{v}])[x/f_{A[\bar{s}_i/\bar{v}]}]$ does not contain final state identifiers, we may apply Lemma 6.2.6 to (6.48), giving

$$(\sigma_0, \sigma_1) \Vdash_{\iota'} Sk(A[\bar{s}_i/\bar{v}])[x/f_{A[\bar{s}_i/\bar{v}]}] \tag{6.58}$$

Also, by the execution (6.52), we know that

$$\langle \mathsf{extract}_{\mathsf{IHL}}(p), \sigma_0 \rangle \hat{\triangleright} \langle \mathsf{a}_1, \sigma_1 \rangle$$

and so, by (6.50),

$$(\sigma_0, \sigma_1) \Vdash \forall x : \mathsf{etype}(A[\bar{s}_i/\bar{v}]) \bullet Sk(A[\bar{s}_i/\bar{v}])[x/f_{A[\bar{s}_i/\bar{v}]}] \Rightarrow Sk(B[\bar{s}_f/\bar{v}])[(a_1\ x)/f_{B[\bar{s}_f/\bar{v}]}] \tag{6.59}$$

Then, by the definition of $\Vdash$, (6.59) and (6.58) yield

$$(\sigma_0, \sigma_1) \Vdash_{\iota'} Sk(B[\bar{s}_f/\bar{v}])[(a_1\ x)/f_{B[\bar{s}_f/\bar{v}]}]$$

By (6.53), σ_1 and σ_2 differ only over $\mathtt{rv}_p$, which does not occur in

$$Sk(B[\bar{s}_f/\bar{v}])[(a_1\ x)/f_{B[\bar{s}_f/\bar{v}]}]$$

So, by Lemma 6.2.4,

$$(\sigma_0, \sigma_2) \Vdash_{\iota'} Sk(B[\bar{s}_f/\bar{v}])[(a_1\ x)/f_{B[\bar{s}_f/\bar{v}]}] \tag{6.60}$$

We then apply Lemma 6.2.2 to (6.60) to give

$$(\sigma_2, \sigma_3) \Vdash_{\iota'} Sk(B[\bar{s}_i/\bar{v}])[(a_1\ x)/f_{B[\bar{s}_i/\bar{v}]}] \tag{6.61}$$

Now, by the execution (6.52), we know that

$$\langle \mathsf{extract}_{\mathsf{IHL}}(q), \sigma_2 \rangle \hat{\triangleright} \langle \mathsf{a}_2, \sigma_3 \rangle$$

and so, by the IH (6.51),

$$(\sigma_2, \sigma_3) \Vdash \forall x : \mathsf{etype}(B[\bar{s}_i/\bar{v}]) \bullet Sk(B[\bar{s}_i/\bar{v}])[x/f_{A[\bar{s}_i/\bar{v}]}] \Rightarrow Sk(C[\bar{s}_f/\bar{v}])[(a_2\ x)/f_{B[\bar{s}_f/\bar{v}]}] \tag{6.62}$$

Then, by the definition of $\Vdash$, (6.62) and (6.61) yield

$$(\sigma_2, \sigma_3) \Vdash_{\iota'} Sk(C[\bar{s}_f/\bar{v}])[(a_2\ (a_1\ x))/f_{C[\bar{s}_f/\bar{v}]}] \tag{6.63}$$

By (6.54), σ_3 and σ_4 differ only over $\mathtt{rv}_q$, which does not occur in

$$Sk(C[\bar{s}_f/\bar{v}])[(a_2\ (a_1\ x))/f_{C[\bar{s}_f/\bar{v}]}]$$

so, by Lemma 6.2.4,

$$(\sigma_2, \sigma_4) \Vdash_{\iota'} Sk(C[\bar{s}_f/\bar{v}])[(a_2\ (a_1\ x))/f_{C[\bar{s}_f/\bar{v}]}] \tag{6.64}$$

But, as initial state values do not occur in

$$Sk(B[\bar{s}_f/\bar{v}])[(a_2\ (a_1\ x))/f_{C[\bar{s}_f/\bar{v}]}]$$

we can use Lemma 6.2.6 on (6.64) to give

$$(\sigma_0, \sigma_4) \Vdash_{\iota'} Sk(C[\bar{s}_f/\bar{v}])[(a_2\ (a_1\ x))/f_{C[\bar{s}_f/\bar{v}]}] \tag{6.65}$$

Now, by (6.57),

$$answer\ x \rhd_{\Sigma_p} (fn\ x : \mathsf{etype}(A) => a_2\ (a_1 x))\ x \rhd_{\Sigma_p} (a_2\ (a_1 x))$$

So we can apply Lemma 6.2.5 to (6.65) to give

$$(\sigma_0, \sigma_4) \Vdash_{\iota'} Sk(C[\bar{s}_f/\bar{v}])[answer\ x/f_{C[\bar{s}_f/\bar{v}]}]$$

This is (6.49), as required.

Cases 3b–3d. The remaining cases are similar in approach to 3a.

Case 4: Proof ends in an application of (ite). Assume that $t^{\mathtt{w} \diamond T}$ is of the form

$$\mathsf{ite}(p, q)^{\mathtt{if\ b\ then\ w_1\ else\ w_2} \diamond C}$$

derived

$$\frac{\vdash p^{\mathtt{w_1} \diamond \mathsf{tologic_i}(\mathtt{b}) = true \Rightarrow C} \quad \vdash q^{\mathtt{w_2} \diamond \mathsf{tologic_i}(\mathtt{b}) = false \Rightarrow C}}{\vdash \mathsf{ite}(p, q)^{\mathtt{if\ b\ then\ w_1\ else\ w_2} \diamond C}}\ ite$$

We need to show that

$$(\sigma, \sigma') \Vdash_{\iota} Sk(C)[answer/f_C] \tag{6.66}$$

Because $\mathsf{tologic_i}(\mathtt{b}) = true$ is Harrop, by the IH

$$\mathsf{extract_{IHL}}(p)\ \ \mathsf{retr}_{\iota}\ \mathsf{tologic_i}(\mathtt{b}) = true \Rightarrow C \tag{6.67}$$

Similarly, $\mathsf{tologic_i}(\mathtt{b}) = false$ is Harrop, by the IH

$$\mathsf{extract_{IHL}}(q)\ \ \mathsf{retr}_{\iota}\ \mathsf{tologic_i}(\mathtt{b}) = false \Rightarrow C \tag{6.68}$$

Using (6.67) and (6.68) we obtain

$$\mathsf{extract_{IHL}}(p)\ \mathsf{retr}_{\iota}\ \mathsf{tologic_i}(\mathtt{b}) = true \Rightarrow C \tag{6.69}$$

and

$$\mathsf{extract_{IHL}}(q)\ \mathsf{retr}_{\iota}\ \mathsf{tologic_i}(\mathtt{b}) = false \Rightarrow C \tag{6.70}$$

So, by the definition of retr and Sk, (6.69) means that, for any states τ, τ',

$$\langle \mathsf{extract_{IHL}}(p), \tau \rangle \hat{\rhd} \langle \mathtt{answer}_p, \tau' \rangle \text{ entails}$$
$$(\tau, \tau') \Vdash_{\iota} \mathsf{tologic_i}(\mathtt{b}) = true \Rightarrow Sk(C)[answer_p/f_C] \tag{6.71}$$

and (6.70) means that, for any states τ, τ'

$$\langle \mathsf{extract}_{\mathsf{IHL}}(q), \tau\rangle \hat{\triangleright} \langle \mathtt{answer}_q, \tau'\rangle \text{ entails } (\tau,\tau') \Vdash_\iota \mathsf{tologic_i}(\mathtt{b}) = false \Rightarrow Sk(C)[answer_q/f_C] \quad (6.72)$$

Either $\sigma(\mathtt{b}) = true$ or $\sigma(\mathtt{b}) = false$. We reason over these two cases to obtain (6.66).

Subcase 4a: $\sigma(\mathtt{b}) = true$. Then

$$\langle \mathtt{b}, \sigma\rangle \triangleright \langle \mathtt{true}, \sigma\rangle$$

and so, by Lemma 6.2.7, this means that

$$(\sigma, \sigma') \Vdash_\iota \mathsf{tologic_i}(\mathtt{b}) = true \quad (6.73)$$

Also, the operational semantics of $\mathsf{extract}_{\mathsf{IHL}}(t)$ demands the following holds:

$$\langle \mathtt{if}\ \mathtt{b}\ \mathtt{then}\ \mathsf{extract}_{\mathsf{IHL}}(p)\ \mathtt{else}\ \mathsf{extract}_{\mathsf{IHL}}(q), \sigma\rangle \hat{\triangleright} \langle \mathtt{answer}, \sigma'\rangle \text{ entails } \langle \mathsf{extract}_{\mathsf{IHL}}(p), \sigma\rangle \hat{\triangleright} \langle \mathtt{answer}, \sigma'\rangle$$

So

$$\langle \mathsf{extract}_{\mathsf{IHL}}(p), \sigma\rangle \hat{\triangleright} \langle \mathtt{answer}, \sigma'\rangle \quad (6.74)$$

Instantiating (6.71) with (6.74) gives

$$(\sigma, \sigma') \Vdash_\iota \mathsf{tologic_i}(\mathtt{b}) = true \Rightarrow Sk(C)[answer/f_C]$$

Instantiating this with (6.73) gives

$$(\sigma, \sigma') \Vdash_\iota Sk(C)[answer/f_C]$$

which establishes (6.66), as required.

Subcase 4b: $\sigma(\mathtt{b}) = false$. Similar reasoning to the previous subcase will establish (6.66).

Case 5: Proof ends in an application of (cons). Assume $t^{\mathtt{w}\diamond T}$ is of the form

$$\mathsf{cons}(p^{\mathtt{w}\diamond P}, q^{P\Rightarrow A})^{\mathtt{w}\diamond A}$$

derived by

$$\frac{\vdash p^{\mathtt{w}\diamond P} \quad \vdash_{\mathsf{Int}} q^{P\Rightarrow A}}{\vdash \mathsf{cons}(p^{\mathtt{w}\diamond P}, q^{P\Rightarrow A})^{\mathtt{w}\diamond A}} \text{(cons)}$$

By the IH, we know that

$$\mathsf{extract}_{\mathsf{IHL}}(p)\ \mathsf{retr}_\iota\ P \quad (6.75)$$

There are two cases, depending on whether P is Harrop or not. We only deal with the latter situation, as the proof for the former is similar.

By Theorem 6.2.1,

$$\texttt{fn x}_\mathsf{v} : \mathsf{etype(P)} \texttt{=>} \mathsf{extract}_{\mathsf{Int}}(q) \ \mathsf{mr}\ P \Rightarrow A \tag{6.76}$$

By Lemma 5.1.1 of Chapter 5 (p. 136), this means that, for any (τ, τ'),

$$(\tau, \tau') \Vdash \forall x_v : \mathsf{etype}(P) \bullet Sk(P)[x_v/f_P] \Rightarrow Sk(A)[(a\ x_v)/f_A] \tag{6.77}$$

for any $a = \mathsf{tologic}(\mathtt{a})$ where

$$\mathtt{a} = \texttt{fn x}_\mathsf{v} : \mathsf{etype(P)} \texttt{=>} \mathsf{extract}_{\mathsf{Int}}(q)[\tau(\bar{s})/\bar{s}_i][\tau'(\bar{s})/\bar{s}_f] \tag{6.78}$$

Now, the execution of $\mathsf{extract}_{\mathsf{IHL}}(t)$ must result in a sequence of four states

$$\sigma = \sigma_0, \sigma_1, \sigma_2, \sigma_3 = \sigma'$$

such that

$$\begin{array}{l} \left\langle \left(\begin{array}{l} \bar{\mathtt{i}} := \bar{\mathtt{s}};\ \mathtt{rv}_p := \mathsf{extract}(p);\ \bar{\mathtt{f}} := \bar{\mathtt{s}}; \\ (\texttt{fn}\ \bar{\mathtt{s}}_\mathtt{i} :: \bar{\mathtt{s}}_\mathtt{f} \Rightarrow \\ \quad \texttt{fn x}_\mathsf{v} : \mathsf{etype(P)} \texttt{=>} \mathsf{extract}_{\mathsf{Int}}(q)) \\ \quad !\bar{\mathtt{i}} :: \bar{\mathtt{f}}\ !\mathtt{rv}_p \end{array} \right), \sigma_0 \right\rangle \\ \rhd \left\langle \begin{array}{l} (\texttt{fn}\ \bar{\mathtt{s}}_\mathtt{i} :: \bar{\mathtt{s}}_\mathtt{f} \Rightarrow \\ \quad \texttt{fn x}_\mathsf{v} : \mathsf{etype(P)} \texttt{=>} \mathsf{extract}_{\mathsf{Int}}(q)), \sigma_3 \\ \quad !\bar{\mathtt{i}} :: \bar{\mathtt{f}}\ !\mathtt{rv}_p \end{array} \right\rangle \\ \hat{\rhd}\ \langle \mathtt{answer}, \sigma_3 \rangle \end{array} \tag{6.79}$$

where

$$\mathtt{answer} = \texttt{fn x}_\mathsf{v} : \mathsf{etype(P)} \texttt{=>} \mathsf{extract}_{\mathsf{Int}}(q)[\sigma_3(\bar{\mathtt{i}})/\bar{s}_i][\sigma_3(\bar{\mathtt{f}})/\bar{s}_f]\ \sigma_3(\mathtt{rv}_p) \tag{6.80}$$

and

$$\begin{array}{r} \langle \bar{\mathtt{i}} := \bar{\mathtt{s}}, \sigma_0 \rangle\ \hat{\rhd}\ \langle \mathtt{a}_1, \sigma_1 \rangle \\ \langle \mathtt{rv}_p := \mathsf{extract}_{\mathsf{IHL}}(p), \sigma_1 \rangle\ \hat{\rhd}\ \langle \mathtt{a}_p, \sigma_2 \rangle \\ \langle \bar{\mathtt{f}} := \bar{\mathtt{s}}, \sigma_2 \rangle\ \hat{\rhd}\ \langle \mathtt{a}_2, \sigma_3 \rangle \end{array} \tag{6.81}$$

so that

$$\sigma_1 = \sigma_0[\bar{\mathtt{i}} \mapsto \sigma_0(\bar{\mathtt{s}})] \sigma_3 = \sigma_2[\bar{\mathtt{f}} \mapsto \sigma_2(\bar{\mathtt{s}})] \sigma_3(\mathtt{rv}_p) = \mathtt{a}_p \tag{6.82}$$

Now, because $\bar{\mathtt{i}}$ do not occur in $\mathsf{extract}_{\mathsf{IHL}}(p)$, (6.82) and inspection of (6.81) reveal that

$$\sigma_3(\bar{\mathtt{i}}) = \sigma_1(\bar{\mathtt{i}}) = \sigma_0(\bar{\mathtt{s}}) \tag{6.83}$$

Also, because the values of $\bar{\mathtt{s}}$ are unchanged in the assignment $\bar{\mathtt{f}} := \bar{\mathtt{s}}$,

$$\sigma_3(\bar{\mathtt{f}}) = \sigma_2(\bar{\mathtt{s}}) = \sigma_3(\bar{\mathtt{s}}) \tag{6.84}$$

So, using (6.82), (6.83) and (6.84) in (6.80) gives

$$\mathtt{answer} = \texttt{fn x}_\mathsf{v} : \mathsf{etype(P)} \texttt{=>} \mathsf{extract}_{\mathsf{Int}}(q)[\sigma(\bar{\mathtt{s}})/\bar{s}_i][\sigma'(\bar{\mathtt{s}})/\bar{s}_f]\ \mathtt{a}_p \tag{6.85}$$

Define

$$\mathtt{a}_q = \mathtt{fn}\ \mathtt{x_v} : \mathsf{etype}(\mathtt{P}) \Rightarrow \mathsf{extract_{Int}}(q)[\sigma(\bar{\mathtt{s}})/\bar{s}_i][\sigma'(\bar{\mathtt{s}})/\bar{s}_f] \tag{6.86}$$

Then

$$\mathtt{answer} = \mathtt{a}_q\ \mathtt{a}_p \tag{6.87}$$

By (6.77), it is the case that

$$(\sigma, \sigma') \Vdash \forall x_v : \mathsf{etype}(P) \bullet Sk(P)[x_v/f_P] \Rightarrow Sk(A)[(a_q\ x_v)/f_A] \tag{6.88}$$

Also, given that

$$\langle \mathtt{rv}_p := \mathsf{extract_{IHL}}(p), \sigma_1 \rangle \hat{\rhd} \langle \mathtt{a}_p, \sigma_2 \rangle$$

we let σ_1' be the state such that

$$\langle \mathsf{extract_{IHL}}(p), \sigma_1 \rangle \hat{\rhd} \langle \mathtt{a}_p, \sigma_1' \rangle$$

Now, recall the IH, (6.75)

$$\mathsf{extract_{IHL}}(p)\ \mathsf{retr}_\iota\ P$$

This means that

$$(\sigma_1, \sigma_1') \Vdash_\iota Sk(P)[a_p/f_P]$$

Note that σ_0 differs from σ_1 only over $\bar{\mathtt{i}}$, and σ_3 differs from σ_1' only over $\bar{\mathtt{f}}$ and $\mathtt{rv}_p$. So, because $\bar{\mathtt{i}}$, $\bar{\mathtt{f}}$ and $\mathtt{rv}_p$ do not occur in $Sk(P)[a_p/f_P]$, by Lemma 6.2.4, we have

$$(\sigma, \sigma') \Vdash_\iota Sk(P)[a_p/f_P] \tag{6.89}$$

We instantiate (6.88) with (6.89) to give

$$(\sigma, \sigma') \Vdash_\iota Sk(A)[(a_q\ a_p)/f_A]$$

But then, by (6.87) and Lemma 6.2.5, we have

$$(\sigma, \sigma') \Vdash_\iota Sk(A)[answer/f_A]$$

as required.

This last case concludes our proof. □

6.3 The Curry–Howard protocol for program synthesis

In this section, we show that our extraction map leads to an effective application of the Curry–Howard protocol for the synthesis of IHL-realizers from proofs of specifications.

6.3.1 Logical and computational type theories

In the Curry–Howard protocol of Chapter 3, we gave a general framework for program synthesis from proofs of specifications, generalizing state-of-the-art proofs-as-programs to new programming languages and logical contexts.

The protocol requires a logical type theory and a computational type theory. We take the logical type theory as the $LTT(\mathsf{IHL})$ of Chapter 5.2 (identified as an LTT for IHL in Section 5.2 p. 149). We take the computational type theory to be IML of Chapter 4. We shall take our computational type theory to be the IML (identified as a CTT in Section 4.2, p. 104).

6.3.2 Conformance to the Curry–Howard protocol

The Curry–Howard protocol (Definition 3.2.5, Chapter 3, p. 87) holds between the $LTT(\mathsf{IHL})$, and IML, for the following reasons

1. There are extraction maps etype from formulae of $LTT(\mathsf{IHL})$ to types of IML and extract from proof-terms of $LTT(\mathsf{IHL})$ to programs of IML,

$$\mathsf{extract}_{\mathsf{IHL}} : PT(LTT(\mathsf{IHL})) \Rightarrow Term(IML)$$
$$\mathsf{etype} : WFF(\Sigma_t) \Rightarrow Type(IML)$$

   such that, given a proof $d \in PT(LTT(\mathsf{IHL}))$ with the property that

$$\vdash_{LTT(\mathsf{IHL})} d^{\mathtt{w} \diamond A}$$

   then $\mathsf{extract}(d)$ is in IML, and is of type $\mathsf{etype}A$. The maps were given in Figs. 6.1 and 6.3. The required typing property was shown in Theorem 6.2.2.
2. There is a realizability relation kr between programs and formulae, such that, for any proof

$$\vdash_{LTT(\mathsf{IHL})} p^{\mathtt{w} \diamond A}$$

   it is true that

$$\mathsf{extract}(p) \;\; \mathsf{kr} \;\; (\mathtt{w} \diamond A)$$

   The realizability relation was identified in Definition 6.1.6. The required property holds by Theorems 6.2.4 and 6.2.5.

6.3.3 Application of the protocol

Recalling the process of protocol application described in Chapter 3, Section 3.3, p. 87, we have sucessfully taken the required steps

1. We defined a signature and a logical calculus that involves the signature in Chapter 4. This involved deriving some properties that were orthogonal to the protocol process itself, but which were necessary for deriving the extraction theorem. Specifically, we provided a semantics for the calculus (in Chapter 4) and proved soundness (in Chapter 5).

2. We defined a logical type theory for the logical calculus in Chapter 5.
3. We Identified a programming language and described it as a computational type theory in Chapter 4.
4. Finally, in this chapter, we completed the process by proving the Curry–Howard protocol to hold over the above domains.

6.4 Example: electronic banking system

We illustrate our approach to program synthesis using the electronic banking example used throughout this part of the book. The system consists of a database of account details, indexed by user identification. In Section 4.5.5 of Chapter 4, we used IHL to develop a program/formula pair that satisfied the following property. By using the program to search through the database, it is possible to obtain a list of all accounts held at the bank by the user, given a user's details. This was shown by the following:

$$\begin{aligned} \vdash\ & \texttt{counter} := 0; \\ & \quad \texttt{while}\ \texttt{y} < (\texttt{length}\ \texttt{db}) - 1\ \texttt{do}\ \texttt{counter} := !\texttt{counter} + 1 \diamond \\ & \exists y : (account\ list) \bullet allAccountsAt(currentUser, y, counter_f) \wedge \\ & \qquad\qquad (counter_f < (length\ db) - 1) = false \end{aligned} \tag{6.90}$$

The formula here is the unSkolemized form of

$$listAllAccounts(currentUser, y, (length\ db))$$

Consequently, when viewed as a specification of a side-effect-free return value, according to our notion of IHL-realizability, the program/formula pair of (6.90) specifies a program that, given a user's details, will search through a database to obtain all accounts held at the bank by the user, and then return this list. The program should exhibit the same side-effects as the program of the pair with respect to the state identifiers used in the formula of the pair.

Section 5.3 of Chapter 5 showed how the proof of (6.90) might be represented in the logical type theory for IHL.

Using Theorems 6.2.4 and 6.2.5, we can take the proof-term of (6.90) given in the logical type theory and obtain the required IHL-realizer.

Before examining the IHL-realizer, we make some observations relating about the subproofs of the theorem and the corresponding subprograms of the realizer.

Note that the proof of (6.90) involved a non-Harrop axiom available to intuitionistic proofs

$$y < (length\ db) - 1 \Rightarrow sub(l, y + 1).owner = u \vee \neg sub(l, y + 1).owner = u$$

By Assumption 6.1, this axiom is presumed to be associated with a side-effect-free program $\texttt{PK}_{5.59}$ that is an intuitionistic modified realizer of (5.59), so that

$$(\forall x : \mathsf{etype}(B) \bullet PK_{5.59} = inl(x) \Rightarrow Sk(B)[x/f_B]) \\ \wedge (\forall y : \mathsf{etype}(C) \bullet PK_{5.59} = inr(y) \Rightarrow Sk(C)[y/f_C])$$

where B and C denote the left and right subformulae of the axiom, respectively.

We use the normalized proof of the theorem. The proof involves a subproof of the form

$$\vdash_{\mathsf{Int}} counter_f = counter_i + 1 \Rightarrow (counter_i < (length\ db) - 1) = true \wedge \\ \exists l : (account\ list) \bullet allAccountsAt(currentUser, l, counter_i) \Rightarrow \\ \exists l : (account\ list) \bullet allAccountsAt(currentUser, l, counter_f) \quad (6.91)$$

The normalized proof-term corresponding to this proof, $p'_{5.63}$, is of the form

$$\mathsf{specific}(\mathsf{abstract}\ r^{counter_f = counter_i + 1}. \\ \mathsf{Schema}(subst, [[(p'_{5.61}); r^{counter_f = counter_i + 1}]; \\ [\forall u : user \bullet (counter_i < (length\ db) - 1) \wedge \\ \exists l : (account\ list) \bullet allAccountsAt(u, l, counter_i) \Rightarrow \\ \exists l : (account\ list) \bullet allAccountsAt(u, l, m)] \\ [counter_i + 1; counter_f]; [int]]), currentUser)$$

where $p'_{5.61}$ is the normalized form of $\mathsf{specific}(p_{5.61}, counter_i)$ ($p_{5.61}$ was defined Section 5.3 of Chapter 5, p. 159),

$$\mathsf{use}\ u : user.\ \mathsf{abstract}\ m^{(counter_i < (length\ db) - 1) \wedge \exists l : (account\ list)}. \\ \mathsf{app}(\mathsf{app}(\mathsf{abstract}\ e^{counter_i < (length\ db) - 1}. \\ \mathsf{abstract}\ i^{\exists l : (account\ list) \bullet allAccountsAt(u, l, counter_i)}. \\ \mathsf{specific}(i, x.f^{allAccountsAt(u, x, counter_i)}.p_{5.60}), \mathsf{fst}(m)), \mathsf{snd}(m))$$

and where $p_{5.60}$ is a proof-term denoting proof by cases

$$\mathsf{case}\ \mathsf{app}(\mathsf{Axiom}(A_{(5.59)}), e)\ \mathsf{of} \\ \mathsf{inl}(g^{sub(l, y+1).owner = u}).\mathsf{show}(sub(db, y + 1) :: x, p_2), \\ \mathsf{inr}(h^{\neg sub(l, y+1).owner = u}).\mathsf{show}(x, p_3)$$

We define $p'_{5.60}$ by

$$\mathsf{case}\ \mathsf{app}(\mathsf{Axiom}(A_{(5.59)}), e)\ \mathsf{of} \\ \mathsf{inl}(g^{sub(l, counter_i + 1).owner = u}).\mathsf{show}(sub(db, counter_i + 1) :: x, p'_2), \\ \mathsf{inr}(h^{\neg sub(l, counter_i + 1).owner = u}).\mathsf{show}(x, p'_3)$$

where p'_2 is

$$\vdash \mathsf{app}(\mathsf{app}(\mathsf{app}(\mathsf{app}(\mathsf{Axiom}(A_{(5.56)}), u), x), counter_i), \langle g, f\rangle)^{allAccountsAt(u, sub(db, counter_i+1)::x, counter_i+1)}$$

and p'_3 is

$$\vdash \mathsf{app}(\mathsf{app}(\mathsf{app}(\mathsf{app}(\mathsf{Axiom}(A_{(5.56)}), u), x), counter_i), \langle g, f\rangle)^{allAccountsAt(u, x, counter_i+1)}$$

We denote the modified realizer of (6.91), $\mathsf{extract}_{\mathsf{Int}}(p_{6.91})$, by

$$\mathtt{PP} : \mathtt{user} \rightarrow (\mathtt{account\ list}) \rightarrow (\mathtt{account\ list})$$

Written in full, PP is

```
(fn u : user =>
  fn m : account list =>
    fn i : account list =>
    (fn x : account list =>
    match PK5.59 with
inl(()) => sub(db, counter_i + 1) :: x | inr(()) => x
     m)
currentUser)
```

Inspection shows that the form of PP mirrors the structure of $p'_{5.63}$, but with simplifications achieved by ignoring proof-terms for Harrop formulae.

The normalized proof-term for the required theorem (6.90) was given as $p'_{5.72}$ in Section 5.3 of Chapter 5,

$$\mathsf{cons}(\mathsf{seq}(\mathsf{cons}(\mathsf{cons}(\mathsf{assign}(counter, 0), p_{5.67}), ptrue), \mathsf{wd}(\mathsf{cons}(\mathtt{assign}(counter, counter + 1), p_{5.63'}))), qtrue)$$

We apply extract to $p'_{5.72}$ and obtain the required program

```
rv1 := fun x:account list => x;
while !counter<(length db) - 1 do
(rv2 := (ic := !counter; counter:=!counter+1;
     if := !counter;
        (fun counter_i => fun counter_f =>
             (PP !counter_i currentUser)) ic if)
rv1 := fun x_2::x_1=> fun x => (x_2 (rv1 x)) !rv2 !rv1;)
!rv1 [];
```

Inspection of this program's execution shows that it is indeed an IHL-realizer for (6.90).

6.5 Example: synthesis of contracts

We illustrate our approach with a second example. We consider the specification and synthesis of a routine for processing orders for books at an online bookstore. The specification involves a disjunction, and, consequently, the return values of the synthesized program are of disjoint union type. We will briefly discuss how our program can be considered equivalent to a program with an *assertion contract* in the sense defined by Bertrand Meyer's principle of design-by-contract [Mey97]. Briefly, an assertion contract is a boolean function that evaluates at the beginning or end of a program for run-time testing. Disjoint unions can be used instead of booleans, and, in fact, carry more useful information to define more complex contracts. This opens up a potentially useful application of our methods — the synthesis of programs with complex contracts to be used in systems built using design-by-contract.

6.5.1 Design-by-contract

Design-by-contract is a method of software development, first proposed by Bertrand Meyer [Mey97]. Its roots are in the Hoare logic and pre- and post-condition specifications of programs. Design-by-contract is incorporated into languages such Eiffel and Oberon, and also in specification languages such as the OCL part of the UML [WK98]. Design-by-contract is often given as a method of object-oriented and structured program development — for our purposes, we shall restrict ourselves to examining design-by-contract for isolated imperative programs.

Briefly, the idea is as follows. When a program is developed, it must be accompanied with two boolean-valued functions, called assertions. These form the so-called contract of the program. The boolean functions are called the pre- and post- condition assertions. Programs are tested at run-time by evaluating the values of the assertions in a dedicated test suite. If the pre-condition assertion evaluates to true before the program is executed, and the post-condition evaluates to false, then the program has an error and the designer is altered by the test suite.

The assertions are defined by the programmer to specify expectations about code. The programmer writes the code independently of the specification in the sense there is no guarantee that the code satisfies the contract assertions. However, testing enables the programmer to systematically check the code's validity against logical expectations. In this way, design-by-contract facilitates a logical, specification-oriented approach to run-time testing of programs.

For example, a program in an Eiffel-like language takes the form:

```
pre:
  Odd(s)
body:
  s:=s+1;
post:
  Even(s)
```

The program consists of a body of code, $\mathtt{s} := \mathtt{s} + 1$ and two boolean assertions $\mathtt{Odd(s)}$ and $\mathtt{Even(s)}$, before and after the body. The former assertion is true if the state identifier s contains an odd integer. The latter is true if the state identifier contains an even integer. The execution of the program proceeds by evaluating $\mathtt{Odd(s)}$, then $\mathtt{s} := \mathtt{s} + 1$ and then $\mathtt{Even(s)}$. The program will never generate an error, because it is always the case that the increment of an odd number is even. However, if the body of the code was replaced with $\mathtt{s} := \mathtt{s} + 2$, any execution would generate an error.

Remark 6.2. Note that, in contrast to the pre- and post-conditions of our version of the Hoare logic, these assertions are *decidable* boolean functions. This is why we refer to these functions as pre- and post-condition *assertions*, to distinguish them from our pre- and post-condition predicate formulae. In the literature, the functions are often simply referred to as pre- and post- conditions.

The assertions of the example are very simple. However, it was noted in [Mey00] that, as programs become more complex, accompanying assertions will also grow in complexity. When this happens, it becomes more likely that an assertion can be incorrectly coded — in the sense that the boolean function does not correctly represent the required specification. The assertions of a program are side-effect-free. Our program synthesis methods are designed to develop programs with side-effects and side-effect-free aspects. It is therefore an interesting question to see if we can use our methods to synthesize programs with provably correct assertions.

6.5.2 Synthesis of post-conditions in *SML*

For our purposes, we will consider only programs that use post-condition assertion contracts, without pre-conditions. We leave a full treatment of contracts for future research.

There is no native support for design-by-contract in *SML*. We will take a very simple approach, that will make it easy for us to apply our program synthesis methods to post-condition contracts.

We will simulate post-conditions as *return values* of disjoint union type. That is to say, to be used in design-by-contract development, a *SML* program code must have evaluation sequences of the form

$$\langle \mathtt{code}, \sigma \rangle \hat{\triangleright} \langle \mathtt{p}, \sigma' \rangle$$

where the return value p is of type $(\mathtt{t}|\mathtt{t})$, for some type t. The post-condition assertion for code is taken to be true if post is of the form $\mathtt{Inl(a)}$ and false if it is of the form `Inr(b)`. We will assume that such programs are evaluated within a test suite that will generate appropriate error reports given false post-conditions.

For example, assume a program that

$$\texttt{s} := !\texttt{s} * 2; \texttt{Even}(!\texttt{s})$$

consists of some imperative code $\texttt{s} := !\texttt{s} * 2$ with return values arising from `Even(!s)`, of type (`Unit`|`Unit`). If the state value `!s` is even, then the return value of the program is `Inl(())`, and `Inr(())` otherwise.

Because post-condition assertions are taken to be return values, we can employ this chapter's synthesis techniques to contracts. The specification of a required post-condition assertion is given by a disjunction of the form $A \vee \neg A$. The disjunction specifies the required post-condition as a return value realizer, of type $(\mathsf{etype}(A)|\mathsf{etype}(\neg A))$. By Theorems 6.2.4 and 6.2.5, given a proof of the form

$$\vdash \mathsf{body} \diamond A \vee \neg A \tag{6.92}$$

we can extract a program that is visibly side-effect equivalent to `body`, with the required post-condition assertion.

6.5.3 Using flawed programs to build new programs

The designer uses the rules of our Hoare logic to make proofs of the form (6.92). As usual, the designer should only use true properties about given programs. However, we permit programs to have faults. Faulty programs are reasoned about as follows. Instead of using formulae that assert the correctness of programs, we use disjunctive statements stating that the program may or may not be faulty.

Consider a program designed to connect to a database, `connectDB`, for example. The program is intended to always result in a successful connection, specified by formula $connected_f = true$. However, the problem has a fault, and sometimes results in an unsuccessful connection. This situation is described truthfully by the disjunction

$$\vdash \texttt{connectDB} \diamond connected_f = true \vee \neg connected_f = true$$

This property is true of the program, and so the program may be used within the Hoare logic to develop a larger program/formula theorem, without jeapordizing the truth of the final result.

6.5.4 Order processing system

We consider a program that processes a number of book orders for an online bookstore. The program first initializes a bulk order in the database. Then the program takes a list of individual orders and successively adds them to the bulk order, until no more individual orders are left. There is a possibility that the program may corrupt the database. We wish to synthesize this program, together with a post-condition assertion that determines if the program has corrupted the database or not.

We make the following domain assumptions:

- We presume there is a *SML* type of databases, `DBT`.
- We use the predicate $DBCorrupt(d)$ to say that a database $\mathtt{d} : \mathtt{DBT}$ is corrupt.
- We take a *SML* state reference `db` of reference type `DBT`, the database to be manipulated by our required program.
- We use a boolean state reference `orderRemaining`, to determine if there are any orders that need to be added to the database `db`.
- There is a program that initializes the bulk order, `init`. This program always leaves the database free of corruption. More formally, we have an axiom

$$\mathtt{init} \diamond \neg DBCorrupt(db_f) \tag{6.93}$$

  We denote this axiom by $A_{6.93}$.
- We are given a black-box program `add` that adds an order request to the database `db`. Each time this program is called, there is a chance that the database may become corrupt. The program should only be called if there are orders to be processed (that is, if `orderRemaining` is `true`). If the status of the database is known, prior to adding a new entry, we can determine if the new entry results in corruption or not.
  This is stated formally by a non-Harrop axiom,

$$\begin{aligned}\mathtt{add} \diamond (orderRemaining_i = true \wedge \\ (\neg DBCorrupt(db_i) \vee DBCorrupt(db_i))) \Rightarrow \\ (\neg DBCorrupt(db_f) \vee DBCorrupt(db_f))\end{aligned} \tag{6.94}$$

  We denote this axiom by $A_{6.94}$.
- Because axiom $A_{6.94}$ is non-Harrop, by Assumption 6.1, $A_{6.94}$ is presumed to be associated with a side-effect-free program `PK`, its IHL-realizer.
  Let P denote

$$\begin{aligned}(orderRemaining_i = true \wedge \\ (\neg DBCorrupt(db_i) \vee DBCorrupt(db_i))) \Rightarrow \\ (\neg DBCorrupt(db_f) \vee DBCorrupt(db_f))\end{aligned}$$

  so that `PK` kr P. Then

$$[\![\mathtt{PK}]\!] \Vdash P \tag{6.95}$$

$$\mathtt{PK} \equiv_P \mathtt{add} \tag{6.96}$$

  and

$$\mathsf{PK}\ \mathsf{retr}\ P \tag{6.97}$$

Condition (6.97) is important. It says that PK is a return value realizer of P. That is, for any σ, σ' and interpretation ι, if

$$\langle \mathtt{PK}, \sigma\rangle \mathrel{\hat{\triangleright}} \langle \mathtt{answer}, \sigma'\rangle$$

then

$$(\sigma, \sigma') \Vdash_\iota Sk(P)[answer/f_F]$$

where $Sk(P)[answer/f_F]$ is of the form

$$\begin{aligned}
\forall x : (Unit|Unit) \bullet (orderRemaining_i = true \wedge \\
((\forall y : Unit \bullet x = inl(y) \Rightarrow \neg DBCorrupt(db_i)) \wedge \\
(\forall y : Unit \bullet x = inr(y) \Rightarrow DBCorrupt(db_i)))) \\
\Rightarrow ((\forall y : Unit \bullet (answer\ x) = inl(y) \Rightarrow \neg DBCorrupt(db_f)) \wedge \\
(\forall y : Unit \bullet (answer\ x) = inr(y) \Rightarrow DBCorrupt(db_f))
\end{aligned}$$

We wish to obtain a program that

- initializes the database, processes each order until no orders remain, and
- provides the required assertion as a return value realizer of

$$\neg DBCorrupt(db_f) \vee DBCorrupt(db_f)$$

From (6.94) we can apply (loop) to obtain

$$\begin{aligned}
&\mathtt{while\ orderRemaining\ do\ add}\diamond \\
&\qquad (\neg DBCorrupt(db_i) \vee DBCorrupt(db_i)) \Rightarrow \\
&((\neg DBCorrupt(db_f) \vee DBCorrupt(db_f)) \wedge orderRemaining_f = true)
\end{aligned} \tag{6.98}$$

The proof-term for this theorem is simply

$$\mathsf{wd}(\mathsf{Axiom}(A_{DBC-1}))$$

Let $True$ be a provable intuitionistic statement, say $\bot \Rightarrow \bot$. It is easy to derive

$$\begin{aligned}
&\vdash_{\mathsf{Int(IHL)}} \neg DBCorrupt(db_f) \Rightarrow \\
&\qquad (True \Rightarrow (\neg DBCorrupt(db_f) \vee DBCorrupt(db_f)))
\end{aligned} \tag{6.99}$$

by (Ass-I), ($\vee_2$-I) and weakening the result with $True$ as hypothesis. But then by applying (cons) to from (6.93) and (6.99) we can derive

$$\mathtt{init} \diamond True \Rightarrow (\neg DBCorrupt(db_f) \vee DBCorrupt(db_f)) \tag{6.100}$$

Written in full, the proof-term for this theorem is

$$\mathsf{cons}(\mathsf{abstract}\ u^{\neg DBCorrupt(db_f)}.\ \mathsf{app}(\mathsf{abstract}\ w^{\neg DBCorrupt(db_f) \vee DBCorrupt(db_f)}.\ \mathsf{abstract}\ x^{True}.\ \mathsf{fst}(\langle w, x\rangle), \mathsf{inl}(u)^{\neg DBCorrupt(db_f) \vee DBCorrupt(db_f)}), \mathsf{Axiom}(A_{DBC-2}))$$

We will denote this proof-term by $p_{6.100}$.

We apply (seq) to (6.100) and (6.98) to obtain

$$\texttt{init; while orderRemaining do add} \diamond True \Rightarrow ((\neg DBCorrupt(db_f) \vee DBCorrupt(db_f)) \wedge orderRemaining_f = true) \quad (6.101)$$

with proof-term

$$\mathsf{seq}(p_{6.100}, \mathsf{wd}(\mathsf{Axiom}(A_{DBC-1})))$$

It is a simple task to derive

$$\vdash_{\mathsf{Int(IHL)}} (True \Rightarrow (\neg DBCorrupt(db_f) \vee DBCorrupt(db_f)) \wedge orderRemaining_f = true) \Rightarrow (\neg DBCorrupt(db_f) \vee DBCorrupt(db_f)) \wedge orderRemaining_f = true \quad (6.102)$$

with proof-term $\mathsf{abstract}\ l.\ \mathsf{app}(l, \mathsf{abstract}\ v^{\perp}.\ v^{\perp})$ where l is of type

$$True \Rightarrow ((\neg DBCorrupt(db_f) \vee DBCorrupt(db_f)) \wedge orderRemaining_f = true)$$

We will abbreviate this proof-term by $p_{6.102}$.

By (cons) applied to (6.101) and (6.102) we have

$$\vdash \texttt{init; while orderRemaining do add} \diamond (\neg DBCorrupt(db_f) \vee DBCorrupt(db_f)) \wedge orderRemaining_f = true$$

Finally, we can apply($\wedge_1$-E) we obtain

$$\texttt{init; while orderRemaining do add} \diamond \neg DBCorrupt(db_f) \vee DBCorrupt(db_f) \quad (6.103)$$

with proof-term

$$\mathsf{fst}(\mathsf{cons}(\mathsf{seq}(p_{6.100}, \mathsf{wd}(\mathsf{Axiom}(A_{DBC-1}))), p_{6.102}))$$

Observe that the program of theorem (6.103) satisfies one of the system requirements: it initializes the database, processes each order until no orders remain. However, this program is not sufficient for our needs. Recall that our system is required to include a post-condition assertion, the return value realizer of

$$\neg DBCorrupt(db_f) \vee DBCorrupt(db_f)$$

The full system, with accompanying post-condition, is synthesized by applying the extraction map to the proof-term for (6.103). This results in an *SML* program of the form:

```
i_1 := !db;
rv_r := (
  rv_p := (
   i_2 := !db; init; f_2 := !db;
   ((fn s_i => fn s_f => Inl(())) i_2 f_2);
  )
  rv_q := (
   rv_1 := (fn x:Unit|Unit => x);
   while OrderRemaining do
    rv_2 := PK;
    rv_1 := (fn x_2 => fn x_1 => fn x:Unit|Unit =>
              x_2 (x_1 x)) !rv_2 !rv_1;
    !rv_1
  )
 (!rv_q !rv_p)
 )
f_1 := !db;
(fn s_i => fn s_f => fn x:Unit|Unit => x) !i_1 !f_1 !rv_r;
```

To understand this program, it is helpful to consider the equivalent, optimised form:

```
rv_r := (
  rv_p := (init; Inl(()))
  rv_q := (
   rv_1 := (fn x:Unit|Unit => x);
   while OrderRemaining do
    rv_2 := PK;
    rv_1 := (fn x_2 => fn x_1 => fn x:Unit|Unit =>
              x_2 (x_1 x)) !rv_2 !rv_1;
    !rv_1;
  )
 (!rv_q !rv_p)
 )
!rv_r;
```

Equivalence of side-effects and of return values is easily shown, because the values of `i_1` and `i_2` are not used anywhere except

```
(fn s_i => fn s_f => fn x:Unit|Unit => x) !i_1 !f_1 !rv_r
```

which, by the operational semantics is equivalent to $\mathtt{rv_r}$. Similarly, the values of i_2 and i_1 are not used anywhere except

```
((fn s_i => fn s_f => Inl(())) i_2 f_2)
```

which is equivalent to Inl(()) under any evaluation.

This program will initialize the database, by calling init and then, using the while loop, it will processes each order until no orders remain. Also, by virtue of our extraction techniques, it forms a return value realizer of

$$\neg DBCorrupt(db_f) \vee DBCorrupt(db_f)$$

Assume an initial state such that only 3 orders need to be processed — that is, OrderRemaining has the value true for 3 calls to PK. Then, by the operational semantics, the extracted program will result in a return value of the form

$$\mathtt{rPK_3(rPK_2(rPK_1 Inl(())))} \tag{6.104}$$

where $\mathtt{rPK_1}$, $\mathtt{rPK_2}$ and $\mathtt{rPK_3}$ are modified realizers of

$$(orderRemaining_i = true \wedge (\neg DBCorrupt(db_i) \vee DBCorrupt(db_i))) \\ \Rightarrow (\neg DBCorrupt(db_f) \vee DBCorrupt(db_f)) \tag{6.105}$$

for states after the first, second, and third calls to PK, respectively.

Condition (6.97) above is equivalent to the following condition on terms ($\mathtt{rPK_i}$ t) ($i = 1, 2, 3$). For each call to PK, if the database is known to be either corrupt (t is of the form Inr(())) or not corrupt (t is of the form Inl(())) *prior* to the call, ($\mathtt{rPK_i}$ t) is guaranteed to tell us if the database is either corrupt (($\mathtt{rPK_i}$ t) evaluating to Inr(())) or not corrupt (($\mathtt{rPK_i}$ t) evaluating to Inl(())) *after* the call. (This is true by the fact that PK is a return value realizer of (6.105).)

Thus, it can be seen that (6.104) will provide a return value realizer of

$$\neg DBCorrupt(db_f) \vee DBCorrupt(db_f)$$

for the initial and final states of the evaluation.

The extraction map produces the required program and post-condition assertion. The assertion will always return Inl(()) or Inr(()), telling us if the program evaluation has resulted in the database becoming corrupt or not.

Remark 6.3. Our work has applied constructive methods to the synthesis of imperative programs, taking realizability as a specification of pure functional return value terms. However, return values are not the only places in an imperative program where pure functional terms may occur.

Also, in the case of languages such as Eiffel side-effect-free boolean assertions are sometimes used for run-time testing of programs. These assertions can be understood as functional return values of boolean range [Mey00]. It would be an interesting and potentially fruitful topic to examine how our methods of return value synthesis could be adapted to the synthesis of such assertions.

6.6 Discussion

This chapter concludes Part III of the monograph. Chapter 4 defined IHL. Chapter 5 discussed issues relating to semantics and gave a soundness theorem, showing that IHL theorems represent truths about *SML* program side-effects. It also showed how IHL can be represented as a logical type theory, in the style of the Curry–Howard isomorphism for constructive logic, where proofs are given as terms and program/formula pairs as types.

In this chapter we have achieved our ultimate goal of adapting proofs-as-programs to IHL, building upon the results of previous chapters. We applied the Curry–Howard protocol of Chapter 3 from Part II. A new notion of realizability was given between IHL program/formula pairs and *SML* programs — where a pair specifies a required program in terms of side-effects and a return-value. We then defined an extraction map from proofs in the logical type theory of Chapter 5, to realizing imperative *SML* programs that are terms of a computational type theory.

These results show a successful and practical approach to merging constructive proofs-as-programs with Hoare logic. We retain the advantages of both methods, using them to target their concerns separately. Hoare logic is retained to reason about and develop a program in terms of side-effects. Constructive realizability is adapted to reason and develop functional return-values. Throughout the extraction process, programs with both aspects are synthesized from proofs.

To the best of our knowledge, this is the first time such an approach has been given.

Nondeterminism and total correctness have long been understood in extensions of the Hoare logic. It would be interesting to examine how our results could be adapted to such extensions.

Our work has applied constructive methods to the synthesis of imperative programs, taking realizability as a specification of pure functional return value terms. However, return values are not the only places in an imperative program where pure functional terms may occur. In the case of languages such as Eiffel [Mey97], side-effect-free boolean assertions are sometimes used for the run-time testing of programs. In general, boolean assertions can contain complex functional aspects, such as higher-order abstractions [Mey00]. We have briefly shown by example how these assertions can be understood as functional return values of boolean range, and how they can be synthesized using our approach. It would be an interesting and potentially fruitful topic to develop these results further to an industrial strength approach to assertion synthesis, for a language such as Eiffel.

We leave such investigations to future research.

The work of this part has been an application of the Curry–Howard protocol of Part II. Its success and utility provide a justification for the protocol as a good framework for generalizing proofs-as-programs. In the next part of the book, we shall provide a second application of the protocol in the domain of structured algebraic specifications and structured program synthesis.

Part IV

Structured Proofs-as-Programs

7

Reasoning about Structured Specifications

In this chapter, we introduce a logical system, called the Structured Specification Logic (SSL), for reasoning constructively about structured specifications. We consider specifications written in the Common Algebraic Specification Language (*CASL*) as defined in the CoFI group's standard [CoF01]. Our logic is compositional in the sense that proofs about a structured specifications are given in a modular fashion, using knowledge about sub-specifications to derive knowledge about composite specifications. This promotes the desirable features of a divide-and-conquer approach and proof reuse.

SSL is based on one of the first compositional proof systems for structured specifications, defined by Martin Wirsing in [Wir91]. (See the introduction of Chapter 1 for a review of related work.) Peterreins, Crossley, and Wirsing [Pet96, WCP98] extended that calculus to a natural deduction system. Their system was concerned with structured specifications in an *ASL*-like kernel language [Wir86, ST88a], involving basic operators for composing specifications (renaming and hiding signatures and taking unions of specifications). That work was given as a natural deduction system and used classical logic.

The novelty of the SSL system is as follows. We simplify the rules of the original calculus of Peterreins, Crossley, and Wirsing, dividing them into structural and logical classes. We use *CASL* syntax for structured specifications instead of *ASL*. Also, we make the calculus constructive.

Theorems of our calculus consist of labelled formulae, of the form

$$\vdash_{\mathsf{SSL}} \textsc{Sp} \diamond P$$

where $\textsc{Sp}$ is a structured specification from *CASL* and P is a many-sorted formula that is derivable from the axioms of $\textsc{Sp}$ via constructive reasoning. It will be shown using soundness that P is true for the models of $\textsc{Sp}$.

Our calculus involves structural and logical rules, permitting us to do two things.

1. *Reasoning about specifications.* This is achieved by means of logical rules that augment the rules of intuitionistic many-sorted logic to deal with speci-

fications. For example, we define the following rule to prove the conjunction $(A \wedge B)$ from the axioms of a specification SP, given that we already have proofs of A and B from SP:

$$\frac{\Gamma_1 \vdash \mathrm{SP} \diamond A \quad \Gamma_2 \vdash \mathrm{SP} \diamond B}{\Gamma_1, \Gamma_2 \vdash \mathrm{SP} \diamond (A \wedge B)} \ (\wedge\text{-I})$$

This rule is the usual constructive rule for $(\wedge$-I$)$, augmented to accommodate specification labels.

2. *Building new specifications.* This is done by adding so-called structural rules that, given a theorem, permit us to change the specification label. In this chapter, structural rules correspond to the standard ways of creating structured specifications as presented in *CASL*: translating, hiding signatures, taking unions, and extending specifications. For instance, we define the following rule,

$$\frac{\Gamma \vdash_{\mathsf{SSL}} \mathrm{SP\_1} \diamond A}{\Gamma \vdash_{\mathsf{SSL}} (\mathrm{SP\_1}\ \mathbf{and}\ \mathrm{SP\_2}) \diamond A} \ (\mathrm{union}_1)$$

which tells us that if A is true about SP_1, then A is also true about the union of SP_1 and SP_2.[1] In Chapter 10, we provide additional structural rules that deal with defining and instantiating generic (parametrized) specifications.

This process of reasoning about and constructing new specifications is similar to the Hoare logic of Part III, which also had a notion of manipulating formulae and the associated information carried by labels. In the case of Hoare logic, the labels were imperative programs. In the present case, they are structured specifications. The analogy is continued in the next chapter, where we define a logical type theory for SSL with labelled formulae treated as types, and in Chapter 9, where we adapt proofs-as-programs to SSL, defining a realizability notion for labelled formulae.

The chapter is organized as follows:

- In Section 7.1, we provide an overview of background knowledge and define the basic specifications used in *CASL*.
- Section 7.2 defines the necessary concepts of *CASL* structured specifications.
- Our logic is presented in Section 7.3.
- Section 7.4 gives a soundness theorem.
- Section 7.5 provides a summary and discussion of our results.

We illustrate our work with a password checking system example, similar to that examined in Chapter 2 of Part II. We show how to specify and reason about this system in a structured fashion. We will continue to use this example in the following two chapters of this part, illustrating how our reasoning can be used to extract correct programs and construct specification refinements.

[1] This is a specialized version of the (union_1) rule defined in this chapter: see particularly Remark 7.16 of Section 7.3, and the list of structural rules in Fig. 7.5, p. 239.

7.1 Specifications

In this section, we outline the important concepts of many-sorted signatures, basic specifications, and models, as provided in the *CASL* CoFI document [CoF01].

The ideas of this section are all from elementary model theory. A basic specification consists essentially of a signature and a set of axioms over a signature. From a computational perspective, basic specifications provide a means of defining simple, unstructured components. The semantics of a basic specification is the class of all models that satisfy the axioms. Semantics is important when specifying components because classes of models can be considered to denote a range of possible implementations.

Later in this chapter we will be representing, and reasoning about, *SML* programs using the *CASL* syntax for basic specifications. We will only be interested in *SML* programs that implement total functions. So, in order to minimize the technical problems of representing such programs in our specifications, we confine ourselves to basic specifications with total functions, although partial functions are permitted in *CASL*.

7.1.1 Many-sorted signatures

We use the COFI document's definition of a many-sorted signature [CoF01, p. 3], but restricted to involve only total function symbols.

Definition 7.1.1 (Many-sorted signature with total functions). A *many-sorted signature* $\Sigma = \langle S, TF, P\rangle$ consists of:

- a set S of sorts;
- sets $TF_{w,s}$ of total function symbols for each function profile (w, s) consisting of a word sequence of argument sorts $w \in S^*$ and a result sort $s \in S$ (constants are treated as functions with no arguments);
- sets P_w of predicate symbols for each predicate profile consisting of a word sequence of argument sorts $w \in S^*$.

We overload $\in$ to denote membership of sorts, functions, or predicates in the appropriate sets of a signature. That is, given a sort symbol s we write $s \in \Sigma$ if $s \in S$. Similarly, for a function symbol f and a predicate symbol A, we write $f \in \Sigma$ and $A \in \Sigma$ if $f \in TF$ and $A \in P$ respectively.

Constants and functions are also referred to as *operations*.

Remark 7.1. *CASL* signatures are the standard notion of signatures. They are similar to those of Definition 2.1.1, Chapter 2 of Part II. However, we do not (yet) include functional sorts — and the terms of our signatures as they will be defined in this chapter are not lambda terms. In Chapter 9, we will extend *CASL* signatures and terms to include functional sorts and also lambda abstraction and application, for the purposes of program extraction. However, for the purposes of this chapter, it is only necessary to understand first-order many-sorted signatures.

Remark 7.2. Conforming to *CASL* syntax, given a signature $\Sigma = \langle S, TF, P\rangle$, we will often denote membership to a set of TF using product and function typing notation. That is, rather than writing

$$f \in TF_{s_1 \ldots s_n, s}$$

we will simply write

$$f : (s_1 \times \ldots \times s_n) \to s$$

for $s_1 \ldots s_n \in S^*$ and a result sort $s \in S$.

As in the previous part, and also following the CoFI standard [CoF01, p. 3], the symbols that identify operations and predicates may be overloaded, occurring in more than one of the sorted sets. Whenever there is ambiguity in sentences, function symbols f and predicate symbols P should be qualified by profiles, written $f_{w,s}$ and p_w respectively. We omit these profiles when there is no ambiguity.

We will require a definition of *signature morphisms.*

Definition 7.1.2 (Many-sorted signature morphism). A many-sorted signature morphism

$$\sigma : \langle S, TF, P\rangle \to \langle S', TF', P'\rangle$$

is a mapping such that

1. $\sigma(S) \subseteq (S')$
2. $\sigma(TF) \subseteq TF'$
3. $\sigma(P) \subseteq P'$
4. if $f \in TF_{w,s}$, then $\sigma(f) \in TF'_{\sigma^*(w),\sigma(s)}$
5. if $Q \in P_w$ then $\sigma(Q) \in P'_{\sigma(w)}$

where (given $w = s_1 \ldots s_n$) $\sigma^*(w) = \sigma(s_1) \ldots \sigma(s_n)$.

Definition 7.1.3. We define **CSig** to be the category with signatures as objects and signature morphisms as morphisms.

We will require several operations on signatures.

Disjoint unions of signatures are defined as follows.

Definition 7.1.4. Given two signatures

$$\Sigma_1 = \langle S_1, TF_1, P_1\rangle \text{ and } \Sigma_2 = \langle S_2, TF_2, P_2\rangle$$

the disjoint union

$$\Sigma_1 \uplus \Sigma_2 = \langle S_1 \uplus S_2, TF_1 \uplus TF_2, P_1 \uplus P_2\rangle$$

where $\uplus$ is the disjoint union for sets, so that, for any symbol t, we have symbols t_l and t_r such that $t_l \in \Sigma_1 \uplus \Sigma_2$ if, and only if, $t \in \Sigma_1$ and $t_r \in \Sigma_1 \uplus \Sigma_2$ if, and only if, $t \in \Sigma_2$.

The amalgamated union of signatures is defined using a pushout construction (following, e.g., [Cen94, pp. 18–21]).

Definition 7.1.5 (Amalgamated unions). Given two signatures,

$$\Sigma_1 = \langle S_1, TF_1, P_1 \rangle \text{ and } \Sigma_2 = \langle S_2, TF_2, P_2 \rangle$$

that share a (possibly empty) sub-signature $\Sigma = \langle S, TF, P \rangle$, we define the disjoint union $\Sigma_1 +_\Sigma \Sigma_2$ to be the pushout in **CSig**,

$$\begin{array}{ccc} \Sigma & \xrightarrow{i_1} & \Sigma_1 \\ {\scriptstyle i_2}\downarrow & & \downarrow{\scriptstyle inl} \\ \Sigma_2 & \xrightarrow{inr} & \Sigma_1 +_\Sigma \Sigma_2 \end{array}$$

where

- i_1 and i_2 are injections of Σ into Σ_1 and Σ_2.
- $\Sigma_1 +_\Sigma \Sigma_2$ is defined as

  $$\langle S_1/S \uplus S_2/S \uplus S, TF_1/TF \uplus TF_2/TF \uplus TF, P_1/P \uplus P_2/P \uplus P \rangle$$

  where $\uplus$ is the disjoint union for sets. For any symbol t, we have symbols t_l, t_r such that $t_l \in \Sigma_1 +_\Sigma \Sigma_2$ if, and only if, $t \in \Sigma_1/\Sigma$, $t_r \in \Sigma_1 +_\Sigma \Sigma_2$ whenever $t \in \Sigma_2/\Sigma$ and $t \in \Sigma_1 +_\Sigma \Sigma_2$ whenever $t \in \Sigma$.
- The pushout morphisms are defined by

  $$inl(t) = \begin{cases} t & \text{if } t \in \Sigma \\ t_l & \text{otherwise} \end{cases}$$
  $$inr(t) = \begin{cases} t & \text{if } t \in \Sigma \\ t_r & \text{otherwise} \end{cases}$$

To define renaming and hiding operations, we require the following concept.

Definition 7.1.6 (Symbol lists and mappings, [CoF01], section 6.4). A *symbol list* is a list of sort, function (and predicate) symbols and a *symbol mapping* ρ is a list of maps of the form

$$SY_1 \mapsto SY_1', ..., SY_n \mapsto SY_n'$$

where each SY_i and SY_i' is a sort, function or predicate symbol. $SY_i \mapsto SY_i'$ denotes a map that takes the sort or symbol SY_i to the sort or symbol SY_i', respectively. A symbol mapping must not map the same symbol to two different symbols.

We write ρ^{-1} for the same symbol mapping with mappings reversed.

A symbol mapping ρ extends to a morphism between signatures,

$$\widehat{\rho} : \Sigma \to \Sigma'$$

so that, for any sort, function or predicate symbol t of Σ, we define

$$\widehat{\rho}(t) = \begin{cases} t' \text{ if } t \mapsto t' \text{ is in } \rho \\ t \text{ otherwise} \end{cases}$$

This morphism is well-defined, *provided* that $\widehat{\rho}(t)$ is a symbol of Σ'. It can easily be seen that if ρ is well-defined, $\widehat{\rho^{-1}}$ is the inverse $(\widehat{\rho})^{-1}$ of $\widehat{\rho}$. When there is no ambiguity, we will overload ρ to denote with a symbol mapping ρ and its associated morphism.

Renaming is an operation on signatures using symbol mappings.

Definition 7.1.7. Given a signature $\Sigma = \langle S, TF, P \rangle$, and a symbol mapping ρ, we define the renaming $\rho(\Sigma)$ to be

$$\langle \rho(S), \rho(TF), \rho(P) \rangle$$

The final important operation on signatures is *hiding*, defined as follows.

Definition 7.1.8 (Hiding). Let SL be a symbol list consisting of (possibly empty) sets of sort symbols S_0, function symbols TF_0 and predicate symbols P_0. Let Σ_1 be the associated signature $\langle S_1, TF_1, P_1 \rangle$.

Then we define Σ_1 *with* SL *hidden*, written Σ_1/SL, to be the signature

$$\langle S_1/S_0, TF_1/TF_0, P_1/P_0 \rangle$$

7.1.2 Terms and formulae

A signature is associated with sets of well-formed terms formed from free variables and the function symbols of the signature. These terms are sorted according to the sorts of the signature. These terms, in turn, together with the predicate and sort symbols of the signature, form a set of well-formed formulae for the signature.

Definition 7.1.9 (Terms of a signature). Let $\Sigma = \langle S, TF, P \rangle$ be a signature. Let X be a set that includes an S-sorted set of free variables, disjoint from the constants in TF (so that X consists of disjoint subsets, X_s, indexed by $s \in S$).

For every sort $s \in S$, *the set* $Term(\Sigma, X)_s$ *of terms of sort* s is the least set containing

1. every $x \in X_s$ of sort s and every nullary operation symbol $f \in TF_{\emptyset,s}$, and
2. every $f(t_1, \ldots, t_n)$ where $f \in TF_{s_1 \ldots s_n, s}$ is a function symbol in TF with range s and every t_i $(i = 1, \ldots, n)$ is a term of sort s_i in $Term(\Sigma, X)_{s_i}$.

Terms without elements of X are referred to as *ground terms* and $Term(\Sigma, \emptyset)_s$ is denoted by $Term(\Sigma)_s$.

If t is a Σ-term, then $FV(t)$ denotes the set of free variables in t.

A *sensible signature* has at least one ground term for each sort.

Definition 7.1.10 (Well-formed formulae of a signature). Let $\Sigma = \langle S, TF, P \rangle$ be a signature.

Let X be a S-sorted set of free variables, disjoint from the constants in TF.

The set of well-formed formulae for a signature, $WFF(\Sigma, X)$ is the least set containing

- every $P(t_1, \ldots, t_n)$ where $P \in P_{s_1 \ldots s_n}$ is a predicate symbol in P and every t_i $(i = 1, \ldots, n)$ is a term of sort s_i in $Term(\Sigma, X)_{s_i}$,
- every formula $(A \wedge B)$, where $A, B \in WFF(\Sigma, X)$,
- every formula $(A \vee B)$, where $A, B \in WFF(\Sigma, X)$,
- every formula $(A \Rightarrow B)$, where $A, B \in WFF(\Sigma, X)$,
- every formula $\forall x : s \bullet F$, where $x \in X_s$ and $F \in WFF(\Sigma, X)$,
- every formula $\exists x : s \bullet F$, where $x \in X_s$ and $F \in WFF(\Sigma, X)$, and
- the formula $\perp$.

We often write $\neg A$ as an abbreviation for $A \Rightarrow \perp$.

Definition 7.1.11. We can inductively extend the signature morphism $\sigma : \Sigma \rightarrow \Sigma'$, to a morphism between formulae of $WFF(\Sigma', X)$ in the obvious way. That is, $\sigma(F) \in WFF(\Sigma', X)$ for $F \in WFF(\Sigma, X)$ is obtained by applying σ to each term and predicate symbol in F.

Assumption 7.1. As in previous parts of this monograph we will always use terms built over a denumerable set of variables, Var. We will assume that Var is large enough to include S-sorted sets of free variables for any signature's list of sorts S.

7.1.3 Structures

As usual, signatures are associated with semantic objects by means of a semantic *interpretation function.* Interpreted signatures without predicates are known as algebras (in universal algebra theory). In our case, where predicate symbols are used, the resulting interpretations are called *structures* for signatures.

A Σ-structure has a carrier set for each sort of Σ, a function on those sets corresponding to each function *symbol* of Σ, and a subset of tuples of carriers for each predicate symbol of Σ.

Definition 7.1.12 (Σ-structure). Let M be a Σ-structure. Let w^M denote the Cartesian product $s_1^M \times \ldots \times s_n^M$ if $w = s_1 \ldots s_n$.

For a many-sorted signature $\Sigma = \langle S, TF, P \rangle$, a *$\Sigma$-structure* M consists of

- non-empty carrier sets s^M for each sort $s \in S$,
- a total function f^M from w^M to s^M for each function symbol $f \in TF_{w,s}$,
- a relation $P^M \subseteq s_1^M \times \ldots \times s_n^M$ for each predicate symbol $P \in P_w$ with $w = s_1 \ldots s_n$.

We write $Struct(\Sigma)$ for the set of all Σ-structures.

A Σ-homomorphism is a map between Σ-structures preserving the operations interpreting the functions symbols of Σ and the relations interpreting the predicate symbols of Σ.

A (weak) many-sorted homomorphism h from M_1 to M_2, with $M_1, M_2 \in Struct(\Sigma)$, $\Sigma = \langle S, TF, PF, P \rangle$, consists of a function $h_s : s^{M_1} \rightarrow s^{M_2}$ for each $s \in S$ preserving not only the values of functions but also their definedness, and preserving the truth of predicates.

Definition 7.1.13 (Reducts, [Wir91, p. 682]). Given a structure $M = Struct(\Sigma')$ and a signature morphism $\sigma : \Sigma \rightarrow \Sigma'$, one can recover the $\Sigma = \langle S, TF, P \rangle$ structure buried inside M — this structure is called the *σ-reduct* of M, written $M|_\sigma$, consisting of

- carrier sets $s^{M|_\sigma} = s^M$ for each $s \in S$,
- a total function $f^{M|_\sigma} = \sigma(f)^M$ for each $f \in TF$, and
- a relation $A^{M|_\sigma} = \sigma(A)^M$ for each $A \in P$.

7.1.4 Interpretations of terms

Given a signature, we can define an interpretation map ι from terms of the signature to a structure for the signature — this map provides a meaning for the terms.

Definition 7.1.14 (Interpretation). Given a signature Σ, we extend an interpretation map ι from terms $Term(\Sigma, X)$ to a structure $M \in Struct(\Sigma)$ via a *variable valuation map* $\hat{\iota} : X \rightarrow M$:

- $\iota(x) = \hat{\iota}(x)$ for every $x \in X_s$,
- $\iota(f(t_1, \ldots, t_n)) = f^M(\iota(t_1), \ldots, \iota(t_n))$

We call interpretation τ the *x-variant* of interpretation τ' when they differ only over a particular variable x.

7.1.5 Formula satisfaction

We take the usual approach to defining when a well-formed formula of a signature is true of a structure for the signature.

Definition 7.1.15 (Satisfaction). Take any signature Σ and a structure M of Σ (from $Struct(\Sigma)$).

For any formula F in $WFF(\Sigma, Var)$ and valuation $\hat{\iota} : X \rightarrow M$, then M *satisfies F under ι*, written

$$M \models_\iota F$$

when

- if F is $P(t_1, \ldots, t_n)$, then $(\iota(t_1), \ldots, \iota(t_n)) \in P^M$
- if F is $(A \wedge B)$ then $M \models_\iota A$ and $M \models_\iota B$,
- if F is $(A \vee B)$ then $M \models_\iota A$ or $M \models_\iota B$,

- if F is $(A \Rightarrow B)$ then, if $M \models_\iota A$ holds, it must be the case that $M \models_\iota B$,
- if F is $\forall x : s \bullet Q$, where $x \in X_s$, then $M \models_{\iota'} Q$ for every x-variant ι' of ι,
- if F is $\exists x : s \bullet Q$, where $x \in X_s$, then $M \models_{\iota'} Q$ for some x-variant ι' of ι,

We require that $M \models_\iota \bot$ never holds.

If $M \models_\iota F$ holds for every valuation ι, then we say that M *satisfies* F and write $M \models F$.

spec NATBOOLEAN =
sorts
  $nat, boolean$
ops $0 : nat; s : nat \to nat; + : nat \times nat \to nat, T : boolean,$
  $F : boolean, ge : nat \times nat \to boolean$
preds
  $\geq: nat \times nat$
axioms
  $\forall x : nat \bullet x + 0 = x$
  $\forall x; y : nat \bullet x + s(y) = s(x + y)$
  $\forall x; y : nat \bullet x + y = y + x$
  $\forall x; y; v; w : nat \bullet x \geq v \wedge y \geq w \Rightarrow x + y \geq v + w \forall x : nat \bullet x \geq 0$
  $\forall x : nat \bullet s(x) \geq x$
  $\forall x; y : nat \bullet x \geq y \Rightarrow ge(x, y) = T$
end

Fig. 7.1. A basic specification of the natural numbers and booleans.

7.1.6 Basic specifications

Ultimately we will be building larger specifications of components from smaller specifications of smaller components. The atomic building blocks of this process are *basic* (otherwise called *simple* or *flat*) specifications. From a computational perspective, basic specifications provide a means of defining primitive, unstructured components for use in defining more complex data types.

For our purposes, we take a basic specification to be as follows.

Definition 7.1.16 (Basic specification). A basic specification is a specification of the form

$$\text{SP} = \langle \Sigma, Ax \rangle$$

where Σ is a signature and Ax is a set *axioms* for the signature, formulae from $WFF(\Sigma, Var)$.

Assumption 7.2. For the purposes of program extraction, described in Chapter 9, we will assume all axioms are Harrop (given by Definition 6.1.1 of Chapter 2, Part II, p. 33).

spec StringBool =
sorts
  $string, bool$
ops $a : string, b : string, \ldots, x : string, y : string, z : string, A : string,$
  $B : string, \ldots, Z : string, space : string, emptystring : string,$
  $concat : string \times string \rightarrow string, toUpper : string \rightarrow string,$
  $toLower : string \rightarrow string, true : bool, false : bool, not : bool \rightarrow bool,$
  $or : bool \times bool \rightarrow bool$
axioms
  $\forall x : string \bullet concat(emptystring, x) = x$
  $\forall x : string \bullet concat(x, emptystring) = x$
  $\forall x; y : string \bullet toUpper(concat(x, y)) = concat(toUpper(x), toUpper(y))$
  $\forall x; y : string \bullet toLower(concat(x, y)) = concat(toLpper(x), toLower(y))$
  $toUpper(a) = A \ldots toUpper(z) = Z$
  $toLower(A) = a \ldots toLower(Z) = z$
  $not(true) = false \quad not(false) = true$
  $or(true, true) = true \quad or(true, false) = true$
  $or(false, true) = true \quad or(false, false) = false$
end

We will often use '$a_1 \ldots a_{n-1}a_n$' for strings of the form $concat(a_1, concat(\ldots, concat(a_{n-1}, a_n) \ldots))$ and write a blank space for the constant *space*.

Fig. 7.2. A basic specification of strings and booleans.

Assumption 7.3. We assume that, for each basic specification $\text{Sp} = \langle \Sigma, Ax \rangle$, $\Sigma = \langle S, TF, P \rangle$ contains a distinguished equality predicate $=_s \in P_{ss}$ for each $s \in S$.

Remark 7.3. We use equality predicates to define the so-called existential equations of *CASL*, of the form

$$t_1 = t_2$$

(we omit the profile subscript in $=_s$ when the sorts of the terms are clear). Intuitively, an existential equation holds in a structure when the interpreted values of both terms t_1 and t_2 are defined and identical.[2]

Assumption 7.4. We assume that, for each basic specification $\text{Sp} = \langle \Sigma, Ax \rangle$, $\Sigma = \langle S, TF, P \rangle$, there is a distinguished set $Cons_s \subseteq (\bigcup_{r \in S} TF_r)$ for each $s \in S$, called the set of constructors for s. If $Cons_s \neq \emptyset$, we call s a *generated sort*.

Remark 7.4. Informally, the constructors of a generated sort form a canonical representation of every element in the sort. From the perspective of our intended

[2] For the sake of simplicity, we do not deal with strong equations, which also hold when the values of both terms are undefined. Our work could easily be adapted to include strong equations.

semantics, when a generated sort is taken as a set of elements, all the elements of the set always represent terms formed from the sort's constructors — see our discussion of loose models for specifications below.

In Chapter 9, we investigate program synthesis from proofs in our logic by applying the Curry–Howard protocol. For reasons to do with program synthesis, we will make the convenient (though not essential) assumption that all axioms of basic specifications are Harrop formulae (see Definition 2.2.1 of Chapter 2 in Part II).

It is possible to write basic specifications using a more lengthy, human-readable syntax employed in *CASL*. For instance, using the longer syntax, the specification $\text{SP} = \langle \Sigma, Ax \rangle$, $\Sigma = \langle S, TF, P \rangle$ can be written in standard *CASL* notation:

spec SP =
sorts
  SortList
ops
  OpList
preds
  PredList
axioms
  AxiomList
end
where

- *SortList* is a list of the sorts of S,
- *OpList* is a list of every operation of TF and their associated sorts, so that, if $t \in TF_{s_1 \ldots s_n, s}$ then

$$t : s_1 \times \ldots \times s_n \to s \in OpList$$

- *PredList* is a list of every predicate of P with associated sorts, so that, if $t \in P_{s_1 \ldots s_n}$ then

$$P : s_1 \times \ldots \times s_n \in PredList$$

- *Axioms* is a list of the axioms Ax

We use this syntax when we wish to clearly present a large specification.

Example 7.1. A basic specification of the natural numbers and booleans is given in Fig. 7.1. The signature of the specification consists of the sort of natural numbers, a successor function s, the addition function, and ordering predicate $\geq$. The first two axioms provide a recursive definition of addition, the third axiom defines addition to be commutative, and the remaining axioms define the $\geq$ predicate as an ordering over the natural numbers. The booleans are given a simple definition, consisting only of constant symbols for truth and falsity. Finally, there is a boolean function over the naturals $ge(x, y)$ that has the value T when $x \geq y$ holds.

A basic specification of strings and booleans is given as StringBool in Fig. 7.2. The only predicates used are the implicitly assumed equality predicates for strings and booleans. The upper and lowercase letters of the alphabet are given as constant symbols. Function symbols consist of concatenation and changing the case of a string. In contrast to the specification NatBool of Fig. 7.1, StringBool contains a more detailed axiomatization of the booleans, with function symbols for negation and disjunction.

7.1.7 Semantics of basic specifications

The semantics of a basic specification consists of a class of structures that satisfy the axioms of the specification. We will consider two ways of defining the semantics — an *algebraic* and a *loose* semantics (we take our definitions from [Wir91, pp. 696–699]).

Algebraic semantics is as follows. Given a basic specification $\text{Sp} = \langle \Sigma, Ax \rangle$, we can associate with it a class of Σ-structures that satisfies all the axioms Ax. We call this class the (algebraic) models of Sp, $Alg(\text{Sp})$ so

$$Alg(\text{Sp}) = \{M \in Struct(\Sigma) \mid M \models A \text{ for every } A \in Ax\} \tag{7.1}$$

This class of structures denotes the range of possible meanings a specification may have. For instance, some of these models could be executable programs that implement the functions defined in the specification.

An algebraic semantics admits *nonstandard* models — those models in which models use elements that are not representable by the terms of Sp. The problem with this semantics is that it includes such "useless," non-specified elements. Computationally, this means the models are not precise enough to represent a concrete definition of the data types used in a program component.

Also, we would like a generated sort to denote a set whose elements are all represented by terms formed from the constructors for the sort. Because we have nonstandard models, this is not possible.

To rectify these problems, the loose semantics is defined by the following restriction on the class of structures used in (7.1). Essentially we constrain our semantics to models whose elements always correspond to terms of the specification. Given a basic specification $\text{Sp} = \langle\langle S, TF, P \rangle, Ax\rangle$, we associate a class of *reachable* Σ-structures that satisfy all the axioms Ax, defined by

$$Loose(\text{Sp}) = \{M \in Struct(\Sigma) \mid M \models A \text{ for every } A \in Ax \text{ and, for each } s \in S \text{ and each } a \in s^M \text{ if } s \text{ is not generated, then there is a corresponding ground term } t \in Term(\langle S, TF, P \rangle, \emptyset)_s, \text{ and if } s \text{ is generated, there is a corresponding ground term formed from the constructors for } s,\ t \in Term(\langle S, Cons_s, P \rangle, \emptyset)_s\} \tag{7.2}$$

Unless otherwise stated, loose semantics will be used here to provide the intended interpretation of basic specifications used in *CASL*.

7.2 Structured specifications

In writing large specifications it is desirable to design specifications in a structural fashion by combining and modifying smaller specifications. This supports modular decomposition, facilitating a divide-and-conquer approach to defining system component requirements. This is desirable because a complex system typically involves many functions and axioms, which become unmanageable when defined using a simple basic specification.

In *CASL* a structured specification is formed by combining specifications in various ways, starting from basic specifications. For instance,

- specifications may be united,
- a specification may be extended with further signature items and/or axioms,
- parts of a signature may be hidden,
- the signature may be translated to use different symbols (with corresponding translation of the sentences) by a signature morphism, and
- models may be restricted to initial models.

We will now provide an overview of how structured specifications are treated in the CoFI standard [CoF01]. For the purposes of this and the next two chapters, we will be concerned with the following specification structuring operations: building unions and extensions, hiding signatures and translations of symbols. We defer treatment of generic (parametrized) and named specifications in *CASL* to Chapter 10.

Example 7.2. We will illustrate our concepts with the following ongoing example of a password checking system example, similar to that used throughout Chapter 2.

The informal domain specification is as follows. We consider a service that hosts email accounts for a number of users. When a user joins the service, he/she is required to define a new numerical password. We make the following assumptions concerning the password correctness functions for a new user joining or logging onto the system:

- Password numbers must be 4 digits long (and so within the range of 0000 to 9999).
- If the number chosen is not of the right length, the system should output a response message, asking the user to select a new number within the correct range.
- If the number is within the correct range, then the system should output a response message to this effect.

We shall model the system within *CASL* by formally specifying these assumptions, defining notions of acceptable lengths of passwords and the correct responses for given passwords.

7.2.1 Specification expressions

In *CASL*, we understand structured specifications by means of the collection **CSpec** of *specification expressions*, denoting the range of possible basic and structured specifications. Later in this section we will introduce the *CASL* operators to build specification expressions.

We require that every specification expression is associated with two maps, namely

$$Sig : \mathbf{CSpec} \to \mathbf{CSig}$$

giving the *visible signature* of the specification, and

$$Mod : \mathbf{CSpec} \to \{M \subseteq Struct(\Sigma) \mid \Sigma \in \mathbf{CSig}\}$$

giving the models of the specification.

7.2.2 Basic specifications

A basic specification

$$\langle \Sigma, Ax \rangle$$

is a specification expression, with

$$Sig(\langle \Sigma, Ax \rangle) = \Sigma$$

and

$$Mod(\langle \Sigma, Ax \rangle) = Loose(\langle \Sigma, Ax \rangle)$$

Remark 7.5. Note that we will sometimes consider an algebraic semantics instead of a loose semantics for basic specifications. In this case we will take

$$Mod(\langle \Sigma, Ax \rangle) = Alg(\langle \Sigma, Ax \rangle)$$

Unless otherwise stated, however, we will employ the loose semantics.

7.2.3 Translation

Syntactically, the translation operation permits us to rename the signature and axioms of a specification to give a new specification that uses renamed symbols. If we consider a specification as specifying component requirements, the renamed specification can be considered as a means of wrapping the component requirements with a new interface.

In *CASL*, given a specification expression SP and a symbol mapping ρ, we will write

$$\rho \bullet \text{SP}$$

for the expression denoting the translation of SP by ρ.

We define

$$Sig(\rho \bullet \text{SP}) = \rho(Sig(\text{SP}))$$

and

$$Mod(\rho \bullet \text{SP}) = \{A|_{\rho^{-1}} \mid A \in Mod(\text{SP})\}$$

Example 7.3. Consider the basic specifications of Example 7.1. Take the symbol mapping

$$BtoB = [boolean \mapsto Bool, T \mapsto true, F \mapsto false]$$

Then the specification

$$BtoB \bullet \text{NATBOOLEAN}$$

is NATBOOLEAN of Fig. 7.1 with the booleans now renamed to have the same sort and constant symbols as those of the specification STRINGBOOL in Fig. 7.2. The axioms and functions are appropriately renamed, so that *ge* is now a function from $Nat \times Nat$ to *Bool*, whose behavior is given according to the renamed axiom:

$$\forall x; y : nat \bullet x \geq y \Rightarrow ge(x, y) = T$$

7.2.4 Union

The union of two specifications is a new specification that retains the meaning of the shared parts of the specifications. In *CASL*, given two specification expressions SP_1 and SP_2, and a signature Σ such that $\Sigma \subseteq Sig(\text{SP\_1})$ and $\Sigma \subseteq Sig(\text{SP\_2})$, the union

$$\text{SP\_1 } \mathbf{and} \text{ SP\_2}$$

is a specification expression with

$$Sig(\text{SP\_1 } \mathbf{and} \text{ SP\_2}) = Sig(\text{SP\_1}) +_{\Sigma} Sig(\text{SP\_2})$$

and

$$Mod(\text{SP\_1 } \mathbf{and} \text{ SP\_2}) = \{C \in Mod(Sig(\text{SP\_1}) +_{\Sigma} Sig(\text{SP\_2})) \mid C|_{inl} \in Mod(\text{SP\_1}) \text{ and } C|_{inr} \in Mod(\text{SP\_2})\}$$

(The signature morphisms *inl* and *inr* are from the pushout construction for $Sig(\text{SP\_1})$ and $Sig(\text{SP\_2})$ and their common signature — see Definition 7.1.5, p. 221 above.)

Example 7.4. Consider the union

$$\text{SNB} = (BtoB \bullet \text{NATBOOLEAN}) \textbf{ and } \text{STRINGBOOL}$$

(where *BtoB* was given in Example 7.3 above). The two sub-specifications ($BtoB \bullet$ NATBOOLEAN) and STRINGBOOL involve a common signature — that with the boolean sort *Bool* and *true* and *false* constant symbols. Consequently, by the nature of the pushout construction, the boolean symbols of both the sub-specifications are interpreted by the same objects in all models of SNB.

7.2.5 Extension

The extension of a specification is a way of adding additional symbols and axioms to a specification to extend the specification, whilst retaining the original meaning.

Extensions are useful when we wish to define new axioms Ax_{ext} using symbols from a given specification SP_1 and possibly new symbols Σ_{ext}. A collection of such new axioms and symbols $\langle \Sigma_{ext}, Ax_{ext} \rangle$ constitutes a so-called *partial specification*. Such a partial specifications has no models in isolation — its axioms require symbols from SP_1 to be interpreted.

For the purposes of this monograph, we will define the meaning of extensions using unions.

If SP_1 is a specification and SP_EXT $= \langle \Sigma_{ext}, Ax_{ext} \rangle$ is a (possibly partial) specification which determines an extension from Sig(SP_1) to a complete signature Σ, then the extension

SP_1 **then** SP_EXT

is a specification expression in *CASL*. We define this expression to be equivalent to a union of the form

SP_1 **and** SP_2

where SP_2 $= \langle \Sigma \cup \Sigma_{ext}, Ax_{ext} \rangle$, the partial specification SP_EXT extended to include the complete signature Σ. It follows that

$$Sig(\text{SP\_1}) = \Sigma$$

and

$$Mod(\text{SP\_1 } \mathbf{then} \text{ SP\_EXT}) = Mod(\text{SP\_1 } \mathbf{and} \text{ SP\_2})$$

Remark 7.6. Extensions add nothing that cannot be expressed using unions. However, they are a convenient, because, if we want to add new information to a specification, we need not define a full specification to add appropriate new axioms, as would have to be done in the case of unions.

Example 7.5. We use extensions to specify the password system outlined in Example 7.2. The specification of the system, PWDCORE, is given in Fig. 7.3.

The specification PWDCORE extends the naturals, strings and booleans given by SNB

$$\text{SNB} = (BtoB \bullet \text{NATBOOLEAN}) \ \mathbf{and} \ \text{STRINGBOOL}$$

We model aspects of the password checking system by adding new functions and predicates.

- We define a new boolean function *inRange*(x) that will output *true* if the password number (x) is within the required range (between the natural numbers 1000 and 9999).

- A new predicate $OkPwd(x)$ holds over a number, if the number is an acceptable password (that is, if $inRange(x) = true$).
- A new predicate $ValidMsg(x, y)$ that holds if a string y is a correct response message for the input of a password number x.

spec PWDCORE = (*BtoB* • NATBOOLEAN) **and** STRINGBOOL **then**
ops $inRange : nat \rightarrow Bool$
preds
$OkPwd : nat,\ ValidMsg : nat \times String$
axioms
$\forall x : nat \bullet ge(x, 0) = true \wedge ge(s^{9999}(0), x) = true \Rightarrow inRange(x) = true$
$\forall x : nat \bullet ge(x, 0) = true \wedge ge(s^{9999}(0), x) = false \Rightarrow inRange(x) = false$
$\forall x : nat \bullet ge(x, 0) = false \wedge ge(s^{9999}(0), x) = true \Rightarrow inRange(x) = false$
$\forall x : nat \bullet ge(x, 0) = false \wedge ge(s^{9999}(0), x) = false \Rightarrow inRange(x) = false$
$\forall x : nat \bullet inRange(x) = true \Rightarrow OkPwd(x)$
$\forall x : nat \bullet inRange(x) = false \Rightarrow \neg OkPwd(x)$
$\forall x : nat \bullet \forall y : string \bullet OkPwd(x) \Rightarrow ValidMsg(x,$ 'Password acceptable')
$\forall x : nat \forall y : string \bullet \neg OkPwd(x) \Rightarrow ValidMsg(x,$ 'Please choose a password in correct range')

end

We write $s^i(0)$ for the successor function applied i times to 0.

Fig. 7.3. Specification of the password checking system.

7.2.6 Hiding

We can hide symbols used by a signature of a specification while retaining the original meaning. If we consider a specification as a requirement of a component, hiding is a means of encapsulating functionality. Hiding permits us to expose only certain important parts of a component interface, but taking other other parts to be black-box workings of the component.

In *CASL*, given a specification expression SP and symbol list SL, SP *hidden by* SL is the specification expression

$$\text{SP } \mathbf{hide}\ SL$$

which is

$$Sig(\text{SP } \mathbf{hide}\ SL) = \Sigma = Sig(\text{SP})/SL$$

and

$$Mod(\text{SP } \mathbf{hide}\ SL) = \{C|_\sigma \mid C \in Mod(\text{SP}) \text{ with } \sigma \text{ the injection from } \Sigma \text{ to } Sig(\text{SP})\}$$

Example 7.6. We give a final specification of the password checking system outlined in Example 7.2, by encapsulating some of the functionality exposed by PwdCore (Example 7.5, Fig. 7.3).

While PwdCore models all the assumptions we require of the the system, it contains some implementation detail that should be encapsulated. In particular we hide the functions *ge* (defined in the sub-specification NatBoolean) and the function *inRange* (defined in the extension part of PwdCore). These functions are closer to implementation detail, because they are concerned with defining when a password is valid.

The resulting specification is

$$\text{PwdSys} = \text{PwdCore}\ \mathbf{hide}\ \{ge, inRange\}$$

When viewed as a component, this specification only exposes the relevant functionality that is related to the validity of the password and axioms defining correct response messages. The component does not expose details about how a password is determined to be valid.

7.2.7 Flattening structured specifications

To understand properties of a specification expression, it will sometimes be useful to consider a *normal form* of the expression, written as a basic specification with hidden symbols. The normal form can be considered a means of "flattening" the structure of the specification to a basic specification with some hidden symbols. The normal form is equivalent to the original specification in that the signature and set of models are the same.

The following theorem shows that a normal form exists for every specification. The proof of the theorem is constructive, showing how to build the normal form *nf*(Sp) for an expression Sp.

Theorem 7.2.1. *For any specification expression* Sp *there is a* normal form *specification expression of the form*

$$nf(\text{Sp}) = \langle \Sigma, Ax \rangle\ \mathbf{hide}\ SL_e$$

where SL_e *is some symbol list, such that*

$$Sig(\text{Sp}) = \Sigma / SL_e$$

and

$$Mod(\text{Sp}) = Mod(nf(\text{Sp}))$$

Proof. Given a specification Sp′ such that

$$Sig(\text{Sp}) = Sig(\text{Sp}')$$

and

$$Mod(\textsc{Sp}) = Mod(\textsc{Sp}')$$

we write

$$\textsc{Sp} \cong \mathit{nf}(\textsc{Sp})$$

If Sp is a basic specification, then SL_e is empty.

If Sp is Sp_1 **and** Sp_2, then by the IH, there are symbol lists SL_e^1 and SL_e^2 such that

$$\begin{aligned}\textsc{Sp\_1} &\cong \mathit{nf}(\textsc{Sp\_1}) = \langle \Sigma_1, Ax_1\rangle \textbf{ hide } SL_e^1\\ \textsc{Sp\_2} &\cong \mathit{nf}(\textsc{Sp\_2}) = \langle \Sigma_2, Ax_2\rangle \textbf{ hide } SL_e^2\end{aligned}$$

It can be shown that

$$\begin{aligned}\textsc{Sp} &\cong \mathit{nf}(\textsc{Sp\_1}) \textbf{ and } \mathit{nf}(\textsc{Sp\_1})\\ &= \langle \Sigma_1 +_\Sigma \Sigma_2, inl(Ax_1) \cup inr(Ax_2)\rangle \textbf{ hide } SL_e\\ &= \mathit{nf}(\textsc{Sp})\end{aligned}$$

where

- Σ is the signature common to Σ_1 and Σ_2
- inl and inr are the maps for the pushout of these signatures
- if Σ_1^e and Σ_2^e are signatures formed from the symbol lists SL_e^1 and SL_e^2, SL_e is the symbol list formed from $\Sigma_1^e +_\Sigma \Sigma_2^e$.

The proof of this is straightforward, but involved. See [Cen94, pp. 85–86] for details.

If Sp is Sp_1 **then** Sp_EXT, then Sp is semantically equivalent to a union of the form Sp_1 **and** Sp_2 where Sp_2 $= \langle \Sigma, Ax_{ext}\rangle$, where Σ is the extension of Sig(Sp_1) by Sig(Sp_EXT). So we can take *nf*(Sp) to be *nf*(Sp_1 **and** Sp_2).

If Sp is Sp_1 **hide** SL, then, by the IH, there is a symbol list SL_e^1 such that

$$\textsc{Sp\_1} \cong \mathit{nf}(\textsc{Sp\_1}) = \langle \Sigma_1, Ax_1\rangle \textbf{ hide } SL_e^1$$

So,

$$\begin{aligned}\textsc{Sp} &\cong \mathit{nf}(\textsc{Sp\_1}) \textbf{ hide } SL\\ &= \langle \Sigma_1, Ax_1)\rangle \textbf{ hide } (SL_e^1, SL)\\ &= \mathit{nf}(\textsc{Sp})\end{aligned}$$

where $SL_e^1, SL = SL_e$ is the concatenation of SL_e^1 and SL.

If Sp is $\rho \bullet$ Sp_1, then, by the IH, there is a symbol list SL_e^1 such that

$$\textsc{Sp\_1} \cong \mathit{nf}(\textsc{Sp\_1}) = \langle \Sigma_1, Ax_1\rangle \textbf{ hide } SL_e^1$$

Again, following [Cen94, pp. 85–86] it can be shown that there is a ρ' such that

$$\rho \bullet (\langle \Sigma_1, Ax_1\rangle \textbf{ hide } SL_e^1) \cong \langle \rho'(\Sigma_1), \rho'(Ax_1)\rangle \textbf{ hide } \rho(SL_e^1)$$

which we take to be *nf*(Sp). □

Definition 7.2.2 (Visible symbols and axioms). Given a specification SP with normal form

$$nf(\text{SP}) = \langle \Sigma, Ax \rangle \ \textbf{hide} \ SL_e$$

we call the symbols of Σ/SL_e the *visible sorts, functions, and predicates of* SP.

The subset of axioms Ax that only involve visible sorts, functions and predicates, is called the *visible axioms* of SP, written

$$Axioms(\text{SP}) = \{A \mid A \in Ax \text{ and } A \in WFF(\Sigma/SL_e, Var)\}$$

7.3 Reasoning about *CASL* specifications

Having understood the basic concepts of structured *CASL* specifications, we are now ready to define a logical calculus for constructing and reasoning about these specifications.

In this section, we develop an intuitionistic version of the natural deduction calculus originally proposed by Wirsing, Peterreins, and Crossley in [WCP98, Pet96] for reasoning about *ASL* specifications. We use *CASL* instead of *ASL* as our specification language. We call our calculus the Structured Specification Logic (SSL). SSL is constructive, extending intuitionistic logic.

This section presents the basic rules of our calculus. Later chapters will investigate additional rules for SSL. (Specifically, Chapter 9 adds a rule to deal with adding new extracted functions to specifications, and Chapter 10 adds rules to develop and reason with parametrized and named specifications from *CASL*.)

7.3.1 Judgements

The formal calculus is presented as a natural deduction system.

We deal with *judgements*, which we write in sequent form as

$$\Gamma \vdash_{\mathsf{SSL}} \text{SP} \diamond A$$

where SP is a *CASL* specification and A is a formula from $WFF(Sig(\text{SP}), Var)$. The *context*, Γ, is a set of assumption formulae from $WFF(Sig(\text{SP}), Var)$. The intended meaning of the judgement is that, *assuming Γ are satisfied by the models of* SP*, then A is also satisfied.*

Throughout this monograph, when the context is clear, we will abbreviate $\vdash_{\mathsf{SSL}}$ by $\vdash$.

As usual, we employ a sequent style presentation of natural deduction, but switch to a proof-tree notation when convenient.

Remark 7.7. We have modified the syntax for judgements from that used in the papers [CPW00] and [PCW02]. There, the specification label appeared as a subscript of the turnstile, so that judgements were written as

$$\Gamma \vdash_{\text{SP}} \diamond A$$

We use the new notation to draw the analogy with our treatment of Hoare logic in Part III. As we will see in the next chapter where we define a logical type theory for our logic, we can take labelled formulae as types, in a similar fashion to our treatment of the logical type theory for IHL given in Chapter 5 of Part III.

7.3.2 Logical rules

The basic rules for SSL are of two kinds: logical and structural.

The *logical rules* of SSL are shown in Fig. 7.4. These rules are essentially the standard rules for many-sorted intuitionistic logic, but with specification labels. Intuitively, a logical rule enables us to do constructive reasoning about the properties of a single specification.

For example, we augment the usual ($\Rightarrow$-I) rule of constructive logic as follows:

$$\frac{\Gamma, A \vdash \text{SP} \diamond B}{\Gamma \vdash \text{SP} \diamond (A \Rightarrow B)} \ (\Rightarrow\text{-I})$$

This rule permits us to prove that the implication $(A \Rightarrow B)$ satisfies the specification SP, given a proof that, assuming A, then B is satisfiable for SP.

The conditions of application for the logical rules are similar to those of intuitionistic logic presented in Chapter 2.

Remark 7.8 *(Substitution for individual variables).* As usual $A[t/x]$ denotes the result of substituting t for all free occurrences of x in A, subject to avoiding clashes of variables, where t and x share the same sort.

We illustrate the motivation of our calculus by considering several rules. The other rules can be understood similarly.

Remark 7.9. Axiom and assumption introduction rules deserve some discussion. These rules allow formulae to be proved about a specification, using the specification axioms and assumptions about the specification. The axiom introduction rule (Ax-I) permits the use the axioms of a basic specification. Assumption introduction (Ass-I) enables assumptions to be made about a specification, using visible symbols of the specification's signature.

Remark 7.10. The rule ($\vee$-I$_1$)

$$\frac{\Gamma \vdash \text{SP} \diamond A}{\Gamma \vdash \text{SP} \diamond (A \vee B)} \ (\vee\text{-I}_1)$$

means that, we know $(A \vee B)$ is true of SP because we know A is true of SP. This is an important principle of constructive systems — a disjunction is known only if the left or right formula of the disjunction is known. Motivation for ($\vee$-I$_2$) is similar.

Take any structured specification expression SP. Let t be a term of $Term(Sig(\mathrm{SP}), X)_s$ with a sort, s, of $\mathrm{Sig}(\mathrm{SP})$.

$$\frac{}{A \vdash \mathrm{SP} \diamond A}\ (\text{Ass-I}) \quad \text{where } \mathrm{Sig}(A) \subseteq Sig(\mathrm{SP})$$

$$\frac{}{\emptyset \vdash \langle \Sigma, Ax \rangle \diamond A}\ (\text{Ax-I}) \quad \text{where } A \in Ax$$

$$\frac{\Gamma, A \vdash \mathrm{SP} \diamond B}{\Gamma \vdash \mathrm{SP} \diamond (A \Rightarrow B)}\ (\Rightarrow\text{-I}) \qquad \frac{\Gamma_1 \vdash \mathrm{SP} \diamond (A \Rightarrow B) \quad \Gamma_2 \vdash \mathrm{SP} \diamond A}{\Gamma_1, \Gamma_2 \vdash \mathrm{SP} \diamond B}\ (\Rightarrow\text{-E})$$

$$\frac{\Gamma_1 \vdash \mathrm{SP} \diamond A \quad \Gamma_2 \vdash \mathrm{SP} \diamond B}{\Gamma_1, \Gamma_2 \vdash \mathrm{SP} \diamond A \wedge B}\ (\wedge\text{-I})$$

$$\frac{\Gamma \vdash \mathrm{SP} \diamond (A_1 \wedge A_2)}{\Gamma \vdash \mathrm{SP} \diamond A_1}\ (\wedge\text{-E}_1) \qquad \frac{\Gamma \vdash \mathrm{SP} \diamond (A_1 \wedge A_2)}{\Gamma \vdash \mathrm{SP} \diamond A_2}\ (\wedge\text{-E}_2)$$

$$\frac{\Gamma \vdash \mathrm{SP} \diamond A}{\Gamma \vdash \mathrm{SP} \diamond (A \vee B)}\ (\vee\text{-I}_1) \qquad \frac{\Gamma \vdash \mathrm{SP} \diamond B}{\Gamma \vdash \mathrm{SP} \diamond (A \vee B)}\ (\vee\text{-I}_2)$$

$$\frac{\Gamma \vdash \mathrm{SP} \diamond (A \vee B) \quad \Gamma_1, A \vdash \mathrm{SP} \diamond C \quad \Gamma_2, B \vdash \mathrm{SP} \diamond C}{\Gamma, \Gamma_1, \Gamma_2 \vdash \mathrm{SP} \diamond C}\ (\vee\text{-E})$$

$$\frac{\Gamma \vdash \mathrm{SP} \diamond A}{\Gamma \vdash \mathrm{SP} \diamond \forall x : s \bullet A}\ (\forall\text{-I}) \qquad \frac{\Gamma \vdash \mathrm{SP} \diamond \forall x : s \bullet A}{\Gamma \vdash \mathrm{SP} \diamond A[t/x]}\ (\forall\text{-E})$$

where x is free in A, not free in Γ

$$\frac{\Gamma \vdash \mathrm{SP} \diamond A[t/x]}{\Gamma \vdash \mathrm{SP} \diamond \exists x : s \bullet A}\ (\exists\text{-I})$$

$$\frac{\Gamma_1 \vdash \mathrm{SP} \diamond \exists x : s \bullet A \quad \Gamma_2, A[z/x] \vdash \mathrm{SP} \diamond C}{\Gamma_1, \Gamma_2 \vdash \mathrm{SP} \diamond C}\ (\exists\text{-E})$$

where z does not occur free in C

$$\frac{\Gamma \vdash \mathrm{SP} \diamond \bot}{\Gamma \vdash \mathrm{SP} \diamond A}\ (\bot\text{-E})$$

provided A is Harrop

Fig. 7.4. The logical rules of SSL.

Remark 7.11. As was the case for intuitionistic logic, the premise formula of ($\bot$-E) is restricted to Harrop formulae. This restriction, though not necessary, aids program extraction, investigated in Chapter 9. The restriction does not affect the power of our logical rules. That is, we can conservatively extend our calculus with a rule of the form

$$\frac{\Gamma \vdash_{\mathsf{Int}} \mathrm{SP} \diamond \bot}{\Gamma \vdash \mathrm{SP} \diamond A}\ (\bot\text{-E}^*)$$

for all formulae A. This rule is derivable from ($\bot$-E) by a similar proof to that of Lemma 2.2.1 in Chapter 2, Part II, p. 33.

Remark 7.12. Our presentation of the logical rules follows that given in [CPW00]. However, we have restricted logical rules to use the same specification in premises and conclusion. This does not affect the logical strength of the system as the structural rules, given below, allow us to derive all the rules in [CPW00] in our present system.

7.3.3 Structural rules

The *structural rules* of SSL are given in Fig. 7.5. These rules allow us to simultaneously build a structured specification and prove properties about the result, given previously known properties about smaller specifications.

For instance, the translation rule (trans)

$$\frac{\Gamma \vdash \textsc{Sp} \diamond A}{\rho`\Gamma \vdash \textsc{Sp}\ \textbf{with}\ \rho \diamond \rho \bullet (A)}\ \text{(trans)}$$

permits us to simultaneously

- rename a specification Sp to Sp **with** ρ, and
- transform a known property A of the specification to a new property $\rho \bullet A$ of the renamed specification.

Remark 7.13. Observe that, for the logical rules, the specification in the conclusion is the same as that in the premises. On the other hand, for the structural rules, the change in structure is reflected in the new specification label of the conclusion.

$$\frac{\Gamma \vdash \textsc{Sp} \diamond A}{\rho`\Gamma \vdash \textsc{Sp}\ \textbf{with}\ \rho \diamond \rho \bullet (A)}\ \text{(trans)}$$

$$\frac{\Gamma \vdash \textsc{Sp} \diamond A}{\Gamma \vdash \textsc{Sp}\ \textbf{hide}\ SL \diamond A}\ \text{(hide)}$$

where $\Gamma \cup \{A\} \subseteq WFF(Sig(\textsc{Sp})/SL, Var)$

$$\frac{\Gamma \vdash \textsc{Sp\_1} \diamond A}{\Gamma \vdash \textsc{Sp\_1}\ \textbf{and}\ \textsc{Sp\_2} \diamond inl(A)}\ (\text{union}_1)$$

$$\frac{\Gamma \vdash \textsc{Sp\_2} \diamond A}{\Gamma \vdash \textsc{Sp\_1}\ \textbf{and}\ \textsc{Sp\_2} \diamond inr(A)}\ (\text{union}_2)$$

$$\frac{\Gamma \vdash \textsc{Sp\_1} \diamond A}{\Gamma \vdash \textsc{Sp\_1}\ \textbf{then}\ \textsc{Sp\_Ext} \diamond inl(A)}\ (\text{ext}_1)$$

$$\frac{\Gamma \vdash \textsc{Sp\_Ext} \diamond A}{\Gamma \vdash \textsc{Sp\_1}\ \textbf{then}\ \textsc{Sp\_Ext} \diamond inr(A)}\ (\text{ext}_2)$$

Fig. 7.5. The structural rules of SSL.

Remark 7.14. We make the following remarks about the structural rules of Fig. 7.5:

- Given a formula F and symbol translation ρ, then $\rho(F)$ denotes the obvious translation of F by replacing function symbols according to ρ.
- The translation of the context Γ by ρ is written $\rho`\Gamma$. This is simply the recursive application of $\rho$ to every formula in $\Gamma$ to obtain a result context $\Gamma' = \rho`\Gamma$.
- The morphisms $inl : Sig(\text{SP\_1}) \rightarrow Sig(\text{SP\_1}) +_\Sigma Sig(\text{SP\_2})$ and $inr : Sig(\text{SP\_2}) \rightarrow Sig(\text{SP\_1}) +_\Sigma Sig(\text{SP\_2})$ in (union$_1$) and (union$_2$) are the pushout morphisms over common signature Σ, as defined in Definition 7.1.5.
- The morphisms $inl : Sig(\text{SP\_1}) \rightarrow Sig(\text{SP\_1}) +_\Sigma Sig(\text{SP\_2})$ and $inr : Sig(\text{SP\_2}) \rightarrow Sig(\text{SP\_1}) +_\Sigma Sig(\text{SP\_2})$ in (ext$_1$) and (ext$_2$) are pushout morphisms over the common signature $\Sigma = Sig(\text{SP\_1})$, where $\text{SP\_2} = \langle Sig(\text{SP\_1}) \cup \Sigma_e, Ax_e\rangle$ is formed by extending the signature of $\text{SP\_EXT} = \langle \Sigma_e, Ax_e\rangle$.

Remark 7.15. Our presentation of the structural rules follows that of [CPW00] but with the additional rules for extensions that were given in [PCW02].

Remark 7.16 *(Standard representation of unions).* In later chapters, following [Pet96, WCP98, CPW00, PCW02], we will use $Sig(\text{SP\_1}) \cup Sig(\text{SP\_2})$ as standard representation of the isomorphism class of $Sig(\text{SP\_1 } \textbf{and} \text{ SP\_2})$. This simplifies the presentation of the (union$_i$), (ext$_i$) ($i = 1, 2$), since we do not have to write the embedding morphisms *inl* and *inr* explicitly.

However, in this present chapter, we will use the full pushout construction for

$$Sig(\text{SP\_1 } \textbf{and} \text{ SP\_2})$$

as given in 7.2.4, Section 7.2, p. 231.

7.3.4 Reasoning with equality

CASL makes the following assumptions to aid reasoning about equality between terms. Every signature contains an equality predicate $=_s$ for each sort s of the signature, and also the following sets of axioms are in every basic specification $\langle \Sigma, Ax\rangle$, $\Sigma = \langle S, TF, P\rangle$:

1. Reflexivity: for every $s \in S$,

$$\forall x : s \bullet x =_s x$$

2. Symmetry: for every $s \in S$,

$$\forall x, y : s \bullet x =_s y \Rightarrow y =_s z$$

3. Transitivity: for every $s \in S$,

$$\forall x, y, z : s \bullet x =_s y \wedge y =_s z \Rightarrow x =_s z$$

4. Substitutivity for terms: for every $s_1, \ldots, s_n, s \in S$ and every term $t \in TF_{s_1 \ldots s_n, s}$,

$$\forall a_1, b_1 : s_1 \bullet \ldots a_n, b_n : s_n \bullet (a_1 = b_1 \wedge \ldots \wedge a_n = b_n) \Rightarrow f(a_1, \ldots, a_n) = f(b_1, \ldots, b_n)$$

5. Substitutivity in predicates: for every $s_1, \ldots, s_n \in S$ and every predicate $F \in P_{s_1 \ldots s_n}$,

$$\forall a_1, b_1 : s_1 \bullet \ldots a_n, b_n : s_n \bullet (a_1 = b_1 \wedge \ldots \wedge a_n = b_n) \Rightarrow F(a_1, \ldots, a_n) \Rightarrow F(b_1, \ldots, b_n)$$

In *CASL*, these axioms are presumed implicit in basic specifications — they do not need to be given by the specification writer.

These assumptions permit us to take the following as a convenient derivable schema, for any specification expression SP:

$$\frac{\Gamma_1 \vdash \mathrm{SP} \diamond x_1 =_s y_1 \quad \ldots \quad \Gamma_n \vdash \mathrm{SP} \diamond x_n =_s y_n}{\Gamma_1, \ldots, \Gamma_n \vdash \mathrm{SP} \diamond (P[x/z] \Rightarrow P[y/z])} \text{(subst}[P]) \tag{7.3}$$

for $s \in Sig(\mathrm{SP})$ and $P \in WFF(Sig(\mathrm{SP}), Var)$.

To see why (7.3) is a derivable schema, we require Lemmata 7.3.1 and 7.3.2 below.

Lemma 7.3.1. *For any specification* SP, *for every* $a : s_2 \in Term(Sig(\mathrm{SP}))$

$$\vdash \mathrm{SP} \diamond \forall x, y : s_1 \bullet \forall x =_{s_1} y \Rightarrow a[x/z] =_{s_2} a[y/z]$$

where formula P *is a formula in* $WFF(Sig(\mathrm{SP}))$.

Proof. We use the axioms of substitutivity for basic specifications.

By induction on the forms of a and of SP. Suppose

$$\vdash \mathrm{SP} \diamond x =_{s_1} y \tag{7.4}$$

Assume SP is basic.

Either z is free in a or not. If the latter holds, then we are done. If the former, then a is of two forms: either

- a is $f(a_1,\ldots,a_n)$, and z is only free in the subset $\{a'_1 : s'_1,\ldots,a'_m : s'_m\}$ of $\{a_1,\ldots,a_n\}$. By the IH, it is possible to prove

$$\vdash \text{SP} \diamond a'_1[x/z] =_{s'_1} a'_1[y/z]$$

$$\ldots$$

$$\vdash \text{SP} \diamond a'_m[x/z] =_{s'_m} a'_m[y/z]$$

  Then, by the assumed axioms for substitutivity of terms,

$$\vdash \text{SP} \diamond f(a_1,\ldots,a_n)[x/z] = f(a_1,\ldots,a_n)[y/z]$$

  is provable
- a is a variable z. In this case, we are done, because we have assumed (7.4).

  If SP is a structured specification expression, we prove

$$\vdash \text{SP\_2} \diamond \forall x, y : s_1 \bullet x =_{s_1} y \Rightarrow a[x/z] =_{s_2} a[y/z]$$

for the sub-specification SP_2 of SP whose signature includes s_1, s_2 as sorts and a as a term. Then we use the structural rules to construct SP. For example, if SP is of the form $\rho \bullet$ SP_2, then, by the IH, it is possible to derive

$$\vdash \text{SP\_2} \diamond \rho^{-1}(\forall x, y : s_1 \bullet \forall x =_{s_1} y \Rightarrow a[x/z] =_{s_2} a[y/z])$$

We then apply the rule (trans) to obtain the required conclusion by translation. □

Lemma 7.3.2. *For any specification* SP, *for sort* $s \in Sig(\text{SP})$

$$\vdash \text{SP} \diamond \forall x, y : s \bullet x =_s y \Rightarrow (P[x/z] \Rightarrow P[y/z])$$

for any formula $P \in WFF(Sig(\text{SP}))$.

Proof. We use the assumed axioms of substitutivity, proceeding by induction on the forms of P and of SP. Suppose

$$\vdash \text{SP} \diamond x =_s y$$

If P is atomic, of the form $F(a_1,\ldots,a_n)$, then either z is free in F or not. If the latter, then we are done. If the former, then z is free only in some non-empty subset $\{a'_1 : s_1,\ldots a'_m : s_m\}$ of $\{a_1,\ldots,a_n\}$. By Lemma (7.3.1), it is possible to prove

$$\vdash a'_1[x/z] =_{s_1} a'_1[x/z]$$

$$\ldots$$

$$\vdash a'_m[x/z] =_{s_m} a'_m[x/z]$$

Using the basic specification axioms of substitutivity of predicates, we have

$$\vdash \text{SP} \diamond F(a_1, \ldots, a_n)[x/z] \Rightarrow F(a_1, \ldots, a_n)[y/z]$$

as required.

Otherwise we proceed over the structure of P.

If SP is a structured specification expression, we prove

$$\vdash \text{SP\_2} \diamond \forall x, y : s \bullet x =_s y \Rightarrow (P[x/z] \Rightarrow P[y/z])$$

for the smallest sub-specification SP_2 of SP whose signature includes s_1, s_2 as sorts and a as a term, and we then use the structural rules to construct SP in the same way as with the proof of Lemma 7.3.1. □

Repeated application of the last lemma is enough to derive any application of the schema (7.3).

7.3.5 Induction

To reason inductively over generated sorts, we include structural induction schemata.

Recall, in Chapter 2 (p. 34), we dealt with specific cases of induction schemata for many-sorted intuitionistic logic. In the case of SSL, we will go further and provide a general induction schemata for any generated sort of any basic specification.

Given a sort in a signature $\Sigma = \langle S, F, P \rangle$, generated by constructors $Cons_s \subseteq F_s$, we permit an induction schema of the form given in Fig. 7.6.

The general form of this schema is complicated. It is best illustrated by examples.

Example 7.7. Consider the specification NATBOOLEAN of Example 7.1, Fig. 7.1, in which the booleans *boolean* are generated by constants

$$\{F : boolean, T : boolean\}$$

The corresponding induction schema *Ind*(*boolean*) produced by the general schema of Fig. 7.6 is

$$\frac{\text{NATBOOLEAN} \diamond P[F/x] \quad \text{NATBOOLEAN} \diamond P[T/x]}{\text{NATBOOLEAN} \diamond \forall x : boolean \bullet P} \text{ (Ind[boolean])}$$

Example 7.8. Given a signature Σ in which the natural numbers *nat* are generated by

$$Cons_{nat} = \{0 : nat, suc : nat \to nat\}$$

The induction schema (Ind[nat]), a special case of Fig. 7.6, is

$$\frac{\langle \Sigma, Ax \rangle \diamond P[0/x] \quad \langle \Sigma, Ax \rangle \diamond \forall y : nat \bullet P[y/x] \Rightarrow P[suc(y)/x]}{\langle \Sigma, Ax \rangle \diamond \forall x : nat \bullet P} \text{ (Ind[nat])}$$

Let $\Sigma = \langle S, F, P \rangle$ be a signature that contains a sort s generated by constructors $Cons_s \subseteq F_s$,

$$Cons_s = \{c_1 : s, \ldots, c_n : s, f_1 : (s_1^1 \times \ldots \times s_{m_1}^1) \to s, \ldots, \\ f_p : (s_1^p \times \ldots \times s_{m_p}^p) \to s\}$$

Let P be a formula of Σ with $x : s$ free.
For each $i = 1, \ldots, p$ we make the following definitions. Take a set of variables $\{x_j^i\}_{j=1,\ldots,m_i}$ corresponding to argument sorts of f_i, we define

$$M(s, \{x_j^i : s_j^i\}_{j=1,\ldots,m_i}) = \{x_j^i \mid s_j^i = s\}$$

(When this set is empty, then f_i does not involve s as an argument sort.)
If this set is empty, then we define

$$P_{f_i} = \forall x_1^i : s_1^i, \ldots, x_{m_i}^i : s_{m_i}^i \bullet P[f_i(x_1^1, \ldots, x_{m_i}^m)/x]$$

Otherwise, we take

$$P_{f_i} = \forall x_1^i : s_1^i, \ldots, x_{m_i}^i : s_{m_i}^i \bullet Q_{f_i} \Rightarrow P[f_i(x_1^i, \ldots, x_{m_i}^i)/x]$$

where Q_{f_i} is formed from $M(s, \{x_j^i : s_j^i\}_{j=1,\ldots,m_i})$ as follows. Assume $M(s, \{x_j^i : s_j^i\}_{j=1,\ldots,m_i}) = \{x_1, \ldots, x_k\}$. If $k = 1$,

$$Q_{f_i} = P(x_1)$$

If $k > 1$,

$$Q_{f_i} = P(x_1) \Rightarrow \ldots \Rightarrow P(x_k)$$

Then we have the following **induction schema** for s in Σ:

$$\frac{\vdash \langle \Sigma, Ax \rangle \diamond P_{c_1} \ \ldots \vdash \langle \Sigma, Ax \rangle \diamond P_{c_n} \qquad \vdash \langle \Sigma, Ax \rangle \diamond P_{f_1} \ \ldots \vdash \langle \Sigma, Ax \rangle \diamond P_{f_p}}{\vdash \langle \Sigma, Ax \rangle \diamond \forall x : s \bullet P} \ (\mathrm{Ind}[s])$$

Fig. 7.6. General structural induction schema.

Example 7.9. Given a signature Σ in which lists of natural numbers $lnat$ are generated by

$$Cons_{lnat} = \{l : nat \to lnat, con : (lnat \times lnat) \to lnat\}$$

The induction schema (Ind[lnat]), a special case of Fig. 7.6, is

$$\frac{\langle \Sigma, Ax \rangle \diamond \forall y : nat \bullet P[l(y)/x] \qquad \langle \Sigma, Ax \rangle \diamond \forall y : lnat \bullet \forall z : lnat \bullet P[y/x] \Rightarrow P[z/x] \Rightarrow P[con(y,z)/x]}{\langle \Sigma, Ax \rangle \diamond \forall x : nat \bullet P} \ (\mathrm{Ind[lnat]})$$

7.3.6 Example: Password checking system

We will now illustrate SSL in practice. We use the calculus to simultaneously reason about and to construct specifications, developing a theorem about the password checking system PWDSYS described throughout Section 7.2.

We will prove that, given any input x of a password, there is always an appropriate response message to be output. The response message will tell the user if the password is of the correct length or not. This requirement is stated as follows

$$\vdash \text{PWDSYS} \diamond \forall x : nat \bullet \exists y : string \bullet ValidMsg(x, r) \tag{7.5}$$

We now derive (7.5).

First we need to prove the following lemma using boolean induction over NATBOOLEAN (this induction schema was defined in Example 7.7 above):

$$\dfrac{\dfrac{\dfrac{}{\text{NATBOOLEAN} \diamond F = F}\,(\text{Ax})}{\text{NATBOOLEAN} \diamond T = T \vee F = F}\,(\vee\text{-I}_2) \quad \dfrac{\dfrac{}{\text{NATBOOLEAN} \diamond T = T}\,(\text{Ax})}{\text{NATBOOLEAN} \diamond T = T \vee T = F}\,(\vee\text{-I}_1)}{\text{NATBOOLEAN} \diamond \forall b : boolean \bullet b = T \vee b = F} \tag{7.6}$$

where with the conclusion obtained by Ind(Boolean) and where the axioms are from the assumed axioms of equality.

We wish to use (7.6) to show that the value $inRange(x)$ of PWDCORE will always be either $true$ or $false$ for any input password x.

In order to use (7.6) in PWDCORE, we must employ structural rules to build PWDCORE from NATBOOLEAN. The function *inRange* is defined with boolean sort symbols of PWDCORE, which are renamed versions of those in NATBOOLEAN. So we proceed by renaming boolean symbols of NATBOOLEAN to their appropriate counterparts used by *inRange*, taking a union with STRINGBOOLEAN and then extending the new symbols and axioms of PWDCORE as follows

$$\dfrac{\dfrac{\dfrac{\begin{array}{c}\vdots\ (7.6)\\ \text{NATBOOLEAN} \diamond \forall b : boolean \bullet b = T \vee b = F\end{array}}{(BtoB \bullet \text{NATBOOLEAN}) \diamond \forall b : bool \bullet b = true \vee b = false}\,(\text{trans})}{\text{SNB} \diamond \forall b : bool \bullet b = true \vee b = false}\,(\text{union}_1)}{\text{PWDCORE} \diamond \forall b : bool \bullet b = true \vee b = false}\,(\text{ext}_1) \tag{7.7}$$

where SNB $=$ $(BtoB \bullet \text{NATBOOLEAN})$ **and** STRINGBOOL and $BtoB$ is the symbol mapping $[boolean \mapsto bool, T \mapsto true, F \mapsto false]$.

Then we can instantiate (7.7) with $inRange(x)$ for b, by application of ($\forall$-E), to give

$$\text{PWDCORE} \diamond inRange(x) = true \vee inRange(x) = false \tag{7.8}$$

We will derive (7.5) by reasoning over the possible cases that $\textsc{PwdCore} \diamond inRange(x) = true$ or $\textsc{PwdCore} \diamond inRange(x) = false$.

First we consider the former case, $\textsc{PwdCore} \diamond inRange(x) = true$.

We apply ($\forall$-E) on two axioms of $\textsc{PwdCore}$, to remove the quantifiers:

$$\dfrac{\dfrac{}{\textsc{PwdCore} \diamond \forall x : nat \bullet inRange(x) = true \Rightarrow OkPwd(x)}\,(\text{Ax-I})}{\textsc{PwdCore} \diamond inRange(x) = true \Rightarrow OkPwd(x)}\,(\forall\text{-E}) \tag{7.9}$$

and

$$\dfrac{\dfrac{}{\begin{array}{c}\textsc{PwdCore} \diamond \forall x : nat \bullet OkPwd(x) \Rightarrow \\ ValidMsg(x, \text{'Password acceptable'})\end{array}}\,(\text{Ax-I})}{\textsc{PwdCore} \diamond OkPwd(x) \Rightarrow ValidMsg(x, \text{'Password acceptable'})}\,(\forall\text{-E}) \tag{7.10}$$

We use these axioms and the assumption to derive (7.5):

$$\dfrac{\dfrac{\dfrac{\dfrac{(7.10)}{\textsc{PwdCore} \diamond A_2} \quad \dfrac{\dfrac{(7.9)}{\textsc{PwdCore} \diamond A_1} \quad \dfrac{}{\textsc{PwdCore} \diamond inRange(x) = true}\,(\text{Ass-I})}{\textsc{PwdCore} \diamond OkPwd(x)}\,(\Rightarrow\text{-E})}{\textsc{PwdCore} \diamond ValidMsg(x, \text{'Password acceptable'})}\,(\Rightarrow\text{-E})}{\textsc{PwdCore} \diamond \exists y : string \bullet ValidMsg(x, y)}\,(\exists\text{-I})}{\textsc{PwdCore} \diamond \forall x : nat \bullet \exists y : string \bullet ValidMsg(x, y)}\,(\forall\text{-I}) \tag{7.11}$$

where A_1 denotes the formula $inRange(x) = true \Rightarrow OkPwd(x)$ in the instantiated axiom of (7.9) and A_2 denotes the formula $OkPwd(x) \Rightarrow ValidMsg(x, \text{'Password acceptable'})$ in the instantiated axiom of (7.10).

Similar reasoning from the assumption $\textsc{PwdCore} \diamond inRange(x) = false$, using the axioms of $\textsc{PwdCore}$ will give a proof of the form

$$\dfrac{\dfrac{\begin{array}{c}\dfrac{}{\textsc{PwdCore} \diamond inRange(x) = false}\,(\text{Ass-I}) \\ \vdots \\ \textsc{PwdCore} \diamond ValidMsg(x, \text{'Please choose a password in correct range'})\end{array}}{\textsc{PwdCore} \diamond \exists y : string \bullet ValidMsg(x, y)}\,(\exists\text{-I})}{\textsc{PwdCore} \diamond \forall x : nat \bullet \exists y : string \bullet ValidMsg(x, y)}\,(\forall\text{-I}) \tag{7.12}$$

We apply ($\vee$-E) using (7.8), (7.11) and (7.12), to obtain

$$\textsc{PwdCore} \diamond \forall x : nat \bullet \exists y : string \bullet ValidMsg(x, y)$$

This proves the *formula* of the required theorem, but over the system specification prior to encapsulation, $\textsc{PwdCore}$.

By hiding $\{ge, inRange\}$ in the result, we obtain the theorem (7.5) as required:

$$\frac{\vdots \atop \text{PwdCore} \diamond \forall x : nat \bullet \exists y : string \bullet ValidMsg(x, y)}{\text{PwdSys} \diamond \forall x : nat \bullet \exists y : string \bullet ValidMsg(x, y)} \text{(hide)}$$

This theorem is a truth about the specification PwdSys, given known properties about its required behavior. It tells us that there is always a correct response message for a given password number.

To build an implementation of the password checking system, it would be useful to find a function for producing such a message for given passwords. In isolation the theorem does not tell us what this function is. However, by utilizing the Curry–Howard protocol to adapt proofs-as-programs to SSL, it is possible to extract such a function from the proof of the theorem. As we develop such a methodology in the next two chapters, we shall return to this example to illustrate our ideas.

In Chapter 8 we will define a logical type theory for representing proofs in SSL. We will show how the proof of this example can be encoded as a term in the type theory (see Section 3.2.1, p. 260).

In Chapter 9 we will use the Curry–Howard protocol to provide a method for extracting correct programs from proofs in SSL. We will show how the proof of theorem (7.5) can be transformed into a password checking function which outputs an appropriate response for a given password number input. Then we will also show how this function can be used to build a consistent extension of PwdSys (Section 9.4, p. 334).

7.4 Soundness

In this section, we outline soundness proofs for SSL, using the semantics of specification expressions and definition of formula satisfaction.

We require the following notion of validity for sequents in SSL.

Definition 7.4.1. Take any specification Sp.

Let $\Gamma = \{G_1, \ldots, G_n\}$ and F be $WFF(Sig(\text{Sp}), Var)$ formulae.

We write

$$\text{Sp}, \Gamma \models F$$

if, for every $M \in Mod(\text{Sp})$, and every interpretation $\hat{\iota} : Var \to M$, assuming

$$M \models_\iota G_i$$

for each $i = 1, \ldots, n$ then

$$M \models_\iota F$$

In this case we say that F is *valid* for Sp, assuming Γ.

If Γ is empty and

$$\text{Sp} \models F$$

then we say the formula F is *valid for the specification.*

Soundness is an important result: it tells us that if $\mathrm{SP} \diamond P$ can be proved, then P is a valid statement for SP.

Theorem 7.4.2 (Soundness). *Given an* SSL *proof*

$$\Gamma \vdash \mathrm{SP} \diamond F$$

Then

$$\mathrm{SP}, \Gamma \models F$$

Proof. We are required to prove that, if, for every $M \in Mod(\mathrm{SP})$, and valuation $\hat{\iota} : Var \to M$, $M \models_\iota G_i$ (each $G_i \in \Gamma$), then $M \models_\iota F$.

We show this by induction on the length of the proof.

Take any $M \in Mod(\mathrm{SP})$ and valuation $\hat{\iota} : X \to M$ and assume

$$M \models_\iota G_i \tag{7.13}$$

for every $G_i \in \Gamma$.

Logical rules. If the proof ends in the application of a logical rule, then the approach is straightforward. We obtain the required conclusion using the definition of satisfaction (Definition 7.1.15, p. 224).

(*Ass*-I). If the proof is of the form

$$\frac{}{\emptyset \vdash \langle \Sigma, Ax \rangle \diamond A} \text{(Ax-I)}$$

then $M \models_\iota A$ by definition of $Mod(\langle \Sigma, Ax \rangle)$ and the fact that $A \in Ax$, and we are done.

(*Ass*-I). If the proof is of the form

$$\frac{Sig(A) \subseteq Sig(\mathrm{SP})}{A \vdash \mathrm{SP} \diamond A} \text{(Ass-I)}$$

Then $\Gamma = \{A\}$ and we have $M \models_\iota A$ by assumption (7.13).

($\Rightarrow$-I). Assume the proof ends in a rule application of the form

$$\frac{\Gamma, A \vdash \mathrm{SP} \diamond B}{\Gamma \vdash \mathrm{SP} \diamond (A \Rightarrow B)} (\Rightarrow\text{-I})$$

The IH and assumption (7.13) dilate that, if $M \models_\iota A$ then $M \models_\iota B$. Then, by definition of satisfaction, we know $A \Rightarrow B$, as required.

($\Rightarrow$-E). Assume the proof ends in a rule application of the form

$$\frac{\Gamma_1 \vdash \mathrm{SP} \diamond A \Rightarrow B \quad \Gamma_2 \vdash \mathrm{SP} \diamond A}{\Gamma_1, \Gamma_2 \vdash \mathrm{SP} \diamond B} (\Rightarrow\text{-E})$$

Then the IH assumption (7.13) dictate that $M \models_\iota A \Rightarrow B$ and $M \models_\iota A$. By definition of satisfaction for $A \Rightarrow B$, we know that $M \models_\iota B$ as required.

The cases of proofs ending in ($\wedge$-I), ($\wedge_1$-E), ($\wedge_2$-E), ($\vee_1$-I), ($\vee_2$-I), ($\vee$-E) and ($\bot$-E) are similar.

($\forall$-I). Assume the proof ends in a rule application of the form

$$\frac{\Gamma \vdash \textsc{Sp} \diamond A}{\Gamma \vdash \textsc{Sp} \diamond \forall x : s \bullet A} \ (\forall\text{-I})$$

Take any $x : s$-variant ι' of ι. By assumption (7.13), and the fact that $x : T \notin FV(\Gamma)$, it is easy to derive

$$M \models_{\iota'} G_i \qquad (7.14)$$

for each $G_i \in \Gamma$. So, by the IH, $M \models_{\iota'} A$ for every $x : T$-variant ι' of ι, as required.

The case of a proof ending in ($\exists$-I) is similar.

($\forall$-E). Assume the proof ends in a rule application of the form

$$\frac{\Gamma \vdash \textsc{Sp} \diamond \forall x : s \bullet A}{\Gamma \vdash \textsc{Sp} \diamond A[t/x]} \ (\forall\text{-E})$$

By assumption (7.13), $M \models_\iota \forall x : s \bullet A$. So, $M \models_{\iota'} A$ for the ι' variant that maps x to $\iota(t)$. It is then easy to derive $M \models_\iota A[t/x]$ as required.

The case of a proof ending in ($\exists$-E) is similar.

Structural rules. We follow [Cen94, p. 88], using the definition of Mod over structured specification expressions.

($union_1$). Assume the proof ends in rule application

$$\frac{\Gamma \vdash \textsc{Sp\_1} \diamond A}{\Gamma \vdash \textsc{Sp\_1}\ \textbf{and}\ \textsc{Sp\_2} \diamond inl(A)} \ (\text{union}_1)$$

By the IH, $N \models A$ for any $N \in Mod(\textsc{Sp\_1})$. Now, $M \in Mod(\textsc{Sp\_1}\ \textbf{and}\ \textsc{Sp\_2})$. By definition of Mod, $M|_{inl} \in Mod(SpecNameSp_1)$. So, $M \models inl(A)$, as required.

($union_2$). This case is symmetric.

(ext_1). Assume the proof ends in rule application

$$\frac{\Gamma \vdash \textsc{Sp\_1} \diamond A}{\Gamma \vdash \textsc{Sp\_1}\ \textbf{then}\ \textsc{Sp\_EXT} \diamond inl(A)} \ (\text{ext}_1)$$

where $\textsc{Sp\_EXT} = \langle \Sigma, Ax \rangle$. Let Σ be the $Sig(\textsc{Sp\_1}\ \textbf{then}\ \textsc{Sp\_EXT})$. By the IH, $N \models A$ for any $N \in Mod(\textsc{Sp\_1})$. Now, $M \in Mod(\textsc{Sp\_1}\ \&\ \textsc{Sp\_2})$ for $\textsc{Sp\_2} = \langle \Sigma, Ax$. By definition of Mod, $M|_{inl} \in Mod(SpecNameSp_1)$. So, $M \models inl(A)$ as required.

(ext_2). This case is symmetric.

The cases of ($hide$) and ($trans$) are trivial.

Induction schemata. Assume that a proof ends in an induction schema:

$$\frac{\begin{array}{ll}\vdash \langle \Sigma, Ax \rangle \diamond P[c_1/x] & \ldots \vdash \langle \Sigma, Ax \rangle \diamond P[c_n/x] \\ \vdash \langle \Sigma, Ax \rangle \diamond P_{f_1} & \ldots \vdash \langle \Sigma, Ax \rangle \diamond P_{f_p}\end{array}}{\vdash \langle \Sigma, Ax \rangle \diamond \forall x : s \bullet P} \ \text{Ind(s)}$$

where

- $\langle \Sigma, Ax \rangle$ is a basic specification with $\Sigma = \langle S, TF, P \rangle$ with constructors $Cons_s \subseteq TF$ for a sort $s \in S$,

$$Cons_s = \{c_1 : s, \ldots, c_n : s, f_1 : (s_1^1 \times \ldots s_{m_1}^1) \to s, \ldots, \\ f_p : (s_1^p \times \ldots s_{m_p}^p) \to s\}$$

- where each P_{f_i} is defined as in Fig. 7.6 (p. 244).

We are required to show that, for every $M \in Mod(\langle \Sigma, Ax \rangle)$ (using loose semantics), for every interpretation $\hat{\imath} : Var \to M$

$$M \models_{\iota'} P \tag{7.15}$$

for every x-variant ι' of ι.

We establish 7.15 by a second induction over the form of $a = \iota'(x) \in s^M$.

Because we use loose semantics and s has constructors, we know that a must be of the following forms

- $a = c_i^M$ for some $i = 1, \ldots, n$. In this case, by the main IH, $M \models_\iota P[c_i/x]$ and so (7.15) holds.
- $a = f_i^M((a_1^i)^M, \ldots, (a_{m_i}^i)^M)$ for some $i = 1, \ldots, m$, with each $(a_j^i)^M \in (s_j^i)^M$ $(j = 1, \ldots, m_i)$
  First, let $M(s, \{x_j^i : s_j^i\}_{j=1,\ldots,m_i})$ be defined as in Fig. 7.6 (p. 244):

$$M(s, \{x_j^i : s_j^i\}_{j=1,\ldots,m_i}) = \{x_j^i : s_j^i \mid s_j^i = s \text{ for } j_i = 1_i, \ldots, m_i\} = \\ \{x_1 : s, \ldots, x_k : s\}$$

  We deal with the more complicated case where $M(s, \{x_j^i : s_j^i\}_{j=1,\ldots,m_i})$ is not empty and $k > 1$. The other cases (where the set is empty or when $k = 1$) are similar.
  We have a set of terms corresponding to $M(s, \{x_j^i : s_j^i\}_{j=1,\ldots,m_i})$

$$M(s, \{a_j^i : s_j^i\}_{j=1,\ldots,m_i}) = \{a_j^i : s_j^i \mid s_j^i = s \text{ for } j_i = 1_i, \ldots, m_i\} = \\ \{a_1 : s, \ldots, a_k : s\}$$

  By the second IH, for each $j = 1, \ldots, k$ it is true that $M \models_{\iota_j} P$ for the x-variant ι_j of ι defined $\iota_j = \iota[x \mapsto (a_j)^M]$. But this means

$$M \models_\iota P[a_j/x] \tag{7.16}$$

  Now,

$$P_{f_i} = \forall x_1^i : s_1^i, \ldots, x_{m_i}^i : s_{m_i}^i \bullet Q_{f_i} \Rightarrow P[f_i(x_1^i, \ldots, x_{m_i}^i)/x]$$

  where Q_{f_i} is

$$Q_{f_i} = P[x_1/x] \Rightarrow \ldots \Rightarrow P[x_k/x]$$

By the main IH,

$$M \models_\iota \forall x_1^i : s_1^i, \ldots x_m^i : s_m^i \bullet Q_{f_i} \Rightarrow P[f_i(x_1^i, \ldots, x_m^i)/x]$$

By repeated instantiation, this means

$$M \models_\iota Q'_{f_i} \Rightarrow P[f_i(a_{1_i}, \ldots, a_{m_i})/x]$$

with

$$Q'_{f_i} = P[a_1/x] \Rightarrow \ldots \Rightarrow P[a_k/x]$$

So we can repeatedly instantiate Q' with (7.16) $j = 1, \ldots, k$ to obtain

$$M \models_\iota P[f_i(a_{1_i}, \ldots, a_{m_i})/x]$$

This gives us (7.15) as required.

□

Remark 7.17. We omit the opposite direction — completeness. Completeness is not as important a property as soundness for the purposes of program extraction.

The proof of completeness would follow along the lines given by Cengarle in [Cen94, pp. 89–91], but would require constructive models for specifications (e.g., Kripke semantics), due to the constructive nature of our logical rules.

Also, as noted by Cengarle, the completeness theorem does not hold for the case where we use loose semantics and where SP involves basic sub-specifications with constructors for sorts. This is because the model classes for such sub-specifications are computation structures, which, by Gödel's Incompleteness Theorem, do not have complete formal systems. Note that only these sub-specifications permit reasoning with structural induction schemata. Thus, our logic with schemata is sound with respect to our loose semantics, but not complete.

7.5 Discussion

This chapter presented the important concepts of structured specifications in *CASL* and defined the logic SSL. The next two chapters will be concerned with application of the Curry–Howard protocol to SSL.

In Chapter 8 we will define a logical type theory for SSL for which the Curry–Howard isomorphism holds. Then, in Chapter 9, we will show how to transform SSL proofs into provably correct functional programs, which may then be used to consistently extend structured specifications.

The logic SSL is extensible. Later in this part of the monograph, we will propose various extensions to the basic SSL. In Chapter 9, we shall extend the language of *CASL* and our logic to specify with and reason about a lambda calculus of the *SML* programming language. This is necessary for us to use the

Curry–Howard protocol to extract lambda terms from proofs of SSL and to obtain executable refinements of *CASL* specifications.

The current chapter has not dealt with generic (parametrized) specifications. Chapter 10 extends SSL to the generic specifications of *CASL*.

The final chapter of this part of the monograph, Chapter 11, will examine how our calculus and the synthesis results can be applied to give methods for structured program synthesis.

8

Proof-theoretic Properties of SSL

According to the Curry–Howard isomorphism, ordinary intuitionistic logic corresponds to a logical type theory. Intuitionistic proofs correspond to terms, formulae to types and the logical rules to type inference rules of the theory. In Chapter 3 of Part II, we defined the Curry–Howard protocol as a framework for generalizing proofs-as-programs to new logics and programming paradigms. Integral to that framework is the identification of a type theory for the target logic so that there is a correspondence in the style of the Curry–Howard isomorphism.

We want to apply the Curry–Howard protocol to synthesize correct functions from proofs about *CASL* specifications. To achieve this objective, we must first investigate a similar version of the Curry–Howard isomorphism between SSL and an associated logical type theory *LTT*(SSL). Then, in the next chapter (Chapter 9), we will use *LTT*(SSL) to achieve the ultimate goal of adapting proofs-as-programs to SSL.

Our logical type theory is a kind of lambda calculus. Its distinguishing feature is that the types are specification/formula pairs (similar to the logical type theory for Hoare logic of Chapter 5 in Part III, where types were program/formulae pairs). As in the case of the type theory for intuitionistic logic there is an associated normalizing relation. This relation corresponds to a proof normalization strategy for simplifying proofs by eliminating redundant application of rules. Owing to the presence of structural rules, we need to consider a more complex normalization strategy than that used by intuitionistic logic. We will derive two important proof-theoretic properties with respect to this reduction relation: the Church–Rosser property and strong normalization.

We proceed as follows:

- Section 8.1 defines a logical type theory for our calculus and explains how the Curry–Howard isomorphism holds for this new context.
- Section 8.2 identifies rules for proof normalization and proves the strong normalization theorem (Theorem 8.2.9).

- Section 8.3 derives the Church–Rosser property for our proof-terms (Theorem 8.3.3).
- Finally, in Section 8.4, we provide a discussion of our results.

In Part III we applied the Curry–Howard protocol to the synthesis of imperative programs with return values. Part of that work involved defining a similar adaptation of the Curry–Howard isomorphism to the Hoare logic and an associated type theory (Chapter 5). This chapter provides a second example of how the isomorphism can be adapted for use with an application of the protocol.

$a, b, c ::=$	Proof-terms, $PT(\mathsf{SSL})$, of $LTT(\mathsf{SSL})$
$\mathsf{ass}(\text{SP}, u^F)$	assumption, where F is a formula, SP is a specification expression , u is in $Var_{PT(\mathsf{SSL})}$
$\mathsf{ax}(\langle \Sigma, Ax\rangle, F)$	axiom, where F is a formula, $\langle \Sigma, Ax\rangle$ is a basic specification expression
$\mathsf{rec}(Cons_s, s, p)$	structural induction, where $Cons_s$ is a list of constructors for sort s and p is a list of proof-terms.
$\mathsf{abstract}\ u^F.\ a$	abstraction, u from $Var_{PT(\mathsf{SSL})}$, F a formula
$\mathsf{app}(a, b)$	application
$\langle a, b\rangle$	pair
$\mathsf{fst}(a)$	first projection
$\mathsf{snd}(a)$	second projection
$\mathsf{use}\ x : s.\ a$	term variable abstraction, $x \in Var$, s a sort symbol
$\mathsf{specific}(a, t)$	term application, t an individual term
$\mathsf{show}(t, a)$	witness, t an individual term
$\mathsf{select}\ (a)\ \mathsf{in}\ x.y.b$	select, x a term variable and y from $Var_{PT(\mathsf{SSL})}$
$\mathsf{case}\ a\ \mathsf{of}\ \mathsf{inl}(x).b,\ \mathsf{inr}(y).c$	case, x and y are variables from $Var_{PT(\mathsf{SSL})}$
$\mathsf{inl}(a)$	in left
$\mathsf{inr}(a)$	in right
$\mathsf{abort}(a, F)$	abort, where F is a formula
$\rho \bullet a$	translation
$\mathsf{hide}(a, SL)$	hide, SL a symbol list
$\mathsf{union}_1(a, \text{SP})$	first union, SP a specification expression
$\mathsf{union}_2(a, \text{SP})$	second union, SP a specification expression
$\mathsf{ext}_1(a, \text{SP})$	first extension, SP a specification expression
$\mathsf{ext}_2(a, \text{SP})$	second extension, SP a specification expression

Fig. 8.1. Syntax for the proof-terms of $LTT(\mathsf{SSL})$.

8.1 A type theory for SSL

To define our logical type theory $LTT(\mathsf{SSL})$, we proceed as we did for our version of Hoare Logic (Chapter 5 of Part III). We define a version of the lambda calculus whose terms (referred to as "proof-terms") represent proofs. We take types of proof-terms to be *pairs* of specification expressions and formulae. Then we define type inference rules so that correct typing of a proof-term corresponds to a valid proof according to the rules of SSL.

8.1.1 Proof-terms

Recall that the rules of SSL extend intuitionistic logic with rules for using structured specifications. Correspondingly our proof-terms consist of the lambda calculus for intuitionistic type theory (Chapter 2 of Part II) extended with additional constructs to handle structured specifications.

The syntax for proof-terms, $PT(\mathsf{SSL})$ is given in Fig. 8.1.

Notation 8.1. Proof-terms are given with respect to two sets of variables:

- $Var_{PT(\mathsf{SSL})}$, a denumerable set of proof-term variables, and
- Var, the set of individual term variables (already used to define terms of SSL in the previous chapter).

Similar to the logical type theory for intuitionistic logic in Chapter 2, our proof-terms involve individual terms (the terms used within the types of the proof-terms). Individual terms come from signatures of specification labels used in rules. However, for the purposes of speaking about proof-terms, it will sometimes be helpful to speak about the set of all individual terms, without reference to the particular signature they come from.

Notation 8.2. The set of all individual terms is the set of terms for *all* signatures,

$$\bigcup_{\Sigma \in \mathbf{CSig}} Term(\Sigma, Var)$$

For the rest of this book, we shall simply refer to this set as *individual terms.*

Remark 8.1. Note that individual terms are distinct from the proof-terms of $LTT(\mathsf{SSL})$. The former are used by formulae of IHL, while the latter are used in the logical type theory to represent SSL proofs. Individual terms occur within proof-terms to denote the use of individual terms as witnesses in instances of ($\exists$-I) and for instantiation in instances of ($\forall$-E).

Notation 8.3. Formulae used in proof-terms are taken from the set of well-formed formulae for *all* signatures,

$$\bigcup_{\Sigma \in \mathbf{CSig}} WFF(\Sigma, Var)$$

For the rest of this book, we shall simply refer to this set as *formulae.*

Remark 8.2. On its own, the translation proof-term

$$\rho \bullet a$$

is pure syntax and has no semantics of evaluation — it does *not* denote the evaluated application of the symbol map ρ to a. However, later, when we define a normalization strategy for proofs, we will simplify translation terms by evaluating symbol maps.

8.1.2 Types

A theorems of SSL involves a specification expression paired with a formula. Because we want types of proof-terms to correspond to proved expressions, a distinguishing feature of our theory is that its types will be specification/formula pairs, from the set

$$\begin{aligned}Pairs(\mathsf{SSL}) = \{\mathrm{S}\textsc{p} \diamond F \mid\ & \mathrm{S}\textsc{p} \text{ is a specification expression and} \\ & F \text{ is a formula from } \textstyle\bigcup_{\Sigma \in \mathbf{CSig}} WFF(\Sigma, Var)\}\end{aligned}$$

Remark 8.3. This treatment of types is similar to our work on adapting the Curry–Howard isomorphism to Hoare logic, where types range over program/formula pairs. It suggests a wider range of application of the Curry–Howard isomorphism to logical systems that involve formulae paired with some kind of labels (so-called labelled deductive systems [Gab96]). We leave exploration of this possibility to future research.

8.1.3 Type inference rules

We use proof-terms to denote SSL proofs. This follows from associating types with proof-terms by the typing relation $(.)^{(.)}$. When $p \in PT(\mathsf{SSL})$ is associated with type $(\mathrm{S}\textsc{p} \diamond F) \in Pairs(\mathsf{SSL})$, we write $p^{\mathrm{S}\textsc{p} \diamond F}$ and say that p *represents* a proof of the pair $\mathrm{S}\textsc{p} \diamond F$.

A proof-term correctly represents a proof when it is typed with the proof's formula and specification according to a set of type inference rules. Each type inference rule corresponds to a rule of SSL. Consequently the division between logical and structural rules is preserved.

The rules for typing proof-terms that denote instances of logical rules are given in Fig. 8.2, while correct typing of proof-terms for structural rules is given in Fig. 8.3.

The general form of the induction schema corresponds to the typing schema given in Fig. 8.4.

These rules define an inference relation $\vdash_{LTT(\mathsf{SSL})}$ that holds between a context Γ and a typed proof-term $p^{\mathrm{S}\textsc{p} \diamond F}$. A context Γ consists of a set of assumption formulae with associated proof-term variables

$$\begin{aligned}\Gamma \in \mathcal{P}(\{u^F \mid u \in Var_{PT(\mathsf{SSL})} & \text{ and} \\ & F \text{ is a formula from } \textstyle\bigcup_{\Sigma \in \mathbf{CSig}} WFF(\Sigma, Var)\})\end{aligned}$$

(We often write (Γ_1, Γ_2) for the union of two contexts $(\Gamma_1 \cup \Gamma_2)$.)

$$\frac{}{u^A \vdash \mathsf{ass}(\mathrm{SP}, u^A)^{\mathrm{SP}\diamond A}} \text{(Ass-I)} \quad \text{where } Sig(A) \subseteq Sig(\mathrm{SP})$$

$$\frac{}{\emptyset \vdash \mathsf{ax}(\langle \Sigma, Ax\rangle, A)^{\langle \Sigma, Ax\rangle \diamond A}} \text{(Ax-I)} \quad \text{where } A \in Ax$$

$$\frac{\Gamma, x : A \vdash d^{\mathrm{SP}\diamond B}}{\Gamma \vdash \lambda x : A.d^{\mathrm{SP}\diamond(A \Rightarrow B)}} (\Rightarrow\text{-I}) \qquad \frac{\Gamma_1 \vdash d^{\mathrm{SP}\diamond A \Rightarrow B} \quad \Gamma_2 \vdash r^{\mathrm{SP}\diamond A}}{\Gamma_1, \Gamma_2 \vdash (dr)^{\mathrm{SP}\diamond B}} (\Rightarrow\text{-E})$$

$$\frac{\Gamma_1 \vdash d^{\mathrm{SP}\diamond A} \quad \Gamma_2 \vdash e^{\mathrm{SP}\diamond B}}{\Gamma_1, \Gamma_2 \vdash \langle d, e\rangle^{\mathrm{SP}\diamond A \wedge B}} (\wedge\text{-I})$$

$$\frac{\Gamma \vdash d^{\mathrm{SP}\diamond A_1 \wedge A_2}}{\Gamma \vdash \mathsf{fst}(d)^{\mathrm{SP}\diamond A_1}} (\wedge\text{-E}_1) \qquad \frac{\Gamma \vdash d^{\mathrm{SP}\diamond A_1 \wedge A_2}}{\Gamma \vdash \mathsf{snd}(d)^{\mathrm{SP}\diamond A_2}} (\wedge\text{-E}_2)$$

$$\frac{\Gamma \vdash d^{\mathrm{SP}\diamond A}}{\Gamma \vdash \mathsf{inl}(d)^{\mathrm{SP}\diamond A \vee B}} (\vee\text{-I}_1) \qquad \frac{\Gamma \vdash e^{\mathrm{SP}\diamond B}}{\Gamma \vdash \mathsf{inr}(e)^{\mathrm{SP}\diamond A \vee B}} (\vee\text{-I}_2)$$

$$\frac{\Gamma \vdash f^{\mathrm{SP}\diamond A \vee B} \quad \Gamma_1, x^A \vdash d^{\mathrm{SP}\diamond C} \quad \Gamma_2, y^B \vdash e^{\mathrm{SP}\diamond C}}{\Gamma, \Gamma_1, \Gamma_2 \vdash \mathsf{case}\ f\ \mathsf{of}\ \mathsf{inl}(x).d,\ \mathsf{inr}(y).e^{\mathrm{SP}\diamond C}} (\vee\text{-E})$$

$$\frac{\Gamma \vdash d^{\mathrm{SP}\diamond A}}{\Gamma \vdash \mathsf{use}\ x : s.\ d^{\mathrm{SP}\diamond \forall x:s \bullet A}} (\forall\text{-I}) \qquad \frac{\Gamma \vdash d^{\mathrm{SP}\diamond \forall x:s \bullet A}}{\Gamma \vdash \mathsf{specific}(d, t)^{\mathrm{SP}\diamond A[t/x]}} (\forall\text{-E})$$

$$\frac{\Gamma \vdash d^{\mathrm{SP}\diamond A[t/x]}}{\Gamma \vdash \mathsf{show}(t, d)^{\mathrm{SP}\diamond \exists x:s \bullet A}} (\exists\text{-I})$$

$$\frac{\Gamma_1 \vdash d^{\mathrm{SP}\diamond \exists x:s \bullet A} \quad \Gamma_2, y^{A[z/x]} \vdash e^{\mathrm{SP}\diamond C}}{\Gamma_1, \Gamma_2 \vdash \mathsf{select}\ (d)\ \mathsf{in}\ z.y^{A[z/x]}.e^{\mathrm{SP}\diamond C}} (\exists\text{-E})$$

$$\frac{\Gamma \vdash d^{\mathrm{SP}\diamond \bot}}{\Gamma \vdash \mathsf{abort}(d, A)^{\mathrm{SP}\diamond A}} (\bot\text{-E})$$

We abbreviate the relation $\vdash_{LTT(\mathsf{SSL})}$ by $\vdash$. These rules have the same conditions of application as the corresponding rules of Fig. 7.4, Chapter 7, p. 238.

Fig. 8.2. Logical rules of SSL represented in the logical type theory $LTT(\mathsf{SSL})$.

Remark 8.4. An assumption variable u from a context can be used to build a larger proof-term by means of the (Ass-I) rule. Applications of that rule will be denoted by subterms of the form $\mathsf{ass}(F, \mathrm{SP}, u)$ in the larger proof-term.

The rule for translation, (trans), requires us to apply symbol mappings to context formulae and to the formula of the conclusion. So we require the following definition.

Definition 8.1.1 (Translation of contexts). The translation of the context Γ by a symbol mapping ρ is written $\rho`\Gamma$, and is defined by

$$\rho`\Gamma = \{u^{\rho(G)} \mid u^G \in \Gamma\}$$

$$\frac{\Gamma \vdash d^{\mathrm{SP} \diamond A}}{\rho`\Gamma \vdash (\rho \bullet d)^{(\mathrm{SP}\ \mathbf{with}\ \rho) \diamond \rho \bullet A}}\ (\text{trans})$$

$$\frac{\Gamma \vdash d^{\mathrm{SP} \diamond A}}{\Gamma \vdash \mathsf{hide}(d, SL)^{(\mathrm{SP}\ \mathbf{hide}\ SL) \diamond A}}\ (\text{hide})$$
where $\Gamma \cup \{A\} \subset WFF(Sig(\mathrm{SP})/SL, Var)$

$$\frac{\Gamma \vdash d^{\mathrm{SP\_1} \diamond A}}{\Gamma \vdash \mathsf{union}_1(d, \mathrm{SP\_2})^{(\mathrm{SP\_1}\ \mathbf{and}\ \mathrm{SP\_2}) \diamond A}}\ (\text{union}_1)$$

$$\frac{\Gamma \vdash d^{\mathrm{SP\_2} \diamond A}}{\Gamma \vdash \mathsf{union}_2(d, \mathrm{SP\_1})^{(\mathrm{SP\_1}\ \mathbf{and}\ \mathrm{SP\_2}) \diamond A}}\ (\text{union}_2)$$

$$\frac{\Gamma \vdash d^{\mathrm{SP\_1} \diamond A}}{\Gamma \vdash \mathsf{ext}_1(d, \mathrm{SP\_EXT})^{(\mathrm{SP\_1}\ \mathbf{then}\ \mathrm{SP\_EXT}) \diamond A}}\ (\text{ext}_1)$$

$$\frac{\Gamma \vdash d^{\mathrm{SP\_EXT} \diamond A}}{\Gamma \vdash \mathsf{ext}_2(d, \mathrm{SP\_EXT})^{(\mathrm{SP\_1}\ \mathbf{then}\ \mathrm{SP\_EXT}) \diamond inr(A)}}\ (\text{ext}_2)$$

We abbreviate the relation $\vdash_{LTT(\mathsf{SSL})}$ by $\vdash$. These rules have the same conditions of application as the corresponding rules of Fig. 7.5, p. 239, Chapter 7.

Fig. 8.3. Type inference rules of $LTT(\mathsf{SSL})$ corresponding to the structural rules of SSL.

8.1.4 The Curry–Howard isomorphism

The Curry–Howard isomorphism for intuitionistic logic states that proofs of formulae can be denoted by correctly typed terms of a typed lambda calculus with dependent products and sums. An analogous property holds between terms of our logical type theory and proofs of theorems in our logic.

Theorem 8.1.2 (Curry–Howard correspondence between LTT(SSL) and SSL). *The following properties hold*

1. *Given a natural deduction proof D of $\vdash_{\mathsf{SSL}} \mathrm{SP} \diamond A$, we can construct a well typed term $\vdash_{LTT(\mathsf{SSL})} f^{\mathrm{SP} \diamond A}$.*
2. *Given a well-typed term $\vdash_{LTT(\mathsf{SSL})} f^{\mathrm{SP} \diamond A}$, we can construct a natural deduction proof D with conclusion $\mathrm{SP} \diamond A$*

$$\begin{array}{c} \vdots \\ \vdash_{LTT(\mathsf{SSL})} \mathrm{SP} \diamond A \end{array}$$

Proof. Item 1) is derived by straightforward induction on the structure of the deduction D. Item 2) is obtained by induction on the structure of the inference $\vdash_{LTT(\mathsf{SSL})} f^{\mathrm{SP} \diamond A}$. □

Let $\Sigma = \langle S, F, P \rangle$ be a signature that contains a sort s generated by constructors $Con_s \subseteq F_s$,

$Cons_s = \{c_1 : s, \ldots, c_n : s, f_1 : (s^1_1 \times \ldots \times s^1_{m_1}) \to s, \ldots, f_p : (s^p_1 \times \ldots \times s^p_{m_p}) \to s\}$

Let P be a formula of Σ with $x : s$ free.
For each $i = 1, \ldots, p$ we make the following definitions. Take a set of variables $\{x^i_j\}_{j=1,\ldots,m_i}$ corresponding to argument sorts of f_i, we define

$$M(s, \{x^i_j : s^i_j\}_{j=1,\ldots,m_i}) = \{x^i_j \mid s^i_j = s\}$$

(When this set is empty, then f_i does not involve s as an argument sort.)
If this set is empty, then we define

$$P_{f_i} = \forall x^i_1 : s^i_1, \ldots, x^i_{m_i} : s^i_{m_i} \bullet P[f_i(x^1_1, \ldots, x^m_{m_i})/x]$$

Otherwise, we take

$$P_{f_i} = \forall x^i_1 : s^i_1, \ldots, x^i_{m_i} : s^i_{m_i} \bullet Q_{f_i} \Rightarrow P[f_i(x^i_1, \ldots, x^i_{m_i})/x]$$

where Q_{f_i} is formed from $M(s, \{x^i_j : s^i_j\}_{j=1,\ldots,m_i})$ as follows. Assume $M(s, \{x^i_j : s^i_j\}_{j=1,\ldots,m_i}) = \{x_1, \ldots, x_k\}$. If $k = 1$,

$$Q_{f_i} = P(x_1)$$

If $k > 1$,

$$Q_{f_i} = P(x_1) \Rightarrow \ldots \Rightarrow P(x_k)$$

Then we have the following induction schema for s in Σ:

$$\frac{a_1^{\langle \Sigma, Ax \rangle \diamond P[c_1/x]} \quad \ldots \quad a_n^{\langle \Sigma, Ax \rangle \diamond P[c_n/x]} \quad b_1^{\langle \Sigma, Ax \rangle \diamond P_{f_1}} \quad \ldots \quad b_p^{\langle \Sigma, Ax \rangle \diamond P_{f_p}}}{\mathsf{rec}(Cons_s, s, [a_1; \ldots; a_n; b_1; \ldots; b_p])^{\langle \Sigma, Ax \rangle \diamond \forall x : s \bullet P}} \; Ind(s, \Sigma)$$

Fig. 8.4. Type inference schema corresponding to the general structural induction schema.

8.1.5 Proof-term information

A proof-term is a very compact representation of information important for formal reasoning about proofs. Given a proof-term d with a derivation

$$\Gamma \vdash_{LTT(\mathsf{SSL})} d^{\mathrm{SP} \diamond F}$$

there are algorithms to compute the following data from d:

1. the current context $\mathrm{con}(d)$,
2. the specification $\mathrm{sp}(d)$ for which d is a derivation, and
3. the derived formula, $\mathrm{for}(d)$.

We define these algorithms in Figs. 8.5 and 8.6, by a modification of those given for the system of [WCP98].

con *is defined by:*

$$
\begin{array}{rl|rl}
\mathrm{con}(\mathsf{ass}(\textsc{Sp}, u^A)) &= \{u^A\} & \mathrm{con}(\mathsf{ax}(\langle\Sigma, Ax\rangle, F)) &= \emptyset \\
\mathrm{con}(\mathsf{abstract}\ u^A.\ d) &= \mathrm{con}(d)/\{u^A\} & \mathrm{con}(\mathsf{app}(d, r)) &= \mathrm{con}(d) \cup \mathrm{con}(r) \\
\mathrm{con}(\mathsf{inl}(d)) &= \mathrm{con}(d) & \mathrm{con}(\mathsf{inr}(e)) &= \mathrm{con}(e) \\
\mathrm{con}(\mathsf{use}\ x : s.\ d) &= \mathrm{con}(d) & \mathrm{con}(\mathsf{specific}(d, t)) &= \mathrm{con}(d) \\
\mathrm{con}(\langle d, e\rangle) &= \mathrm{con}(d) \cup \mathrm{con}(e) & \mathrm{con}(\mathsf{abort}(d, A)) &= \mathrm{con}(d) \\
\mathrm{con}(\mathsf{fst}(d)) &= \mathrm{con}(d) & \mathrm{con}(\mathsf{snd}(d)) &= \mathrm{con}(d)
\end{array}
$$

$$
\begin{array}{rl}
\mathrm{con}(\mathsf{case}\ f\ \mathsf{of}\ \mathsf{inl}(x).d,\ \mathsf{inr}(y).e) &= \mathrm{con}(d) \cup \mathrm{con}(e) \cup \mathrm{con}(f) \\
\mathrm{con}(\mathsf{select}\ (d)\ \mathsf{in}\ z.y.e) &= \mathrm{con}(d) \cup \mathrm{con}(e)
\end{array}
$$

$$
\begin{array}{rl|rl}
\mathrm{con}(\rho \bullet d) &= \rho`(\mathrm{con}(d)) & \mathrm{con}(\mathsf{hide}(d, SL)) &= \mathrm{con}(d) \\
\mathrm{con}(\mathsf{union}_1(d, \textsc{Sp\_2})) &= \mathrm{con}(d) & \mathrm{con}(\mathsf{union}_2(d, \textsc{Sp\_1})) &= \mathrm{con}(d) \\
\mathrm{con}(\mathsf{ext}_1(d, \textsc{Sp\_2})) &= \mathrm{con}(d) & \mathrm{con}(\mathsf{ext}_2(d, \textsc{Sp\_1})) &= \mathrm{con}(d)
\end{array}
$$

Fig. 8.5. The definition of con.

8.1.6 Example: Password checking system

Recall the password checking system example given in the previous chapter. In Section 7.3, pp. 245–247, we developed a theorem in SSL about the system specification PWDSYS: given any input x of a password, there is always an appropriate response message to be output, explaining if the password is of the correct length or not. This is the theorem

$$\vdash_{LTT(\mathsf{SSL})} \textsc{PwdSys} \diamond \forall x : nat \bullet \exists y : string \bullet ValidMsg(x, y) \tag{8.1}$$

We will now show how the proof of this theorem can be encoded as a proof-term in our logical type theory.

We require the following lemma, proved by boolean induction over NATBOOLEAN, as identified in Example 7.7 of the previous chapter. Observe that a proof by induction corresponds to application of the induction schema proof-term constructor.

$$\dfrac{\dfrac{\overline{\mathsf{ax}(\textsc{NatBoolean}, F = F)^{\textsc{NatBoolean} \diamond F = F}}\ (\vee\text{-}\mathrm{I}_2)}{p_1^{\textsc{NatBoolean} \diamond F = T \vee F = F}} \qquad \dfrac{\overline{\mathsf{ax}(\textsc{NatBoolean}, T = T)^{\textsc{NatBoolean} \diamond T = T}}\ (\vee\text{-}\mathrm{I}_1)}{p_2^{\textsc{NatBoolean} \diamond T = T \vee T = F}}}{\mathsf{rec}([T, F], boolean, [p_1, p_2])^{\textsc{NatBoolean} \diamond \forall b : boolean \bullet b = T \vee b = F}}\ \mathrm{Ind(Boolean)} \tag{8.2}$$

where $p_1 = \mathsf{inr}(\mathsf{ax}(\textsc{NatBoolean}, F = F))$ and $p_2 = \mathsf{inl}(\mathsf{ax}(\textsc{NatBoolean}, T = T))$.

In the proof of the last chapter, Section 7.3, pp. 245–247, the property (8.2) was used to show that the function $inRange(x)$ of PWDCORE is either *true*

sp *is defined by:*

$$\begin{array}{rcl|rcl}
\mathrm{sp}(\mathsf{ass}(\mathrm{SP}, u^A)) & = & \mathrm{SP} & \mathrm{sp}(\mathsf{ax}(\langle \Sigma, Ax\rangle, A)) & = & \langle \Sigma, Ax\rangle \\
\mathrm{sp}(\mathsf{abstract}\ x.\ d^A) & = & \mathrm{sp}(d) & \mathrm{sp}(\mathsf{app}(d, r)) & = & \mathrm{sp}(d) \\
\mathrm{sp}(\mathsf{use}\ x : s.\ d) & = & \mathrm{sp}(d) & \mathrm{sp}(\mathsf{specific}(d, t)) & = & \mathrm{sp}(d) \\
\mathrm{sp}(\langle d, e\rangle) & = & \mathrm{sp}(d) & \mathrm{sp}(\mathsf{fst}(d)) & = & \mathrm{sp}(d) \\
\mathrm{sp}(\mathsf{snd}(d)) & = & \mathrm{sp}(d) & & & \\
\mathrm{sp}(\mathsf{show}(t, d)) & = & \mathrm{sp}(d) & \mathrm{sp}(\mathsf{abort}(d, F)) & = & \mathrm{sp}(d) \\
\mathrm{sp}(\mathsf{inl}(d)) & = & \mathrm{sp}(d) & \mathrm{sp}(\mathsf{inr}(e)) & = & \mathrm{sp}(e)
\end{array}$$

$$\mathrm{sp}(\mathsf{case}\ f\ \mathsf{of}\ \mathsf{inl}(x).d,\ \mathsf{inr}(y).e) = \mathrm{sp}(d)$$

$$\mathrm{sp}(\mathsf{select}\ (d)\ \mathsf{in}\ z.y.e) = \mathrm{sp}(d)$$

$$\begin{array}{rcl|rcl}
\mathrm{sp}(\rho \bullet d) & = & \mathrm{sp}(d)\ \textbf{with}\ \rho & \mathrm{sp}(\mathsf{hide}(d, SL)) & = & \mathrm{sp}(d)\ \textbf{hide}\ SL \\
\mathrm{sp}(\mathsf{union}_1(d, \mathrm{SP_2})) & = & \mathrm{sp}(d)\ \textbf{and}\ \mathrm{SP_2} & \mathrm{sp}(\mathsf{union}_2(d, \mathrm{SP_1})) & = & \mathrm{sp}(d)\ \textbf{and}\ \mathrm{SP_2} \\
\mathrm{sp}(\mathsf{ext}_1(d, \mathrm{SP_2})) & = & \mathrm{sp}(d)\ \textbf{and}\ \mathrm{SP_2} & \mathrm{sp}(\mathsf{ext}_2(d, \mathrm{SP_1})) & = & \mathrm{sp}(d)\ \textbf{and}\ \mathrm{SP_2}
\end{array}$$

for *is defined by:*

$$\begin{array}{rcl|rcl}
\mathrm{for}(\mathsf{ass}(\mathrm{SP}, u^A)) & = & A & \mathrm{for}(\mathsf{ax}(\langle \Sigma, Ax\rangle, A)) & = & A \\
\mathrm{for}(\mathsf{abstract}\ x^A.\ d) & = & (A \Rightarrow \mathrm{for}(d)) & \mathrm{for}(\langle d, e\rangle) & = & \mathrm{for}(d) \wedge \mathrm{for}(e) \\
\mathrm{for}(\mathsf{inl}(d)) & = & (\mathrm{for}(d) \vee B) & \mathrm{for}(\mathsf{inr}(e)) & = & (A \vee \mathrm{for}(e)) \\
\mathrm{for}(\mathsf{use}\ x : s.\ d) & = & \forall x : s \bullet \mathrm{for}(d) & \mathrm{for}(\mathsf{show}(t, d)) & = & \exists x : s \bullet \mathrm{for}(d)
\end{array}$$

$$\begin{array}{rcl}
\mathrm{for}(\mathsf{app}(d, e)) & = & B \text{ where } \mathrm{for}(d) = (A \to B) \text{ and } \mathrm{for}(r) = A \\
\mathrm{for}(\mathsf{fst}(d)) & = & A_1 \text{ where } \mathrm{for}(d) = (A_1 \vee A_2) \\
\mathrm{for}(\mathsf{snd}(d)) & = & A_2 \text{ where } \mathrm{for}(d) = (A_1 \vee A_2) \\
\mathrm{for}(\mathsf{specific}(d, t)) & = & A[t/x] \text{ where } \mathrm{for}(d) = \forall x : s \bullet A \\
\mathrm{for}(\mathsf{abort}(d, A)) & = & A
\end{array}$$

$$\mathrm{for}(\mathsf{case}\ f\ \mathsf{of}\ \mathsf{inl}(x).d,\ \mathsf{inr}(y).e) = \mathrm{for}(d)$$

$$\mathrm{for}(\mathsf{select}\ (d)\ \mathsf{in}\ z.y.e) = \mathrm{for}(e)$$

$$\begin{array}{rcl|rcl}
\mathrm{for}(\rho \bullet d) & = & \rho(\mathrm{for}(d)) & \mathrm{for}(\mathsf{hide}(d, SL)) & = & \mathrm{for}(d) \\
\mathrm{for}(\mathsf{union}_1(d, \mathrm{SP_2})) & = & \mathrm{for}(d) & \mathrm{for}(\mathsf{union}_2(d, \mathrm{SP_1})) & = & \mathrm{for}(d) \\
\mathrm{for}(\mathsf{ext}_1(d, \mathrm{SP_2})) & = & \mathrm{for}(d) & \mathrm{for}(\mathsf{ext}_2(d, \mathrm{SP_1})) & = & \mathrm{for}(d)
\end{array}$$

Fig. 8.6. The definitions of sp and for.

or $false$ for any input password x. This required the use of structural rules to build PwdCore from NatBoolean. The resulting proof-term makes use of proof-term constructors corresponding to these rules.

$$\dfrac{\dfrac{\dfrac{\begin{array}{c}\vdots\ (8.2)\\ p_3^{\mathrm{NATBOOLEAN} \diamond \forall b:boolean \bullet b=T \vee b=F}\end{array}}{BtoB \bullet p_3^{(BtoB \bullet \mathrm{NATBOOLEAN}) \diamond \forall b:bool \bullet b=true \vee b=false}}\ (\mathrm{trans})}{\mathsf{union}_1(BtoB \bullet p_3, \mathrm{STRINGBOOL})^{\mathrm{SNB} \diamond \forall b:bool \bullet b=true \vee b=false}}\ (\mathrm{union}_1)}{\mathsf{ext}_1(\mathsf{union}_1(BtoB \bullet p_3, \mathrm{STRINGBOOL}), \langle SExt, AExt\rangle)^{\mathrm{PWDCORE} \diamond \forall b:bool \bullet b=true \vee b=false}}\ (\mathrm{ext}_1) \quad (8.3)$$

where $p_3 = \mathsf{rec}([T, F], boolean, [p_1, p_2])$,

$$\mathrm{SNB} = (BtoB \bullet \mathrm{NATBOOLEAN})\ \textbf{and}\ \mathrm{STRINGBOOL}$$

and where $BtoB$ is the symbol mapping $[Boolean \mapsto Bool, T \mapsto true, F \mapsto false]$ and $\langle SExt, AExt \rangle$ is the signature and axiom extension required to form PwdCore from SNB.

In the SSL proof of the previous chapter, we instantiated (8.3) with $inRange(x)$ for b, by application of ($\forall$-E).

That proof corresponds to the following proof-term construction

$$p_4^{\text{PwdCore} \diamond inRange(x)=true \vee inRange(x)=false} \tag{8.4}$$

where $p_4 = \mathsf{specific}(\mathsf{ext}_1(\mathsf{union}_1(BtoB \bullet p_3, \text{StringBool}), \langle SExt, AExt \rangle), inRange(x))$.

We derive (8.1) by reasoning over the possible cases that either $\text{PwdCore} \diamond inRange(x) = true$ or $\text{PwdCore} \diamond inRange(x) = false$.

Assuming the first case corresponds to the use of a proof-term variable

$$u^{\text{PwdCore} \diamond inRange(x)=true}$$

in a derivation of the form

$$\dfrac{\dfrac{\begin{matrix}\vdots\ (7.10)\\ p_7^{\text{PwdCore}\diamond A_2}\end{matrix} \quad \dfrac{\begin{matrix}\vdots\ (7.9)\\ p_6^{\text{PwdCore}\diamond A_1}\end{matrix} \quad \dfrac{u^{inRange(x)=true}}{p_5^{\text{PwdCore}\diamond inRange(x)=true}}\ (\text{Ass-I})}{\mathsf{app}(p_6, p_5)^{\text{PwdCore}\diamond OkPwd(x)}}\ (\to\text{-E})}{\mathsf{app}(p_7, \mathsf{app}(p_6, p_5))^{\text{PwdCore}\diamond ValidMsg(x,\text{`Password acceptable'})}}\ (\to\text{-E})}{\mathsf{show}(\text{`Password acceptable'}, \mathsf{app}(p_7, \mathsf{app}(p_6, p_5)))^{\text{PwdCore}\diamond \exists y:string \bullet ValidMsg(x,y)}}\ (\exists\text{-I}) \tag{8.5}$$

where $p_5 = \mathsf{ass}(SpecNamePwdCore, u^{inRange(x)=true})$. The proof-term $p_6^{\text{PwdCore}\diamond A_1}$ corresponds to an instantiated axiom of PwdCore, which can be written in full as

$$\mathsf{specific}(\mathsf{ax}(\text{PwdCore}, \forall x : nat \bullet inRange(x) = true \to OkPwd(x)), x)$$

with type $\text{PwdCore} \diamond inRange(x) = true \to OkPwd(x)$. The proof-term $p_7^{\text{PwdCore}\diamond A_2}$ also corresponds to an instantiated axiom of PwdCore, which when written in full is

$$\begin{aligned}\mathsf{specific}(\mathsf{ax}(&\text{PwdCore}, \forall x : nat \bullet OkPwd(x) \to \\ &ValidMsg(x, \text{`Password acceptable'})), x)\end{aligned}$$

with type $\text{PwdCore} \diamond OkPwd(x) \to ValidMsg(x, \text{`Password acceptable'})$

In the second case, similar reasoning over an assumption variable $v^{\text{PwdCore}\diamond inRange(x)=false}$, using the axioms of PwdCore will give a proof-term derivation of the form

$$\dfrac{\begin{matrix}\dfrac{v^{inRange(x)=true}}{\mathsf{ass}(\text{PwdCore}, v^{inRange(x)=true})^{\text{PwdCore}\diamond inRange(x)=false}}\ (\text{Ass-I})\\ \vdots \\ p_8^{\text{PwdCore}\diamond ValidMsg(x,\text{`Please choose a password in correct range'})}\end{matrix}}{\mathsf{show}\left(\begin{pmatrix}\text{`Please choose a}\\ \text{password in correct range'}\end{pmatrix}, p_8\right)^{\text{PwdCore}\diamond \exists y:string \bullet ValidMsg(x,y)}}\ (\exists\text{-I}) \tag{8.6}$$

here p_8 is a proof-term involving manipulation of PwdCore axioms.

By applying ($\vee$-E) using (8.4), (8.5) and (8.6), followed by ($\forall$-I) and (hide) will give (8.1), as required

$$\vdash_{LTT(\mathsf{SSL})} p^{\text{PwdSys}\diamond\forall x:nat\bullet\exists y:string\bullet ValidMsg(x,y)}$$

with proof-term

$$\begin{aligned}p = \mathsf{hide}(&\mathsf{use}\ x : nat.\ \mathsf{case}\ p_4\ \mathsf{of}\\ &\quad \mathsf{inl}(u).\mathsf{show}(\text{`Password acceptable'}, \mathsf{app}(p_7, \mathsf{app}(p_6, p_5))),\\ &\quad \mathsf{inr}(v).\mathsf{show}(\text{`Please choose a password in correct range'}, p_8),\\ &\qquad \{ge, inRange\})\end{aligned}$$

Fully expanded, the proof-term is as follows

$$\begin{aligned}p = \mathsf{hide}(&\mathsf{use}\ x : nat.\ \mathsf{case}\\ &\quad (\mathsf{specific}(\mathsf{ext}_1(\mathsf{union}_1(BtoB \bullet (\mathsf{rec}([T, F], boolean,\\ &\qquad [(\mathsf{inr}(\mathsf{ax}(\text{NatBoolean}, T = T))),\\ &\quad (\mathsf{inl}(\mathsf{ax}(\text{NatBoolean}, T = T)))])), \text{StringBool}), \langle SExt, AExt\rangle),\\ &\qquad inRange(x)))\\ &\quad \mathsf{of}\ \mathsf{inl}(u).\mathsf{show}(\text{`Password acceptable'}, \mathsf{app}(p_7, \mathsf{app}(p_6, p_5))),\\ &\quad \mathsf{inr}(v).\mathsf{show}(\text{`Please choose a password in correct range'}, p_8),\\ &\qquad \{ge, inRange\})\end{aligned}$$

By our version of the Curry–Howard isomorphism (Theorem 8.1.2 above), the entire SSL proof can be retrieved from this proof-term. In particular, the proof-term encodes constructive information obtained from the ($\exists$-I) steps of (8.5) and (8.6). The witness strings for the variable y are always valid messages, such that $ValidMsg(x, y)$ given a password number x.

In Chapter 9 we will use the Curry–Howard protocol to provide a method for extracting correct programs from proofs in SSL. We will show how the constructive information in proof-term p can be used to extract a password checking program, which outputs an appropriate response for any given password number input. Then we will also show how this program can be treated as a specification of a function to build a consistent extension of PwdSys.

However, before we investigate program extraction, we will investigate some simplifications corresponding to proof normalization that can be made to proof-terms. These simplifications will help in extracting programs, because they will yield more optimal programs.

8.1.7 Full form of the logical type theory

In full, the logical type theory $LTT(\mathsf{SSL})$ is defined as a tuple, following the general definition of Definition 3.2.3, Chapter 3, p. 84:

$$LTT(\mathsf{SSL}) = \langle PT(\mathsf{SSL}), Pairs(\mathsf{SSL}), (.)^{(.)}, \vdash_{LTT(\mathsf{SSL})}, PTR(LTT(\mathsf{SSL})), \rhd_{\mathsf{SSL}} \rangle$$

consisting of:

- a set of proof-terms $PT(\mathsf{SSL})$, described in 8.1.1 and Fig. 8.1,
- a set of types, taken as $Pairs(\mathsf{SSL})$,
- a typing relation $(.)^{(.)}$ between proof-terms and types, so that if $p \in PT(\mathsf{SSL})$ has type $(\mathrm{SP} \diamond F) \in Pairs(\mathsf{SSL})$, we write $p^{\mathrm{SP} \diamond F}$,
- a type inference relation given by $\vdash_{LTT(\mathsf{SSL})}$ with rules $PTR(LTT(\mathsf{SSL}))$, explained in 8.1.3, and
- a normalizing relation $\rhd_{\mathsf{SSL}}$, described in the next section (Section 8.2, p. 264).

8.2 Normalization and proof-term reduction

We define a normalization strategy for removing redundant parts of SSL proofs, based on proof normalization for intuitionistic logic.

In intuitionistic logic, normalization consists of repeatedly deleting matching applications of introduction and elimination rules (see, e.g., [Gen69] or [GLT89] and also Chapter 2). As we have seen, the logical rules of SSL correspond to intuitionistic rules, with introduction and elimination rules for the connectives. Consequently, we can define a similar kind of normalization.

As was the case for Int in Chapter 2 (pp. 39–41) and for Hoare logic in Chapter 5 (pp. 155–156), we define our proof normalization strategy in terms of a normalizing relation over proof-terms of our logical type theory.

The relation $\rhd_{\mathsf{SSL}}$ corresponding to normalization is given inductively by the rules of Figs. 8.7 and 8.9. As usual, the LHD and the RHD of a rule are called the *redex* and the *reduct* of the rule, respectively.

We write

$$p \mathrel{\hat{\rhd}_{\mathsf{SSL}}} p'$$

when p' may be obtained from p by the transitive closure of $\rhd_{LTT(\mathsf{SSL})}$ as defined in Figs. 8.7 and 8.9. When $p \mathrel{\hat{\rhd}_{\mathsf{SSL}}} p'$ holds, then p' is obtainable from p by a sequence of replacements of subterms using the rules of Figs. 8.7 and 8.9. In this case, we say that p is *reducible* to p'.

The rules fall into two categories:

- *Reductions of logical rules.* These reductions follow intuitionistic normalization, and form extension of β-reduction over the proof-terms that are corresponding to sub-proofs in which a connective is introduced and then immediately removed. Fig. 8.7 gives these rules.
- *Moving of structural rules.* Redundances can also occur due to application of structural rules in between two logical introduction and elimination rules that may be further simplified. To deal with this, we introduce reduction rules for moving structural rules up a proof tree before logical introduction rules. These reduction rules are given in Fig. 8.9.

$$
\begin{array}{rcl}
\mathsf{app}(\mathsf{abstract}\ x^A.\ a^{\mathrm{SP}\diamond B}, b^{\mathrm{SP}\diamond A}) & \rhd_{\mathsf{SSL}} & a[b/x]^{\mathrm{SP}\diamond B} \\
\mathsf{specific}(\mathsf{use}\ x : s.\ a^{\mathrm{SP}\diamond\forall x:s\bullet A}, v) & \rhd_{\mathsf{SSL}} & a[v/i]^{\mathrm{SP}\diamond A[v/x]} \\
\mathsf{fst}(\langle a, b\rangle^{\mathrm{SP}\diamond A\wedge B}) & \rhd_{\mathsf{SSL}} & a^{\mathrm{SP}\diamond A} \\
\mathsf{snd}(\langle a, b\rangle^{\mathrm{SP}\diamond A\wedge B}) & \rhd_{\mathsf{SSL}} & b^{\mathrm{SP}\diamond B} \\
\mathsf{case}\ \mathsf{inl}(a)^{\mathrm{SP}\diamond A\vee B}\ \mathsf{of}\ \mathsf{inl}(x).b^{\mathrm{SP}\diamond C},\ \mathsf{inr}(y).c^{\mathrm{SP}\diamond C} & \rhd_{\mathsf{SSL}} & b[a/x]^{\mathrm{SP}\diamond C} \\
\mathsf{case}\ \mathsf{inr}(a)^{\mathrm{SP}\diamond A\vee B}\ \mathsf{of}\ \mathsf{inl}(x).b^{\mathrm{SP}\diamond C},\ \mathsf{inr}(y).c^{\mathrm{SP}\diamond C} & \rhd_{\mathsf{SSL}} & c[a/y]^{\mathrm{SP}\diamond C} \\
\mathsf{select}\ (\mathsf{show}(v, a)^{\mathrm{SP}\diamond\exists y:s\bullet P})\ \mathsf{in}\ x.y.b^{\mathrm{SP}\diamond C} & \rhd_{\mathsf{SSL}} & b[a/x][v/y]^{\mathrm{SP}\diamond C}
\end{array}
$$

Fig. 8.7. Logical reduction rules that inductively define $\rhd_{\mathsf{SSL}}$.

The rest of this section provides some motivation for these two categories. Also, at the end of this section, we briefly discuss further permutations of rules that are possible, but which are not included in our normalization strategy. These permutations were identified in [WCP98, Pet96] for the calculus of that work, but are also applicable here.

8.2.1 Reductions of logical rules

We provide some motivation for the rules of Fig. 8.7. Our rules define reduction as a relation $\rhd_{\mathsf{SSL}}$ that holds between proof-terms.

The simplest cases of logical reduction are where a connective is introduced and then immediately removed.

Here we give two simple examples.

Example 8.1. The proof

$$
\dfrac{\dfrac{\begin{matrix}\vdots \\ \Gamma_1 \vdash_{LTT(\mathsf{SSL})} d^{\mathrm{SP}\diamond A}\end{matrix} \quad \begin{matrix}\vdots \\ \Gamma_2 \vdash_{LTT(\mathsf{SSL})} e^{\mathrm{SP}\diamond B}\end{matrix}}{\Gamma_1, \Gamma_2 \vdash_{LTT(\mathsf{SSL})} \langle d, e\rangle^{\mathrm{SP}\diamond(A\wedge B)}}\ (\wedge\text{-I})}{\Gamma_1, \Gamma_2 \vdash_{LTT(\mathsf{SSL})} \mathsf{fst}(\langle d, e\rangle)^{\mathrm{SP}\diamond A}}\ (\wedge\text{-E})
$$

can be reduced to the very simple proof

$$
\begin{matrix}\vdots \\ \Gamma_1 \vdash_{LTT(\mathsf{SSL})} d^{\mathrm{SP}\diamond A}\end{matrix}
$$

because the introduction of the conjunct B ultimately adds nothing to the proof.

To define the full range of possible logical reductions using proof-terms, we need to define substitution, in the obvious way.

Definition 8.2.1 (Substitution). The substitution of the proof-term r for the proof-term variable x in the proof-term d, written $d[r/x]$, is defined recursively, as in Fig. 8.8.

1) Logical rules

$$\mathsf{ass}(\mathrm{SP}, y^A)[r/x] = \begin{cases} r & \text{if } y = x \\ \mathsf{ass}(\mathrm{SP}, y^A) & \text{otherwise} \end{cases}$$

$$\mathsf{ax}(\langle \Sigma, Ax\rangle, A)[r/x] = \mathsf{ax}(\langle \Sigma, Ax\rangle, A)$$

$$\mathsf{abstract}\ y^B.\ d[r/x] = \begin{cases} \mathsf{abstract}\ y^B.\ d & \text{if } y:B = x:A \\ \mathsf{abstract}\ y^B.\ d[r/x] & \text{if } y:B \neq x:A, \\ & y \text{ not free in } r, \\ \mathsf{abstract}\ y^B.\ d[z/y][r/x] & \text{otherwise, where } z \text{ is new} \end{cases}$$

$$\begin{aligned}
\langle d, e\rangle[r/x] &= \langle d[r/x], e[r/x]\rangle \\
(\mathsf{inl}(d))[r/x] &= \mathsf{inl}(d[r/x]) \\
(\mathsf{inr}(d))[r/x] &= \mathsf{inr}(d[r/x]) \\
(\mathsf{use}\ y : s.\ d[r/x] &= \mathsf{use}\ y : s.\ (d[r/x]) \\
\mathsf{show}(t, d)[r/x] &= \mathsf{show}(t, d[r/x]) \\
\mathsf{app}(d, e)[r/x] &= \mathsf{app}((d[r/x]), (e[r/x])) \\
(\mathsf{fst}(d))[r/x] &= \mathsf{fst}(d[r/x]) \\
(\mathsf{snd}(d))[r/x] &= \mathsf{snd}(d[r/x]) \\
\mathsf{specific}(d, t)[r/x] &= \mathsf{specific}(d[r/x], t)
\end{aligned}$$

$$\begin{aligned}
\mathsf{case}\ f\ \mathsf{of}\ \mathsf{inl}(z).d,\ \mathsf{inr}(y).e[r/x] &= \mathsf{case}\ f[r/x]\ \mathsf{of}\ \mathsf{inl}(z).d[r/x],\ \mathsf{inr}(y).e[r/x] \\
\mathsf{select}\ (d)\ \mathsf{in}\ z.y.e[r/x] &= \mathsf{select}\ (d[z/x])\ \mathsf{in}\ z.y.e[z/x]
\end{aligned}$$

2) Structural rules

$$\begin{aligned}
(\rho \bullet d)[r/x] &= (\rho \bullet d[r/x]) \\
(\mathsf{hide}(d, SL))[r/x] &= \mathsf{hide}(d[r/x], SL) \\
\mathsf{union}_1(d,\ \mathrm{SP_2})[r/x] &= \mathsf{union}_1(d[r/x],\ \mathrm{SP_2}) \\
\mathsf{union}_2(d,\ \mathrm{SP_1})[r/x] &= \mathsf{union}_2(d[r/x],\ \mathrm{SP_1}) \\
\mathsf{ext}_1(d,\ \mathrm{SP_2})[r/x] &= \mathsf{ext}_1(d[r/x],\ \mathrm{SP_2}) \\
\mathsf{ext}_2(d,\ \mathrm{SP_1})[r/x] &= \mathsf{ext}_2(d[r/x],\ \mathrm{SP_1})
\end{aligned}$$

Fig. 8.8. Definition of substitution of proof-terms for proof-term variables.

Example 8.2. Substitution is used to define reductions involving the $\Rightarrow$ connective. To see this, consider the proof

$$\dfrac{\dfrac{\begin{matrix}\{x^A\} \vdash_{LTT(\mathsf{SSL})} x^A \\ \vdots \\ \Gamma_1, \{x^A\} \vdash_{LTT(\mathsf{SSL})} e^{\mathrm{SP}\diamond B}\end{matrix}}{\Gamma_1 \vdash_{LTT(\mathsf{SSL})} \mathsf{abstract}\ x^A.\ e^{\mathrm{SP}\diamond(A\Rightarrow B)}}\ (\Rightarrow\text{-I}) \qquad \begin{matrix}\vdots \\ \Gamma_2 \vdash_{LTT(\mathsf{SSL})} d^{\mathrm{SP}\diamond A}\end{matrix}}{\Gamma_2, \Gamma_1 \vdash_{LTT(\mathsf{SSL})} \mathsf{app}(\mathsf{abstract}\ x^A.\ e, d)^{\mathrm{SP}\diamond B}}\ (\Rightarrow\text{-E})$$

This too can be reduced to the much simpler proof:

$$\begin{array}{c} \vdots \\ \Gamma_2 \vdash_{LTT(\mathsf{SSL})} d^{\mathrm{SP} \diamond A} \\ \vdots \\ \Gamma_1, \Gamma_2 \vdash_{LTT(\mathsf{SSL})} e[d/x]^{\mathrm{SP} \diamond B} \end{array}$$

where the proof

$$\begin{array}{c} \vdots \\ \Gamma_2 \vdash_{LTT(\mathsf{SSL})} d^{\mathrm{SP} \diamond A} \end{array}$$

has replaced the assumption

$$\{x^A\} \vdash_{LTT(\mathsf{SSL})} x^{\mathrm{SP} \diamond A}$$

Example 8.3. Here is the logical reduction for a particular ($\vee$-I$_1$) followed by a ($\vee$-E). Note that, for simplicity, we have assumed that the contexts for the two proofs of C are the same. (The case of a ($\vee$-I$_2$) followed by a ($\vee$-E) is similar.)

$$\dfrac{\begin{array}{c} \vdots \\ \Gamma_2, \{x^A\} \vdash_{LTT(\mathsf{SSL})} d^{\mathrm{SP} \diamond C} \end{array} \quad \begin{array}{c} \vdots \\ \Gamma_2, \{y^B\} \vdash_{LTT(\mathsf{SSL})} e^{\mathrm{SP} \diamond C} \end{array} \quad \dfrac{\begin{array}{c} \vdots \\ \Gamma_1 \vdash_{LTT(\mathsf{SSL})} g^{\mathrm{SP} \diamond A} \end{array}}{\Gamma_1 \vdash_{LTT(\mathsf{SSL})} \mathsf{inl}(g)^{\mathrm{SP} \diamond (A \vee B)}} (\vee\text{-I}_1)}{\Gamma_2, \Gamma_1 \vdash (\mathsf{case}\ \mathsf{inl}(g)\ \mathsf{of}\ \mathsf{inl}(x).d,\ \mathsf{inr}(y).e)^{\mathrm{SP} \diamond C}} (\vee\text{-E})$$

This reduces to application of a new, derivable rule (Ass-E):

$$\dfrac{\begin{array}{c} \vdots \\ \Gamma_2, \{x^A\} \vdash_{LTT(\mathsf{SSL})} d^{\mathrm{SP} \diamond C} \end{array} \quad \Gamma_1 \vdash_{LTT(\mathsf{SSL})} g^{\mathrm{SP} \diamond A}}{\Gamma_2, \Gamma_1 \vdash_{LTT(\mathsf{SSL})} d[g/x]^{\mathrm{SP} \diamond C}} (\text{Ass-E})$$

The corresponding proof-term reductions are as follows:

$$(\mathsf{case}\ \mathsf{inl}(g)\ \mathsf{of}\ \mathsf{inl}(x).d,\ \mathsf{inr}(y).e)^{\mathrm{SP} \diamond C} \rhd_{\mathsf{SSL}} d[g/x]^{\mathrm{SP} \diamond C}$$

8.2.2 Moving structural rules

We now provide some motivation for the rules of Fig. 8.9. Under certain conditions, it is also possible to move structural rules up and down proofs. This can lead to matching of previously separated introduction and elimination rules and, consequently, to further logical reductions.

Example 8.4. Consider the following example.

$$\dfrac{\dfrac{\dfrac{\Gamma, \{u^A\} \vdash_{LTT(\mathsf{SSL})} d^{\mathrm{SP} \diamond B}}{\Gamma \vdash_{LTT(\mathsf{SSL})} (\mathsf{abstract}\ u^A.\ d)^{\mathrm{SP} \diamond A \Rightarrow B}} (\Rightarrow\text{-I})}{\rho`(\Gamma) \vdash_{LTT(\mathsf{SSL})} (\rho \bullet \mathsf{abstract}\ u^A.\ d)^{(\mathrm{SP}\ \mathbf{with}\ \rho) \diamond \rho(A) \Rightarrow \rho(B)}} (\text{trans}) \quad \Gamma_1 \vdash_{LTT(\mathsf{SSL})} r^{(\mathrm{SP}\ \mathbf{with}\ \rho) \diamond \rho(A)}}{\rho`(\Gamma), \Gamma_1 \vdash_{LTT(\mathsf{SSL})} (\mathsf{app}(\rho \bullet \mathsf{abstract}\ u^A.\ d, r))^{(\mathrm{SP}\ \mathbf{with}\ \rho) \diamond \rho(B)}}$$

$$\begin{aligned}
S(\mathsf{abstract}\ u^A.\ d) &\rhd_{\mathsf{SSL}} \mathsf{abstract}\ u^A.\ S(d)\\
S(\mathsf{use}\ x : s.\ d) &\rhd_{\mathsf{SSL}} \mathsf{use}\ x : s.\ S(d)\\
S(\mathsf{show}(t, d)) &\rhd_{\mathsf{SSL}} \mathsf{show}(t, S(d))\\
S(\langle d, e\rangle) &\rhd_{\mathsf{SSL}} \langle S(d), S(e)\rangle\\
S(\mathsf{inl}(d)) &\rhd_{\mathsf{SSL}} \mathsf{inl}(S(d))\\
S(\mathsf{inr}(d)) &\rhd_{\mathsf{SSL}} \mathsf{inr}(S(d))
\end{aligned}$$

where $S(p)$ denotes any of the following possible operations on a proof-term p: $\mathsf{union_1}(p, \mathrm{SP})$, $\mathsf{union_2}(p, \mathrm{SP})$ $\mathsf{ext_1}(p, \mathrm{SP})$ or $\mathsf{ext_2}(p, \mathrm{SP})$.

$$\begin{aligned}
\rho \bullet (\mathsf{abstract}\ u^A.\ d) &\rhd_{\mathsf{SSL}} \mathsf{abstract}\ u^{\rho(A)}.\ S(d)\\
\rho \bullet (\mathsf{use}\ x : s.\ d) &\rhd_{\mathsf{SSL}} \mathsf{use}\ x : \rho(s).\ (\rho \bullet d)\\
\rho \bullet (\mathsf{show}(t, d)) &\rhd_{\mathsf{SSL}} \mathsf{show}(\rho(t), (\rho \bullet d))\\
\rho \bullet (\langle d, e\rangle) &\rhd_{\mathsf{SSL}} \langle (\rho \bullet d), (\rho \bullet e)\rangle\\
\rho \bullet (\mathsf{inl}(d)) &\rhd_{\mathsf{SSL}} \mathsf{inl}((\rho \bullet d))\\
\rho \bullet (\mathsf{inr}(d)) &\rhd_{\mathsf{SSL}} \mathsf{inr}((\rho \bullet d))
\end{aligned}$$

$$\begin{aligned}
\mathsf{hide}(\mathsf{abstract}\ u^A.\ d, SL) &\rhd_{\mathsf{SSL}} \mathsf{abstract}\ u^A.\ S(d)\\
\mathsf{hide}(\mathsf{use}\ x : s.\ d, SL) &\rhd_{\mathsf{SSL}} \mathsf{use}\ x : s.\ \mathsf{hide}(d, SL)\\
\mathsf{hide}(\langle d, e\rangle, SL) &\rhd_{\mathsf{SSL}} \langle \mathsf{hide}(d, SL), \mathsf{hide}(e, SL)\rangle\\
\mathsf{hide}(\mathsf{inl}(d), SL) &\rhd_{\mathsf{SSL}} \mathsf{inl}(\mathsf{hide}(d, SL))\\
\mathsf{hide}(\mathsf{inr}(d), SL) &\rhd_{\mathsf{SSL}} \mathsf{inr}(\mathsf{hide}(d, SL))
\end{aligned}$$

Fig. 8.9. Structural reduction rules that inductively define $\rhd_{\mathsf{SSL}}$.

where the last step of the proof is an instance of ($\Rightarrow$-E). Observe that the lambda abstraction and application in the proof-term $(\mathsf{app}(\rho \bullet \mathsf{abstract}\ u.\ d, r))$ cannot be matched for reduction, because of the renaming over the abstraction. However, if we *swap* the order of the rule applications ($\Rightarrow$-I) and (trans), we obtain a correctly typed proof-term, for which a logical reduction can be applied:

$$\dfrac{\dfrac{\dfrac{\Gamma, \{u^A\} \vdash_{LTT(\mathsf{SSL})} d^{\mathrm{SP} \diamond B}}{\rho`(\Gamma), \{u^{\rho(A)}\} \vdash_{LTT(\mathsf{SSL})} (\rho \bullet d)^{(\mathrm{SP}\ \mathbf{with}\ \rho) \diamond \rho(B)}}\ (\text{trans})}{\rho`(\Gamma) \vdash_{LTT(\mathsf{SSL})} (\mathsf{abstract}\ u^A.\ \rho \bullet d)^{\mathrm{SP} \diamond \rho(A) \Rightarrow \rho(B)}}\ (\Rightarrow\text{-I}) \quad \Gamma_1 \vdash_{LTT(\mathsf{SSL})} r^{(\mathrm{SP}\ \mathbf{with}\ \rho) \diamond \rho(A)}}{\rho`(\Gamma), \Gamma_1 \vdash_{LTT(\mathsf{SSL})} (\rho \bullet \mathsf{app}(\mathsf{abstract}\ u^A.\ \rho \bullet d, r))^{(\mathrm{SP}\ \mathbf{with}\ \rho) \diamond \rho(B)}}\ (\Rightarrow\text{-E})$$

The resulting proof-term reduces as follows

$$\mathsf{app}(\mathsf{abstract}\ u^A.\ \rho \bullet d, r) \rhd_{\mathsf{SSL}} (\rho \bullet d)[r/u]$$

Clearly it is in our interest to systematically move structural rules above introduction rules, when possible, in order to eliminate further redundancies in a proof. We do this by extending $\rhd_{\mathsf{SSL}}$ with rules for swapping proof-terms for structural and introduction rules. Fig. 8.9 gives these additional rules.

(composition)	$\rho_2 \bullet \rho_1 \bullet d^{\text{SP with } \rho_1 \text{ with } \rho_2 \diamond \rho_2 \bullet (\rho_1 \bullet (A))}$
	$\rho_* \bullet d^{\text{SP with } \rho_* \diamond \rho_* \bullet (A)}$
(union$_1$) trivializing	$\mathsf{hide}(\mathsf{union}_1(d, \text{SP\_2}), \Sigma_1)^{((\text{SP\_1 and SP\_2}) \text{ hide } \Sigma_1) \diamond A}$
	$\rhd_{\text{triv}} d^{\text{SP\_2} \diamond A}$
(union$_2$) trivializing	$\mathsf{hide}(\mathsf{union}_2(d, \text{SP\_1}), \text{SP\_2})^{((\text{SP\_1 and SP\_2}) \text{ hide } \Sigma_1) \diamond A}$
	$\rhd_{\text{triv}} d^{\text{SP\_1} \diamond A}$
(ext$_1$) trivializing	$\mathsf{hide}(\mathsf{ext}_1(d, \text{SP\_2}), \Sigma_1)^{((\text{SP\_1 and SP\_2}) \text{ hide } \Sigma_1) \diamond A}$
	$\rhd_{\text{triv}} d^{\text{SP\_1} \diamond A}$
(ext$_2$) trivializing	$\mathsf{hide}(\mathsf{ext}_2(d, \text{SP\_1}), \Sigma_2)^{((\text{SP\_1 and SP\_2}) \text{ hide } \Sigma_2) \diamond A}$
	$\rhd_{\text{triv}} d^{\text{SP\_2} \diamond A}$

SP_1 and SP_2 are arbitrary specification expressions. Σ_1 denotes Sig(SP_1) and Σ_2 denotes Sig(SP_2). Note that ρ^* is the composition of ρ_1 followed by ρ_2.

Fig. 8.10. Trivializing structural reductions.

Remark 8.5. Such interchange is not possible for (hide) and ($\exists$-I) rules because an $\exists$ introduction may be applied with respect to a witness term that is later hidden.

In this case, the rules cannot be reversed. For example:

$$\dfrac{\dfrac{\Gamma \vdash_{LTT(\mathsf{SSL})} d^{\text{SP} \diamond A[t/x]}}{\Gamma \vdash_{LTT(\mathsf{SSL})} \mathsf{show}(t, d)^{\text{SP} \diamond \exists x:s \bullet A}}\ (\exists\text{-I})}{\Gamma \vdash_{LTT(\mathsf{SSL})} \mathsf{hide}(\mathsf{show}(t, d), t)^{(\text{SP } \mathbf{hide}\ t) \diamond \exists x:s \bullet A}}\ (\text{hide})$$

cannot be replaced by a sequence

$$\dfrac{\dfrac{\Gamma \vdash_{LTT(\mathsf{SSL})} d^{\text{SP} \diamond A[t/x]}}{\Gamma \vdash_{LTT(\mathsf{SSL})} \mathsf{hide}(d, t)^{(\text{SP } \mathbf{hide}\ t) \diamond A[t/x]}}\ (\text{hide})}{\Gamma \vdash_{LTT(\mathsf{SSL})} \mathsf{show}(t, \mathsf{hide}(d, t))^{(\text{SP } \mathbf{hide}\ t) \diamond \exists x:s \bullet A}}\ (\exists\text{-I})$$

The hiding inference is not permitted because the hidden term t is then used as a witness by the existential introduction.

8.2.3 Reduction preserves derivability

We have the following important lemma about the reduction relation $\rhd_{\mathsf{SSL}}$

Lemma 8.2.1. *Let p and p' be proof-terms such that $p \rhd_{\mathsf{SSL}} p'$. If we have a derivation $\Gamma \vdash_{LTT(\mathsf{SSL})} p^{\text{SP} \diamond A}$ then we can construct a derivation $\Gamma \vdash_{LTT(\mathsf{SSL})} p'^{\text{SP} \diamond A}$.*

Proof. By a straightforward induction on the possible forms of p. □

Remark 8.6. Our main concern is the extraction of programs from proofs. As discussed in Chapter 3 of Part II and seen in the cases of intuitionistic logic (Chapter 2 of Part II) and Hoare logic (Chapter 5 of Part III), proof-normalization is a useful means of simplifying proof-terms prior to transformation into required programs. Application of a normalization strategy will yield simpler programs, because, as we will see in the next chapter, the size of an extracted program often reflects the size of the proof.

Note, however, we are generalizing state-of-the-art proofs-as-programs, where the importance of normalization is devalued from naïve methods for constructive program extraction. In those methods, proofs are treated as programs with proof-normalization considered to be an operational semantics. Our work stands in contrast to that treatment, by virtue of our adherence to the Curry–Howard protocol. *Normalization does not correspond to executing programs, because, in the protocol, proofs are not considered to be programs.* (See Remark 2.14 of Chapter 2, p. 2.14.) Proofs are transformed into programs via an extraction map. Potentially, we could extract correct programs from proofs that are not normal. So, from the perspective of program extraction according to the protocol, normalization should be seen as a pre-processing strategy to be carried out prior to extraction for the purpose of yielding more optimal programs.

8.2.4 Further possible reductions

The normalization process defined by Figs. 8.7 and 8.9 does not eliminate all possible redundancies in a proof. Following [WCP98] and [Pet96], we can define further *trivializing reductions* over structural rules. However, these reductions result in changing the specification of the conclusion. This is not the case of the normalization reductions discussed above, where the formula and specification expression remain the same, and only the proof is simplified.

Example 8.5. As an example of a structural rule reduction, two translations can be consolidated into one. In particular translating by ρ followed by ρ^{-1} can be regarded as redundant. In general we have the reduction:

$$\dfrac{\dfrac{\Gamma \vdash_{LTT(\mathsf{SSL})} d^{\mathrm{SP} \diamond A}}{\rho_1^{\text{‘}}(\Gamma) \vdash_{LTT(\mathsf{SSL})} (\rho_1 \bullet d)^{\mathrm{SP}\ \mathbf{with}\ \rho_1 \diamond \rho_1(A)}}\ (\text{trans})}{\rho_2^{\text{‘}}(\rho_1^{\text{‘}}(\Gamma)) \vdash_{LTT(\mathsf{SSL})} (\rho_2 \bullet \rho_1 \bullet d)^{\mathrm{SP}\ \mathbf{with}\ \rho_1\ \mathbf{with}\ \rho_2 \diamond \rho_2(\rho_1(A))}}\ (\text{trans})$$

reduces to

$$\dfrac{\Gamma \vdash_{LTT(\mathsf{SSL})} d^{\mathrm{SP} \diamond A}}{\rho^{*\,\text{‘}}(\Gamma) \vdash_{LTT(\mathsf{SSL})} (\rho^* \bullet d)^{(\mathrm{SP}\ \mathbf{with}\ \rho^*) \diamond \rho^*(A)}}\ (\text{trans}) \tag{8.7}$$

where $\rho^* = \rho_2 \circ \rho_1$ (the concatenation of the symbol maps ρ_2 and ρ_1). In the special case that $\rho_2 = \rho_1^{-1}$ or *vice versa*, the translation of (8.7) becomes a triviality and can be omitted.

Example 8.6. Another example of a trivial, reducible proof is taking the union with a specification whose signature is Σ and then hiding Σ, or *vice versa.* The pushout for the union can be regarded as an introduction rule and hiding as an elimination rule in the case where the signature goes from SP_1 to (SP_1 **and** SP_2) and then back (by hiding) to SP_1. Thus we can remove the following pair of rules (and similarly for (union_1) in a proof:

$$\dfrac{\dfrac{\Gamma \vdash_{LTT(\mathsf{SSL})} d^{\text{SP\_1}\diamond A}}{\Gamma \vdash_{LTT(\mathsf{SSL})} \mathsf{union}_1(d, \text{SP\_2})^{(\text{SP\_1 } \mathbf{and} \text{ SP\_2})\diamond A}}\ (\text{union}_1)}{\Gamma \vdash_{LTT(\mathsf{SSL})} \mathsf{hide}(\mathsf{union}_1(d, \text{SP\_2}), \text{Sig}(\text{SP\_2}))^{(\text{SP\_1 } \mathbf{and} \text{ SP\_2}) \text{ } \mathbf{hide} \text{ } \text{Sig}(\text{SP\_2})\diamond A}}\ (\text{hide})$$

The list of trivializing structural reductions can be formalized by a reduction relation $\rhd_{\text{triv}}$ over proof-terms, displayed in Fig. 8.9. The list is modified from that given in [Pet96], to account for the different syntax of our proof-terms.

Extending $\rhd_{\mathsf{SSL}}$ to include $\rhd_{\text{triv}}$ will conserve the strong normalization and Church–Rosser properties (to be proved for $\rhd_{\mathsf{SSL}}$ below). The reader is referred to [Pet96] for a proof that can be readily adapted to our system.

Remark 8.7. For the purposes of program extraction, these structural reductions are not as important as the logical reductions. Therefore we do not use them for the remainder of this monograph. The reasons for this is as follows. In the following chapter, we will define an extraction map from proof-terms to *SML* functions that will ignore all structural proof-terms (except for renamings). So the simplification of redundant structural rules is not a vital consideration. Logical reductions, in contrast, are important for program extraction, because the size of logical proof-terms can affect the size of extracted *SML* programs.

8.2.5 Strong normalization

We now prove *strong normalization* for our normalization reductions: that is, we show that any sequence of normalizing reductions obtained by Figs. 8.7 and 8.9 will always terminate. We will prove this in Theorem 8.2.9 and Corollary 8.2.1. The proof follows that given by Crossley and Shepherdson [CS93] for intuitionistic logic (based in part on the reducibility methods of Tait and of Girard [GLT89]).

Recall that SSL consists of both logical and structural rules. The logical part of the formal calculus is essentially intuitionistic logic with the same specification expression used to label premisses and conclusions. The proof-terms, and reduction rules, for that part of the calculus are essentially identical to those of intuitionistic logic. It follows that the logical part of SSL, considered separate from the structural part, is strongly normalizing with respect to the logical reduction rules of Fig. 8.7.

However, in order to prove strong normalization for the whole calculus we have to include the possibility of structural reductions.

We require the following notation and definitions in our proof.

Definition 8.2.2. We define the relation $a \rhd_1 b$ to hold if b is obtained from a by a single application of one of the reduction rules of Figs. 8.7 and 8.9 to a redex of a.

If $a \rhd_1 b$ holds, we say that b is *immediately reducible* from a.

Definition 8.2.3. We say that a proof-term is *normal* if it contains no redex. A normal proof-term is *irreducible.*

Definition 8.2.4. Given a proof-term t, we let $\mathcal{N}(t)$ denote the least upper bound of lengths of reduction sequences for t. We say that t is *strongly normalizable* if all reduction sequences are finite.

Remark 8.8. It is the case that, if t is strongly normalizable, then $\mathcal{N}(t)$ must be finite. The converse also holds, by König's Lemma.

Definition 8.2.5 (Neutral proof-terms). A proof-term is *neutral* if it is a $\mathsf{LTT}_{\mathsf{SSL}}$ variable or is of one of the following forms:

$\mathsf{ass}(\mathrm{SP}, u^A)$	$\mathsf{ax}(\langle \Sigma, Ax\rangle, A)$	$\mathsf{app}(a, b)$	
$\mathsf{specific}(a, v)$	$\mathsf{fst}(a)$	$\mathsf{snd}(b)$	$\mathsf{case}\ a\ \mathsf{of}\ \mathsf{inl}(x).b,\ \mathsf{inr}(y).c$
$\mathsf{abort}(d, A)$	$\mathsf{select}\ (a)\ \mathsf{in}\ x.y.b$	$\rho \bullet d$	$\mathsf{hide}(d, SL)$
$\mathsf{union}_1(d, \mathrm{SP})$	$\mathsf{union}_1(d, \mathrm{SP})$	$\mathsf{ext}_1(d, \mathrm{SP})$	$\mathsf{ext}_2(d, \mathrm{SP})$

Remark 8.9. A proof-term is *not* neutral if it is of one of the following forms:

	$\mathsf{abstract}\ x.\ a$		$\mathsf{use}\ i : s.\ a$
$\langle a, b\rangle$	$\mathsf{inl}(a)$	$\mathsf{inr}(b)$	$\mathsf{show}(v, a)$

Neutral proof-terms satisfy the following lemma.

Lemma 8.2.2. *Let p, q, r and s be proof-terms. Assume that p is neutral. Then the following properties are true:*

- *Every immediate reduct of* $\mathsf{app}(p, q)$ *is obtained by reducing p or q. That is, the immediate reduct must be of the form* $\mathsf{app}(p', q)$ *or* $\mathsf{app}(p, q')$*, where p' is an immediate reduct of p and q' is an immediate reduct of q.*
- *Every immediate reduct of* $\mathsf{select}\ (p)\ \mathsf{in}\ x.y.q$ *is obtained by reducing p or q.*
- *Every immediate reduct of* $\mathsf{case}\ p\ \mathsf{of}\ \mathsf{inl}(x).q,\ \mathsf{inr}(y).r$ *is obtained by reducing p, q or r.*

Also, every immediate reduct of $\mathsf{use}\ x : s.\ p$, $\mathsf{abort}(p, A)$, $\mathsf{specific}(p, v)$, $\mathsf{fst}(p)$, $\mathsf{snd}(p)$, $(\rho \bullet p)$, $\mathsf{hide}(p, SL)$, $\mathsf{union}_1(p, \mathrm{SP})$, $\mathsf{union}_2(p, \mathrm{SP})$, $\mathsf{ext}_1(p, \mathrm{SP})$ *and* $\mathsf{ext}_2(p, \mathrm{SP})$ *is obtained by reducing p.*

Proof. The proof follows easily from the definition of $\rhd_{\mathsf{SSL}}$. □

Our definition of candidates for reducibility (CR) is similar to that for the logical type theory for intuitionistic logic, following [CS93] and Girard [GLT89]. Candidates for reducibility are sets of strongly normalizing proof-terms of a common type. We will use CR to prove strong normalizability, by showing that every derivable proof-term is in a CR. Because our types now range over pairs of specification expressions and formulae, we associate CR with such pairs (rather than single formulae, as in the intuitionistic case). This is necessary to define well-formed operations over CR. We define CR for specification/formulae pairs as follows.

Definition 8.2.6 (Candidate for reducibility). A candidate for reducibility (CR) of formula A is a set C of proof-terms of type $\textsc{Sp} \diamond A$ such that
CR1. If $t \in C$, then t is strongly normalizable.
CR2. If $t \in C$ and $t \rhd_1 t'$, then $t' \in C$.
CR3. If t is neutral and all immediate reducts t' of t are in C, then $t \in C$.

Lemma 8.2.3. *Assume that $t^{\textsc{Sp}\diamond A}$ is a proof-term and C is a CR of type $\textsc{Sp}\diamond A$. If t is neutral and normal, then $t \in C$.*

Proof. This is a direct consequence of CR3. □

Definition 8.2.7 (Operations on CR). We define operations on CR corresponding to the connectives used to construct types, as follows.

Suppose that C_1, C_2 are CR of types $(\textsc{Sp} \diamond A_1)$ and $(\textsc{Sp} \diamond A_2)$ respectively. Then define

- $(C_1 \Rightarrow C_2)$ as the set of all proof-terms t of type $\textsc{Sp} \diamond (A_1 \Rightarrow A_2)$ such that, for every proof-term $u \in C_1$, it is true that $\mathsf{app}(t, u) \in C_2$.
- $(C_1 \wedge C_2)$ as the set of all proof-terms t of type $\textsc{Sp} \diamond (A_1 \wedge A_2)$ such that $\mathsf{fst}(t) \in C_1$ and $\mathsf{snd}(t) \in C_2$.
- $(C_1 \vee C_2)$ as the set of all proof-terms t of type $\textsc{Sp} \diamond (A_1 \vee A_2)$ such that the following holds. Assume C is any CR of some type $\textsc{Sp} \diamond P$, $f_1 \in (C_1 \Rightarrow C)$ and $f_2 \in (C_2 \Rightarrow C)$, where u^{C_1} is not free in f_1 and v^{C_2} is not free in f_2. Then,
$$\mathsf{case}\ t\ \mathsf{of}\ \mathsf{inl}(u).\mathsf{app}(f_1, u),\ \mathsf{inr}(v).\mathsf{app}(f_2, v) \in C$$

Lemma 8.2.4. *A CR can be obtained by applying any of the operations of Definition 8.2.7 to other CR.*

Proof. We essentially follow the proof given in [CS93], adapted for our system. For the operations $\Rightarrow$, $\wedge$ and $\vee$ we must verify CR1, CR2 and CR3.

Assume that C_1 and C_2 are CR of types $(\textsc{Sp}\diamond A_1)$ and $(\textsc{Sp}\diamond A_2)$ respectively.

($\Rightarrow$ *case*). $(C_1 \Rightarrow C_2)$ is a CR because of the following facts.

CR1 Assuming t is in $(C_1 \Rightarrow C_2)$ then, by Lemma 8.2.3 applied to C_1, the variable u^{C_1} is in C_1, so $\mathsf{app}(t, u)$ is in C_2. It is the case that $\mathsf{app}(t, u)$ is strongly normalizable, by CR1 for C_1. Consequently, $\mathcal{N}(\mathsf{app}(t, u))$ is finite. Then, as $\mathcal{N}(\mathsf{app}(t, u)) \geq \mathcal{N}(t)$, $\mathcal{N}(t)$ is finite and t is strongly normalizable.

CR2 Assume t is in $(C_1 \Rightarrow C_2)$ and t is immediately reducible to t'. If u is in C_1 then the proof-term $\mathsf{app}(t, u)$ in C_2, is immediately reducible to $\mathsf{app}(t', u)$. By CR2 for C_2, $\mathsf{app}(t', u)$ is in C_2. Hence t' is in $(C_1 \Rightarrow C_2)$.

CR3 Suppose t is neutral and all immediate reducts t' of t are in $(C_1 \Rightarrow C_2)$. We wish to show that, if u is in C_1, then $\mathsf{app}(t, u)$ in C_2. We proceed by induction on $\mathcal{N}(u)$. By CR3 for C_2, since $\mathsf{app}(t, u)$ is neutral, we need only prove that every immediate reduct of $\mathsf{app}(t, u)$ is in C_2. Because t is neutral, such a proof-term can be of two forms:
  a) t can be of the form $\mathsf{app}(t', u)$, where t' is an immediate reduct of t. Observe that t' is in $(C_1 \Rightarrow C_2)$. As a consequence $\mathsf{app}(t', u)$ is in C_2.
  b) t can be of the form $\mathsf{app}(t, u')$, where u' is an immediate reduct of u. But here $\mathcal{N}(u') < \mathcal{N}(u)$, so the result follows by the induction hypothesis.

($\wedge$ *case*). $(C_1 \wedge C_2)$ is a CR because we can verify the following conditions.

CR1 Assume that t is in $(C_1 \wedge C_2)$. This means $\mathsf{fst}(t)$ is in C_1. Because it can be shown that $\mathcal{N}(t) \leq \mathcal{N}(\mathsf{fst}(t)) + \mathcal{N}(\mathsf{snd}(t))$, we are done.

CR2 Assume t is in $(C_1 \wedge C_2)$ and t is immediately reducible to t'. Then it is the case that $\mathsf{fst}(t)$ is in C_1 and is immediately reducible to $\mathsf{fst}(t')$. Consequently $\mathsf{fst}(t')$ is in C by CR2 for C_1. Similarly we have that $\mathsf{snd}(t')$ is in C_2. So, t' is in $(C_1 \wedge C_2)$.

CR3 Assuming t is neutral and all immediate reducts t' of t are in $C_1 \wedge C_2$, we have that $\mathsf{fst}(t)$ is neutral and all immediate reducts of $\mathsf{fst}(t)$ are of the form $\mathsf{fst}(t')$, where t' is an immediate reduct of t. Because any such $\mathsf{fst}(t')$ is in C_1, $\mathsf{fst}(t)$ is in C_1 by CR3. Similarly $\mathsf{snd}(t)$ is in C_2. This gives us that t is in $C_1 \wedge C_2$.

($\vee$ *case*). $(C_1 \vee C_2)$ is a CR due to the following reasoning. First, suppose t is in $C_1 \vee C_2$. Take an arbitrary type $(\mathrm{SP} \diamond P)$, and a CR, called C, of type $(\mathrm{SP} \diamond P)$. Take arbitrary proof-terms $f_1 \in (C_1 \Rightarrow C_2)$ and $f_2 \in (C_2 \Rightarrow C)$, where u^{C_1} is not free in f_1 and v^{C_2} is not free in f_2.

Then the following conditions for CR are satisfied.

CR1 It is the case that

$$\mathcal{N}(\mathsf{case}\ t\ \mathsf{of}\ \mathsf{inl}(u).\mathsf{app}(f_1, u),\ \mathsf{inr}(v).\mathsf{app}(f_2, v)) \geq \mathcal{N}(t)$$

But $\mathsf{case}\ t\ \mathsf{of}\ \mathsf{inl}(u).\mathsf{app}(f_1, u),\ \mathsf{inr}(v).\mathsf{app}(f_2, v)$ is in C and hence strongly normalizable by CR1 for C. So $\mathcal{N}(t)$ is finite, as required

CR2 If t is immediately reducible to t' then

$$\mathsf{case}\ t\ \mathsf{of}\ \mathsf{inl}(u).\mathsf{app}(f_1, u),\ \mathsf{inr}(v).\mathsf{app}(f_2, v)$$

is immediately reducible to $\mathsf{case}\ t'\ \mathsf{of}\ \mathsf{inl}(u).\mathsf{app}(f_1, u),\ \mathsf{inr}(v).\mathsf{app}(f_2, v)$. By CR2 for C, $\mathsf{case}\ t'\ \mathsf{of}\ \mathsf{inl}(u).\mathsf{app}(f_1, u),\ \mathsf{inr}(v).\mathsf{app}(f_2, v)$ is in C. As a consequence, t' is in $(C_1 \vee C_2)$, as required.

CR3 Assume that t is neutral and every immediate reduct t' of t is in $(C_1 \vee C_2)$. Then

$$\mathsf{case}\ t'\ \mathsf{of}\ \mathsf{inl}(u).\mathsf{app}(f_1, u),\ \mathsf{inr}(v).\mathsf{app}(f_2, v) \in C$$

We are required to show that $t \in (C_1 \vee C_2)$. This is tantamount to deriving

$$\mathsf{case}\ t\ \mathsf{of}\ \mathsf{inl}(u).\mathsf{app}(f_1, u),\ \mathsf{inr}(v).\mathsf{app}(f_2, v)$$

is in C. This proof-term is neutral, so we need only show that all its immediate reducts are in C.
We can rephrase this requirement. We fix t and define $g_1 = \mathsf{app}(f_1, u^{C_1})$ and $g_2 = \mathsf{app}(f_2, v^{C_2})$. We are then required to prove that

$$r = \mathsf{case}\ t\ \mathsf{of}\ \mathsf{inl}(u^{C_1}).g_1,\ \mathsf{inr}(v^{C_2}).g_2$$

is in C, assuming that g_1 and g_2 are in C, t is neutral and, for each immediate reduct t' of t, $\mathsf{case}\ t'\ \mathsf{of}\ \mathsf{inl}(u^{C_1}).g_1,\ \mathsf{inr}(v^{C_2}).g_2$ is in C.
We prove this by induction over $\mathcal{N}(g_1)+\mathcal{N}(g_2)$. Since r is neutral it is enough to show all its immediate reducts are in C. But t is neutral, so these reducts must be of one of three forms:

a) $r_1 = \mathsf{case}\ t'\ \mathsf{of}\ \mathsf{inl}(u^{C_1}).g_1,\ \mathsf{inr}(v^{C_2}).g_2$ where $t \rhd_1 t_1$. By the induction hypothesis, this is in C.
b) $r_2 = \mathsf{case}\ t\ \mathsf{of}\ \mathsf{inl}(u^{C_1}).g_1',\ \mathsf{inr}(v^{C_2}).g_2$ where g_1' is an immediate reduct of g_1. By CR2 for C, g_1' is in C and $\mathcal{N}(g_1') < \mathcal{N}(g_1)$. Also, if $t \rhd_1 t'$, then

$$\begin{aligned}&\mathsf{case}\ t'\ \mathsf{of}\ \mathsf{inl}(u^{C_1}).g_1,\ \mathsf{inr}(v^{C_2}).g_2 \rhd_1\\ &\qquad \mathsf{case}\ t'\ \mathsf{of}\ \mathsf{inl}(u^{C_1}).g_1',\ \mathsf{inr}(v^{C_2}).g_2\\ &\qquad \Rightarrow \mathsf{case}\ t'\ \mathsf{of}\ \mathsf{inl}(u^{C_1}).g_1',\ \mathsf{inr}(v^{C_2}).g_2 \in C \quad \text{by CR2}\end{aligned}$$

So, r_2 is in C by the induction hypothesis.
c) $r_3 = \mathsf{case}\ t\ \mathsf{of}\ \mathsf{inl}(u^{C_1}).g_1,\ \mathsf{inr}(v^{C_2}).g_2'$. This case follows similarly to r_2.

This last case concludes the proof. □

Definition 8.2.8 (Canonical CR). For every labelled formula $\text{SP} \diamond A$ we define a canonical CR, $C_{\text{SP}\diamond A}$ as in Fig. 8.11.

Lemma 8.2.5. *Consider formulae of the forms* $\text{SP} \diamond \forall x : s \bullet A$ *and* $\text{SP} \diamond \exists x : s \bullet A$. *Then it is the case that* $C_{\text{SP}\diamond\forall x:s\bullet A}$ *and* $C_{\text{SP}\diamond\exists x:s\bullet A}$ *are CR.*

Proof. Our proof adapts that given in [CS93]. As in Lemma 8.2.4 we must verify the conditions for CR (CR1, CR2 and CR3) hold over $C_{\text{SP}\diamond\forall x:s\bullet A}$ and $C_{\text{SP}\diamond\exists x:s\bullet A}$.

($C_{\text{SP}\diamond\forall x:s\bullet A}$ *case*). We show $C_{\text{SP}\diamond\forall x:s\bullet A}$ is a CR by the fact that the conditions for CR are satisfied.

CR1 Assume $t \in C_{\text{SP}\diamond\forall x:s\bullet A}$. Then $\mathsf{specific}(t, x : s)$ is in $C_{\text{SP}\diamond A}$. But because $\mathcal{N}(\mathsf{specific}(t, x : s)) \geq \mathcal{N}(t)$ we have that t is strongly normalizing as required.
CR2 If $t \rhd_1 t'$, then $\mathsf{specific}(t, a)$ ($a \in Term(\text{Sig}(\text{SP}), Var)$) is immediately reducible to $\mathsf{specific}(t', a)$. Assume $\mathsf{specific}(t, a)$ is in $C_{A[a/x]}$. Then, by CR2, $\mathsf{specific}(t', a)$ is also in $C_{A[a/x]}$. Consequently t' is in $C_{\text{SP}\diamond\forall x:s\bullet A}$.

Formula A	Canonical CR $C_{\text{SP}\diamond A}$
Atomic	All strongly normalizable proof-terms of type $\text{SP} \diamond A$
$\perp$	All strongly normalizable proof-terms of type $\text{SP} \diamond \perp$
$A \Rightarrow B$	$C_{\text{SP}\diamond A} \Rightarrow C_{\text{SP}\diamond B}$
$A_1 \wedge A_2$	$C_{\text{SP}\diamond A_1} \wedge C_{\text{SP}\diamond A_2}$
$A_1 \vee A_2$	$C_{\text{SP}\diamond A_1} \vee C_{\text{SP}\diamond A_2}$
$\forall x : s \bullet P$	The set of all proof-terms t of type $(\text{SP} \diamond \forall x : s \bullet P)$ such that $\mathsf{specific}(t, a) \in C_{\text{SP}\diamond P[a/x]}$ for every proof-term a in $Term(\text{Sig}(\text{SP}), Var)$ of sort s.
$\exists x : s \bullet P$	The set of all proof-terms t of type $(\text{SP}\diamond\exists x : s \bullet P)$ satisfying the following condition. Take any type $(\text{SP} \diamond G)$, with x not free in G, and an arbitrary CR, called D, of type $(\text{SP} \diamond G)$. Take any proof-term g of type $(\text{SP} \diamond A \Rightarrow G)$ satisfying $$g[t/x] \in C_{\text{SP}\diamond A[t/x]} \Rightarrow D$$ for any $t \in Term(\text{Sig}(\text{SP}), Var)$. It must be the case that the proof-term $$\mathsf{select}\ (t)\ \mathsf{in}\ x : s.y^A.\mathsf{app}(g, y)$$ is in D.

Fig. 8.11. Definition of canonical CR for a labelled formula $\text{SP} \diamond A$.

CR3 Assume t is neutral and every immediate reduct t' of t is in $C_{\text{SP}\diamond\forall x:s\bullet A}$. It follows that, for any $a \in Term(\text{Sig}(\text{SP}), Var)$, the proof-term $\mathsf{specific}(t', a)$ is in $\text{SP} \diamond C_{A[a/x]}$. So $\mathsf{specific}(t, a)$ is neutral and any immediate reduct of $\mathsf{specific}(t, a)$ is of the form $\mathsf{specific}(t', a)$ where $t \rhd_1 t'$. By assumption these proof-terms are in $C_{\text{SP}\diamond A[a/x]}$. It follows by CR3 for $C_{\text{SP}\diamond A[a/x]}$ that $\mathsf{specific}(t, a)$ is in $C_{\text{SP}\diamond A[a/x]}$. Thus t is in $C_{\text{SP}\diamond\forall x:s\bullet A}$.

($C_{\exists x:s\bullet A}$ *case*). It is true that $C_{\text{SP}\diamond\exists x:s\bullet A}$ is a CR.

To see this, we first make some assumptions. Take an arbitrary type $(\text{SP}\diamond G)$ in which $x : s$ does not occur free. Let D be an arbitrary CR of type $(\text{SP} \diamond G)$. Take any $t \in C_{\exists x:s\bullet A}$. Then we know that

$$\mathsf{select}\ (t)\ \mathsf{in}\ x.y^A.\mathsf{app}(g, y) \in D$$

for any proof-term g of type $(\text{SP} \diamond A \Rightarrow G)$ which satisfies

$$g[a/x] \in C_{\text{SP}\diamond A[a/x]} \Rightarrow D$$

for any $a \in Term(\text{Sig}(\text{SP}), Var)$.

Then we can derive the conditions for CR.

CR1 Assume t is in $C_{\exists x:s\bullet A}$. We abbreviate the proof-term $\mathsf{abstract}\ v^A.\ \mathsf{ass}(\text{SP}, z^G)$ by g. This is a proof-term of type $\text{SP} \diamond (A \Rightarrow G)$.

First we show that, for each $a \in Term(\text{Sig}(\text{SP}), Var)$, it is the case that

$$g[a/x] = \mathsf{abstract}\ v^{A[a/x]}.\ z^{\text{SP}\diamond A[a/x]} \tag{8.8}$$

is in $C_{\text{SP}\diamond(A[a/x]} \Rightarrow D)$.
We rephrase (8.8) as follows: we wish to prove that, for every $u \in C_{\text{SP}\diamond A[a/x]}$, it is the case that

$$s = \mathsf{app}(g[a/x], u^{\text{SP}\diamond A[a/x]}) = \mathsf{app}(\mathsf{abstract}\ v^{A[a/x]}.\ , \mathsf{ass}(\text{SP}, z^G))u^{\text{SP}\diamond A[a/x]} \in D$$

This is derived by induction on $\mathcal{N}(u)$. Since s is neutral, it is enough to show that all its immediate reducts are in D. These are of two possible forms

a) $s_1 = \mathsf{ass}(\text{SP}, z^G)[u/v] = \mathsf{ass}(\text{SP}, z^G)$. Because this is neutral and normal, s_1 is in D, as required.
b) $s_2 = \mathsf{app}(\mathsf{abstract}\ v^{A[a/x]}.\ , \mathsf{ass}(\text{SP}, z^G))u'^{\text{SP}\diamond A[a/x]}$ where $u \rhd_1 u'$, so $\mathcal{N}(u') < \mathcal{N}(u)$ and this is in D by the induction hypothesis.

If (8.8) holds, then the term $r = \mathsf{select}\ (t)\ \mathsf{in}\ x : s.y^A.\mathsf{app}(g, y)$ must be in D. Because $\mathcal{N}(r) \geq \mathcal{N}(t)$, t is strongly normalizable, as required.

CR2 Assume t is in $C_{\text{SP}\diamond\exists x:s\bullet A}$. If t is immediately reducible to t', then $\mathsf{select}\ (t)\ \mathsf{in}\ .x : s.y^A\mathsf{app}(g, y)$ is immediately reducible to $\mathsf{select}\ (t')\ \mathsf{in}\ x : s.y^A.\mathsf{app}(g, y)$. Thus we have that t' is in $C_{\text{SP}\diamond\exists x:s\bullet A}$ by property CR2 for D.

CR3 Assume t is a neutral proof-term such that every immediate reduct t' of t is in $C_{\text{SP}\diamond\exists x:s\bullet A}$. The proof-term $\mathsf{select}\ (t')\ \mathsf{in}\ x : s.y^A.\mathsf{app}(g, y)$ is therefore in D. We are required to prove $\mathsf{select}\ (t)\ \mathsf{in}\ x : s.y^A.\mathsf{app}(g, y)$ is in D.
We can rephrase this requirement as follows. We define h to denote $\mathsf{app}(g, y)$, so that h is in D. We need to prove $r = \mathsf{select}\ (t)\ \mathsf{in}\ x : s.y^A.h$ is in D, given that the proof-term $\mathsf{select}\ (t')\ \mathsf{in}\ x : s.y^A.\mathsf{app}(g, y)$ is in D for any $t \rhd_1 t'$.
We proceed by induction on $\mathcal{N}(h)$. Since r is neutral, it is enough to show that all its immediate reducts are in D. Due to the fact that t is neutral, these reducts can be of only two forms:

a) The first possible form is $\mathsf{select}\ (t')\ \mathsf{in}\ x : s.y^A.h$ where $t \rhd_1 t'$. This proof-term is in D by hypothesis.
b) The second possible form is $r_2 = \mathsf{select}\ (t)\ \mathsf{in}\ x : s.y^A.h'$ where $h \rhd_1 h'$. By the property CR2 for D, it is true that $h' \in D$ and $\mathcal{N}(h') < \mathcal{N}(h)$. Also, if t' is an immediate reduct of t, then $\mathsf{select}\ (t')\ \mathsf{in}\ x : s.y^A.h'$ is an immediate reduct of $\mathsf{select}\ (t')\ \mathsf{in}\ x : s.y^A.h \in D$ and therefore is in D. So, r_2 is in D by the induction hypothesis.

This last case concludes the proof. □

We now show that every proof-term corresponding to an SSL proof is strongly normalizable. We do this by first showing that such proof-terms are always contained in a CR (Theorem 8.2.9). Strong normalization follows as a simple corollary (Corollary 8.2.1).

Theorem 8.2.9 (Strong normalization for SSL proofs). *Each derivable proof-term f of type $\text{SP} \diamond A$ is in $C_{\text{SP}\diamond A}$.*

Proof. We adapt the proof of strong normalization for intuitionistic logic given by Crossley and Shepherdson in [CS93]. The adaptation is straightforward: additional complication is due to the additional proof-terms corresponding to structural proofs.

We proceed by induction on the structure of f. To aid the derivation, we strengthen the induction hypothesis to the following:

> Let $\bar{z} = z_1, \ldots, z_s$ be a list of distinct individual variables, $\bar{t} = t_1, \ldots, t_s$ a list of individual terms, $\bar{y} = y_1^{G_1}, \ldots y_r^{G_r}$ a list of distinct proof-term variables, and $\bar{p}' = p_1^{\mathrm{SP\_1} \diamond G_1'}, \ldots p_k^{\mathrm{SP\_R} \diamond G_r'}$ a list of proof-terms in $C_{\mathrm{SP\_1} \diamond G_1'}, \ldots C_{\mathrm{SP\_R} \diamond G_r'}$ respectively, where each $G_i' = G_i[\bar{t}/\bar{z}]$ so that $\mathrm{Sig}(G_i) \subseteq \mathrm{Sig}(\mathrm{SP\_I})$ and $\mathrm{Sig}(G_i') \subseteq \mathrm{Sig}(\mathrm{SP\_I})$. Let $\bar{y}' = y_1^{G_1'}, \ldots, y_r^{G_r'}$. Then,
> $$(f^{\mathrm{SP} \diamond A})[\bar{t}/\bar{z}][\bar{p}'/\bar{y}']$$
> is a proof-term of type $\mathrm{SP} \diamond A'$ and is in $C_{\mathrm{SP} \diamond A'}$, where $A' = A[\bar{t}/\bar{z}]$.

Without loss of generality, we will assume bound variables of f are not equivalent to any $\bar{y}'$ or any variable in $\bar{p}'$.

(*Base case*). If f is an assumption $\mathsf{ass}(\mathrm{SP}, x^A)$, then either

- x is not y_i, for every $i \in \{1, \ldots, r\}$, and so $f^{\mathrm{SP} \diamond A}[\bar{t}/\bar{z}][\bar{p}'/\bar{y}']$ is $\mathsf{ass}(\mathrm{SP}, x^{A'})$, which belongs to $C_{\mathrm{SP} \diamond A'}$, or
- x is y_i, some $i \in \{1, \ldots, r\}$, and so $f^{\mathrm{SP} \diamond A}[\bar{t}/\bar{z}][\bar{p}'/\bar{y}']$ is $p_i^{\mathrm{SP} \diamond A'}$, which belongs to $C_{\mathrm{SP} \diamond A'}$ by the induction hypothesis.

($\Rightarrow$-I). We are required to prove that, if $f^{\mathrm{SP} \diamond B}$ satisfies the induction hypothesis, then so does $(\mathsf{abstract}\ x^A.\ f^{\mathrm{SP} \diamond B})$. This is equivalent to proving r is in $C_{\mathrm{SP} \diamond A' \Rightarrow B'}$, where r is defined to be $(\mathsf{abstract}\ x^A.\ f^{\mathrm{SP} \diamond B}[\bar{t}/\bar{z}][\bar{p}'/\bar{y}'])$.

Observe that, by definition, r can be rewritten $\mathsf{abstract}\ x^{A'}.\ (f^{\mathrm{SP} \diamond B}[\bar{t}/\bar{z}])[\bar{p}'/\bar{y}']$. We can rewrite this again as

$$r = \mathsf{abstract}\ x^{A'}.\ (f^{\mathrm{SP} \diamond B}[\bar{t}/\bar{z}][\bar{p}'/\bar{y}'])$$

because we can assume $x^{A'}$ does not occur in $\bar{y}'$ or $\bar{p}'$.

We define $g^{\mathrm{SP} \diamond B'}$ to be $f^{\mathrm{SP} \diamond B}[\bar{t}/\bar{z}][\bar{p}'/\bar{y}']$, so that $r = \mathsf{abstract}\ x^{A'}.\ g^{\mathrm{SP} \diamond B'}$.

Noting that $C_{\mathrm{SP} \diamond A' \Rightarrow B'} = C_{\mathrm{SP} \diamond A'} \Rightarrow C_{\mathrm{SP} \diamond B'}$, we have to show that, for all $u \in C_{\mathrm{SP} \diamond A'}$, it is true that $\mathsf{app}(r, u) = \mathsf{app}(\mathsf{abstract}\ x^{A'}.\ g^{\mathrm{SP} \diamond B'}, u) \in C_{\mathrm{SP} \diamond B'}$.

By the induction hypothesis,

$$f^{\mathrm{SP} \diamond B}[\bar{t}/\bar{z}][\bar{p}' :: q/\bar{y}' :: x^{A'}] \in C_{\mathrm{SP} \diamond B'}$$

since $x^{A'}$ is not equivalent to any $\bar{y}'$ or any variable $\bar{p}'$. Consequently, we may infer

$$
\begin{array}{ll}
f^{\mathrm{SP}\diamond B}[\bar{t}/\bar{z}][\bar{p}' :: q/\bar{y}' :: x^{A'}] \in C_{\mathrm{SP}\diamond B'} & \text{since } x^{A'} \text{ is not equivalent to any } \bar{y}' \text{ or any variable } \bar{p}'. \\
\Rightarrow f^{\mathrm{SP}\diamond B}[\bar{t}/\bar{z}][\bar{p}'/\bar{y}'][q/x^{A'}] \in C_{\mathrm{SP}\diamond B'} & \text{because } C_{\mathrm{SP}\diamond B'} \text{ is closed under equivalence of proof-terms.} \\
\Rightarrow f^{\mathrm{SP}\diamond B}[\bar{t}/\bar{z}][\bar{p}'/\bar{y}'][q/x^{A'}] \in C_{\mathrm{SP}\diamond B'} & \\
g^{\mathrm{SP}\diamond B'}[q/x^{A'}] \in C_{\mathrm{SP}\diamond B'} &
\end{array}
\qquad (8.9)
$$

for any $q \in C_{\mathrm{SP}\diamond A'}$.

The proof-term $\mathsf{app}(r, u)$ is neutral. So, by condition CR3 for $C_{\mathrm{SP}\diamond B'}$, it is enough to show that all immediate reducts are in $C_{\mathrm{SP}\diamond B'}$.

We proceed to prove this by subsidiary induction on $\mathcal{N}(g)+\mathcal{N}(u)$, given that

$$g^{\mathrm{SP}\diamond B'} \in C_{\mathrm{SP}\diamond B'}$$

and

$$u^{\mathrm{SP}\diamond A'} \in C_{\mathrm{SP}\diamond A'}$$

In the base case, when $\mathcal{N}(g)+\mathcal{N}(u) = 0$, $\mathsf{app}(r, u)$ is in normal form. Because $\mathsf{app}(r, u)$ is neutral, $\mathsf{app}(r, u)$ must be in $C_{\mathrm{SP}\diamond B'}$ by Lemma 8.2.3.

For the subsidiary induction step, consider that the reducts of r can be of only three possible forms:

1. The first possible reduct form is $\mathsf{app}(\mathsf{abstract}\ x^{A'} .\ g^{\mathrm{SP}\diamond B'}, u')$, where $u \rhd_1 u'$. So, by CR2, $u' \in C_{\mathrm{SP}\diamond A'}$. Also, $\mathcal{N}(u') < \mathcal{N}(u)$. So, the reduct is in $C_{\mathrm{SP}\diamond B'}$ by the subsidiary induction hypothesis.
2. The second possible reduct form is $\mathsf{app}(\mathsf{abstract}\ x^{A'} .\ g'^{\mathrm{SP}\diamond B'}, u)$, where $g \rhd_1 g'$. So, by CR2, $g' \in C_{\mathrm{SP}\diamond B'}$. Also, $\mathcal{N}(g') < \mathcal{N}(g)$. This reduct is in $C_{\mathrm{SP}\diamond B'}$ by the subsidiary induction hypothesis.
3. The third possible form is $g^{\mathrm{SP}\diamond B'}[u/x^{\mathrm{SP}\diamond A'}]$, which is in $C_{\mathrm{SP}\diamond B'}$, by (8.9).

($\Rightarrow$-E). We have to show that $\mathsf{app}(f^{\mathrm{SP}\diamond A \Rightarrow B}, g^{\mathrm{SP}\diamond A})$ satisfies the induction hypothesis. That is, we must show that $\mathsf{app}(f^{\mathrm{SP}\diamond A \Rightarrow B}[\bar{t}/\bar{z}][\bar{p}'/\bar{y}'], g^{\mathrm{SP}\diamond A}[\bar{t}/\bar{z}][\bar{p}'/\bar{y}'])$ is in $C_{\mathrm{SP}\diamond B'}$. This follows from the induction hypothesis for $f^{\mathrm{SP}\diamond A \Rightarrow B}[\bar{t}/\bar{z}][\bar{p}'/\bar{y}']$ and $g^{\mathrm{SP}\diamond A}[\bar{t}/\bar{z}][\bar{p}'/\bar{y}']$ and the definition of $C_{\mathrm{SP}\diamond A' \Rightarrow B'}$.

($\wedge$-I). We define r to be $\langle f^{\mathrm{SP}\diamond A}[\bar{t}/\bar{z}][\bar{p}'/\bar{y}'], g^{\mathrm{SP}\diamond B}[\bar{t}/\bar{z}][\bar{p}'/\bar{y}']\rangle$ We are required to prove r is in $C_{\mathrm{SP}\diamond A' \wedge B'}$. That is, we must prove that $\mathsf{fst}(r) \in C_{\mathrm{SP}\diamond A'}$ and $\mathsf{snd}(r) \in C_{\mathrm{SP}\diamond B'}$.

Because $\mathsf{fst}(r)$ and $\mathsf{snd}(r)$ are neutral, we can proceed in a similar way to the case for ($\Rightarrow$-I) above, showing all immediate reducts of these proof-terms are in $C_{\mathrm{SP}\diamond A'}$ and $C_{\mathrm{SP}\diamond B'}$, respectively. We use a subsidiary induction on $\mathcal{N}(f^{\mathrm{SP}\diamond A}[\bar{t}/\bar{z}][\bar{p}'/\bar{y}']) + \mathcal{N}(g^{\mathrm{SP}\diamond B}[\bar{t}/\bar{z}][\bar{p}'/\bar{y}'])$.

For instance, in the case of $\mathsf{fst}(r)$ one of the possible immediate reducts of $\mathsf{fst}(r)$ is $f^{\mathrm{SP}\diamond A}[\bar{t}/\bar{z}][\bar{p}'/\bar{y}']$. This is in $C_{\mathrm{SP}\diamond A'}$, by the main induction hypothesis. The other immediate reducts obtained by reducing f or g follow easily.

($\wedge$-E$_i$) ($i = 1, 2$). When $i = 1$, we have to show $r = \mathsf{fst}(f^{\mathrm{SP}\diamond A\wedge B})[\bar{t}/\bar{z}][\bar{p}'/\bar{y}']$ is in $C_{\mathrm{SP}\diamond A'}$. This follows from the definition of $C_{\mathrm{SP}\diamond(A'\wedge B')}$ and the inductive assumption that $f^{\mathrm{SP}\diamond(A\wedge B)}[\bar{t}/\bar{z}][\bar{p}'/\bar{y}']$ is in $C_{\mathrm{SP}\diamond A'\wedge B'}$. We proceed similarly when $i = 2$ for $\mathsf{snd}(r)$.

($\forall$-I). We show that $(\mathsf{use}\ x : s.\ f^{\mathrm{SP}\diamond A})[\bar{t}/\bar{z}][\bar{p}'/\bar{y}']$ is in $C_{\mathrm{SP}\diamond\forall x:s\bullet A'}$. We can rewrite the proof-term as $\mathsf{use}\ x : s.\ (f^{\mathrm{SP}\diamond A}[\bar{t}/\bar{z}][\bar{p}'/\bar{y}'])$, where x does not occur in $\bar{p}'$, nor in the types of $\bar{y}'$.

So, we are required to prove that, for all $t_0 : s \in Term(\mathrm{Sig}(\mathrm{SP}), Var)$,

$$r = \mathsf{specific}(\mathsf{use}\ x : s.\ f^{\mathrm{SP}\diamond A}[\bar{t}/\bar{z}][\bar{p}'/\bar{y}'], t_0)$$

is in $C_{A'[t_0/x]}$. This is proved by induction on $\mathcal{N}(f^{\mathrm{SP}\diamond A}[\bar{t}/\bar{z}][\bar{p}'/\bar{y}'])$.

Since r is neutral we need only consider its immediate reducts. These are of two forms:

1. The first form of reduct derives from a reduction of f. By induction on $\mathcal{N}(f[\bar{t}/\bar{z}][\bar{p}'/\bar{y}'])$ we can show that this reduct is in $C_{A'[t_0/x]}$ as required.
2. The second reduct is of the form $f[\bar{t}/\bar{z}][\bar{p}'/\bar{y}'][t_0/x]$. This may be rewritten

$$f[\bar{t}/\bar{z}][\bar{p}'/\bar{y}'][t_0/x]$$

Because $x : s$ does not occur in $\bar{p}'$ or $\bar{y}'$, this proof-term is equal to

$$f[\bar{t}/\bar{z}][x/t_0][\bar{p}'/\bar{y}']^{\mathrm{SP}\diamond A'[t_0/x]}$$

which is in $C_{\mathrm{SP}\diamond A'[t_0/x]}$ by the main induction hypothesis.

($\forall$-E). We have to show $\mathsf{specific}(f^{\mathrm{SP}\diamond\forall x:s\bullet A}[\bar{t}/\bar{z}][\bar{p}'/\bar{y}'], t_0)$ is in $C_{\mathrm{SP}\diamond A'[t_0/x]}$ if, for all $t_0 : s$,

$$f^{\mathrm{SP}\diamond\forall x.A'}[\bar{t}/\bar{z}][\bar{p}'/\bar{y}'] \in C_{\forall x:s\bullet A'}$$

This follows from the definition of $C_{\mathrm{SP}\diamond\forall x:s\bullet A'}$.

($\vee$-I$_1$). We have to show $\mathsf{inl}(f^{\mathrm{SP}\diamond A})[\bar{t}/\bar{z}][\bar{p}'/\bar{y}']^{\mathrm{SP}\diamond A'\vee B'}$ is in $C_{\mathrm{SP}\diamond A'\vee B'}$, assuming that $f^{\mathrm{SP}\diamond A}[\bar{t}/\bar{z}][\bar{p}'/\bar{y}']$ is in $C_{\mathrm{SP}\diamond A'}$.

Take arbitrary types $(\mathrm{SP}\diamond C)$, all CR C of type $(\mathrm{SP}\diamond C)$, and proof-terms f_1, f_2 in $C_{\mathrm{SP}\diamond A'\Rightarrow C}$ and $C_{\mathrm{SP}\diamond B'\Rightarrow C}$ respectively (with $x^{\mathrm{SP}\diamond A'}$ not free in f_1 and $y^{\mathrm{SP}\diamond B'}$ not free in f_2). We must prove that the proof-term

$$\begin{aligned} r = \mathsf{case}\ \mathsf{inl}(f^{\mathrm{SP}\diamond A}[\bar{t}/\bar{z}][\bar{p}'/\bar{y}'])\ \mathsf{of}\ &\mathsf{inl}(x^{A'}).\mathsf{app}(f_1, \mathsf{ass}(\mathrm{SP}, x^{A'})),\\ &\mathsf{inr}(y^{B'}).\mathsf{app}(f_2, \mathsf{ass}(\mathrm{SP}, x^{B'})) \end{aligned}$$

is in C.

We reformulate this requirement. First we define g_1 to be $\mathsf{app}(f_1, x)$, g_2 to be $\mathsf{app}(f_2, y)$ and g_3 to be $f[\bar{t}/\bar{z}][\bar{p}'/\bar{y}']$. We have to show that the proof-term

$$r = \mathsf{case}\ \mathsf{inl}(g_3)\ \mathsf{of}\ \mathsf{inl}(x).g_1,\ \mathsf{inr}(y).g_2$$

is in C, given that g_1 and g_2 are in $C_{\mathrm{SP}\diamond C}$, g_3 is in $C_{\mathrm{SP}\diamond A'}$, and $g_1[g_3/x]$ is in C. (Note that $g_1[g_3/x]$ is $\mathsf{app}(f_1, g_3)$, which is in C by the definition of $C_{\mathrm{SP}\diamond A'} \Rightarrow C$.)

We proceed by subsidiary induction on $\mathcal{N}(G_1) + \mathcal{N}(G_2) + \mathcal{N}(G_3)$, using the fact that r is neutral, examining immediate reducts of r.

There are four possible immediate reducts of r.

The first possible immediate reduct of r is of the form $g_1[g_3/x]$. This is in C by hypothesis.

Each alternative possible immediate reducts r' is obtained by reducing one of g_1, g_2 or g_3. Observe that $\mathcal{N}(r')$ is less than $\mathcal{N}(g_1) + \mathcal{N}(g_2) + \mathcal{N}(g_3)$. So we only have to verify that each reduct will still leave

1. the reduced g_1 or g_2 in $C_{\text{SP}\diamond C}$ or the reduced g_3 in $C_{\text{SP}\diamond A'}$, and
2. the reduced version of $g_1[g_3/x]$ still in C.

The first of these requirements follows by CR2 for $C_{\text{SP}\diamond C}$ and $C_{\text{SP}\diamond A'}$. The second follows by CR2 for C by the following reasoning. If g_3' is an immediate reduct of g_3, then $g_1[g_3'/x]$ is a (not necessarily immediate) reduct of $g_1[g_3/x]$. Also, if g_1' is an immediate reduct of g_1 then $g_1'[g_3/x]$ is an immediate reduct of $g_1[g_3/x]$. This concludes the proof for this subsidiary induction and for this case.

($\vee$-I$_2$). Similar to the previous case.

($\vee$-E). We have to show

$$\mathsf{case}\ h^{\text{SP}\diamond A\vee B}[\bar{t}/\bar{z}][\bar{p}'/\bar{y}']\ \mathsf{of}\ \mathsf{inl}(x^{A'}).f_1^{\text{SP}\diamond G}[\bar{t}/\bar{z}][\bar{p}'/\bar{y}'],\ \mathsf{inr}(y^{B'}).f_2^{\text{SP}\diamond G}[\bar{t}/\bar{z}][\bar{p}'/\bar{y}']$$

is in $C_{\text{SP}\diamond G'}$. This proof-term is neutral. As in previous cases, it is enough to show all immediate reducts of the proof-term are in $C_{\text{SP}\diamond G'}$.

We proceed by induction on

$$\mathcal{N}(h^{\text{SP}\diamond A\vee B}[\bar{t}/\bar{z}][\bar{p}'/\bar{y}']) + \mathcal{N}(f_1^{\text{SP}\diamond G}[\bar{t}/\bar{z}][\bar{p}'/\bar{y}']) + \mathcal{N}(f_2^{\text{SP}\diamond G}[\bar{t}/\bar{z}][\bar{p}'/\bar{y}'])$$

There are several possible immediate reducts. The reducts obtained by reducing h, f_1 or f_2 are then in $C_{\text{SP}\diamond G'}$ by this subsidiary induction. As a consequence,

$$\mathsf{case}\ h^{\text{SP}\diamond A\vee B}[\bar{t}/\bar{z}][\bar{p}'/\bar{y}']\ \mathsf{of}\ \mathsf{inl}(x^{A'}).f_1^{\text{SP}\diamond G}[\bar{t}/\bar{z}][\bar{p}'/\bar{y}'],\ \mathsf{inr}(y^{B'}).f_2^{\text{SP}\diamond G}[\bar{t}/\bar{z}][\bar{p}'/\bar{y}']$$

is in $C_{\text{SP}\diamond G'}$.

There are two further possible reducts, depending on whether h is $\mathsf{inl}(k)$ or $\mathsf{inr}(k)$. We deal with the former case as the latter is similar. If h is $\mathsf{inl}(k)$ there is another immediate reduct,

$$f_1^{\text{SP}\diamond G}[\bar{t}/\bar{z}][\bar{p}'/\bar{y}'][(k[\bar{t}/\bar{z}][\bar{p}'/\bar{y}'])/x]$$

Because we can assume $x^{A'}$ is not equivalent to any free proof-term variable in $\bar{y}'$, this proof-term is equivalent to

$$f_1^{\text{SP}\diamond G}[\bar{t}/\bar{z}][\bar{p}' :: (k[\bar{t}/\bar{z}][\bar{p}'/\bar{y}'])/\bar{y}' :: x^{\text{SP}\diamond A'}]$$

which is in $C_{\text{SP}\diamond G'}$, the main induction hypothesis, as required.

($\exists$-I). A proof-term obtained by ($\exists$-I) is of the form $\mathsf{show}(u, f^{\text{SP}\diamond A[u/x]})^{\text{SP}\diamond\exists x:s\bullet A}$. We can assume that $\bar{z}$ denotes the list of all free individual variables in this proof-term. We write $u_0 = u[\bar{t}/\bar{z}]$.

So, we have to show $\mathsf{show}(u_0, f^{\text{SP}\diamond A[u/x]}[\bar{t}/\bar{z}][\bar{p}'/\bar{y}'])^{\text{SP}\diamond\exists x:s\bullet A'}$ is in $C_{\text{SP}\diamond\exists x:s\bullet A'}$. Take any type $(\text{SP} \diamond G)$ with x not free in G, any CR D of type $(\text{SP} \diamond G)$, and any proof-term g of type $(\text{SP} \diamond A' \Rightarrow G)$ satisfying, for any individual term u of sort s,

$$g[u/x] \in C_{\text{SP}\diamond A'[u/x]} \Rightarrow D$$

We must show that the proof-term

$$r = \mathsf{select}\ (\mathsf{show}(u_0, f^{\text{SP}\diamond A[u/x]})[\bar{t}/\bar{z}][\bar{p}'/\bar{y}'])\ \mathsf{in}\ x : s.y^{A'}.\mathsf{app}(g, y)^{\text{SP}\diamond G}$$

is in D.

As in previous cases, we reformulate the requirement.

First we set g_1 to be $\mathsf{app}(g, y)$ and f_1 to be $f^{\text{SP}\diamond A[u/x]}[\bar{t}/\bar{z}][\bar{p}'/\bar{y}']$. So we can easily see that g_1 is in D, f_1 is in $C_{\text{SP}\diamond A'[u_0/x]}$, x is not free in any type superscripts in g_1 (except perhaps $y^{\text{SP}\diamond A'}$) and $g_1[u_0/x][f_1/y]$ is in D.

We are required to prove $r = \mathsf{select}\ (\mathsf{show}(u_0, f_1))\ \mathsf{in}\ x : s.y.g_1$ is in D.

We prove this by a subsidiary induction in $\mathcal{N}(g_1) + \mathcal{N}(f_1)$. As r is neutral we need only prove that its immediate reducts are in D. These reducts can be of three kinds:

1. The first form is $r_1' = \mathsf{select}\ (\mathsf{show}(u_0, f_1))\ \mathsf{in}\ x : s.y^{A'}.g_1'$ where g_1' is an immediate reduct of g_1. Since $\mathcal{N}(g_1') < \mathcal{N}(g_1)$, the subsidiary induction hypothesis tells us that this is in D, provided that $g_1'[u_0/x][f_1/y]$ is in D. To see this, we observe that $g_1'[u_0/x][f_1/y]$ is an immediate reduct of $g_1[u_0/x][f_1/y]$, which is in D by the hypothesis.
2. The second form is $r_2' = \mathsf{select}\ (\mathsf{show}(u_0, f_1'))\ \mathsf{in}\ x : s.y^{A'}.g_1$ where f_1' is an immediate reduct of f_1. The subsidiary induction hypothesis tells us that this is in D, because $g_1[u_0/x][f_1/y]$ is in D. This follows from CR2 for D, using the easily verifiable fact that $g_1[u_0/x][f_1'/y]$ is a reduct of $g_1[u_0/x][f_1/y]$.
3. The third form is $r_3' = g_1[u_0/x][f_1/y]$, which is in D by the hypothesis.

($\exists$-E). A proof-term obtained by ($\exists$-E) is of the form

$$(\mathsf{select}\ (h^{\text{SP}\diamond\exists x:s\bullet A})\ \mathsf{in}\ x : s.y^A.g^{\text{SP}\diamond G})^{\text{SP}\diamond G}$$

where $x : s$ is not free in G nor in any type of any free variable of g. So, we have to show that $r = (\mathsf{select}\ (h^{\text{SP}\diamond\exists x:s\bullet A}[\bar{t}/\bar{z}][\bar{p}'/\bar{y}'])\ \mathsf{in}\ x : s.y^{A'}.g^{\text{SP}\diamond G}[\bar{t}/\bar{z}][\bar{p}'/\bar{y}']$ is in $C_{\text{SP}\diamond G'}$. We assume that the $\bar{y}'$ do not have $x : s$ occurring free in the type of any of their free variables.

Since r is neutral, it is enough to show its immediate reducts are in $C_{\text{SP}\diamond G'}$. This is proved by induction on $\mathcal{N}(h^{\text{SP}\diamond\exists x:s\bullet A}[\bar{t}/\bar{z}][\bar{p}'/\bar{y}']) + \mathcal{N}(g^{\text{SP}\diamond G}[\bar{t}/\bar{z}][\bar{p}'/\bar{y}'])$.

The reducts obtained by reducing g or h are in $C_{\text{SP}\diamond G'}$ by the subsidiary induction hypothesis.

If h is of the form $\mathsf{show}(u : s, f_1^{\mathrm{SP} \diamond A[u/x]})$ there is another reduct,

$$g^{\mathrm{SP} \diamond G}[\bar{t}/\bar{z}][\bar{p}'/\bar{y}'][u_0/x][(f_1[\bar{t}/\bar{z}][\bar{p}'/\bar{y}'])/y]$$

where $u_0 = u[\bar{t}/\bar{z}]$. This proof-term is equivalent to

$$r' = g^{\mathrm{SP} \diamond G}[\bar{t} :: u_0/\bar{z} :: x][\bar{p}' :: (f_1[\bar{t}/\bar{z}][\bar{p}'/\bar{y}'])/\bar{y}' :: y]$$

because (as in the case of ($\forall$-I) above) the $\bar{t}$ do not contain x free, and x does not occur free in the type of any $\bar{z}$ which is free in g, or of any of the $\bar{p}'$. We are done, as the proof-term $r'^{\mathrm{SP} \diamond A'}$ is in $C_{\mathrm{SP} \diamond A'}$ by the main induction hypothesis.

($\bot$-E). We must show $r = \mathsf{abort}(a^{\mathrm{SP} \diamond \bot}, A)^{\mathrm{SP} \diamond A}[\bar{t}/\bar{z}][\bar{p}'/\bar{y}']$ is in $C_{\mathrm{SP} \diamond A'}$. We know that $a^{\mathrm{SP} \diamond \bot}$ is in $C_{\mathrm{SP} \diamond \bot}$. We can rewrite r as $r = \mathsf{abort}(a[\bar{t}/\bar{z}][\bar{p}'/\bar{y}']^{\mathrm{SP} \diamond \bot})^{\mathrm{SP} \diamond A'}$. We can prove this is in $C_{\mathrm{SP} \diamond A'}$, by induction on $\mathcal{N}(a[\bar{t}/\bar{z}][\bar{p}'/\bar{y}']^{1 \diamond \bot})$ using CR3 for $C_{\mathrm{SP} \diamond A'}$ and the fact that r is neutral.

(trans). We must show $r = (\rho \bullet c^{\mathrm{SP} \diamond A})[\bar{t}/\bar{z}][\bar{p}'/\bar{y}']$ is in $C_{(\mathrm{SP}\ \mathbf{with}\ \rho) \diamond \rho(A)'}$, where $\rho(A)' = \rho(A)[\bar{t}/\bar{z}]$. By the induction hypothesis, $c^{\mathrm{SP} \diamond A}[\bar{t}/\bar{z}][\bar{p}'/\bar{y}']$ is in C_{SP}.

As this proof-term is neutral, it is enough to show that every immediate reduct $r \rhd_1 r'$ is in $C_{(\mathrm{SP}\ \mathbf{with}\ \rho) \diamond \rho(A)'}$. This is done by a subsidiary induction on $\mathcal{N}(c)$. In the base case, $\mathcal{N}(c) = 0$, and so $(\rho \bullet c^{\mathrm{SP} \diamond A})[\bar{t}/\bar{z}][\bar{p}'/\bar{y}']$ is in normal form and must be in $C_{(\mathrm{SP\_1}\ \mathbf{with}\ \rho) \diamond \rho(A)'}$

For the inductive step, we reason as follows.

Suppose $r' = (\rho \bullet c')$ where $c[\bar{t}/\bar{z}][\bar{p}'/\bar{y}'] \rhd_1 c'$. In this case, $\mathcal{N}(c') < \mathcal{N}(c)$ and we are done.

Otherwise, c must have one of the following forms: $\mathsf{abstract}\ u^B.\ d^{\mathrm{SP} \diamond D}$, $(\mathsf{use}\ x : s.\ d^{\mathrm{SP} \diamond D})$, $(\langle d^{\mathrm{SP} \diamond D}, e^{\mathrm{SP} \diamond E} \rangle)$, $\mathsf{inl}(d^{\mathrm{SP} \diamond D})$ or $\mathsf{inr}(d^{\mathrm{SP} \diamond D})$. In each of these cases r' must be obtained by a structural reduction rule and take the form

$$\begin{gathered}
\mathsf{abstract}\ u^{\rho(B)}.\ (\rho \bullet d^{\mathrm{SP} \diamond D})[\bar{t}/\bar{z}][\bar{p}'/\bar{y}'] \\
\mathsf{use}\ x : \rho(s).\ (\rho \bullet d^{\mathrm{SP} \diamond D})[\bar{t}/\bar{z}][\bar{p}'/\bar{y}'] \\
\mathsf{show}(\rho(t), (\rho \bullet d^{\mathrm{SP} \diamond D}))[\bar{t}/\bar{z}][\bar{p}'/\bar{y}'] \\
\langle (\rho \bullet d^{\mathrm{SP} \diamond D}), (\rho \bullet e^{\mathrm{SP} \diamond E}) \rangle [\bar{t}/\bar{z}][\bar{p}'/\bar{y}'] \\
\mathsf{inl}((\rho \bullet d^{\mathrm{SP} \diamond D}))[\bar{t}/\bar{z}][\bar{p}'/\bar{y}'] \\
\mathsf{inr}((\rho \bullet d^{\mathrm{SP} \diamond D}))[\bar{t}/\bar{z}][\bar{p}'/\bar{y}']
\end{gathered}$$

respectively. Each of these is in $C_{(\mathrm{SP}\ \mathbf{with}\ \rho) \diamond A}$ by the main induction hypothesis.

(union_1). We must show $r = \mathsf{union}_1(c^{\mathrm{SP\_1} \diamond A}, \mathrm{SP\_2})[\bar{t}/\bar{z}][\bar{p}'/\bar{y}']$ is in $C_{(\mathrm{SP\_1}\ \mathbf{and}\ \mathrm{SP\_2}) \diamond A'}$, given that, by the induction hypothesis, $c[\bar{t}/\bar{z}][\bar{p}'/\bar{y}']^{\mathrm{SP\_1} \diamond A'}$ is in $C_{\mathrm{SP\_1}}$.

As this proof-term is neutral, it is enough to show that every immediate reduct $r \rhd_1 r'$ is in $C_{(\mathrm{SP\_1}\ \mathbf{and}\ \mathrm{SP\_2})) \diamond A}$. This is done by a subsidiary induction on $\mathcal{N}(c)$. For the base case $\mathcal{N}(c) = 0$, and so $\mathsf{union}_1(c^{\mathrm{SP\_1} \diamond A}, \mathrm{SP\_2})[\bar{t}/\bar{z}][\bar{p}'/\bar{y}']$ is in normal form and must be in $C_{\mathrm{SP\_1}\ \mathbf{and}\ \mathrm{SP\_2} \diamond A'}$

For the inductive step, we reason as follows.

Suppose $r' = \mathsf{union}_1(c', \mathrm{SP\_2})$ where $c[\bar{t}/\bar{z}][\bar{p}'/\bar{y}'] \rhd_1 c'$. In this case, $\mathcal{N}(c') < \mathcal{N}(c)$ and we are done.

Otherwise, c must take one of the following forms: $\mathsf{abstract}\ u^A.\ d^{\mathrm{SP\_1} \diamond B}$, $(\mathsf{use}\ x : s.\ d^{\mathrm{SP\_1} \diamond B})$, $(\langle d^{\mathrm{SP\_1} \diamond B}, e^{\mathrm{SP\_1} \diamond B} \rangle)$, $\mathsf{inl}(d^{\mathrm{SP\_1} \diamond B})$ or $\mathsf{inr}(d^{\mathrm{SP\_1} \diamond B})$. In each of these cases r' must be obtained by a structural reduction rule and take the form

$$\begin{array}{c}
\mathsf{abstract}\ u^A.\ \mathsf{union}_1(d^{\mathrm{SP_1} \diamond B}, \mathrm{SP_2})[\bar{t}/\bar{z}][\bar{p}'/\bar{y}'] \\
\mathsf{use}\ x : s.\ \mathsf{union}_1(d^{\mathrm{SP_1} \diamond B})[\bar{t}/\bar{z}][\bar{p}'/\bar{y}'] \\
\mathsf{show}(t, \mathsf{union}_1(d^{\mathrm{SP_1} \diamond B}, \mathrm{SP_2}))[\bar{t}/\bar{z}][\bar{p}'/\bar{y}'] \\
\langle \mathsf{union}_1(d^{\mathrm{SP_1} \diamond B}, \mathrm{SP_2}), \mathsf{union}_1(e^{\mathrm{SP_1} \diamond C}, \mathrm{SP_2}) \rangle [\bar{t}/\bar{z}][\bar{p}'/\bar{y}'] \\
\\
\mathsf{inl}(\mathsf{union}_1(d^{\mathrm{SP_1} \diamond B}, \mathrm{SP_2}))[\bar{t}/\bar{z}][\bar{p}'/\bar{y}'] \\
\mathsf{inr}(\mathsf{union}_1(d^{\mathrm{SP_1} \diamond B}, \mathrm{SP_2}))[\bar{t}/\bar{z}][\bar{p}'/\bar{y}']
\end{array}$$

respectively. Each of these is in $C_{\mathrm{SP\_1}\ \mathbf{and}\ \mathrm{SP\_2} \diamond A}$ by the main induction hypothesis.

The cases for proof-terms for (union_1), (ext_1) and (ext_2) and (hide) are similar. □

Strong normalization for our system follows at once from this theorem.

Corollary 8.2.1. *Each proof-term is strongly normalizing.*

Proof. Using Theorem 8.2.9 and CR1. □

8.3 The Church–Rosser Property

The Church–Rosser property says that divergent proof normalization sequences always eventually converge to yield the same proof. We have already discussed the Church–Rosser property for the type theories of intuitionistic logic (Chapter 2, Part II) and intuitionistic Hoare logic (Chapter 5, Part III).

As in the previous parts of this monograph we formalize this notion using the Curry–Howard correspondence, proving the Church–Rosser property in terms of the logical type theory and the normalization relation $\rhd_{\mathsf{SSL}}$. We did not include detailed proofs of the property for the previous two logics discussed. The property is well known for intuitionistic logic. Because intuitionistic Hoare logic did not add any new normalizing reduction rules to proof-terms besides those of intuitionistic logic, the property followed trivially for intuitionistic Hoare logic,

However, the reduction rules for SSL are a non-trivial extension of those for intuitionistic logic, and so it is of interest to show the Church–Rosser property holds for this new logic. We do this by adapting the proof presented in [Bar84, pp. 59–62] for intuitionistic logic.

We define the *diamond property* of relations as before.

Definition 8.3.1. A relation $\#$ over a set S satisfies the *diamond property* when

$$\text{for all } x, x_1, x_2 \text{ in } S\ (x \# x_1 \text{ and } x \# x_2 \Rightarrow \text{ there is a } x_3 \text{ such that } (x_1 \# x_3 \text{ and } x_2 \# x_3))$$

$$
\begin{array}{rrcl}
 & a & \triangleright_* & a \\
a \triangleright_* a' \Rightarrow & \mathsf{abstract}\ x.\ a & \triangleright_* & \mathsf{abstract}\ x.\ a' \\
a \triangleright_* a' \text{ and } b \triangleright_* b' \Rightarrow & \mathsf{app}(a, b) & \triangleright_* & \mathsf{app}(a', b') \\
a \triangleright_* a' \Rightarrow & \mathsf{use}\ i.\ a & \triangleright_* & \mathsf{use}\ i.\ a' \\
a \triangleright_* a' \Rightarrow & \mathsf{show}(v, a) & \triangleright_* & \mathsf{show}(v, a') \\
a \triangleright_* a' \Rightarrow & \mathsf{specific}(a, v) & \triangleright_* & \mathsf{specific}(a', v) \\
a \triangleright_* a' \text{ and } b \triangleright_* b' \Rightarrow & \langle a, b \rangle & \triangleright_* & \langle a', b' \rangle \\
a \triangleright_* a' \Rightarrow & \mathsf{fst}(a) & \triangleright_* & \mathsf{fst}(a') \\
b \triangleright_* b' \Rightarrow & \mathsf{snd}(b) & \triangleright_* & \mathsf{snd}(b') \\
a \triangleright_* a' \Rightarrow & \mathsf{inl}(a) & \triangleright_* & \mathsf{inl}(a') \\
b \triangleright_* b' \Rightarrow & \mathsf{inr}(b) & \triangleright_* & \mathsf{inr}(b') \\
a \triangleright_* a' \text{ and } b \triangleright_* b' \text{ and } c \triangleright_* c' \Rightarrow & \mathsf{case}\ a\ \mathsf{of}\ \mathsf{inl}(x).b,\ \mathsf{inr}(y).c & \triangleright_* & \mathsf{case}\ a'\ \mathsf{of}\ \mathsf{inl}(x).b',\ \mathsf{inr}(y).c' \\
a \triangleright_* a' \Rightarrow & \mathsf{abort}(a) & \triangleright_* & \mathsf{abort}(a') \\
a \triangleright_* a' \Rightarrow & \mathsf{show}(v, a) & \triangleright_* & \mathsf{show}(v, a') \\
a \triangleright_* a' \text{ and } b \triangleright_* b' \Rightarrow & \mathsf{select}\ (a)\ \mathsf{in}\ x.y.b & \triangleright_* & \mathsf{select}\ (a')\ \mathsf{in}\ x.y.b' \\
a \triangleright_* a' \Rightarrow & \mathsf{union}_1(a, \mathrm{SP}) & \triangleright_* & \mathsf{union}_1(a', \mathrm{SP}) \\
a \triangleright_* a' \Rightarrow & \mathsf{union}_2(a, \mathrm{SP}) & \triangleright_* & \mathsf{union}_2(a', \mathrm{SP}) \\
a \triangleright_* a' \Rightarrow & \mathsf{ext}_1(a, \mathrm{SP}) & \triangleright_* & \mathsf{ext}_1(a', \mathrm{SP}) \\
a \triangleright_* a' \Rightarrow & \mathsf{ext}_2(a, \mathrm{SP}) & \triangleright_* & \mathsf{ext}_2(a', \mathrm{SP}) \\
a \triangleright_* a' \Rightarrow & \mathsf{hide}(a, SL) & \triangleright_* & \mathsf{ext}_2(a', SL) \\
a \triangleright_* a' \Rightarrow & \rho \bullet a & \triangleright_* & \rho \bullet a' \\
a \triangleright_* a' \text{ and } b \triangleright_* b' \Rightarrow & \mathsf{app}(\mathsf{abstract}\ x.\ a, b) & \triangleright_* & a'[b'/x] \\
a \triangleright_* a' \Rightarrow & \mathsf{specific}(\mathsf{use}\ x : s.\ a, v) & \triangleright_* & a'[v/x] \\
a \triangleright_* a' \Rightarrow & \mathsf{fst}(\langle a, b \rangle) & \triangleright_* & a' \\
b \triangleright_* b' \Rightarrow & \mathsf{snd}(\langle a, b \rangle) & \triangleright_* & b' \\
a \triangleright_* a' \text{ and } b \triangleright_* b' \Rightarrow & \mathsf{case}\ \mathsf{inl}(a)\ \mathsf{of}\ \mathsf{inl}(x).b,\ \mathsf{inr}(y).c & \triangleright_* & b'[a'/x] \\
a \triangleright_* a' \text{ and } c \triangleright_* c' \Rightarrow & \mathsf{case}\ \mathsf{inr}(a)\ \mathsf{of}\ \mathsf{inl}(x).b,\ \mathsf{inr}(y).c & \triangleright_* & c'[a'/y] \\
a \triangleright_* a' \text{ and } b \triangleright_* b' \Rightarrow & \mathsf{select}\ (\mathsf{show}(v, a))\ \mathsf{in}\ x.y.b & \triangleright_* & b'[a'/x][v/y]
\end{array}
$$

Fig. 8.12. Axioms defining $\triangleright_*$.

We also remind the reader of the following lemma.

Lemma 8.3.1. *Let # be a binary relation over a set and let $\hat{\#}$ be its transitive closure. If # satisfies the diamond property then so does $\hat{\#}$.*

Definition 8.3.2 (Church–Rosser property). Formally, we say that our normalization reduction rules satisfy the Church–Rosser property when $\hat{\triangleright}_{\mathsf{SSL}}$ satisfies the diamond property.

We define the relation $\triangleright_*$ by the axioms of Figs. 8.12 and 8.13. This relation is a single step reflexive closure of the rules presented in Figs. 8.7 and 8.9. Consequently, this relation (when considered as a set of pairs) is a subset of

$$
\begin{array}{rrl}
d \rhd_* d' \Rightarrow & S(\mathsf{abstract}\ u^A.\ d) & \rhd_*\ \mathsf{abstract}\ u^A.\ S(d') \\
d \rhd_* d' \Rightarrow & S(\mathsf{use}\ x : s.\ d) & \rhd_*\ \mathsf{use}\ x : s.\ S(d') \\
d \rhd_* d' \Rightarrow & S(\mathsf{show}(t, d)) & \rhd_*\ \mathsf{show}(t, S(d')) \\
d \rhd_* d' \text{ and } e \rhd_* e' \Rightarrow & S(\langle d, e\rangle) & \rhd_*\ \langle S(d'), S(e')\rangle \\
d \rhd_* d' \Rightarrow & S(\mathsf{inl}(d)) & \rhd_*\ \mathsf{inl}(S(d')) \\
d \rhd_* d' \Rightarrow & S(\mathsf{inr}(d)) & \rhd_*\ \mathsf{inr}(S(d'))
\end{array}
$$

where $S(p)$ denotes the following possible operations over proof-term p: $\mathsf{union}_1(p, \mathrm{SP})$, $\mathsf{union}_2(p, \mathrm{SP})$ $\mathsf{ext}_1(p, \mathrm{SP})$, $\mathsf{ext}_2(p, \mathrm{SP})$.

$$
\begin{array}{rrl}
d \rhd_* d' \Rightarrow & \rho \bullet (\mathsf{abstract}\ u^A.\ d) & \rhd_*\ \mathsf{abstract}\ u^{\rho(A)}.\ S(d') \\
d \rhd_* d' \Rightarrow & \rho \bullet (\mathsf{use}\ x : s.\ d) & \rhd_*\ \mathsf{use}\ x : \rho(s).\ (\rho \bullet d') \\
d \rhd_* d' \Rightarrow & \rho \bullet (\mathsf{show}(t, d)) & \rhd_*\ \mathsf{show}(\rho(t), (\rho \bullet d')) \\
d \rhd_* d' \text{ and } e \rhd_* e' \Rightarrow & \rho \bullet (\langle d, e\rangle) & \rhd_*\ \langle(\rho \bullet d'), (\rho \bullet e')\rangle \\
d \rhd_* d' \Rightarrow & \rho \bullet (\mathsf{inl}(d)) & \rhd_*\ \mathsf{inl}((\rho \bullet d')) \\
d \rhd_* d' \Rightarrow & \rho \bullet (\mathsf{inr}(d)) & \rhd_*\ \mathsf{inr}((\rho \bullet d')) \\
d \rhd_* d' \Rightarrow & \mathsf{hide}(\mathsf{abstract}\ u^A.\ d, SL) & \rhd_*\ \mathsf{abstract}\ u^{rho(A)}.\ S(d') \\
d \rhd_* d' \Rightarrow & \mathsf{hide}(\mathsf{use}\ x : s.\ d, SL) & \rhd_*\ \mathsf{use}\ x : s.\ \mathsf{hide}(d', SL) \\
d \rhd_* d' \text{ and } e \rhd_* e' \Rightarrow & \mathsf{hide}(\langle d, e\rangle, SL) & \rhd_*\ \langle \mathsf{hide}(d', SL), \mathsf{hide}(e', SL)\rangle \\
d \rhd_* d' \Rightarrow & \mathsf{hide}(\mathsf{inl}(d), SL) & \rhd_*\ \mathsf{inl}(\mathsf{hide}(d', SL)) \\
d \rhd_* d' \Rightarrow & \mathsf{hide}(\mathsf{inr}(d), SL) & \rhd_*\ \mathsf{inr}(\mathsf{hide}(d', SL))
\end{array}
$$

Fig. 8.13. Axioms defining $\rhd_*$ (cont.).

the relation $\rhd_{\mathsf{SSL}}$ and it can be easily verified that $\rhd_{\mathsf{SSL}}$ is the transitive closure of $\rhd_*$.

Thus, by Lemma 8.3.1, we can show $\hat{\rhd}_{\mathsf{SSL}}$ satisfies the Church–Rosser property by proving that $\rhd_*$ satisfies the diamond property.

First, we establish the following lemma.

Lemma 8.3.2. *Let x be a proof-term variable, y be an individual term variable, and assume v is an individual term of the same sort as y.*

If $a \rhd_* a'$ and $b \rhd_* b'$ then

1. $a[v/y] \rhd_* a'[v/y]$
2. $a[b/x] \rhd_* a'[b'/x]$
3. $a[b/x][v/y] \rhd_* a'[b'/x][v/y]$

Proof. We prove this by induction on the definition of $a \rhd_* a'$. The first item is straightforward. The third item follows from the second item, which we now prove.

Case: Assume that $a \rhd_* a'$ is $a \rhd_* a$.

Then we must show that $a[b/x] \rhd_* a[b'/x]$. This follows by subsidiary induction over the structure of a:

- Assume that a is a proof-term variable x. Then $a[b'/x]$ is b' and $a[b/x] \rhd_* a[b'/x]$ is the same as saying $b \rhd_* b'$. This is true, so we are done.

- Assume a is the proof-term variable z, not equivalent to x. Then both $a[b/x]$ and $a[b'/x]$ are equivalent to y. Thus $a[b/x] \rhd_* a[b'/x]$ is the same as saying $y \rhd_* y$. This is true, so the we are done.
- Assume a is the proof-term $\mathsf{app}(p, q)$. Then $a[b/x]$ is equivalent to $\mathsf{app}(p[b/x], q[b/x])$ and $a[b'/x]$ is equivalent to $\mathsf{app}(p[b'/x], q[b'/x])$. By the induction hypothesis and the definition of $\rhd_*$, we are done.

Case: Assume that $a \rhd_* a'$ is of the form $(\mathsf{abstract}\ z.\ p) \rhd_* (\mathsf{abstract}\ z.\ p')$, and is a consequence of $p \rhd_* p'$.

By the induction hypothesis, it must be the case that $p[b/x] \rhd_* p'[b'/x]$. But then $\mathsf{abstract}\ z.\ p[b/x] \rhd_* \mathsf{abstract}\ z.\ p'[b'/x]$ holds, as required.

Case: Assume that $a \rhd_* a'$ is of the form $\mathsf{app}(p, q) \rhd_* \mathsf{app}(p', q')$, and is a consequence of $p \rhd_* p'$ and $q \rhd_* q'$. Then we know that $a[b/x]$ is the same as writing $\mathsf{app}(p[b/x], q[b/x]) \rhd_* \mathsf{app}(p'[b'/x], q[b'/x])$ by the induction hypothesis and the definition of $\rhd_*$ is a'[b'/x] as required.

Case: Assume that $a \rhd_* a'$ is of the form $\mathsf{union}_1(p, \mathrm{SP}) \rhd_* \mathsf{union}_1(p', \mathrm{SP})$, and is a consequence of $p \rhd_* p'$.

By the induction hypothesis, it must be the case that $p[b/x] \rhd_* p'[b'/x]$. But then $\mathsf{union}_1(p[b/x], \mathrm{SP}) \rhd_* \mathsf{union}_1(p'[b'/x], \mathrm{SP})$ holds, as required.

Similar cases. We reason similarly (by the induction hypothesis and the definition of $\rhd_*$) for the cases when $a \rhd_* a'$ is of any of the following forms:

$$
\begin{aligned}
\mathsf{abstract}\ x.\ p &\rhd_* \mathsf{abstract}\ x.\ p' \\
\mathsf{app}(p, q) &\rhd_* \mathsf{app}(p', q') \\
\mathsf{use}\ i.\ p &\rhd_* \mathsf{use}\ i.\ p' \\
\mathsf{specific}(p, v) &\rhd_* \mathsf{specific}(p', v) \\
\langle p, q \rangle &\rhd_* \langle p', q' \rangle \\
\mathsf{fst}(p) &\rhd_* \mathsf{fst}(p') \\
\mathsf{snd}(q) &\rhd_* \mathsf{snd}(q') \\
\mathsf{inl}(p) &\rhd_* \mathsf{inl}(p') \\
\mathsf{inr}(q) &\rhd_* \mathsf{inr}(q')
\end{aligned}
$$

$$
\begin{aligned}
\mathsf{case}\ p\ \mathsf{of}\ \mathsf{inl}(x).q,\ \mathsf{inr}(y).r &\rhd_* \mathsf{case}\ p'\ \mathsf{of}\ \mathsf{inl}(x).q',\ \mathsf{inr}(y).r' \\
\mathsf{show}(v : s, p) &\rhd_* \mathsf{show}(v : s, p') \\
\mathsf{select}\ (p)\ \mathsf{in}\ x.y : s.q &\rhd_* \mathsf{select}\ (p')\ \mathsf{in}\ x.y : s.q' \\
\mathsf{union}_2(p, \mathrm{SP}) &\rhd_* \mathsf{union}_2(p', \mathrm{SP}) \\
\mathsf{ext}_1(p, \mathrm{SP}) &\rhd_* \mathsf{ext}_1(p', \mathrm{SP}) \\
\mathsf{ext}_2(p, \mathrm{SP}) &\rhd_* \mathsf{ext}_2(p', \mathrm{SP}) \\
\mathsf{hide}(p, SL) &\rhd_* \mathsf{hide}(p', SL) \\
\rho \bullet p &\rhd_* \rho \bullet p'
\end{aligned}
$$

Case: Assume $a \rhd_* a'$ is $\mathsf{app}(\mathsf{abstract}\ z.\ p, q) \rhd_* p'[q'/z]$ and is a direct consequence of $p \rhd_* p'$ and $q \rhd_* q'$.

Then we reason as follows.

$$\begin{array}{rll} a[b/x] & \text{is} & \mathsf{app}(\mathsf{abstract}\ z.\ (p[b/x]), q[b/x]) \\ & \rhd_* & p'[b'/x][q'[b'/x]/z] \\ & & \text{by the induction hypothesis and the definition of } \rhd_* \\ & \text{is} & p'[q'/z][b'/x] \\ & \text{is} & a'[b'/x] \end{array}$$

as required.

Case: Assume $a \rhd_* a'$ *is* $\mathsf{specific}(\mathsf{use}\ z : s.\ p, v) \rhd_* p'[v/z]$ *and is a direct consequence of* $p \rhd_* p'$.

Then we reason as follows.

$$\begin{array}{rll} a[b/x] & \text{is} & (\mathsf{specific}(\mathsf{use}\ x : s.\ p, v))[b/x] \\ & \rhd_* & p'[b'/x][v/z] \\ & & \text{by the induction hypothesis and the definition of } \rhd_* \\ & \text{is} & a'[v/z][b'/x] \\ & \text{is} & a'[b'/x] \end{array}$$

as required.

Case: Asssume $a \rhd_* a'$ *is* $\mathsf{fst}(\langle p, q\rangle) \rhd_* p'$ *and is a direct consequence of* $p \rhd_* p'$. Then we reason as follows.

$$\begin{array}{rll} a[b/x] & \text{is} & \mathsf{fst}(\langle p, q\rangle)[b/x] \\ & \rhd_* & p'[b'/x] \\ & & \text{by the induction hypothesis and the definition of } \rhd_* \\ & \text{is} & a'[b'/x] \end{array}$$

as required.

The case when $a \rhd_* a'$ is $\mathsf{snd}(\langle p, q\rangle) \rhd_* q'$, a consequence of $q \rhd_* q'$ is similar.

Case: Assume $a \rhd_* a'$ *is* $(\mathsf{case}\ \mathsf{inl}(p)\ \mathsf{of}\ \mathsf{inl}(z_1).q,\ \mathsf{inr}(z_2).r) \rhd_* q'[p'/z_1]$ *and is a direct consequence of* $p \rhd_* p'$ *and* $q \rhd_* q'$.

Then we reason as follows.

$$\begin{array}{rll} a[b/x] & \text{is} & (\mathsf{case}\ \mathsf{inl}(p)\ \mathsf{of}\ \mathsf{inl}(z_1).q,\ \mathsf{inr}(z_2).r)[b/x] \\ & \rhd_* & q'[b'/x][p'[b'/x]/z_1] \\ & & \text{by the induction hypothesis and the definition of } \rhd_* \\ & \text{is} & q'[p'/z_1][b'/x] \\ & \text{is} & a'[b'/x] \end{array}$$

as required.

We proceed in a similar way for the case where $a \rhd_* a'$ is of the form $\mathsf{case}\ \mathsf{inr}(p)\ \mathsf{of}\ \mathsf{inl}(z_1).q,\ \mathsf{inr}(z_2).r \rhd_* r'[p'/z_2]$ and is a direct consequence of $p \rhd_* p'$ and $r \rhd_* r'$.

Case: Assume $a \rhd_* a'$ *is* $\mathsf{select}\ (\mathsf{show}(v, p))\ \mathsf{in}\ z.y.q \rhd_* p'[q'/z][v/y]$ *and is a direct consequence of* $p \rhd_* p'$ *and* $r \rhd_* r'$.

Then we reason as follows.

$$\begin{array}{rll} a[b/x] & \text{is} & \mathsf{select}\ (\mathsf{show}(v, p))\ \mathsf{in}\ z.y.q[b/x] \\ & \rhd_* & p'[b'/x][q'[b'/x]/z][v/y] \\ & & \text{by the induction hypothesis and the definition of } \rhd_* \\ & \text{is} & p'[q'/z][v/y][b'/x] \\ & \text{is} & a'[b'/x] \end{array}$$

as required.

Let $S(p)$ denote any of the following following: $\mathsf{union}_1(p, \mathrm{SP})$, $\mathsf{union}_2(p, \mathrm{SP})$ $\mathsf{ext}_1(p, \mathrm{SP})$, $\mathsf{ext}_2(p, \mathrm{SP})$.

Case: Assume $a \rhd_* a'$ *is* $S(\mathsf{abstract}\ u^A.\ p) \rhd_* \mathsf{abstract}\ u^A.\ S(p')$. *and is a direct consequence of* $p \rhd_* p'$.

Then we reason as follows.

$$\begin{array}{rl} a[b/x] & = S(\mathsf{abstract}\ u^A.\ p)[b/x] \\ & \rhd_* \mathsf{abstract}\ u^A.\ S(p'[b'/x]) \\ & \quad \text{by the induction hypothesis and the definition of } \rhd_* \\ & \text{is } a'[b'/x] \end{array}$$

as required.

The remaining cases of the proof are similar. □

Lemma 8.3.3. *Let a, b, c, p, q and r be arbitrary well-typed proof-terms. Let z_1, z_2, x be well-typed proof-term variables. Let y and i be individual variables and let v be an arbitrary individual term. Then $\rhd_*$ possesses the following properties.*

1. *It is the case that A implies B for the following cases*

A	B
$\mathsf{use}\ i : s.\ a \rhd_* b$	b *is* $\mathsf{use}\ i : s.\ a'$ *with* $a \rhd_* a'$.
$\mathsf{inl}(a) \rhd_* b$	b *is* $\mathsf{inl}(a')$ *with* $a \rhd_* a'$.
$\mathsf{inr}(a) \rhd_* b$	b *is* $\mathsf{inr}(a')$ *with* $a \rhd_* a'$.
$\mathsf{abstract}\ x.\ a \rhd_* b$	b *is* $\mathsf{abstract}\ x.\ a'$ *with* $a \rhd_* a'$.
$\langle a, c\rangle \rhd_* b$	b *is* $\langle a', b'\rangle$ *with* $a \rhd_* a'$ *and* $c \rhd_* c'$.
$\mathsf{show}(v : s, a) \rhd_* b$	b *is* $\mathsf{show}(v : s, a')$ *with* $a \rhd_* a'$.
$\mathsf{abort}(a) \rhd_* b$	b *is* $\mathsf{abort}(a')$ *with* $a \rhd_* a'$.

2. $\mathsf{app}(a, b) \rhd_* c$ *entails that either*
   - c *is* $\mathsf{app}(a', b')$ *with* $a \rhd_* a'$ *and* $b \rhd_* b'$, *or*
   - a *is* $\mathsf{abstract}\ x.\ p$ *and* c *is* $p'[b'/x]$ *where* $p \rhd_* p'$ *and* $b \rhd_* b'$.
3. $\mathsf{specific}(a, v : s) \rhd_* c$ *entails that either*
   - c *is* $\mathsf{specific}(a', v)$ *with* $a \rhd_* a'$, *or*
   - a *is* $\mathsf{use}\ z : s.\ p$ *and* c *is* $p'[v/z]$ *where* $p \rhd_* p'$.
4. $\mathsf{fst}(a) \rhd_* c$ *entails that either*
   - c *is* $\mathsf{fst}(a')$ *with* $a \rhd_* a'$, *or*
   - a *is* $\langle p, q\rangle$ *and* c *is* p' *where* $p \rhd_* p'$.
5. $\mathsf{snd}(a) \rhd_* c$ *entails that either*
   - c *is* $\mathsf{snd}(a')$ *with* $a \rhd_* a'$, *or*
   - a *is* $\langle p, q\rangle$ *and* c *is* q' *where* $q \rhd_* q'$.
6. $\mathsf{case}\ p\ \mathsf{of}\ \mathsf{inl}(z_1).q,\ \mathsf{inr}(z_2).r \rhd_* c$ *entails that either*
   - c *is* $\mathsf{case}\ p'\ \mathsf{of}\ \mathsf{inl}(z_1).q',\ \mathsf{inr}(z_2).r'$ *with* $p \rhd_* p'$, $q \rhd_* q'$ *and* $r \rhd_* r'$,
   - p *is* $\mathsf{inl}(a)$ *and* c *is* $q'[p'/z_1]$ *where* $p \rhd_* p'$ *and* $q \rhd_* q'$, *or*
   - p *is* $\mathsf{inr}(a)$ *and* c *is* $q'[p'/z_2]$ *where* $p \rhd_* p'$ *and* $q \rhd_* q'$.
7. $\mathsf{select}\ (p)\ \mathsf{in}\ x.y.q \rhd_* c$ *entails that either*

- *c is* select (p') in $x.y.q'$ *with* $p \rhd_* p'$, $q \rhd_* q'$, *or*
- *p is* show(v, a) *and c is* $q'[p'/x][v/y]$ *where* $p \rhd_* p'$ *and* $q \rhd_* q'$.

8. *Let $S(a)$ denote the following possible operations over proof-term a:* union$_1(a, \text{SP})$, union$_2(a, \text{SP})$ ext$_1(a, \text{SP})$, ext$_2(a, \text{SP})$ *or* hide(a, SL). *Then $S(a) \rhd_* c$ entails that either*
   - $c = S(a', \text{SP})$ *with* $a \rhd_* a'$
   - $c =$ abstract u^A. $S(a')$ *with* $a \rhd_* a'$
   - $c =$ use $x : s$. $S(a')$ *with* $a \rhd_* a'$,
   - $c =$ show$(t, S(a'))$ *with* $a \rhd_* a'$, *provided $S(a)$ does not denote* hide(a, SL),
   - $c = \langle S(a'), S(b') \rangle$ *with* $a \rhd_* a'$ *and* $b \rhd_* b'$,
   - $c =$ inl$(S(a'))$ *with* $a \rhd_* a'$, *or*
   - $c =$ inr$(S(a'))$ *with* $a \rhd_* a'$.
9. $\rho \bullet a \rhd_* c$ *entails that either*
   - $c = \rho \bullet a'$ *with* $a \rhd_* a'$
   - $c =$ abstract $u^{\rho(A)}$. $\rho \bullet (a')$ *with* $a \rhd_* a'$
   - $c =$ use $x : s$. $\rho \bullet (a')$ *with* $a \rhd_* a'$,
   - $c =$ show$(\rho(t), \rho \bullet (a'))$ *with* $a \rhd_* a'$,
   - $c = \langle \rho \bullet (a'), \rho \bullet (b') \rangle$ *with* $a \rhd_* a'$ *and* $b \rhd_* b'$,
   - $c =$ inl$(\rho \bullet (a'))$ *with* $a \rhd_* a'$, *or*
   - $c =$ inr$(\rho \bullet (a'))$ *with* $a \rhd_* a'$.

Proof. By induction on the definition of $\rhd_*$. □

Theorem 8.3.3 (Church–Rosser property for SSL). *The relation $\rhd_*$ satisfies the diamond property (and therefore $\rhd_{\text{SSL}}$, as the transitive closure of $\rhd_*$, satisfies the diamond property).*

Proof. We proceed by induction on the definition of $a \rhd_* a_1$. We show that for all $a \rhd_* a_2$ there is an a_3 such that $a_1 \rhd_* a_3$ and $a_2 \rhd_* a_3$.

Case: Assume that $a \rhd_* a_1$ is of the form $a \rhd_* a$. We may take a_3 to be the same as a_2.

Case: Assume that $a \rhd_* a_1$ is of the form app$(p, q) \rhd_*$ app(p', q'), *and is a consequence of $p \rhd_* p'$ and $q \rhd_* q'$.* By Lemma 8.3.3 (2), there are two cases:

- a_2 is app(p'', q'') with $p \rhd_1 p''$ and $q \rhd_1 q''$. Then, using the induction hypothesis, we may take a_3 to be app(p''', q'''), where $p' \rhd_* p'''$ and $p'' \rhd_* p'''$ and similarly $q' \rhd_* q'''$ and $q'' \rhd_* q'''$.
- p is abstract x. p_1 and a_2 is $p_1''[q''/x]$ where $p_1 \rhd_* p_1''$ and $q \rhd_* q''$. By Lemma 8.3.3 (1), it is the case that p' is abstract x. p_1' with $p_1 \rhd_* p_1'$. By the induction hypothesis, we can take a_3 to be $p_1'''[q'''/x]$, where $p' \rhd_* p'''$ and $p'' \rhd_* p'''$ and similarly $q' \rhd_* q'''$ and $q'' \rhd_* q'''$.

Case: Assume that $a \rhd_* a_1$ *is of the form* $\mathsf{specific}(p, v) \rhd_* \mathsf{specific}(p', v)$*, and is a consequence of* $p \rhd_* p'$*.*

By Lemma 8.3.3 (3), there are two cases:

- a_2 is $\mathsf{specific}(p'', v)$ with $p \rhd_1 p''$. Then, using the induction hypothesis, we may take a_3 to be $\mathsf{specific}(p''', v)$, where $p' \rhd_* p'''$ and $p'' \rhd_* p'''$.
- p is $\mathsf{show}(v, p_1)$ and a_2 is $p_1''[v/i]$ where $p_1 \rhd_* p_1''$. By Lemma 8.3.3 (1), it is the case that p' is $\mathsf{use}\ i.\ p_1'$ with $p_1 \rhd_* p_1'$. By the induction hypothesis, we can take a_3 to be $p_1'''[v/x]$, where $p' \rhd_* p'''$ and $p'' \rhd_* p'''$.

Case: Assume that $a \rhd_* a_1$ *is of the form* $\mathsf{fst}(p) \rhd_* \mathsf{fst}(p')$ *and is a consequence of* $p \rhd_* p'$*.* By Lemma 8.3.3 (4), there are two cases.

- a_2 is $\mathsf{fst}(p'')$ with $p \rhd_* p''$. Then, using the induction hypothesis, we may take a_3 is $\mathsf{fst}(p''')$, where $p' \rhd_* p'''$ and $p'' \rhd_* p'''$.
- p is $\langle p_1, q\rangle$ and a_2 is p_1'' with $p_1 \rhd_* p_1''$. By Lemma 8.3.3 (1). Note that p' must be of the form $\langle p_1', q'\rangle$ with $p_1 \rhd_* p_1'$ and $q \rhd_* q'$. Then, by the induction hypothesis, we can take a_3 as p_1''' with $p_1' \rhd_* p_1'''$ (so that $\mathsf{fst}(p')$ is $\mathsf{fst}(\langle p_1', q'\rangle) \rhd_* p_1'''$) and $p'' \rhd_* p_1'''$.

We proceed similarly for when $a \rhd_* a_1$ is of the form $\mathsf{snd}(p) \rhd_* \mathsf{snd}(p')$ and is a consequence of $p \rhd_* p'$.

Case: Assume that $a \rhd_* a_1$ *is of the form*

$$(\mathsf{case}\ p\ \mathsf{of}\ \mathsf{inl}(x).q,\ \mathsf{inr}(y).r) \rhd_* (\mathsf{case}\ p'\ \mathsf{of}\ \mathsf{inl}(x).q',\ \mathsf{inr}(y).r')$$

and is a consequence of $p \rhd_* p'$*,* $q \rhd_* q'$ *and* $r \rhd_* r'$*.* By Lemma 8.3.3 (6), there are three cases.

- a_2 is $(\mathsf{case}\ p''\ \mathsf{of}\ \mathsf{inl}(x).q'',\ \mathsf{inr}(y).r'')$ with $p \rhd_1 p''$, $q \rhd_1 q''$ and $r \rhd_1 r''$. Then, using the induction hypothesis, we may take a_3 to be $(\mathsf{case}\ p'''\ \mathsf{of}\ \mathsf{inl}(x).q''',\ \mathsf{inr}(y).r''')$, where $p' \rhd_* p'''$, $p'' \rhd_* p'''$, $q' \rhd_* q'''$, $q'' \rhd_* q'''$, $r' \rhd_* r'''$ and $r'' \rhd_* r'''$.
- p is $\mathsf{inl}(p_1)$ and a_2 is $q''[p_1''/x]$ where $p_1 \rhd_* p_1''$. By Lemma 8.3.3 (1), it is the case that p' is $\mathsf{inl}(p_1')$ with $p_1 \rhd_* p_1'$. Then, by the induction hypothesis, we can take a_3 to be $q'''[p_1'''/x]$, where $p' \rhd_* p'''$, $p'' \rhd_* p'''$, $q' \rhd_* q'''$ and $q'' \rhd_* q'''$.
- p is $\mathsf{inr}(p_1)$ and a_2 is $r''[p_1''/y]$ where $p_1 \rhd_* p_1''$ and $r \rhd_* r''$. By Lemma 8.3.3 (1), it is the case that p' is $\mathsf{inr}(p_1')$ with $p_1 \rhd_* p_1'$. Then, by the induction hypothesis, we can take a_3 to be $r'''[p_1'''/y]$, where $p' \rhd_* p'''$, $p'' \rhd_* p'''$, $r' \rhd_* r'''$ and $r'' \rhd_* r'''$.

Case: Assume that $a \rhd_* a_1$ *is of the form* $\mathsf{select}\ (p)\ \mathsf{in}\ x.y.q \rhd_* \mathsf{select}\ (p')\ \mathsf{in}\ x.y.q'$*, and is a consequence of* $p \rhd_* p'$ *and* $q \rhd_* q'$*.* By Lemma 8.3.3 (7), there are two cases.

- a_2 is select (p'') in $x.y.q''$ with $p \rhd_1 p''$ and $q \rhd_1 q''$. Then, using the induction hypothesis, we may take a_3 is select (p''') in $x.y.q'''$, where $p' \rhd_* p'''$, $p'' \rhd_* p'''$, $q' \rhd_* q'''$, $q'' \rhd_* q'''$.
- p is $\mathsf{show}(v, p_1)$ and a_2 is $q''[p_1''/x][v/y]$ where $p_1 \rhd_* p_1''$. By Lemma 8.3.3 (1), it is the case that that p' is $\mathsf{show}(v, p_1')$ with $p_1 \rhd_* p_1'$. Then, by the induction hypothesis, we can take a_3 to be $q'''[p_1'''/x][v/y]$, where $p' \rhd_* p'''$, $p'' \rhd_* p'''$, $q' \rhd_* q'''$ and $q'' \rhd_* q'''$.

We proceed similarly for the cases where $a \rhd_* a_1$ is of the form

$$(\mathsf{abstract}\ z.\ p) \rhd_* (\mathsf{abstract}\ z.\ p')$$

or

$$\mathsf{use}\ i.\ p \rhd_* \mathsf{use}\ i.\ p'$$

a consequence of $p \rhd_* p'$.

Case: Assume that $a \rhd_* a_1$ *is of the form* $\mathsf{specific}(p, v) \rhd_* \mathsf{specific}(p', v)$, *and is a consequence of* $p \rhd_* p'$. Then a_2 is $\mathsf{specific}(p'', v)$. Using the induction hypothesis and Lemma 8.3.3 (1), we can take a_3 is $\mathsf{specific}(p''', v)$, where $p' \rhd_* p'''$ and $p'' \rhd_* p'''$.

Similar cases. The cases when $a \rhd_* a_1$ is of the form

$$\begin{aligned} \langle a, b\rangle &\rhd_* \langle a', b'\rangle, \\ \mathsf{inl}(a) &\rhd_* \mathsf{inl}(a'), \\ \mathsf{inr}(b) &\rhd_* \mathsf{inr}(b'), \\ \mathsf{show}(v, a) &\rhd_* \mathsf{show}(v, a') \end{aligned}$$

follow similarly (by applications of the induction hypothesis and Lemma 8.3.3 (1)).

Case: Assume that $a \rhd_* a_1$ *is of the form* $\mathsf{app}(\mathsf{abstract}\ x.\ p, q) \rhd_* p'[q'/x]$ *and is a consequence of* $p \rhd_* p'$ *and* $q \rhd_* q'$. By Lemma 8.3.3 (2) we know that either

- a_2 is $\mathsf{app}(\mathsf{abstract}\ x.\ p'', q'')$ with $p \rhd_* p''$ and $q \rhd_* q''$. By induction hypothesis there are proof-terms p''' and q''' with $p' \rhd_* p'''$ and $p'' \rhd_* p'''$ and similarly $q' \rhd_* q'''$ and $q'' \rhd_* q'''$. By Lemma 8.3.2 (1) we can take a_3 to be $p'''[q'''/x]$.
- a_2 is $p''[q''/x]$ where $p \rhd_* p''$ and $q \rhd_* q''$. Then, using the induction hypothesis, we can take a_3 to be $p'''[q'''/x]$ where $p' \rhd_* p'''$ and $p'' \rhd_* p'''$ and similarly $q' \rhd_* q'''$ and $q'' \rhd_* q'''$.

Case: Assume $a \rhd_* a_1$ *is* $\mathsf{specific}(\mathsf{use}\ z : s.\ p, v) \rhd_* p'[v/z]$ *and is a direct consequence of* $p \rhd_* p'$. Then, by Lemma 8.3.3 (3) we know that either

- a_2 is $\mathsf{specific}(\mathsf{use}\ z : s.\ p'', v)$ with $p \rhd_* p''$. By induction hypothesis there is a proof-term p''' with $p' \rhd_* p'''$ and $p'' \rhd_* p'''$. Then, by Lemma 8.3.2 (1) we can take a_3 to be $p'''[v/z]$.

- a_2 is $p''[v/z]$ where $p \rhd_* p''$. Then, using the induction hypothesis, we can take a_3 to be $p'''[v/z]$ where $p' \rhd_* p'''$ and $p'' \rhd_* p'''$.

Case: Asssume $a \rhd_* a_1$ is $\mathsf{fst}(\langle p, q\rangle) \rhd_* p'$ and is a direct consequence of $p \rhd_* p'$. Then, by Lemma 8.3.3 (4) we know that either:

- a_2 is $\mathsf{fst}(\langle p'', q''\rangle)$ with $p \rhd_* p''$ and $q \rhd_* p''$. By the induction hypothesis there are proof-terms p''' and q''' such that $p' \rhd_* p'''$ and $p'' \rhd_* p'''$ and $q' \rhd_* q'''$ and $q'' \rhd_* q'''$. Then we can take a_3 to be p'''. Or,
- a_2 is p'' where $p \rhd_* p''$. Then, using the induction hypothesis, we can take a_3 to be p''' where $p' \rhd_* p'''$ and $p'' \rhd_* p'''$.

We proceed similarly for the case where $a \rhd_* a_1$ is of the form $\mathsf{snd}(\langle p, q\rangle) \rhd_* q'$, a direct consequence of $q \rhd_* q'$.

Case: Assume $a \rhd_* a_1$ is $(\mathsf{case}\ \mathsf{inl}(p)\ \mathsf{of}\ \mathsf{inl}(z_1).q,\ \mathsf{inr}(z_2).r) \rhd_* q'[p'/z_1]$ and is a direct consequence of $p \rhd_* p'$ and $q \rhd_* q'$. Then, by Lemma 8.3.3 (6) we know that either:

- a_2 is $\mathsf{case}\ \mathsf{inl}(p'')\ \mathsf{of}\ \mathsf{inl}(z_1).q'',\ \mathsf{inr}(z_2).r''$ with $p \rhd_* p''$, $q \rhd_* q''$ and $r \rhd_* r''$. By the induction hypothesis there are proof-terms p''', q''' and r''' such that $p' \rhd_* p'''$, $p'' \rhd_* p'''$, $q' \rhd_* q'''$, $q'' \rhd_* q'''$, $r' \rhd_* r'''$ and $r'' \rhd_* r'''$. Then, by Lemma 8.3.2 (2), we can take a_3 to be $q'''[p'''/z_1]$. Or,
- a_2 is $q''[p''/z_1]$ where $q \rhd_* q''$ and $p \rhd_* p''$. Then, using the induction hypothesis, we can take a_3 to be q''' where $p' \rhd_* p'''$, $p'' \rhd_* p'''$, $q' \rhd_* q'''$ and $q'' \rhd_* q'''$.

We proceed similarly for the case when $a \rhd_* a_1$ is of the form

$$(\mathsf{case}\ \mathsf{inr}(p)\ \mathsf{of}\ \mathsf{inl}(z_1).q,\ \mathsf{inr}(z_2).r) \rhd_* r'[p'/z_2]$$

and is a direct consequence of $p \rhd_* p'$ and $r \rhd_* r'$.

Case: Assume $a \rhd_* a_1$ is $(\mathsf{select}\ (\mathsf{show}(v, p))\ \mathsf{in}\ z.y : s.q) \rhd_* p'[q'/z][v/y]$ and is a direct consequence of $p \rhd_* p'$ and $r \rhd_* r'$.

Then, by Lemma 8.3.3 (7) we know that either:

- a_2 is $(\mathsf{select}\ (\mathsf{show}(v, p''))\ \mathsf{in}\ z.y : s.q'')$ with $p \rhd_* p''$ and $q \rhd_* q''$. By the induction hypothesis there are proof-terms p''' and q''' such that $p' \rhd_* p'''$, $p'' \rhd_* p'''$, $q' \rhd_* q'''$ and $q'' \rhd_* q'''$. Then, by Lemma 8.3.2 (3), we can take a_3 to be $p'''[q'''/z][v/y]$. Or,
- a_2 is $p''[q''/z][v/y]$ where $p \rhd_* p''$ and $q \rhd_* q''$. Then, using the induction hypothesis, we can take a_3 is $p'''[q'''/z]$ where $p' \rhd_* p'''$, $p'' \rhd_* p'''$, $q' \rhd_* q'''$ and $q'' \rhd_* q'''$.

Let $S(p)$ denote any of the following possible operations over a proof-term p: $\mathsf{union}_1(p, \textsc{Sp})$, $\mathsf{union}_2(p, \textsc{Sp})$ $\mathsf{ext}_1(p, \textsc{Sp})$, $\mathsf{ext}_2(p, \textsc{Sp})$ or $\mathsf{hide}(p, SL)$, for some specification expression $\textsc{Sp}$ and symbol list SL.

Case: Assume $a \rhd_* a_1$ is $S(\mathsf{abstract}\ u^A.\ p) \rhd_* \mathsf{abstract}\ u^A.\ S(p')$ and is a direct consequence of $p \rhd_* p'$.

Then, by Lemma 8.3.3 (8), we know that either

- $a_2 = S(\mathsf{abstract}\ u^A.\ p'')$ with $p \rhd_* p''$. By the induction hypothesis there is a proof-term p''' such that $p' \rhd_* p'''$ and $p'' \rhd_* p'''$. So we let $a_3 = \mathsf{abstract}\ u^A.\ S(p''')$.
- $a_2 = \mathsf{abstract}\ u^A.\ S(p'')$ with $p \rhd_* p''$. By the induction hypothesis there is a proof-term p''' such that $p' \rhd_* p'''$ and $p'' \rhd_* p'''$. So we let $a_3 = \mathsf{abstract}\ u^A.\ S(p''')$.

We deal similarly with the cases where $a \rhd_* a_1$ is of the form

$$\begin{aligned} S(\mathsf{use}\ x : s.\ p) &\rhd_* \mathsf{use}\ x : s.\ S(p') \\ S(\mathsf{inl}(p)) &\rhd_* \mathsf{inl}(S(p')) \\ S(\mathsf{inr}(p)) &\rhd_* \mathsf{inr}(S(p')) \\ S(\mathsf{show}(t,p)) &\rhd_* \mathsf{show}(t, S(p')) \end{aligned}$$

and is a direct consequence of $p \rhd_* p'$, or of the form

$$S(\langle p, q\rangle) \rhd_* \langle S(p'), S(q')\rangle$$

and is a direct consequence of $p \rhd_* p'$ and $q \rhd_* q'$.

Case: Assume $a \rhd_* a_1$ is $\rho \bullet (\mathsf{show}(t,p)) \rhd_* \mathsf{show}(\rho(t), (\rho \bullet p'))$ and is a direct consequence of $p \rhd_* p'$. Then, by Lemma 8.3.3 (9), we know that either:

- $a_2 = S(\mathsf{show}(t, p''))$ with $p \rhd_* p''$. By the induction hypothesis there is a proof-term p''' such that $p' \rhd_* p'''$ and $p'' \rhd_* p'''$. So we let $a_3 = \mathsf{show}(\rho(t), (\rho \bullet p'''))$. Or,
- $a_2 = \mathsf{show}(\rho(t), (\rho \bullet p''))$ with $p \rhd_* p''$. By the induction hypothesis there is a proof-term p''' such that $p' \rhd_* p'''$ and $p'' \rhd_* p'''$. So we let $a_3 = \mathsf{show}(\rho(t), (\rho \bullet p'''))$.

We deal similarly with the cases where $a \rhd_* a_1$ is of the form

$$\begin{aligned} \rho \bullet (\mathsf{abstract}\ u^A.\ p) &\rhd_* \mathsf{abstract}\ u^{\rho(A)}.\ S(p') \\ \rho \bullet (\mathsf{use}\ x : s.\ p) &\rhd_* \mathsf{use}\ x : \rho(s).\ (\rho \bullet p') \\ \rho \bullet (\mathsf{inl}(p)) &\rhd_* \mathsf{inl}((\rho \bullet p')) \\ \rho \bullet (\mathsf{inr}(p)) &\rhd_* \mathsf{inr}((\rho \bullet p')) \end{aligned}$$

and is a direct consequence of $p \rhd_* p'$, or of the form

$$\rho \bullet \langle p, q\rangle \rhd_* \langle (\rho \bullet p'), (\rho \bullet q')\rangle$$

and is a direct consequence of $p \rhd_* p'$ and $q \rhd_* q'$.

This last case concludes the proof. □

8.4 Discussion

This chapter presented some important proof-theoretic properties of SSL necessary for the application of the Curry–Howard protocol to SSL.

We have defined a logical type theory for SSL for which the Curry–Howard isomorphism holds. The proof-terms of our theory are more complex than those for ordinary intuitionistic logic. We have seen that this is due to the presence of additional proof-term constructors corresponding to structural rules. We examined further normalizing rules that can be given over the new proof-terms. This makes the proof of strong normalization a non-trivial extension of the proof for the intuitionistic case.

In the next chapter, we will show how to transform the proof-terms of this chapter into provably correct functional programs, which may then be used to consistently extend structured specifications. Then, in Chapter 10, we will extend our calculus and its logical type theory to accommodate parametrized specifications. We will show how to extract programs from that augmented calculus. Finally, in Chapter 11, we will examine how our calculus and the synthesis results can be applied to give methods for structured program synthesis.

9

Structured Proofs-as-Programs

In this chapter we show how to synthesize correct *SML* programs from proofs about *CASL* specifications according to the Curry–Howard protocol of Chapter 3. We also provide a method for incorporating extracted programs back into a *CASL* specification, to develop an executable extension. We refer to these techniques as *structured proofs-as-programs*.

Because we follow the protocol, the process of proving a specification and then extracting a realizing program is similar to that given for intuitionistic logic and *SML* programs in Chapter 2 and for IHL and imperative *SML* in Chapter 6 of Part III. We define an extraction map over proof-terms of the logical type theory $LTT(\mathsf{SSL})$ (described in Chapter 8) to elements of a computational type theory, which consists of the lambda calculus of *SML*.

Briefly, our method is as follows. We define a notion of realizability between terms of the *CTT* and the specification/formula pairs of our calculus. A realizer of a pair $\text{Sp}\diamond A$ is taken as a Skolem function for the Skolem form of the formula, true for some extension Sp' of Sp. Then, given a proof

$$\vdash \text{Sp} \diamond A$$

with a corresponding proof-term, $p^{\text{Sp}\diamond A}$, in the *LTT*, we define an extraction map $\mathsf{extract}_{\mathsf{SSL}}$ so that the program $\mathsf{extract}_{\mathsf{SSL}} p$ realizes $\text{Sp} \diamond A$. For instance, given $p^{\text{Sp}\diamond\forall x:s\bullet\exists y:s\bullet A(x,y)}$ we extract the term $\mathsf{extract}_{\mathsf{SSL}}(p)$ as a realizer, standing as a Skolem function for the proved formula, so that

$$\text{Sp}' \models \forall x : s \bullet A(x, (\mathsf{extract}_{\mathsf{SSL}}(p)\ x))$$

holds for an extension Sp' of Sp.

The extraction map extends the map for intuitionistic logic given in Chapter 2 of Part II, with additions to deal with structural rules. In extraction, most of the structural rules are dealt with easily: unions and extension do not affect the extracted program, and renamings define corresponding renamings of programs. However, we find that we cannot immediately extract realizing programs from proofs that involve instances of the rule for hiding symbols in a specification.

We approach this problem by refining a given specification (that may include hide) to a specification that contain sour extracted, realizing functions.

In order to achieve these results, we require *CASL* specifications to accommodate the lambda calculus of *SML*.

We proceed as follows.

- In Section 9.1, we define the computational type theory into which we shall extract programs. This is the lambda calculus fragment of *SML*, constructed over function symbols of *CASL* signatures. We discuss how to extend *CASL* to accommodate these *SML* lambda terms in specifications.
- In Section 9.2, we define notions of realizability between our *CTT* and an *LTT*.
- The extraction map is identified in Section 9.3. We use this map to extract programs from what we call modular proofs (proofs without occurrences of the (hide) rule.
- In Section 9.4, we describe how to overcome some of the complications that arise from hiding, to extract realizing terms from all proofs. (In Chapter 11, we describe how to use these techniques to obtain recursively successive executable extensions of a given specification that contain programs for each declared function symbol of SP.)
- Section 9.5 returns to the password checking system example used in previous chapters, to show how we can extract a correct program from our proof and consistently refine the system specification.
- In section 9.6, we review how our methods of synthesis are an effective application of the Curry–Howard protocol as it was defined in Chapter 3.
- Section 9.7 provides a brief discussion of our results.

9.1 Specifying and reasoning about *SML* programs

We wish to use *CASL* and our calculus to specify, reason about, and synthesize functional, *SML* programs. This is achieved by extending *CASL* specifications by

- extending the set of terms for any signature to include the lambda calculus of *SML*, and
- assuming implicit axioms in every *CASL* specification, enabling SSL proofs for reasoning about lambda terms.

This is similar to Chapter 2 of Part II and Chapter 4 of Part III, where we used *SML* programs within formulae by treating signature terms as extended by the lambda calculus of *SML*.

9.1.1 Extended signatures

To represent *SML* programs in our specifications, we need to extend our signatures with functional, disjoint union and product sort constructors, corresponding to their respective type constructors in *SML*. This is done by replacing our definition of signatures with that given in Definition 2.1.1 of Chapter 2, Part II, p. 27, repeated here for completeness.

Definition 9.1.1 (Many-sorted signature with total functions). A *many-sorted signature* $\Sigma = \langle S, TF, P \rangle$ consists of:

- A set, S, of sorts. Sorts are generated from a set of *basic sorts*, $B(S)$, according to the following inductive definition. First, $B(S) \subseteq S$. Also, if s_1 and s_2 are in S, then so are
  — the function sort $(s_1 \rightarrow s_2)$
  — the product sort $(s_1 * s_2)$
  — the disjoint union $(s_1 | s_2)$
  We assume that $B(S)$ includes a special sort, called $Unit$.
- Sets $TF_{w,s}$ of total function symbols, for each *function profile* (w, s). A function profile (w, s) is a pair of words, consisting of a sequence of argument sorts $w \in S^*$ and a result sort $s \in S$. Constants are treated as functions with no arguments. The length of w is called the *arity* of function symbols in $TF_{w,s}$. We assume that $TF_{\emptyset,Unit}$ contains a unit symbol, written () (this denotes the single inhabitant of the sort $Unit \in B(S)$).
- Sets P_w of predicate symbols, for each *predicate profile* w. A predicate profile consists of a sequence of argument sorts $w \in S^*$. The length of w is called the *arity* of predicate symbols in P_w. For each basic sort $s \in B(S)$, there is a distinguished equality predicate $=_s \in P_{ss}$.

9.1.2 Lambda terms

Our *SML* programs will be represented as terms in *CASL* specifications. Because we are only concerned with the pure fragment of *SML* corresponding to the lambda calculus, we need to extend the terms generated from a signature by the lambda calculus. This is done by following the same pattern as in Fig. 2.1 of Chapter 2, Part II, p. 28, inductively extending the usual terms with operators for lambda abstraction, application, pairing and making disjoint unions. However, we will also add a new operator for recursion.

We redefine the *terms* for a signature $\Sigma = \langle S, TF, P \rangle$ generated over variables Var, $Term(\Sigma, Var)$, as in Fig. 9.1.

Remark 9.1. The recursion operator $rec(s, \overline{cons}, \overline{arg})$ can be written in *SML* syntax as follows. Given the lists of constructors

$$\begin{aligned} \overline{cons} &= [c_1 : s, \ldots, c_n : s, f_1 : (s_{1_1} \times \ldots \times s_{m_1}) \rightarrow s, \ldots, \\ &\quad f_p : (s_{1_p} \times \ldots \times s_{m_p}) \rightarrow s\} \\ \overline{args} &= [a_1, \ldots, a_n, b_1, \ldots, b_p] \end{aligned}$$

$rec(x, s, \overline{cons}, \overline{arg})$ can be thought of as shorthand for a recursive function application

$$(\texttt{rec\_s}\ \mathtt{a}_1\ \ldots\ \mathtt{a}_n\ \mathtt{b}_1\ \ldots\ \mathtt{b}_n)$$

with *rec_s* defined in *SML* as follows

$a, b, c ::=$	elements of $Term(\Sigma, Var)$
$f(a_1, \ldots, a_n)$	$f \in TF_{w,s}$, w of arity n and $(a_1, \ldots, a_n)$ a (possibly empty) list of elements of $Term(\Sigma)$
x	a variable $x \in Var$
$Inl(a)$	in left
$Inr(a)$	in right
$match\ a\ with\ Inl(x) \Rightarrow b \mid Inr(y) \Rightarrow c$	match case, $x, y \in Var$
$fn\ x : s \Rightarrow b$	lambda abstraction, s is a sort of Σ
$(a\ b)$	application
(a, b)	pair
$fst(a)$	first projection
$snd(a)$	second projection
$rec(s, \overline{cons}, \overline{arg})$	recursion, where s is a sort in Σ and $\overline{cons}$ a list of constructors for s, and $\overline{arg}$ is a list of terms

Fig. 9.1. Syntax terms of $Term(\Sigma, Var)$.

$$
\begin{array}{ll}
\mathtt{fun\ rec\_s}\ \mathtt{a_1}\ \ldots\ \mathtt{a_n}\ \mathtt{b_1}\ \ldots\ \mathtt{b_p}\ \mathtt{c_1} & = \mathtt{a_1} \\
\ \ldots & \\
\ \mid \mathtt{rec\_s}\ \mathtt{a_1}\ \ldots\ \mathtt{a_n}\ \mathtt{b_1}\ \ldots\ \mathtt{b_p}\ \mathtt{c_n} & = \mathtt{a_n} \\
\ \mid \mathtt{rec\_s}\ \mathtt{a_1}\ \ldots\ \mathtt{a_n}\ \mathtt{b_1}\ \ldots\ \mathtt{b_p}\ \mathtt{f_1}(\mathtt{x_{1,1}}, \ldots, \mathtt{x_{m,1}}) & = \mathtt{b_1}\mathtt{x_{1,1}}\ \ldots\ \mathtt{x_{m,1}}\ \langle \mathtt{x_{1,1}}, \ldots, \mathtt{x_{m,1}} \rangle \\
\ \ldots & \\
\ \mid \mathtt{rec\_s}\ \mathtt{a_1}\ \ldots\ \mathtt{a_n}\ \mathtt{b_1}\ \ldots\ \mathtt{b_p}\ \mathtt{f_p}(\mathtt{x_{1,p}}, \ldots, \mathtt{x_{m,p}}) & = \mathtt{b_p}\mathtt{x_{1,1}}\ \ldots\ \mathtt{x_{m,1}}\ \langle \mathtt{x_{1,p}}, \ldots, \mathtt{x_{m,p}} \rangle
\end{array}
$$

where

$$\langle \mathtt{x_{1,1}}, \ldots, \mathtt{x_{m,1}} \rangle$$

denotes

$$(\mathtt{rec\_s}\ \mathtt{a_1}\ \ldots\ \mathtt{a_n}\ \mathtt{b_1}\ \ldots\ \mathtt{b_p}\ \mathtt{x_1})\ \ldots\ (\mathtt{rec\_s}\ \mathtt{a_1}\ \ldots\ \mathtt{a_n}\ \mathtt{b_1}\ \ldots\ \mathtt{b_p}\ \mathtt{x_k})$$

for $[\mathtt{x_{1,j}}, \ldots, \mathtt{x_{k,j}}]$ the list obtained from $[\mathtt{x_{1,j}}, \ldots, \mathtt{x_{m,j}}]$ by removing all $x_{i,j}$ such that $s_{i,j} = s$.

9.1.3 Sorting

Our extended lambda terms are associated with sorts of their signature, according to the sort inference rules provided in Fig. 9.2. These identical to those given in Fig. 2.2, Chapter 2 of Part II, p. 29, except to include a sorting rule for the new recursion terms. As usual, we use the words "sorts" and "types" for terms interchangeably, depending on whether we consider terms as elements of a *CASL* specification or programs in *SML*, respectively.

Remark 9.2. The sort inference rule for recursion terms, (Rec), is weaker than that actually used in *SML*. We omit the full rule, as this rule will suffice for our purposes of illustrating program extraction. See, e.g., [MTH90] for the full typing specification for *SML*.

$$\frac{}{\Gamma, x : s \vdash_\Sigma x : s} \text{(Ass)}$$

$$\frac{f \in TF_{(s_1 \dots s_n),s} \quad \Gamma_1 \vdash a_1 : s_1 \dots \Gamma_n \vdash a_n : s_n}{\Gamma, \Gamma_1, \dots, \Gamma_n \vdash_\Sigma f(a_1, \dots, a_n) : s} \text{(Fn)}$$

$$\frac{\Gamma \vdash_\Sigma a : s_1}{\Gamma \vdash_\Sigma \mathit{Inl}(a) : (s_1|s_2)} (\text{Union}_1) \qquad \frac{\Gamma \vdash_\Sigma a : s_2}{\Gamma \vdash_\Sigma \mathit{Inr}(a) : (s_1|s_2)} (\text{Union}_2)$$

$$\frac{\Gamma_1 \vdash_\Sigma a : s_1 \quad \Gamma_2 \vdash_\Sigma b : s_2}{\Gamma_1, \Gamma_2 \vdash_\Sigma (a, b) : (s_1 * s_2)} \text{(Prod)}$$

$$\frac{\Gamma \vdash_\Sigma a : (s_1 * s_2)}{\Gamma \vdash_\Sigma \mathit{fst}(a) : s_1} (\text{Proj}_1) \qquad \frac{\Gamma \vdash_\Sigma a : (s_1 * s_2)}{\Gamma \vdash_\Sigma \mathit{snd}(a) : s_2} (\text{Proj}_2)$$

$$\frac{\Gamma, x : s_1 \vdash_\Sigma a : s_2}{\Gamma \vdash_\Sigma \mathit{fn}\; x : s_1 => a : s_1 \rightarrow s_2} \text{(Abs)} \qquad \frac{\Gamma_1 \vdash_\Sigma a : s_1 \quad \Gamma_2 \vdash_\Sigma b : (s_1 \rightarrow s_2)}{\Gamma_1, \Gamma_2 \vdash_\Sigma (b\; a) : s_2} \text{(App)}$$

$$\frac{\Gamma_1 \vdash_\Sigma a : (s_1|s_2) \quad \Gamma_2, x : s_1 \vdash_\Sigma b : s \quad \Gamma_3, y : s_2 \vdash_\Sigma c : s}{\Gamma_1, \Gamma_2, \Gamma_3 \vdash_\Sigma \mathit{match}\; a\; \mathit{with}\; \mathit{Inl}(x) => b \mid \mathit{Inr}(y) => c : s} \text{(Case)}$$

$$\frac{\begin{array}{c} \Gamma_1 \vdash c_1 : s \;\dots\; \Gamma_n \vdash c_n : s \\ \Delta_1 \vdash f_1 : (s_{1_1} \times \dots \times s_{m_1}) \rightarrow s \;\dots \\ \Delta_p \vdash f_p : (s_{1_p} \times \dots \times s_{m_p}) \rightarrow s \\ \Theta_1 \vdash a_1 : t \dots \Theta_n \vdash a_n : t \\ \text{the result sort of } b_1, \dots, b_p \text{ is } t \end{array}}{\Gamma_1 \dots, \Theta_n \vdash \mathit{rec}(s, \overline{\mathit{cons}}, \overline{\mathit{arg}}) : t} \text{(Rec)}$$

Fig. 9.2. Sort inference rules for terms of Σ.

9.1.4 Computational type theory

By the remarks in Section 2.4 of Chapter 2, our extended signatures and lambda terms, together with the operational semantics, constitute an effective representation of *SML*.

That is, we can consider our set of terms and sorts to be a computational type theory:

$$\mathsf{C}(SML) = \langle Terms(SML), Sorts(SML), \vdash_{SML}, TIR(SML)\rangle$$

where

- $Terms(SML)$ is a set of terms, built from all $Terms(\Sigma, Var)$ for each signature Σ.
- $Types(SML)$ is a set of types built from all sorts available for every signature.
- $TIR(SML)$ is a set of typing rules that define the type inference relation $\vdash_{SML}$ that holds between contexts and typed terms, built from the collection of all sorting rules for each signature.

We will use our *SML* terms within formulae to reason about programs. As usual we use `typewriter` font to distinguish terms when used as programs, as opposed to terms in formulae.

9.1.5 Operational semantics

We provide an operational semantics for lambda terms by a reduction relation $\triangleright_{SML}$, which is given in Figs. 9.3 and 9.4. These are the usual reduction rules for lambda calculus, but using the syntax of *SML*.

Remark 9.3. For the purposes of this chapter, we do not provide an operational semantics for the function symbols that occur in lambda terms. Instead, when we use a lambda term in a specification, we assume that, when used as an *SML* program, its function symbols correspond to functions which, when executed by a standard *SML* compiler, evaluate according to the specification. In particular, evaluation preserves equality. That is to say, if `f` evaluates to `p`, then $f = p$ is true in the models for the specification.

Example 9.1. For example, take the lambda term $fn\ y : Nat => x + y$ in the basic specification:

spec NAT_0 =
sorts
  Nat
ops $0 : Nat; s : Nat \to Nat; + : Nat \times Nat \to Nat; f : Nat \to Nat$
preds
  $\geq: Nat \times Nat$
axioms
  $\forall x : Nat \bullet x + 0 = x \forall x; y : Nat \bullet x + s(y) = s(x + y) \forall x : Nat \bullet f(x) =$
    $fn\ y : Nat => x + y$
end

We presume that, when treated as an *SML* program, $fn\ y : Nat => x + y$ will behave according to the reduction rules and that the function symbol + will be according to the axioms of the specification.

Remark 9.4. A standard *SML* compiler is equipped with a denotational semantics that is compatible with these rules (see, e.g., [MTH90]).

$$
\begin{array}{r}
(fn\, x : s => p)\ a \rhd_{SML} p[a/x] \\
match\ Inl(a)\ with\ Inl(x) => b \mid Inr(y) => c \rhd_{SML} b[a/x] \\
match\ Inr(a)\ with\ Inl(x) => b \mid Inr(y) => c \rhd_{SML} c[a/y] \\
fst(a, b) \rhd_{SML} a \\
snd(a, b) \rhd_{SML} b
\end{array}
$$

Fig. 9.3. The operational semantics of lambda calculus.

Assume a list of constructors for a sort s, $\overline{cons}$, defined

$$
[c_1 : s, \ldots, c_n : s, \quad f_1 : (s_{1_1} \times \ldots \times s_{m_1}) \to s, \ldots, \quad f_p : (s_{1_p} \times \ldots \times s_{m_p}) \to s]
$$

and a list of terms, $\overline{arg}$, defined

$$[a_1, \ldots, a_n, b_1, \ldots, b_p]$$

Then $rec(\overline{cons}, \overline{arg})$ reduces as follows.

$$rec(\overline{cons}, \overline{arg})c_i \rhd_{SML} a_i$$

for each $c_1, \ldots, c_n$.
For $f_i : (s_{1_i} \times \ldots \times s_{m_i})$ such that each $s_{j_i} \neq s$,

$$rec(\overline{cons}, \overline{arg}) f_i(d_{1_i}, \ldots, d_{m_i}) \rhd_{SML} b_i d_{1_i} \ \ldots \ d_{m_i}$$

For $f_i : (s_{1_i} \times \ldots \times s_{m_i})$ with $[d_1, \ldots, d_k]$ the ordered list obtained from

$$\{d_{j_i} \mid s_{j_i} = s\}_{j_i = 1_i, \ldots, m_i}$$

$$
rec(\overline{cons}, \overline{arg}) f_i(d_{1_i}, \ldots, d_{m_i}) \rhd_{SML} b_i d_{1_i} \ \ldots \ d_{m_i} (rec(\overline{cons}, \overline{arg}) d_1) \ldots (rec(\overline{cons}, \overline{arg}) d_k)
$$

Fig. 9.4. The operational semantics of recursion operators in the lambda calculus.

9.1.6 Using *SML* lambda terms in *CASL*

The *CASL* specification document [CoF01] does not use higher-order sorts. That is, we are not permitted to have a function symbol that accepts another function (non-constant) symbol as an argument.

For the remainder of this monograph we will relax this requirement.

It will also be necessary to add extra-logical axioms to *CASL* specifications, so that we can reason about lambda terms. For convenience, we will use the schemata of Fig. 9.5 to generate these axioms.

$$\frac{}{\vdash_{\mathsf{Int}} u =_s r \Rightarrow r =_s u} \text{(ref)}$$

where s is a basic sort

$$\frac{P[r/y] \wedge u =_s r}{P[u/y]} \text{(subst)}[[P];[u;r];[s]]$$

where u and r are well-sorted of *basic sort* s and y is the only free variable in P

$$\frac{\vdash_{\mathsf{Int}} \forall y_1 : s_1 \bullet P[\mathit{Inl}(y_1)/x] \wedge \forall y_2 : s_2 \bullet P[\mathit{Inr}(y_2)/x]}{\vdash_{\mathsf{Int}} \forall x : s_1|s_2 \bullet P} \text{(disj-ind)}[P;[s_1;s_2]]$$

$$\frac{\vdash_{\mathsf{Int}} \mathit{Inl}(u) = \mathit{Inl}(r)}{\vdash_{\mathsf{Int}} u =_{s_1} r} \text{(union=}_1\text{)}[[u;r];[s_1;s_2]]$$

where $\mathit{Inl}(u)$ and $\mathit{Inl}(r)$ are well-sorted terms of sort $(s_1|s_2)$

$$\frac{\vdash_{\mathsf{Int}} \mathit{Inr}(u) = \mathit{Inr}(r)}{\vdash_{\mathsf{Int}} u =_{s_1} r} \text{(union=}_2\text{)}[[u;r];[s_1;s_2]]$$

where $\mathit{Inr}(u)$ and $\mathit{Inr}(r)$ are well-sorted terms of sort $(s_1|s_2)$

$$\frac{}{\vdash_{\mathsf{Int}} \mathit{Inl}(u) = \mathit{Inr}(r) \Rightarrow \bot} \text{(union}\neq\text{)}[[u;r];[s_1;s_2]]$$

where u and r are well-sorted terms of sorts s_1 and s_2 respectively

Fig. 9.5. Equality schemata and schemata for reasoning about disjoint unions.

Remark 9.5. Because formulae may now involve lambda terms from and higher-order types, we have equalities between functions. For example, we can have a specification with an axiom

$$f =_{nat \to nat} \mathit{fn}\ x : nat => x + x$$

We will assume that the semantics for these equalities is extensional. For example, we take the above axiom to mean that $f(y)$ denotes the same value as $(\mathit{fn}\ x : nat => x+x\ y)$ for every natural number y. These equalities will be particularly useful for providing definitions of functions in terms of *SML* lambda terms.

9.1.7 Semantics of extended specifications

CASL specifications extended with *SML* terms can still be viewed as using first-order signatures [HS96, pp. 47–48]. This is possible by translating our extended signatures into first-order signatures that involve combinators (see, e.g.,

[Sim00, pp. 43–45]). We represent the lambda term formation operators by appropriately typed I, S, K and application operators, with appropriate combinator axioms added to every specification [HS86]. Our lambda calculus and its axioms are then taken as syntactic sugar for the resulting system.

From the perspective of semantics, our extensions mean that we now assume any models for a *CASL* specification are also models for a lambda calculus, so that equality between interpretations of lambda terms coincides with $\triangleright_{SML}$ reduction. There are several possible approaches to modelling the lambda calculus. We do not detail a specific approach here, but instead leave the choice open: the results to follow are not dependent on the choice of models for the lambda calculus. See, e.g., [Bar90, pp. 337–347] for a discussion on denotational semantics for the lambda calculus.

Because we can view our extensions as still retaining first-order signatures, our understanding of specification building operations need not be altered. More complex higher-order models can be developed, but it is not necessary to consider these for our program extraction purposes.

9.2 Realizability

We now define a notion of realizability. A *SML* term is *correct* with respect to a specification/formula pair $\text{SP} \diamond A$ when it is a realizer for SP and A. In the next two sections, we will be concerned with the correct synthesis of *SML* terms, by extraction of realizers.

We will define two kinds of realizers: modular realizers and extended realizers. The latter kind subsumes the former kind.

Modular realizability is essentially the same concept as modified realizability for intuitionistic logic, but we now need to include *CASL* specifications. A lambda term is a modular realizer of a formula and specification when it can stand for the Skolem function of the Skolem form of the formula in a proof about the specification.

In the next section, we derive an important extraction result: that we can extract a modular realizer for $\text{SP} \diamond A$ from a proof of this pair. However, as we shall see, that result only works for a subset of SSL proofs, due to complications that arise from application of the (hide) rule.

To provide a more general extraction result for all proofs, we require extended realizability. This weakens modular realizability for $\text{SP} \diamond A$: a lambda term an extended realizer of a formula and specification when it is a modified realizer for the formula and an extension of the specification. In section 9.4, we shall use the result on extracting modular realizers to extract refinement realizers from *all* SSL proofs.

9.2.1 Skolemization

To define our notions of realizability, we use the Skolem form of formulae.

F	$\mathsf{etype}(F)$
any Harrop formula	$Unit$
$(A \wedge B)$	$\begin{cases} \mathsf{etype}(A) & \text{if not } H(B) \\ \mathsf{etype}(B) & \text{if not } H(A) \\ \mathsf{etype}(A) * \mathsf{etype}(B) & \text{otherwise} \end{cases}$
$(A \vee B)$	$\mathsf{etype}(A)\vert\mathsf{etype}(B)$
$(A \Rightarrow B)$	$\begin{cases} \mathsf{etype}(B) & \text{if not } H(A) \\ \mathsf{etype}(A) \to \mathsf{etype}(B) & \text{otherwise} \end{cases}$
$(\forall x : s \bullet A)$	$s \to \mathsf{etype}(A)$
$(\exists x : s \bullet A)$	$\begin{cases} s & \text{if } H(A) \\ s * \mathsf{etype}(A) & \text{otherwise} \end{cases}$
$\bot$	$Unit$

P is an atomic predicate.

Fig. 9.6. Inductive definition of etype.

We need the definition of Harrop formulae, given by Definition 6.1.1 of Chapter 2, Part II, p. 33.

We also need to define a sort extraction map xsort from formulae to sorts of $\mathsf{C}(SML)$. This the same map as given by in Fig. 2.9 of Chapter 2, Part II. For completeness, we repeat the definition in Fig. 9.6.

An analogous result to Theorem 6.2.2 of Chapter 6, Part III holds.

Theorem 9.2.1. *Take any proof*

$$\vdash_{LTT(\mathsf{SSL})} d^{\mathsf{SP} \diamond A}$$

Then

$$\vdash_{SML} \mathsf{extract}_{\mathsf{mod}}(d) : \mathsf{etype} A$$

is a correct type inference.

Proof. The proof is by induction on the possible forms of d, and is similar to that for Theorem 6.2.2 of Chapter 6, Part III. □

We define the Skolem form of a formula, in the same way as we did for formulae of Chapter 2. The definition is repeated here for completeness.

Definition 9.2.2 (Skolem form and Skolem functions). Given a closed formula A, we define the *Skolem form* of A to be the Harrop formula $Sk(A) = Sk'(A, \emptyset)$, where $Sk'(A, AV)$ is defined as follows.

A unique function letter f_A, called the *Skolem function*, is associated with each such formula A, of sort $\mathsf{etype}(A)$. AV represents a list of application variables for A (that is, the variables that will be arguments of f_A). If AV is $\{x_1 : s_1, \ldots, x_n : s_n\}$ then $f(AV)$ stands for the function application $app(f, (x_1, \ldots, x_n))$.

1. If A is Harrop, then $Sk'(A, AV) = A$.
2. If $A = (B \vee C)$, then

$$\begin{aligned} Sk'(A, AV) = {} & (\forall x : \mathsf{etype}(B) \bullet f_A(AV) = Inl(x) \Rightarrow Sk'(B, AV)[x/f_B]) \\ & \wedge(\forall y : \mathsf{etype}(C) \bullet f_A(AV) = Inr(y) \Rightarrow Sk'(C, AV)[y/f_C]) \end{aligned}$$

3. If $A = (B \wedge C)$, then
   a) If B is Harrop and C is not Harrop,

$$Sk'(A, AV) = B \wedge Sk'(C, AV)[snd(f_A)/f_C]$$

   b) If B is not Harrop and C is Harrop,

$$Sk'(A, AV) = (Sk'(B, AV)[fst(f_A)/f_B] \wedge C)$$

   c) If B and C are not Harrop,

$$Sk'(A, AV) = (Sk'(B, AV)[fst(f_A)/f_B] \wedge Sk'(C, AV)[snd(f_A)/f_C])$$

4. If $A = (B \Rightarrow C)$, then
   a) If B is Harrop,

$$Sk'(A, AV) = (B \Rightarrow Sk'(C, AV)[f_A/f_C])$$

   b) If B is not Harrop and C is not Harrop,

$$\begin{aligned} Sk'(A, AV) = {} & \forall x : \mathsf{etype}(B) \bullet (Sk'(B, AV)[x/f_B] \Rightarrow \\ & Sk'(C, AV)[(f_A x)/f_C]) \end{aligned}$$

5. If $A = \exists y : s \bullet P$, then
   a) when P is Harrop, $Sk'(A, AV) = Sk'(P, AV)[f_A(AV)/y]$.
   b) when P is not Harrop,

$$Sk'(A, AV) = Sk'(P, AV)[fst(f_A(AV))/y][snd(f_A(AV))/f_P]$$

6. If $A = \forall x : s \bullet P$, then $Sk'(A, AV) = \forall x : s \bullet Sk'(P, AV)[(f_A x)/f_P]$.

We will require the following Lemma.

Lemma 9.2.1. *For any formula G, any term t and any specification* SP, *if* $Sig(\text{SP})$ *contains* $Sig(G)$, *we can prove*

$$Sk(G)[t/f_G] \vdash_{\mathsf{SSL}} \text{SP} \diamond G$$

Proof. By induction on the form of G. □

9.2.2 Modular realizability

As in the case of modified realizability for intuitionistic logic, a lambda term is a realizer for a formula when it can be used in place of the Skolem function in the Skolem form of the formula, with the result being true (see Definition 2.5.2 of Chapter 2, Part II, p. 46). The only difference is that we now need to take into account the fact that our proofs are abóut *CASL* specifications. So we define a notion of SSL-realizability that is given over pairs of *CASL* specifications and formulae.

Definition 9.2.3 (Modular realizer). Given a specification SP and a formula A, r is a *modular realizer* of $\text{SP} \diamond A$ if, and only if, r can be used for the Skolem function f_A in the Skolem form $Sk(A)$, with the result being true of the models of SP

$$\text{SP} \models Sk(A)[r/f_A]$$

Remark 9.6. A modular realizer r for $\text{SP} \diamond A$ must be a term of $Term(\text{Sig}(\text{SP}), Var)$. This is because $\text{SP} \models Sk(A)[r/f_A]$ is true, and this statement is well-formed only when all terms in $Sk(A)[r/f_A]$ symbols in Sig(SP) (This follows from Definition 7.1.15 of Chapter 7, p. 224, where $\models$ is defined to hold only between models and formulae of the same signature.)

Remark 9.7. This definition of realizability is similar to that of modified realizability for intuitionistic logic, but now extended to proofs about specifications.

Because we extend modified realizers from single formulae to pairs of specifications and formulae, our approach is comparable to that of Chapter 6 in Part III, where modified realizability was adapted to define return value realizers for pairs of programs and formulae of Hoare logic. However, note that, just as with return value realizability, we define our realizers with respect to semantic truth. This is in contrast to modified realizability of intuitionistic logic, which was given with respect to provability of the Skolem form within the calculus. Our reason for this is that it facilitates an easier approach to proving correctness of extraction from proofs that involve induction.

9.2.3 Extended realizers

To define our notions of extended realizability, we require a notion of specification extensions. We choose a simple notion of model inclusion for extension.

Definition 9.2.4 (Specification extension). A specification SP_1 is said to be an *extension* of a specification SP (written $\text{SP} \gg \text{SP\_1}$) if all models of SP_1 (restricted to Sig(SP)) are also models of SP — that is, when every $C \in Mod(\text{SP\_1})$ is such that $C|_{\text{Sig(SP)}} \in Mod(\text{SP})$.

We say an extension $\text{SP} \gg \text{SP\_1}$ is *relatively consistent* if, assuming SP is consistent, then so is SP_1.

We now define a second concept of realizability, based on modular realizability for extensions of specifications.

Definition 9.2.5 (Extended realizer). Given a specification SP and a formula A, r is an *extended realizer* of $\text{SP} \diamond A$ if, and only if,

$$\text{SP}' \models Sk(A)[r/f_A]$$

is true for some relatively consistent extension SP' of SP.

In this case, we write r extr $\text{SP} \diamond A$.

9.3 Extracting modular realizers

In this section, we provide one of the main results of this part of the book: there is an extraction map from proof-terms of the logical type theory to the *SML* lambda terms of the computational type theory, generating modular realizers from proofs of specifications. This is the idea of structured proofs-as-programs.

Because our logical type theory is an extension of intuitionistic type theory, the extraction map extends the extraction map for intuitionistic proof-terms. Proof-terms corresponding to structural rules require some care.

In the case of unions, we can discard the structural rule altogether — it does not affect the constructive content.

Renamings must be carried out over extracted programs.

However, in the case of hiding, we need to be careful. Our notion of modular realizability means that a realizer of $\text{SP} \diamond P$ is written as a term of Sig(SP) (see Remark 9.6). Therefore we cannot extract programs that use hidden symbols of a specification, because these programs cannot be spoken about in the specification.

Consequently, in this section, we will prove that we can extract modular realizers for a subset of SSL proofs, called modular proofs. In the next section, we will describe a method for extracting *extended* realizers from all SSL proofs. That method is based upon the definitions and results of this section.

9.3.1 The extraction map

The extraction map $\mathsf{extract}_{\mathsf{SSL}}$ is defined in Fig. 9.7. It extracts *SML* lambda terms from proof-terms of $LTT(\mathsf{SSL})$.

The map presumes a set of variables in Var, each corresponding to a proof-term variable from $Var_{PT(\mathsf{SSL})}$,

$$\{x_u \mid u \in Var_{PT(\mathsf{Int})}\}$$

Because the proof-terms for logical rules are similar to those for intuitionistic logic, our extraction map is based on the map for intuitionistic logic in Chapter 2.

A lambda term is a modular realizer when it provides the constructive, computational content for a formula. Thus the idea of our map is to extract

p^P	$\mathsf{extract}_{\mathsf{mod}}(p^P)$
any proof-term where $H(P)$	`()`
u^A	x_u not $H(A)$ `()` $H(A)$
$\mathsf{abstract}\ u^A.\ a^B$	`fn` $\mathtt{x_u}$ `=>` $\mathsf{extract}_{\mathsf{mod}}(a)$ not $H(A)$ $\mathsf{extract}_{\mathsf{mod}}(a)$ $H(A)$
$\mathsf{app}(c^{A \Rightarrow B}, a^A)$	$\mathsf{extract}_{\mathsf{mod}}(c)$ $H(A)$ $(\mathsf{extract}_{\mathsf{mod}}(c)\ \mathsf{extract}_{\mathsf{mod}}(a))$ not $H(A)$
$\mathsf{use}\ x : s.\ a^A$	`fn x : s =>` $\mathsf{extract}_{\mathsf{mod}}(a)$
$\mathsf{specific}(a^{\forall x:s \bullet A}, v)$	$(\mathsf{extract}_{\mathsf{mod}}(a)\ \mathtt{v})$
$\langle a^A, b^B \rangle$	$(\mathsf{extract}_{\mathsf{mod}}(a), \mathsf{extract}_{\mathsf{mod}}(b))$
$\mathsf{case}\ a^{A \vee B}\ \mathsf{of}\ \mathsf{inl}(t^A).b^C,\ \mathsf{inr}(u^B).c^C$	`match` $\mathsf{extract}_{\mathsf{mod}}(a)$ `with` $\mathtt{Inl}(\mathtt{x_t})$ `=>` $\mathsf{extract}_{\mathsf{mod}}(b)$, $\mathtt{Inr}(\mathtt{x_u})$ `=>` $\mathsf{extract}_{\mathsf{mod}}(c)$
$\mathsf{show}(v, a^A)$	`v` $H(A)$ $(\mathtt{v}, \mathsf{extract}_{\mathsf{mod}}(a))$ not $H(A)$
$\mathsf{select}\ (a^{\exists y \bullet A})\ \mathsf{in}\ x.u^{A[x/y]}.b^B$	(`fn x =>` $\mathsf{extract}_{\mathsf{mod}}(b)$) $\mathsf{extract}_{\mathsf{mod}}(a)$ } $H(A)$ (`fn x =>` `fn` $\mathtt{x_u}$ `=>` $\mathsf{extract}_{\mathsf{mod}}(b)$) `fst`($\mathsf{extract}_{\mathsf{mod}}(a)$) `snd`($\mathsf{extract}_{\mathsf{mod}}(a)$) } not $H(A)$
$\mathsf{inl}(a)$	`Inl`($\mathsf{extract}_{\mathsf{mod}}(a)$)
$\mathsf{inr}(a)$	`Inr`($\mathsf{extract}_{\mathsf{mod}}(a)$)
$\mathsf{fst}(a)$	`fst`($\mathsf{extract}_{\mathsf{mod}}(a)$)
$\mathsf{snd}(a)$	`snd`($\mathsf{extract}_{\mathsf{mod}}(a)$)
$\mathsf{abort}(a^\perp)$	`()`
$\mathsf{rec}(Cons_s, s, [a_1; \ldots; a_n; b_1; \ldots; b_p])$	$rec(\overline{cons}, \overline{arg})$
$\mathsf{union}_1(d^{\mathrm{SP} \diamond D}, \mathrm{SP})$	$\mathsf{extract}_{\mathsf{mod}}(d)$
$\mathsf{union}_2(d^{\mathrm{SP} \diamond D}, \mathrm{SP})$	$\mathsf{extract}_{\mathsf{mod}}(d)$
$\mathsf{ext}_1(d^{\mathrm{SP} \diamond D}, \mathrm{SP})$	$\mathsf{extract}_{\mathsf{mod}}(d)$
$\mathsf{ext}_2(d^{\mathrm{SP} \diamond D}, \mathrm{SP})$	$\mathsf{extract}_{\mathsf{mod}}(d)$
$\rho \bullet d^{\mathrm{SP} \diamond D}$	$\rho(\mathsf{extract}_{\mathsf{mod}}(d))$
$\mathsf{hide}(d^{\mathrm{SP} \diamond D}, SL)$	$\mathsf{extract}_{\mathsf{mod}}(d)$

Fig. 9.7. The extraction map $\mathsf{extract}_{\mathsf{SSL}}$, defined over the proof-terms of $LTT(\mathsf{SSL})$ to the lambda terms of $\mathsf{C}(SML)$.

this content from the proof-term of a proved specification/formula pair. Harrop formulae have no computational content and so are systematically ignored, while computational content for non-Harrop formulae is mapped to lambda terms. The result yields terms in *SML* that are simply typed (with disjoint unions and products).

Also, we remove much structural information from the proof-term. In particular:

1. Translations of symbols is carried out explicitly,
2. the structural information of taking unions and extending specifications can be ignored, and
3. applications of the hiding rule are ignored — but see below for a discussion on the problem extraction of realizers from proof-terms with hiding, and Section 9.4 for a solution to this problem.

Remark 9.8. Because we have assumed that all axioms available in a specification are Harrop, we always extract the term () from applications of an axiom introduction $\mathsf{ax}(\text{SP}, A)^{\text{SP} \diamond A}$ (by the first case of the table in Fig. 9.7).

Remark 9.9. Recall that, following [Pet96, WCP98, CPW00, PCW02], we use $Sig(\text{SP\_1}) \cup Sig(\text{SP\_2})$ as the standard representation of the isomorphism class of $Sig(\text{SP\_1 } \mathbf{and} \text{ SP\_2})$. This simplifies the presentation of the (union_i), (ext_i) $(i = 1, 2)$ cases, since we do not have to write the embedding morphisms *inl* and *inr* explicitly. So, we do not need these morphisms in our extraction map or in our proof that extraction produces modular realizers.

However, if we wished to, we could avoid this representation, and redefine extraction over union proof-terms as follows:

$$\mathsf{extract}_{\mathsf{mod}}(\mathsf{union}_1(d^{\text{SP\_1} \diamond A}, \text{SP\_2})) = inl(\mathsf{extract}_{\mathsf{mod}}(d))$$
$$\mathsf{extract}_{\mathsf{mod}}(\mathsf{union}_2(d^{\text{SP\_2} \diamond B}, \text{SP\_1})) = inr(\mathsf{extract}_{\mathsf{mod}}(d))$$

where *inl* and *inr* are the embedding morphisms for the pushout construction for amalgamated unions of signatures of SP_1 and SP_2.

9.3.2 Extracting modular realizers from modular proofs

We would like to synthesize a modular realizer of a specification/formula pair from a proof of the pair by means of our extraction map.

Unfortunately, this is not possible for all proofs. We cannot obtain modular realizers for certain applications of the hide rule.

Example 9.2. For instance, recall the password checking system of the previous two chapters. An initial specification of the system's password requirements, PWDCORE, was given in Example 7.6, Chapter 7, p. 234. The function symbols $\{ge, inRange\}$ of PWDCORE were relevant to the internal specification of the system. However, it is not desirable to expose these symbols when specifying the external functionality of the system. By hiding these functions, we restricted the specification of the system to relevant functionality, obtaining a final, encapsulating specification PWDSYS.

We proved that, given any numerical password entered to the system, an appropriate message can be output. The theorem was obtained by proving

$$\vdash \text{PWDCORE} \diamond \forall x : nat \bullet \exists y : string \bullet ValidMsg(x, y)$$

and then applying (hide) to obtain

$$\vdash \textsc{PwdSys} \diamond \forall x : nat \bullet \exists y : string \bullet ValidMsg(x, y)$$

When encoded in the logical type theory, we obtained a proof-term for the final theorem of the form

$$\begin{aligned}
p = {} & \mathsf{hide}(\mathsf{use}\ x : nat.\ \mathsf{case} \\
& \quad (\mathsf{specific}(\mathsf{ext}_1(\mathsf{union}_1(BtoB \bullet (\mathsf{rec}([T, F], boolean, \\
& \quad\quad [(\mathsf{inr}(\mathsf{ax}(\textsc{NatBoolean}, T = T))), \\
& \quad (\mathsf{inl}(\mathsf{ax}(\textsc{NatBoolean}, T = T)))])), \textsc{StringBool}), \\
& \quad\quad \langle SExt, AExt \rangle), inRange(x))) \\
& \quad \mathsf{of}\ \mathsf{inl}(u).\mathsf{show}(\text{`Password acceptable'}, \mathsf{app}(p_7, \mathsf{app}(p_6, p_5))), \\
& \quad \mathsf{inr}(v).\mathsf{show}(\text{`Please choose a password in correct range'}, p_8), \\
& \quad\quad \{ge, inRange\})
\end{aligned}$$

If we were to extract the program for this term using the extraction map as it is defined, ignoring the application of hide, we would obtain the term

$$\begin{aligned}
& k = fn\ x : nat \Rightarrow \\
& match\ rec([true, false], [inr(()), inl(())])inRange(x)\ with \\
& \quad Inl(x_u) \Rightarrow \text{`Password acceptable'}, \\
& \quad Inr(x_v) \Rightarrow \text{`Please choose a password in correct range'}
\end{aligned}$$

Now *inRange* is not in the specification PwdSys, and therefore is not a valid modular realizer of

$$\textsc{PwdSys} \diamond \forall x : nat \bullet \exists y : string \bullet ValidMsg(x, y)$$

because it is not possible to obtain the Skolem form

$$\textsc{PwdSys} \models \forall x : nat \bullet ValidMsg(x, (k\ x))$$

with k for the Skolem function.

We prove that extraction of realizers is guaranteed to be possible if the proof-term is modular according to the following definition.

Definition 9.3.1. A proof-term d is said to be *non-modular* with respect to a symbol list SL if

- d is of the form $\mathsf{hide}(e^{\textsc{Sp} \diamond A}, SL)$ and
- if the term e depends on symbols that are in SL: that is, if $\mathrm{Sig}(\textsc{Sp}) \cap SL \neq \emptyset$.

A proof-term is *modular* if it contains no non-modular subterms.

Remark 9.10. In practice, non-modular proof-terms often occur in the construction of specifications in SSL. This holds because we use functions from other specifications to prove a required result. These functions are often hidden, to aid comprehensibility and encapsulation in the resulting specification signature. Consequently, it is important to find a means of extracting correct programs from non-modular proof-terms. We do this in the next section.

We now prove that we can extract realizers from modular proof-terms.

Theorem 9.3.2. *Take any set of typed proof-terms* $\Gamma = \{u_1^{G_1}, \ldots, u_n^{G_n}\}$. *We define* Γ' *to the corresponding set of Skolemized formulae*

$$\Gamma' = \{Sk(G_1)[x_{u_1}/f_{G_1}], \ldots, Sk(G_n)[x_{u_n}/f_{G_n}]\}$$

If there is a well-formed modular *proof-term* p *for the proof of* $\text{SP} \diamond P$

$$\Gamma \vdash_{LTT(\mathsf{SSL})} p^{\text{SP}\diamond P}$$

then

$$\text{SP}, \Gamma' \models Sk(P)[\mathsf{extract}_{\mathsf{mod}}(p)/f_P]$$

Proof. By induction on the length of the proof.

We can assume that P is not Harrop, because if it is, then $Sk(P)$ is P and we are done. This covers the case of axiom introduction, as, the conclusion formula P is Harrop (by assumption 7.2) and so $Sk(P)[\mathsf{extract}_{\mathsf{mod}}(p)/f_P]$ is simply P.

We proceed as follows for proofs ending in structural rules.

Case: (trans). Assume $p^{\text{SP}\diamond P}$ is of the form $\rho \bullet d^{(\text{SA } \mathbf{with}\ \rho)\diamond(\rho\bullet(A))}$, derived by an application of (trans):

$$\frac{\Delta \vdash d^{\text{SA}\diamond A}}{\rho^{`}(\Delta) \vdash \rho \bullet d^{(\text{SA } \mathbf{with}\ \rho)\diamond(\rho\bullet(A))}}\ (\text{trans})$$

so that Δ must be of the form $\{u_1^{D_1}, \ldots, u_n^{D_n}\}$ with Γ set to be $\{u_1^{\rho\bullet D_1}, \ldots, u_n^{\rho\bullet D_n}\}$. This means that Γ' is

$$\rho^{`}(\Delta') = \{\rho \bullet Sk(D_1)[x_{u_1}/f_{D_1}], \ldots, \rho \bullet Sk(D_n)[x_{u_n}/f_{D_n}]\}$$

By the IH,

$$\text{SA}, \Delta' \models Sk(A)[\mathsf{extract}_{\mathsf{mod}}(d)/f_A]$$

By the semantics of translation, it can be seen that this entails

$$(\text{SA } \mathbf{with}\ \rho), \rho^{`}(\Delta') \models \rho \bullet Sk(A)[\mathsf{extract}_{\mathsf{mod}}(d)/f_A]$$

Because Γ' is $\rho^{`}(\Delta')$ and $Sk(P)[\mathsf{extract}_{\mathsf{mod}}(p)/f_P]$ is

$$Sk(\rho \bullet A)[\rho(\mathsf{extract}_{\mathsf{mod}}(d))/f_A] = \rho \bullet Sk(A)[\mathsf{extract}_{\mathsf{mod}}(d)/f_A]$$

we are done.

Case: (union$_1$). Assume $p^{\mathrm{SP}\diamond P}$ is of the form $\mathsf{union}_1(d, \mathrm{SP\_2})^{\mathrm{SP\_1}\ \mathbf{and}\ \mathrm{SP\_2}\diamond A}$, derived from a proof of the form

$$\frac{\Gamma \vdash d^{\mathrm{SP\_1}\diamond A}}{\Gamma \vdash \mathsf{union}_1(d, \mathrm{SP\_2})^{(\mathrm{SP\_1}\ \mathbf{and}\ \mathrm{SP\_2})\diamond A}}\ union_1$$

By the IH,

$$\mathrm{SP\_1}, \Gamma' \models Sk(A)[\mathsf{extract}_{\mathsf{mod}}(d)/f_A] \tag{9.1}$$

But, by definition, $\mathsf{extract}_{\mathsf{mod}}(p)$ is $\mathsf{extract}_{\mathsf{mod}}(\mathsf{union}_1(d, \mathrm{SP\_2})) = \mathsf{extract}_{\mathsf{mod}}(d)$, and so (9.1) is the same as writing

$$\mathrm{SP\_1}, \Gamma' \models Sk(A)[\mathsf{extract}_{\mathsf{mod}}(p)/f_A] \tag{9.2}$$

By the semantics for unions it then follows that

$$(\mathrm{SP\_1}\ \mathbf{and}\ \mathrm{SP\_2}), \Gamma' \models Sk(A)[\mathsf{extract}_{\mathsf{mod}}(p)/f_A]$$

as required.

The cases of (union$_2$), (ext$_1$), (ext$_2$) (*hide*) are similar to (union$_1$). Because the proof-term is modular, there are no complications with (*hide*).

We proceed as follows for proofs ending in logical rules. Because these cases involve proof-terms that are the same as those of $LTT(\mathsf{Int})$ of Chapter 2, and the extracted programs are identical, we use similar arguments for Int, only augmented to accommodate structured specifications in our notion of validity of Skolem forms:

$$\mathrm{SP}, \Gamma' \models Sk(P)[\mathsf{extract}_{\mathsf{mod}}(p)/f_P]$$

Case: (Ass-I). Assume that $p^{\mathrm{SP}\diamond P}$ is of the form $u^{\mathrm{SP}\diamond A}$, obtained by an application of (Ass-I):

$$\frac{}{u^A \vdash u^{\langle \mathrm{Sig(A)}, \emptyset\rangle \diamond A}}\ (\text{Ass-I})$$

So $\Gamma' = \{Sk(A)[x_u/f_A]\}$ and we can prove

$$\frac{}{Sk(A)[x_u/f_A] \vdash \langle \mathrm{Sig}(Sk(A)[x_u/f_A]), \emptyset\rangle \diamond Sk(A)[x_u/f_A]}\ (\text{Ass-I})$$

It is then easy to show that

$$\mathrm{sp}(u') = \langle \mathrm{Sig}(Sk(A)[x_u/f_A]), \emptyset\rangle = \langle \mathrm{Sig}(A), \emptyset\rangle = \mathrm{sp}(u)$$

Then, by soundness, we are done.

Case: ($\wedge$-I). Assume that $p^{\mathrm{SP}\diamond P}$ is of the form

$$\langle a, b\rangle^{\mathrm{SP}\diamond A\wedge B}$$

obtained by an application of ($\wedge$-I):

$$\frac{\Gamma_1 \vdash a^{\mathrm{SP}\diamond A} \quad \Gamma_2 \vdash b^{\mathrm{SP}\diamond A}}{\Gamma_1, \Gamma_2 \vdash \langle a, b\rangle^{\mathrm{SP}\diamond A\wedge B}}\ (\wedge\text{-I})$$

so that $\Gamma' = \Gamma_1' \cup \Gamma_2'$.

Because we assume that P is not Harrop, either

1. A and B are both non-Harrop.
2. A is Harrop and B is non-Harrop.
3. A is non-Harrop and B is Harrop.

We deal only with the first case, as the other two cases are similar. Here, $\mathsf{extract_{mod}}(p)$ is $(\mathsf{extract_{mod}}(a), \mathsf{extract_{mod}}(b))$ and $Sk(P)[\mathsf{extract_{mod}}(p)/f_P]$ is

$$Sk(A)[Sk(A)[fst(\mathsf{extract_{mod}}(p))/f_A] \wedge Sk(B)[snd(\mathsf{extract_{mod}}(p))/f_A]$$

So, by the IH, we know

$$\begin{array}{l} \mathrm{S}\mathrm{P}, \Gamma_1' \models Sk(A)[\mathsf{extract_{mod}}(a)/f_A] \\ \mathrm{S}\mathrm{P}, \Gamma_2' \models Sk(B)[\mathsf{extract_{mod}}(b)/f_B] \end{array}$$

By the definition of $\models$, it follows from this that

$$\mathrm{S}\mathrm{P}, \Gamma' \models Sk(A)[\mathsf{extract_{mod}}(a)/f_A] \wedge Sk(B)[\mathsf{extract_{mod}}(b)/f_B] \tag{9.3}$$

Because we assume all specifications models include a model of the lambda calculus with equality preserving reduction, we know that

$$\mathrm{S}\mathrm{P} \models fst(\mathsf{extract_{mod}}(a), \mathsf{extract_{mod}}(b)) = \mathsf{extract_{mod}}(a) \tag{9.4}$$

$$\mathrm{S}\mathrm{P} \models snd(\mathsf{extract_{mod}}(a), \mathsf{extract_{mod}}(b)) = \mathsf{extract_{mod}}(b) \tag{9.5}$$

So, (9.4), (9.5) and (9.3) give us

$$\begin{array}{l} \mathrm{S}\mathrm{P}, \Gamma' \models Sk(A)[fst(\mathsf{extract_{mod}}(a), \mathsf{extract_{mod}}(b))] \wedge \\ \qquad\qquad Sk(B)[snd(\mathsf{extract_{mod}}(a), \mathsf{extract_{mod}}(b))/f_A] \end{array}$$

This is the required conclusion, because

$$\begin{array}{l} Sk(A)[fst(\mathsf{extract_{mod}}(a), \mathsf{extract_{mod}}(b))/f_A] \wedge \\ \qquad\qquad Sk(B)[fst(\mathsf{extract_{mod}}(a), \mathsf{extract_{mod}}(b))/f_B] \end{array}$$

is the same as writing

$$Sk(A)[fst(\mathsf{extract_{mod}}(p))/f_A] \wedge Sk(B)[snd(\mathsf{extract_{mod}}(p))/f_B]$$

Case: ($\wedge$-E$_1$). Assume that $p^{\mathrm{S}\mathrm{P} \diamond P}$ is of the form

$$\mathsf{fst}(q)^{\mathrm{S}\mathrm{P} \diamond A}$$

obtained by an application of ($\wedge$-E$_1$):

$$\frac{\Gamma \vdash_{\mathsf{SSL}} q^{\mathrm{S}\mathrm{P} \diamond A \wedge B}}{\Gamma \vdash_{\mathsf{SSL}} \mathsf{fst}(q)^{\mathrm{S}\mathrm{P} \diamond A}}\ (\wedge\text{-E}_1)$$

We are required to prove $\mathrm{S}\mathrm{P}, \Gamma' \models Sk(A)[\mathsf{extract_{mod}}(p)/f_A]$.

There are two possible cases: either B is Harrop or B is not Harrop. We reason over these cases.

1. Assume that B is Harrop, so that $Sk(B) = B$. Then, $\mathsf{extract}_{\mathsf{mod}}(p)$ is $\mathsf{extract}_{\mathsf{mod}}(q)$ and we are required to prove $\mathrm{SP}, \Gamma' \models Sk(A)[\mathsf{extract}_{\mathsf{mod}}(q)/f_A]$. By the IH and the fact that $Sk(B) = B$, we know

$$\mathrm{SP}, \Gamma' \models Sk(A)[\mathsf{extract}_{\mathsf{mod}}(q)/f_A] \wedge B$$

By the semantics $\wedge$, it easily follows that

$$\mathrm{SP}, \Gamma' \models Sk(A)[\mathsf{extract}_{\mathsf{mod}}(q)/f_A]$$

as required.

2. Assume that B is not Harrop. Then, $\mathsf{extract}_{\mathsf{mod}}(p)$ is $fst(\mathsf{extract}_{\mathsf{mod}}(q))$ and we are required to show $\mathrm{SP}, \Gamma' \models Sk(A)[fst(\mathsf{extract}_{\mathsf{mod}}(q))/f_A]$.
By the IH, we know

$$\mathrm{SP}, \Gamma' \models Sk(A)[fst(\mathsf{extract}_{\mathsf{mod}}(q))/f_A] \wedge Sk(B)[snd(\mathsf{extract}_{\mathsf{mod}}(q))/f_B]$$

From this, by the semantics for $\wedge$, we obtain the required conclusion

$$\mathrm{SP}, \Gamma' \models Sk(A)[fst(\mathsf{extract}_{\mathsf{mod}}(q))/f_A]$$

Case: ($\wedge$-E$_2$). Similar to the case ($\wedge$-E$_1$) above.
Case: ($\vee$-I$_1$). Assume that $p^{\mathrm{SP} \diamond P}$ is of the form

$$\mathsf{inl}(a)^{\mathrm{SP} \diamond A \vee B}$$

obtained by an application of ($\vee$-I$_1$)

$$\frac{\Gamma \vdash_{\mathsf{SSL}} a^{\mathrm{SP} \diamond A}}{\Gamma \vdash_{\mathsf{SSL}} \mathsf{inl}(a)^{\mathrm{SP} \diamond A \vee B}} \; ((\vee\text{-I}_1)$$

so that $\mathsf{extract}_{\mathsf{mod}}(p)$ is $\mathtt{Inl}(\mathsf{extract}_{\mathsf{mod}}(a))$.

We are required to show that $\mathrm{SP}, \Gamma' \models Sk(P)[\mathsf{extract}_{\mathsf{mod}}(p)/f_P]$ is true. That is, we must prove

$$\mathrm{SP}, \Gamma' \models (\forall x : \mathsf{etype}(A) \bullet Inl(\mathsf{extract}_{\mathsf{mod}}(a)) = Inl(x) \Rightarrow Sk(A)[x/f_A]) \wedge \\ (\forall y : \mathsf{etype}(B) \bullet Inl(\mathsf{extract}_{\mathsf{mod}}(a)) = Inr(y) \Rightarrow Sk(B)[y/f_B])$$

To show this, we take any model $M \in Mod(\mathrm{SP})$ and any interpretation $\hat{\iota} : Var \to M$ such that

$$M \models_\iota G'$$

for each $G' \in \Gamma'$. We try to show

$$M \models_\iota (\forall x : \mathsf{etype}(A) \bullet Inl(\mathsf{extract}_{\mathsf{mod}}(a)) = Inl(x) \Rightarrow Sk(A)[x/f_A]) \wedge \\ (\forall y : \mathsf{etype}(B) \bullet Inl(\mathsf{extract}_{\mathsf{mod}}(a)) = Inr(y) \Rightarrow Sk(B)[y/f_B]) \quad (9.6)$$

By the IH,

$$M \models_\iota Sk(A)[\mathsf{extract_{mod}}(a)/f_A] \tag{9.7}$$

We prove the left hand side of the conjunction (9.6) — the right hand side is similar.

Take any $x : \mathsf{etype}(A)$-variant ι' of ι, and assume

$$M \models_{\iota'} Inl(\mathsf{extract_{mod}}(a)) = Inl(x) \tag{9.8}$$

Because we use a loose semantics (Chapter 7, p. 228, that models the lambda calculus (p. 304 of this chapter), this must mean that

$$M \models_{\iota'} \mathsf{extract_{mod}}(a) = x \tag{9.9}$$

So, (9.7) may be rewritten

$$M \models_{\iota'} Sk(A)[x/f_A] \tag{9.10}$$

Because the assumption (9.8) entails (9.10) for any $x : \mathsf{etype}(A)$-variant ι' of ι we know that

$$M \models_\iota \forall x : \mathsf{etype}(A) \bullet Inl(\mathsf{extract_{mod}}(a)) = Inl(x) \Rightarrow Sk(A)[x/f_A] \tag{9.11}$$

This gives us the left hand side of the required conjunction.

The right hand side of (9.6) is deduced similarly.

Case: ($\vee$-I$_2$). Similar to the ($\vee$-I$_1$) case above.

Case: ($\vee$-E). Assume that $p^{\textsc{Sp}\diamond P}$ is of the form

$$\mathsf{case}\ e\ \mathsf{of}\ \mathsf{inl}(x).a,\ \mathsf{inr}(y).b^{\textsc{Sp}\diamond C}$$

obtained by an application of ($\vee$-E)

$$\frac{\Gamma_1 \vdash_{\mathsf{SSL}} \textsc{Sp} \diamond A \vee B \quad \Gamma_2, u^A \vdash_{\mathsf{SSL}} \textsc{Sp} \diamond C \quad \Gamma_3, v^B \vdash_{\mathsf{SSL}} \textsc{Sp} \diamond C}{\Gamma_1, \Gamma_2, \Gamma_3 \vdash \mathsf{case}\ e\ \mathsf{of}\ \mathsf{inl}(u).a,\ \mathsf{inr}(v).b^{\textsc{Sp}\diamond C}}\ (\vee\text{-E})$$

so that $\Gamma' = \Gamma_1' \cup \Gamma_2 \cup \Gamma_3$, and $\mathsf{extract_{mod}}(p)$ is defined as

$$\begin{array}{l} \mathit{match}\ \mathsf{extract_{mod}}(e)\ \mathit{with} \\ \quad Inl(x_u)\ \Rightarrow \mathsf{extract_{mod}}(a), \\ \quad Inr(x_v)\ \Rightarrow \mathsf{extract_{mod}}(b) \end{array} \tag{9.12}$$

By the IH,

$$\textsc{Sp}, \Gamma_1' \models Sk(A \vee B)[\mathsf{extract_{mod}}(e)/f_{A\vee B}] \tag{9.13}$$

$$\textsc{Sp}, \Gamma_2' \cup \{Sk(A)[x_u/f_A]\} \models Sk(C)[\mathsf{extract_{mod}}(a)/f_C] \tag{9.14}$$

$$\textsc{Sp}, \Gamma_3' \cup \{Sk(B)[x_v/f_B]\} \models Sk(C)[\mathsf{extract_{mod}}(b)/f_C] \tag{9.15}$$

By definition of $Sk(A \vee B)$, (9.13) may be rewritten as:

$$\textsc{Sp}, \Gamma_1' \models N \tag{9.16}$$

where N is

$$(\forall x : \mathsf{etype}(A) \bullet \mathsf{extract}_{\mathsf{mod}}(e) = Inl(x) \Rightarrow Sk(A)[x/f_A]) \wedge \\ (\forall y : \mathsf{etype}(B) \bullet \mathsf{extract}_{\mathsf{mod}}(e) = Inr(y) \Rightarrow Sk(B)[y/f_B])$$

Take any model $M \in Mod(\text{SP})$ and interpretation $\hat{\iota} : Var \to D$ such that $M \models G'$ for each $G' \in \Gamma'$. We want to show that

$$M \models_\iota Sk(C)[\mathsf{extract}_{\mathsf{mod}}(p)/f_C] \tag{9.17}$$

Because we use loose semantics and our models must always model the lambda calculus, our models must model disjoint union sorts given by Inl and Inr constructors. So, we know that, either there is a term v_l such that

$$\iota(\mathsf{extract}_{\mathsf{mod}}(e)) = \iota(Inl(v_l)) \tag{9.18}$$

or else there is a term v_r such that

$$\iota(\mathsf{extract}_{\mathsf{mod}}(e)) = \iota(Inr(v_r)) \tag{9.19}$$

We reason over these two possible cases to establish (9.17)

1. Assume (9.18) holds. It follows then, by instantiating the first conjunct of (9.16) with (9.12), that

$$M \models_\iota Sk(A)[v_l/f_A] \tag{9.20}$$

But then (9.20) and (9.14) give us

$$M \models_\iota Sk(C)[\mathsf{extract}_{\mathsf{mod}}(a)[v_l/x_u]/f_C] \tag{9.21}$$

Because our models use a loose semantics and must preserve the lambda calculus, we have that

$$\iota(\mathsf{extract}_{\mathsf{mod}}(p)) = \iota\begin{pmatrix} match\ \mathsf{extract}_{\mathsf{mod}}(e)\ with \\ \quad Inl(x_u)\ \ => \mathsf{extract}_{\mathsf{mod}}(a), \\ \quad Inr(x_v)\ \ => \mathsf{extract}_{\mathsf{mod}}(b) \end{pmatrix} =$$

$$\iota\begin{pmatrix} match\ Inl(v_l)\ with \\ \quad Inl(x_u)\ \ => \mathsf{extract}_{\mathsf{mod}}(a), \\ \quad Inr(x_v)\ \ => \mathsf{extract}_{\mathsf{mod}}(b) \end{pmatrix} = \iota(\mathsf{extract}_{\mathsf{mod}}(a)[v_l/x_u]) \tag{9.22}$$

So (9.22) and (9.21) entail (9.17)

$$M \models_\iota Sk(C)[\mathsf{extract}_{\mathsf{mod}}(p)/f_C]$$

as required.
2. The case when (9.19) holds is similar.

Case: ($\exists$-I). Assume that the proof-term $p^{\textrm{SP}\diamond P}$ is of the form $\exists x : s \bullet A$ obtained by an application of ($\exists$-I)

$$\frac{\Gamma \vdash a^{\textrm{SP}\diamond A[v/x]}}{\Gamma \vdash \mathsf{show}(v,a)^{\textrm{SP}\diamond \exists x:s\bullet A}}\ (\exists\text{-I}) \tag{9.23}$$

There are two cases, dependent on whether A is Harrop or not.

1. Assume A is Harrop. Then $\mathsf{extract}_{\mathsf{mod}}(p)$ is defined

$$\mathsf{extract}_{\mathsf{mod}}(p) = \mathrm{v} \tag{9.24}$$

Also, because A is Harrop, we know that $Sk(\exists x : s \bullet A)$ is A. This means $Sk(\exists x : s \bullet A)[\mathsf{extract}_{\mathsf{mod}}(p)/f_P]$ is $A[v/f_A]$, and so it is the case that $Sk(\exists x : s \bullet A)[f_A/x][\mathsf{extract}_{\mathsf{mod}}(p)/f_A] = A[v/x]$. Consequently, by (9.23),

$$\textrm{SP}, \Gamma \models Sk(\exists x : s \bullet A)[\mathsf{extract}_{\mathsf{mod}}(p)/f_{\exists x:s\bullet A}]$$

By application of Lemma 9.2.1, we have

$$\textrm{SP}, \Gamma' \models G_i$$

for each $G_i \in \Gamma$ $i = 1, \ldots, n$.
So, we have

$$\textrm{SP}, \Gamma' \models Sk(\exists x : s \bullet A)[\mathsf{extract}_{\mathsf{mod}}(p)/f_{\exists x:s\bullet A}]$$

2. Assume A is not Harrop. Then

$$\mathsf{extract}_{\mathsf{mod}}(p) = (\mathrm{v}, \mathsf{extract}_{\mathsf{mod}}(a)) \tag{9.25}$$

Because A is not Harrop, $Sk(\exists x : s \bullet A)[\mathsf{extract}_{\mathsf{mod}}(p)/f_P]$ is

$$Sk(A)[fst(\mathsf{extract}_{\mathsf{mod}}(p))/x][snd(\mathsf{extract}_{\mathsf{mod}}(p))/f_A]$$

Now, by the IH

$$\textrm{SP}, \Gamma' \models Sk(A)[v/x][\mathsf{extract}_{\mathsf{mod}}(a)/f_A] \tag{9.26}$$

Because our models preserve the lambda calculus

$$\textrm{SP} \models fst(\mathsf{extract}_{\mathsf{mod}}(p)) = v \tag{9.27}$$

and

$$\textrm{SP} \models snd(\mathsf{extract}_{\mathsf{mod}}(p)) = \mathsf{extract}_{\mathsf{mod}}(a) \tag{9.28}$$

The required conclusion follows from (9.26), (9.27) and (9.28)

$$\textrm{SP}, \Gamma' \models Sk(A)[fst(\mathsf{extract}_{\mathsf{mod}}(p))/x][\mathsf{extract}_{\mathsf{mod}}(a)/f_A] \tag{9.29}$$

Case: ($\exists$-E). Assume that $p^{\text{SP}\diamond P}$ is of the form $\mathsf{select}\ (a)\ \mathsf{in}\ x.u.b^{\text{SP}\diamond C}$, obtained by an application of ($\exists$-E)

$$\frac{\Gamma_1 \vdash a^{\text{SP}\diamond \exists y:s \bullet A} \quad \Gamma_2, u^{\text{SP}\diamond A[v/y]} \vdash b^{\text{SP}\diamond C}}{\Gamma_1, \Gamma_2 \vdash \mathsf{select}\ (a)\ \mathsf{in}\ x.u.b^{\text{SP}\diamond C}}\ (\exists\text{-E}) \tag{9.30}$$

So, $\Gamma' = \Gamma_1' \cup \Gamma_2'$.

We take any model $M \in Mod(\text{SP})$ and any interpretation $\hat{\iota} : Var \to M$ such that

$$M \models_\iota G'$$

for each $G' \in \Gamma'$, and show

$$M \models_\iota Sk(C)[\mathsf{extract}_{\mathsf{mod}}(p)/f_C] \tag{9.31}$$

There are two cases, dependent on whether A is Harrop or not.

1. If A is Harrop, then $\mathsf{extract}_{\mathsf{mod}}(p)$ is

$$(fn\ v \Rightarrow \mathsf{extract}_{\mathsf{mod}}(b))\ \mathsf{extract}_{\mathsf{mod}}(a)$$

Because A is Harrop, $Sk(A[v/y])$ is $A[v/y]$. So, by the IH,

$$\text{SP}, \Gamma_2' \cup \{A[v/y]\} \models Sk(C)[\mathsf{extract}_{\mathsf{mod}}(b)/f_C]$$

and so, in particular,

$$M \models_\iota \forall v : s \bullet A[v/y] \Rightarrow Sk(C)[\mathsf{extract}_{\mathsf{mod}}(b)/f_C] \tag{9.32}$$

Because A is Harrop,

$$\begin{aligned} & Sk(\exists y : s \bullet A)[\mathsf{extract}_{\mathsf{mod}}(a)/f_{\exists y:s\bullet A}] \\ = {} & A[f_{\exists y:s\bullet A}/y][\mathsf{extract}_{\mathsf{mod}}(a)/f_{\exists y:s\bullet A}] \\ = {} & A[\mathsf{extract}_{\mathsf{mod}}(a)/y] \end{aligned}$$

So, by the IH

$$\text{SP}, \Gamma_1' \models A[\mathsf{extract}_{\mathsf{mod}}(a)/y]$$

and, in particular,

$$M \models_\iota A[\mathsf{extract}_{\mathsf{mod}}(a)/y] \tag{9.33}$$

By the definition of $\models$, (9.33) and (9.32) entail

$$M \models_\iota Sk(C)[\mathsf{extract}_{\mathsf{mod}}(b)/f_C][\mathsf{extract}_{\mathsf{mod}}(a)/v] \tag{9.34}$$

Because, by definition of the ($\exists$-E) rule, v cannot occur in C, (9.34) means

$$M \models_\iota Sk(C)[(\mathsf{extract}_{\mathsf{mod}}(b)[\mathsf{extract}_{\mathsf{mod}}(a)/v])/f_C] \tag{9.35}$$

Because our models preserve the lambda calculus, and

$$fn\ v => \mathsf{extract_{mod}}(b))\ \mathsf{extract_{mod}}(a) \rhd_{SML} \mathsf{extract_{mod}}(b)[\mathsf{extract_{mod}}(a)/v]$$

we know that

$$\iota(\mathsf{extract_{mod}}(p)) = \iota((fn\ v => \mathsf{extract_{mod}}(b))\ \mathsf{extract_{mod}}(a)) = \iota(\mathsf{extract_{mod}}(b)[\mathsf{extract_{mod}}(a)/v]) \quad (9.36)$$

The required conclusion (9.31) follows from (9.35) and (9.36).

2. If A is not Harrop, then $\mathsf{extract_{mod}}(p)$ is

$$(fn\ v => fn\ x_u => \mathsf{extract_{mod}}(b))\ fst(\mathsf{extract_{mod}}(a))\ snd(\mathsf{extract_{mod}}(a))$$

By the IH,

$$M \models_\iota \forall v : s \bullet \forall x_u : \mathsf{etype}(A) \bullet Sk(A)[v/y][x_u/f_A] \Rightarrow Sk(C)[\mathsf{extract_{mod}}(b)/f_C] \quad (9.37)$$

Because A is not Harrop,

$$\begin{aligned} & Sk(\exists y : s \bullet A)[\mathsf{extract_{mod}}(a)/f_{\exists y:s\bullet A}] \\ = \ & Sk(A)[fst(f_{\exists y:s\bullet A})/y][snd(f_{\exists y:s\bullet A})/f_A][\mathsf{extract_{mod}}(a)/f_{\exists y:s\bullet A}] \\ = \ & Sk(A)[fst(\mathsf{extract_{mod}}(a))/y][snd(\mathsf{extract_{mod}}(a))/f_A] \end{aligned}$$

So, by the IH

$$M \models_\iota Sk(A)[fst(\mathsf{extract_{mod}}(a))/y][snd(\mathsf{extract_{mod}}(a))/f_A] \quad (9.38)$$

We take (9.37), set v to $fst(\mathsf{extract_{mod}}(a))$, x_u to $snd(\mathsf{extract_{mod}}(a))$, and then instantiating with (9.38) to obtain

$$M \models_\iota Sk(C)[\mathsf{extract_{mod}}(b)/f_C][fst(\mathsf{extract_{mod}}(a))/v][snd(\mathsf{extract_{mod}}(a))/x_u] \quad (9.39)$$

Because, by definition of the ($\exists$-E) rule, v cannot occur in C, and also x_u cannot occur in C,[1]

$$Sk(C)[\mathsf{extract_{mod}}(b)/f_C][fst(\mathsf{extract_{mod}}(a))/v][snd(\mathsf{extract_{mod}}(a))/x_u]$$

is the same formula as

$$Sk(C)[((\mathsf{extract_{mod}}(b)[fst(\mathsf{extract_{mod}}(a))/v][snd(\mathsf{extract_{mod}}(a))/x_u]))/f_C]$$

and so (9.39) means

[1] Recall that all variables of the form x_a ($a \in Var_{PT(LTT_{\mathsf{SSL}})}$) are assumed not to occur in any formulae prior to use.

$$M \models_\iota$$
$$Sk(C)[((\mathsf{extract}_{\mathsf{mod}}(b)[fst(\mathsf{extract}_{\mathsf{mod}}(a))/v][snd(\mathsf{extract}_{\mathsf{mod}}(a))/x_u]))/f_C] \tag{9.40}$$

Because our models preserve the lambda calculus, and

$$(fn\ v => fn\ x_u => \mathsf{extract}_{\mathsf{mod}}(b))$$
$$fst(\mathsf{extract}_{\mathsf{mod}}(a))\ snd(\mathsf{extract}_{\mathsf{mod}}(a)) \rhd_{SML}$$
$$\mathsf{extract}_{\mathsf{mod}}(b)[fst(\mathsf{extract}_{\mathsf{mod}}(a))/v][snd(\mathsf{extract}_{\mathsf{mod}}(a))/x_u]$$

we know that

$$\iota((fn\ v => fn\ x_u => \mathsf{extract}_{\mathsf{mod}}(b))$$
$$fst(\mathsf{extract}_{\mathsf{mod}}(a))\ snd(\mathsf{extract}_{\mathsf{mod}}(a))) =$$
$$\iota(\mathsf{extract}_{\mathsf{mod}}(b)[fst(\mathsf{extract}_{\mathsf{mod}}(a))/v][snd(\mathsf{extract}_{\mathsf{mod}}(a))/x_u]) \tag{9.41}$$

So (9.41) and (9.40) gives us the required conclusion (9.31).

Case: ($\Rightarrow$-I). Assume that $p^{\mathrm{SP} \diamond P}$ is of the form $A \Rightarrow B$ obtained by an application of ($\Rightarrow$-I)

$$\frac{\Gamma, u^A \vdash b^{\mathrm{SP} \diamond B}}{\Gamma \vdash \mathsf{abstract}\ u.\ b^{\mathrm{SP} \diamond C}}\ (\Rightarrow\text{-I})$$

There are two cases, dependent on whether A is Harrop or not.

1. Assume that A is Harrop. Then $\mathsf{extract}_{\mathsf{mod}}(p)$ is $\mathsf{extract}_{\mathsf{mod}}(b)$. By the IH,

$$\mathrm{SP}, \Gamma \cup \{A\} \models Sk(B)[\mathsf{extract}_{\mathsf{mod}}(b)/f_B]$$

because $Sk(A)$ is A. By the semantics of implication, this means

$$\mathrm{SP}, \Gamma \models A \Rightarrow Sk(B)[\mathsf{extract}_{\mathsf{mod}}(b)/f_B]$$

This is the required conclusion, because $\mathsf{extract}_{\mathsf{mod}}(p) = \mathsf{extract}_{\mathsf{mod}}(b)$ and because

$$Sk(A \Rightarrow B)[\mathsf{extract}_{\mathsf{mod}}(p)/f_{A \Rightarrow B}]$$

is the same formula as the conclusion

$$A \Rightarrow Sk(B)[\mathsf{extract}_{\mathsf{mod}}(b)/f_B]$$

2. Assume that A is not Harrop. Then $\mathsf{extract}_{\mathsf{mod}}(p)$ is $fn\ x_u => \mathsf{extract}_{\mathsf{mod}}(b)$. By the IH, we know that there is a proof of the form

$$\mathrm{SP}, \Gamma' \cup \{Sk(A)[x_u/f_A]\} \models Sk(B)[\mathsf{extract}_{\mathsf{mod}}(b)/f_B] \tag{9.42}$$

Take any model $M \in Mod(\mathrm{SP})$ and any interpretation $\hat{\iota} : Var \to M$ such that

$$M \models_\iota G'$$

for each $G' \in \Gamma'$.
Because our models respect the lambda calculus with equality preserving reduction,

$$\iota((fn\ x_u \Rightarrow \mathsf{extract_{mod}}(b))\ x_u) = \iota(\mathsf{extract_{mod}}(b)) \tag{9.43}$$

Now (9.43) and (9.42) entail that

$$M \models_\iota \{Sk(A)[x_u/f_A]\} \text{ entails } M \models_\iota Sk(B)[((fn\ x_u \Rightarrow \mathsf{extract_{mod}}(b))\ x_u)/f_B]$$

So, by the semantics for implication

$$M \models_\iota Sk(A)[x_u/f_A] \Rightarrow Sk(B)[(\mathsf{extract_{mod}}(p)\ x_u)/f_B] \tag{9.44}$$

The semantics for quantification permits us to abstract over x_u, to give

$$M \models_\iota \forall x_u : \mathsf{etype}(A) \bullet Sk(A)[x_u/f_A] \Rightarrow Sk(B)[\mathsf{extract_{mod}}(p)\ x_u/f_B] \tag{9.45}$$

This is the required conclusion.

Case: ($\Rightarrow$-E). Assume that $p^{\mathrm{SP}\diamond P}$ is of the form C, obtained by an application of ($\Rightarrow$-E)

$$\frac{\Gamma_1 \vdash a^{\mathrm{SP}\diamond B \Rightarrow C} \quad \Gamma_2 \vdash b^{\mathrm{SP}\diamond B}}{\Gamma_1, \Gamma_2 \vdash \mathsf{app}(a, b)^{\mathrm{SP}\diamond C}} \ (\Rightarrow\text{-E})$$

so that $\Gamma' = \Gamma_1' \cup \Gamma_2'$.
There are two cases, dependent on whether B is Harrop or not.

1. Assume that B is Harrop. Then $\mathsf{extract_{mod}}(p) = \mathsf{extract_{mod}}(a)$.
   Also, by the IH,

$$\mathrm{SP}, \Gamma_1' \models_\iota B \Rightarrow Sk(C)[\mathsf{extract_{mod}}(a)/f_C] \tag{9.46}$$

   and

$$\mathrm{SP}, \Gamma_2' \models_\iota B \tag{9.47}$$

   The semantics for implication permits us to instantiate (9.46) with (9.47) to give

$$\mathrm{SP}, \Gamma \models_\iota Sk(C)[\mathsf{extract_{mod}}(a)/f_C]$$

   Because $\mathsf{extract_{mod}}(p) = \mathsf{extract_{mod}}(a)$, the conclusion of this proof is the same as stating $Sk(C)[\mathsf{extract_{mod}}(p)/f_C]$, as required.
2. Assume that B is not Harrop. Then $\mathsf{extract_{mod}}(p)$ is $(\mathsf{extract_{mod}}(a)\ \mathsf{extract_{mod}}(b))$. Also, by the IH, we know

$$\mathrm{SP}, \Gamma_1' \models \forall x : \mathsf{etype}(B) \bullet Sk(B)[x/f_B] \Rightarrow Sk(C)[\mathsf{extract_{mod}}(a)\ x/f_C] \tag{9.48}$$

   and

$$\mathrm{SP}, \Gamma_2' \models Sk(B)[\mathsf{extract_{mod}}(b)/f_B] \tag{9.49}$$

The semantics for quantification and implication allows us to take (9.46), instantiate x with $\mathsf{extract_{mod}}(b)$, and then instantiate with (9.47) to give

$$\text{Sp}, \Gamma' \models Sk(C)[(\mathsf{extract_{mod}}(a)\ \mathsf{extract_{mod}}(b))/f_C]$$

Because $\mathsf{extract_{mod}}(p)$ is defined to be $(\mathsf{extract_{mod}}(a)\ \mathsf{extract_{mod}}(b))$, this is the required conclusion.

Case: (∀-I). Assume that $p^{\text{Sp}\diamond P}$ is of the form $\forall x : s \bullet A$, obtained by an application of (∀-I)

$$\frac{\Gamma \vdash a^{\text{Sp}\diamond A}}{\Gamma \vdash \mathsf{use}\ x : s.\ a^{\text{Sp}\diamond \forall x:s \bullet A}}\ (\forall\text{-I})$$

Because we have assumed that P is not Harrop (so $\forall x : s \bullet A$ is not Harrop), A must not be Harrop, and $\mathsf{extract_{mod}}(p)$ is $\mathit{fn}\ x \Rightarrow \mathsf{extract_{mod}}(a)$

Take any model $M \in Mod(\text{Sp})$ and any interpretation $\hat{\iota} : Var \to M$ such that

$$M \models_\iota G'$$

for each $G' \in \Gamma'$

By the IH,

$$M \models_\iota Sk(A)[\mathsf{extract_{mod}}(a)/f_A] \tag{9.50}$$

Because our models take reducible lambda terms as equal (because they satisfy axioms generated by schemata of Fig. 9.5)

$$\iota(\mathit{fn}\ x \Rightarrow \mathsf{extract_{mod}}(a)\ x) = \iota(\mathsf{extract_{mod}}(a)) \tag{9.51}$$

Now (9.51) and (9.50) entail

$$M \models_\iota Sk(A)[(\mathsf{extract_{mod}}(p)\ x)/f_A] \tag{9.52}$$

By the semantics for quantification we can abstract over x in (9.52) to give us the required conclusion

$$M \models_\iota \forall x : \mathsf{etype}(A) \bullet Sk(A)[\mathsf{extract_{mod}}(p)\ x/f_A]$$

Case: (∀-E). Assume that $p^{\text{Sp}\diamond P}$ is of the form $\mathsf{specific}(a,t)^{\text{Sp}\diamond A[t/x]}$ obtained by an application of (∀-E)

$$\frac{\Gamma \vdash a^{\text{Sp}\diamond \forall x:s \bullet A}}{\Gamma \vdash \mathsf{specific}(a,t)^{\text{Sp}\diamond A[t/x]}}\ (\forall\text{-E})$$

Because we have assumed that P is not Harrop, this means that A must not be Harrop, and $\mathsf{extract_{mod}}(p)$ is $(\mathsf{extract_{mod}}(a)\ t)$

By the IH,

$$\text{Sp}, \Gamma' \models \forall x : s \bullet Sk(A)[(\mathsf{extract_{mod}}(a)\ x)/f_A] \tag{9.53}$$

To obtain the required conclusion, we need only instantiate 9.53) with t to give

$$\text{Sp}, \Gamma' \models Sk(A)[(\mathsf{extract}_{\mathsf{mod}}(a)\ t)/f_A] \tag{9.54}$$

This is the required result, because

$$Sk(A)[\mathsf{extract}_{\mathsf{mod}}(a)\ t/f_A]$$

is the same formula as

$$Sk(A)[\mathsf{extract}_{\mathsf{mod}}(p)/f_A]$$

Case: $(Ind(s, \Sigma))$. Assume that $p^{\text{Sp}\diamond P}$ is of the form

$$\mathsf{rec}(Cons_s, s, [a_1; \ldots; a_n; b_1; \ldots; b_p])^{\langle \Sigma, Ax\rangle \diamond \forall x:s \bullet A}$$

obtained by an induction:

$$\frac{a_1^{\langle \Sigma, Ax\rangle \diamond P[c_1/x]} \quad \ldots \quad a_n^{\langle \Sigma, Ax\rangle \diamond P[c_n/x]} \quad b_1^{\langle \Sigma, Ax\rangle \diamond P_{f_1}} \quad \ldots \quad b_p^{\langle \Sigma, Ax\rangle \diamond P_{f_p}}}{\mathsf{rec}(Cons_s, s, [a_1; \ldots; a_n; b_1; \ldots; b_p])^{\langle \Sigma, Ax\rangle \diamond \forall x:s \bullet P}}\ Ind(s, \Sigma)$$

where

- $\langle \Sigma, Ax\rangle$ is a basic specification where $\Sigma = \langle S, TF, P\rangle$ with constructors $Cons_s \subseteq TF$ for a sort $s \in S$,

$$Cons_s = \{c_1 : s, \ldots, c_n : s, f_1 : (s_1^1 \times \ldots \times s_{m_1}^1) \to s, \ldots,$$
$$f_p : (s_1^p \times \ldots \times s_{m_p}^p) \to s\}$$

  and
- where each P_{f_i} is defined as in Fig. 7.6 of Chapter 7 (p. 244).

In the non-Harrop case,

$$\mathsf{extract}_{\mathsf{mod}}(p) = rec(\overline{cons}, \overline{arg})$$

where

$$\overline{cons} = [c_1; \ldots; c_n; f_1; \ldots; f_p] \quad \text{and}$$
$$\overline{arg} = [\mathsf{extract}_{\mathsf{mod}}(F_1); \ldots; \mathsf{extract}_{\mathsf{mod}}(f_p)]$$

We are required to show that

$$M \models_\iota \forall x : s \bullet Sk(A)[(\mathsf{extract}_{\mathsf{mod}}(p)\ x)/f_A]$$

for $M \in Mod(\langle \Sigma, Ax\rangle)$ and every interpretation ι.

That is, we need to show that, for every x-variant $\iota'(x) \in s^M$,

$$M \models_{\iota'} Sk(A)[(\mathsf{extract}_{\mathsf{mod}}(p)\ x)/f_A] \tag{9.55}$$

Let ι'' be the x-variant defined by

$$\iota'' = \iota'[f_P \mapsto (\iota'(\mathsf{extract}_{\mathsf{mod}}(p)\ x))]$$

Because we use a loose semantics and s has constructors $Cons_s$, we know that $\iota'(x)$ must be of one of the following possible forms

$$\begin{array}{c}\iota'(c_1), \ldots, \iota'(c_n),\\ \iota'(f_1(d_1^1, \ldots, d_{m_1}^1))\\ \ldots\\ \iota'(f_p(d_1^p, \ldots, d_{m_p}^p))\end{array}$$

where each $d_1^j \in T(\Sigma, \emptyset)_{s_1^j}, \ldots, d_{m_j}^j \in T(\Sigma, \emptyset)_{s_{m_j}^j}$ $(j = 1, \ldots, p)$.

To establish (9.55), we use a secondary induction. We reason over the possible form of $\iota'(x)$.

- Assume $\iota'(x) = \iota'(c_i)$, for some $i = 1, \ldots, n$.
  By the main IH

  $$M \models Sk(A[c_i/x])[\mathsf{extract}_{\mathsf{mod}}(p_i)/f_A]$$

  So, in particular,

  $$\models_{\iota'} Sk(A)[\mathsf{extract}_{\mathsf{mod}}(p_i)/f_A] \tag{9.56}$$

  But, by the definition of $\triangleright_{SML}$ over recursion terms, and because $\mathsf{extract}_{\mathsf{mod}}(p) = rec(\overline{cons}, \overline{arg})$,

  $$(\mathsf{extract}_{\mathsf{mod}}(p)\ c_i) \triangleright_{SML} \mathsf{extract}_{\mathsf{mod}}(p_i)$$

  and so, because our models take reducibility to entail equality,

  $$\iota'(\mathsf{extract}_{\mathsf{mod}}(p)\ x) = \iota'(\mathsf{extract}_{\mathsf{mod}}(p_i)) \tag{9.57}$$

  So, using (9.56) and (9.57), we can deduce

  $$M \models_{\iota'} Sk(A)[(\mathsf{extract}_{\mathsf{mod}}(p)\ x)/f_A]$$

  as required.
- Assume $\iota'(x) = \iota'(f_i(d_1^i, \ldots, d_{m_i}^i))$ for some $i = 1, \ldots, n$, and that $s_j^i \neq s$ for each $j = 1, \ldots, m_i$.
  In this case, because $\mathsf{extract}_{\mathsf{mod}}(p) = rec(\overline{cons}, \overline{arg})$,

  $$(\mathsf{extract}_{\mathsf{mod}}(p)\ f_i(d_1^i, \ldots, d_{m_i}^i)) \triangleright_{SML} (\mathsf{extract}_{\mathsf{mod}}(p_i) d_1^i\ \ldots\ d_{m_i}^i)$$

  and so, because our models take reducibility to entail equality,

  $$\iota'(\mathsf{extract}_{\mathsf{mod}}(p)\ x) = \iota'(\mathsf{extract}_{\mathsf{mod}}(p_i)\ d_1^i\ \ldots\ d_{m_i}^i) \tag{9.58}$$

  By the main IH,

$$M \models_\iota \forall x_{1_i} : s_{1_i}, \ldots x_{m_i} : s_{m_i} \bullet Sk(A)[f_i(x_1^i, \ldots, x_{m_i}^i)/x] \\ [(\mathsf{extract}_{\mathsf{mod}}(p_i)\ x_1^i\ \ldots\ x_{m_i}^i)/f_A]$$

So, by repeated instantiation,

$$M \models_\iota Sk(A)[f_i(d_1^i, \ldots, d_{m_i}^i)/x][(\mathsf{extract}_{\mathsf{mod}}(p_i)\ d_1^i\ \ldots\ d_{m_i}^i)/f_A]$$

and thus, by the definition of ι'

$$M \models_{\iota'} Sk(A)[(\mathsf{extract}_{\mathsf{mod}}(p_i)\ d_1^i\ \ldots\ d_{m_i}^i)/f_A] \tag{9.59}$$

Then, by (9.58) and (9.59), we have

$$M \models_{\iota'} Sk(A)[(\mathsf{extract}_{\mathsf{mod}}(p)\ x)/f_A]$$

as required.

- Assume $\iota'(x) = \iota'(f_i(d_{1_i}, \ldots, d_{m_i}))$ for some $i = 1, \ldots, n$, where the set $M(s, \{x_j^i : s_j^i\}_{j=1,\ldots,m_i})$ is non-empty, defined as in Fig. 7.6 of Chapter 7 (p. 244):

$$M(s, \{x_j^i : s_j^i\}_{j=1,\ldots,m_i}) = \\ \{x_j^i : s_j^i \mid s_j^i = s \text{ for } j_i = 1_i, \ldots, m_i\} = \{x_1 : s, \ldots, x_k : s\}$$

We deal with the more complicated case where $k > 1$. The other case ($k = 1$) is similar. Note that we have a set of terms corresponding to $M(s, \{x_j^i : s_j^i\}_{j=1,\ldots,m_i})$

$$M(s, \{d_j^i : s_j^i\}_{j=1,\ldots,m_i}) = \\ \{d_j^i : s_j^i \mid s_j^i = s \text{ for } j_i = 1_i, \ldots, m_i\} = \{d_1 : s, \ldots, d_k : s\}$$

In this case, because $\mathsf{extract}_{\mathsf{mod}}(p) = rec(\overline{cons}, \overline{arg})$,

$$(\mathsf{extract}_{\mathsf{mod}}(p)\ f_i(d_1^i, \ldots, d_{m_i}^i)) \rhd_{SML} \\ \mathsf{extract}_{\mathsf{mod}}(p_i)\ d_1^i\ \ldots\ d_{m_i}^i\ (\mathsf{extract}_{\mathsf{mod}}(p)\ d_1) \ldots (\mathsf{extract}_{\mathsf{mod}}(p) d_k)$$

and so

$$\iota'(\mathsf{extract}_{\mathsf{mod}}(p)\ x) = \\ \iota'(\mathsf{extract}_{\mathsf{mod}}(p_i)\ d_1^i\ \ldots\ d_{m_i}^i\ (\mathsf{extract}_{\mathsf{mod}}(p)\ d_1) \ldots (\mathsf{extract}_{\mathsf{mod}}(p) d_k)) \tag{9.60}$$

By the secondary IH, for each $l = 1, \ldots, k$, because $d_l \in T(\Sigma, \emptyset)_s$,

$$M \models_{\iota_l} Sk(A)[(\mathsf{extract}_{\mathsf{mod}}(p)\ x)/f_A]$$

for $\iota_l = \iota[x \mapsto \iota(d_l)]$. This means

$$M \models_\iota Sk(A)[(\mathsf{extract}_{\mathsf{mod}}(p)\ d_l)/f_A] \tag{9.61}$$

By the main IH,

$$\begin{aligned} M \models_\iota \forall x^i_1 : s^i_1, \ldots, x^i_{m_i} : s^i_{m_i} \bullet \\ \forall y_1 : s \bullet Sk(A)[x_1/x][y_j/f_A] \Rightarrow \\ \ldots \\ \forall y_k : s \bullet Sk(A)[x_k/x][y_k/f_A] \Rightarrow \\ Sk(A)[f_i(x^i_1, \ldots, x^i_m)/x][(\mathsf{extract}_{\mathsf{mod}}(p_i)x_{1_i}\ \ldots\ x_{m_i}\ y_1\ \ldots\ y_k)/f_A] \end{aligned} \tag{9.62}$$

By repeatedly instantiating (9.62) with (9.61), we can obtain

$$\begin{aligned} M \models_\iota Sk(A)[f_i(d^i_1, \ldots, d^i_{m_i})/x] \\ [(\mathsf{extract}_{\mathsf{mod}}(p_i)d^i_1 \ldots d^i_{m_i}\ (\mathsf{extract}_{\mathsf{mod}}(p)d_1) \ldots (\mathsf{extract}_{\mathsf{mod}}(p)d_k))/f_A] \end{aligned}$$

Then, by the definition of ι' and (9.60), we have

$$M \models_{\iota'} Sk(A)[(\mathsf{extract}_{\mathsf{mod}}(p)\ x)/f_A]$$

as required.

This last case concludes the proof. □

Because any *CASL* specification is a trivial extension of itself, we have the following corollary to Theorem 9.3.2.

Corollary 9.3.1. *If there is a well-formed* modular *proof-term p for the proof of* $\mathrm{SP} \diamond P$

$$\vdash_{LTT(\mathsf{SSL})} p^{\mathrm{SP} \diamond P}$$

then $\mathsf{extract}_{\mathsf{mod}}(p)$ *is an extended realizer of* $\mathrm{SP} \diamond P$.

9.4 Extracting extended realizers

By adding additional axioms and symbols to a specification, we can add extracted programs back into a specification to form a conservative extension while preserving consistency. This yields executable extensions of specifications. This is a useful result for two reasons:

- This permits a systematic approach to consistent extension (extending a specification while retaining consistency) and the reuse of previously extracted programs
- We need this result in order to synthesize provably correct *SML* terms from *all* SSL proofs (including non-modular proofs). This is achieved by extracting extended realizers.

We use the results of the previous section to transform non-modular proofs of the form

$$\vdash_{\mathsf{SSL}} \textsc{Sp} \diamond A$$

into modular proofs of the form

$$\vdash_{\mathsf{SSL}} \textsc{Sp}' \diamond A$$

where $\textsc{Sp}'$ is a relatively consistent extension of $\textsc{Sp}$. This transformation will require us to extend SSL and $LTT(\mathsf{SSL})$ conservatively with an additional rule and proof-term construct. Then, by extracting a modular realizer from the resulting proof, we have the required extended realizer of $\textsc{Sp} \diamond A$.

9.4.1 Extensions via extraction of modular realizers

We shall first show how relatively consistent extensions of specifications can be given using the extraction of modular realizers.

When we extract a term $\mathsf{extract_{mod}}(d)$ from a modular proof

$$\vdash d^{\textsc{Sp}\diamond A}$$

the equality $f_A = \mathsf{extract_{mod}}(d)$ and the formula $Sk(A)$ can be added to $\textsc{Sp}$ as an axiom, and f_A can be added to $\mathrm{Sig}(\textsc{Sp})$. This will give a larger specification $\textsc{Sp}'$ that is a relatively consistent extension of $\textsc{Sp}$.

This is formalized by the following theorem, which follows from the soundness of SSL (Theorem 7.4.2 of Chapter 7) and our proof that modular proofs yield realizers (Theorem 9.3.2 above).

Theorem 9.4.1. *Given a proof*

$$\emptyset \vdash d^{\textsc{Sp}\diamond A}$$

such that $e = \mathsf{extract_{mod}}(d)$*, then we have that* $\textsc{Sp}' = NewSpec(\textsc{Sp}, A, e)$ *is a consistent extension of* $\textsc{Sp}$*, where* $NewSpec(\textsc{Sp}, A, e)$ *is defined by*

$$\textsc{Sp}\ \mathbf{then}\ \{\langle f_A : \mathsf{etype}(A)\}, \emptyset, \emptyset\rangle, \{f_A = e, Sk(A)\}\}$$

Proof. We can assume that Skolem function symbol f_A does not occur in the specification $\textsc{Sp}$ (because, if it did occur, we could rename f_A to something else). The result then follows easily because

- the equation $f_A = e$ is a conservative extension of the original specification, not affecting consistency.
- we have a proof that $\vdash\ \textsc{Sp} \diamond Sk(A)[e/f_A]$. This means, for every model $M \in Mod(\textsc{Sp})$,

$$M \models Sk(A)[e/f_A]$$

It can then easily be seen that, for any model C such that $C|_{\mathrm{Sig}(\textsc{Sp})} \in Mod(\textsc{Sp})$ where

$$C \models f_A = e \tag{9.63}$$

it must be the case that

$$C \models Sk(A)$$

So, because $Mod(NewSpec(\textsc{Sp}, A, e))$ is the smallest set of models C such that $C|_{\mathrm{Sig}(\textsc{Sp})} \in Mod(\textsc{Sp})$ we know that these models are always consistent, provided the models of $Mod(\textsc{Sp})$ are consistent.

□

Remark 9.11. Note that f_A and e can be functions, in which case the equation $f_A = e$ is a higher-order equation. This is permissible, because we assume *CASL* is now extended to permit higher-order statements. If we did not permit this, we would have to add an equation of the form

$$f_A(\overrightarrow{x}) = e(\overrightarrow{x})$$

where $\overrightarrow{x}$ is the list all abstraction variables for the lambda term e in the specification.

9.4.2 New rules for consistent extension

We will require our logic and logical type theory to be extended with a new rule, an additional constructor unextract, and new type inference rules. The new rule provides a means of consistently extending a specification with an equational definition for a function and a new axiom for defining behavior of the function. In this way, the rule permits consistent extension by and reuse of extracted terms in further proofs and synthesis.

The following is a new rule of SSL

$$\boxed{\frac{}{\vdash NewSpec(\textsc{Sp}, A, e) \diamond A}\ (\mathrm{Sk})} \tag{9.64}$$

where e is a modular realizer of $\textsc{Sp} \diamond A$.

Also,

$$\mathsf{unextract}(\textsc{Sp}, A, e)$$

is a new proof-term to $LTT(\mathsf{SSL})$. for *SML* term e and specification/formula pair $\textsc{Sp} \diamond A$. We add

$$\mathsf{extract}_{\mathsf{mod}}(\mathsf{unextract}(\textsc{Sp}, A, e)) = f_A$$

to the definition of $\mathsf{extract}_{\mathsf{SSL}}$.

If e is a modular realizer of $\textsc{Sp} \diamond A$, then we can apply a the following new type inference rule of $LTT(\mathsf{SSL})$ corresponding to the rule (Sk) in SSL

$$\boxed{\frac{}{\vdash \mathsf{unextract}(\textsc{Sp}, A, e)^{NewSpec(\textsc{Sp}, A, e) \diamond A}}\ (\mathrm{Sk})}$$

These extensions given by the rule are conservative, in the sense that all the important results about SSL and LTT(SSL) are preserved as shown by the following lemma.

Lemma 9.4.1 (Preservation of known properties). *The rules of LTT(SSL) with the* unextract *construct, the additional type inference rule (Sk) and the new extended definition of* $\mathsf{extract}_{\mathsf{mod}}$ *preserve the following theorems about LTT(SSL):*

1. *Soundness for* SSL *(Theorem 7.4.2 of Chapter 7).*
2. *The Curry–Howard isomorphism (Theorem 5.2.5 of Chapter 8):*
   - *Given a natural deduction proof D of $\vdash_{\mathsf{SSL}} \textsc{Sp} \diamond A$, we can construct a well typed proof-term $f^{\textsc{Sp}\diamond A}$.*
   - *Given a well-typed proof-term $f^{\textsc{Sp}\diamond A}$, we can construct a natural deduction proof D of*

$$\begin{array}{c}\vdots\\ \vdash \textsc{Sp} \diamond A\end{array}$$

3. *Strong normalization and the Church–Rosser property (Theorems 8.2.9 and 2.3.7 of Chapter 8):*
4. *Extraction of modular realizers (Theorem 9.3.2 above).*

Proof. Proof of item 1. We need only show that soundness holds for the new rule of SSL, assuming soundness for the original rules. The rule (Sk) is derivable from the (ext$_2$) rule:

$$\frac{\vdash \textsc{Sp} \diamond A}{\textsc{Sp}\ \mathbf{then}\ \{\langle f_A : \mathsf{etype}(A)\}, \emptyset, \emptyset\rangle, \{f_A = e\}\}}\ ext_2$$

and so soundness follows trivially.

Proof of item 2. The first part of Item (2) follows trivially from Theorem 5.2.5 of Chapter 8, because we do not need to use unextract in the derivation of f. The second part of Item (3) follows because (Sk) has a matching rule in SSL.

Proof of item 3. Strong normalization is preserved because we treat unextract proof-terms in the same way as we treat the ax proof-terms — as constants that cannot be reduced. The Church–Rosser property is preserved because, as with ax proof-terms, there are no critical pairs introduced by unextract.

Proof of item 4. The new construct unextract still preserves extraction of modular realizers. We add the following case to the proof of Theorem 9.3.2. Assume we have a modular proof ending in (Sk),

$$\frac{}{\vdash \mathsf{unextract}(\textsc{Sp}, A, e)^{NewSpec(\textsc{Sp},A,e)\diamond A}}\ (\mathrm{Sk})$$

We have that $\mathsf{extract}_{\mathsf{mod}}(\mathsf{unextract}(\textsc{Sp}, A, e)) = f_A$. So, we need to show

$$NewSpec(\textsc{Sp}, A, e) \models Sk(A)$$

Now, as e is a modular realizer, we already know that

$$\text{SP} \models Sk(A)[e/f_A]$$

By the semantics for extension, we can infer from this that

$$\text{SP } \mathbf{then} \ \{\langle f_A : \mathsf{etype}(A)\}, \emptyset, \emptyset\rangle, \{f_A = e\}\} \models Sk(A)[e/f_A]$$

which means

$$NewSpec(\text{SP}, A, e) \models Sk(A)[e/f_A] \tag{9.65}$$

Now, by definition of $NewSpec$, we have $f_A = e$ as an axiom and so

$$NewSpec(\text{SP}, A, \mathsf{extract}_{\mathsf{mod}}(d)) \models f_A = e \tag{9.66}$$

It follows from (9.65) and (9.66) that

$$NewSpec(\text{SP}, A, \mathsf{extract}_{\mathsf{mod}}(d)) \models Sk(A)$$

as required. □

9.4.3 Making proofs modular

We now show how to transform a proof

$$\vdash d^{\text{SP} \diamond A}$$

into a *modular* proof

$$\vdash d'^{\text{SP}' \diamond A}$$

by using Theorem 9.4.1. This result will then be used to extract extended realizers from any proof.

The transformation proceeds according to the proof of the following lemma.

Lemma 9.4.2. *Given any term* $\emptyset \vdash d^{\text{SP} \diamond A}$*, there is a proof-term* $\vdash \mathsf{modular}(d)^{\text{SP}' \diamond A}$ *such that* $\mathsf{modular}(d)$ *contains no non-modular subterms and* $\text{SP} \gg \text{SP}'$.

Proof. We give a recursive definition of $\mathsf{modular}(d)$ using a depth-first traversal of the proof tree encoded by d. (Recall that d is a proof-term and therefore represents a natural deduction proof tree.)

Let $n(t)$ be the total number of non-modular subterms in the proof-term t.

We define a terminating sequence of proof terms $d = d_0, ..., d_k = \mathsf{modular}(d)$.

Given d_i, we determine d_{i+1} as follows.

Case 1. If d_i does not contain any non-modular subterm, then $d_{i+1} = \mathsf{modular}(d) = d_i$ (that is, the sequence terminates).

Case 2. Otherwise, normalize d_i to give a proof-term d'. As long as d_i has no assumptions, the normalized proof-term d' will contain no subterms of the form $\mathsf{hide}((\lambda x : A.p), \Sigma)$.

1. Take the leftmost innermost non-modular subterm of the form $t = \mathsf{hide}(e : B, \Sigma)$ in d'. That is, take the first non-modular subterm t in d' which does not contain any non-modular subterms. So, e itself contains no non-modular subterms.
2. Extract a modular realizer from e to yield a new *SML* term $f = \mathsf{extract}_{\mathsf{mod}}(e)$. Let $\text{SP} \equiv NewSpec(sp(e), A, \mathsf{extract}_{\mathsf{mod}}(e))$. Then, in t, replace all occurrences of e by $\mathsf{unextract}(\text{SP}, A, \mathsf{extract}_{\mathsf{mod}}(e))$, to give $t' = t[\mathsf{unextract}(\text{SP}, A, \mathsf{extract}_{\mathsf{mod}}(e))/e]$.
   Note that t' proves the same formula as t by the typing of unextract and Lemma 9.4.1, but with a refined specification (adding the definition of $f_A = \mathsf{extract}_{\mathsf{mod}}(e)$). The term t' is modular because hidden symbols now do not hide the function f_A, which can be used in $\mathsf{extract}_{\mathsf{mod}}(t')$.
3. Replace t by t' in d', to give $d_{i+1} = d'[t'/t]$. Then d_{i+1} has at least one less non-modular subterm than d_i, and proves the same theorem as d_i.

Since $n(d_{i+1}) < n(d_i)$, this process yields a k such that $n(d_k) = 0$. Then we take $\mathsf{modular}(d) = d_k$, which is a proof-term with no non-modular subterms.

Note that SP' is a conservatively correct extension of SP by Lemma 9.4.1. So the the final specification will be a conservatively correct extension of sp(d). □

9.4.4 Extraction of extended realizers

The previous results can be used to extract correct *SML* terms from any SSL proof. These terms are correct in the sense that they are extended realizers of the proved specification/formula pairs.

Theorem 9.4.2 (Extraction of extended realizers). *There is an extraction map* extract$_{\mathsf{SSL}}$ *from proof-terms to* SML *terms such that, given any proof*

$$\vdash d^{\text{SP} \diamond A}$$

it is the case that extract$_{\mathsf{SSL}}(d)$ *is an extended realizer of* $\text{SP} \diamond A$*,*

$$\mathsf{extract}_{\mathsf{SSL}}(d) \ \mathsf{extr} \ \text{SP} \diamond A$$

Proof. The map extract$_{\mathsf{SSL}}$ is defined according to the following sequence

1. Take any proof
$$\vdash d^{\text{SP} \diamond A}$$
2. Apply Lemma 9.4.2, to transform d into a *modular* proof
$$\vdash d'^{\text{SP}' \diamond A}$$
   where $\text{SP} \gg \text{SP}'$

3. Let $\mathsf{extract}_{\mathsf{SSL}}(d) = \mathsf{extract}_{\mathsf{mod}}(d')$. By Theorem 9.3.2 it is a modular realizer of d'. So
$$\mathrm{SP}' \diamond Sk(A)[\mathsf{extract}_{\mathsf{SSL}}(d)/f_A]$$

4. Because $\mathrm{SP} \gg \mathrm{SP}'$, $\mathsf{extract}_{\mathsf{SSL}}(d)$ is the required *extended realizer* of $\mathrm{SP} \diamond A$. □

Theorem 9.4.3. *Take any proof*
$$\vdash_{LTT(\mathsf{SSL})} d^{\mathrm{SP}\diamond A}$$
Then
$$\vdash_{SML} \mathsf{extract}_{\mathsf{SSL}}(d) : \mathsf{etype}A$$
is a correct type inference.

Proof. The theorem follows easily from Theorem 6.2.2 and the construction of $\mathsf{extract}_{\mathsf{SSL}}$. □

9.5 Example: password checking system

We now demonstrate the extraction of modular and extended realizers using the password checking system example of the previous two chapters.

For reference, we briefly summarize the domain description. In Chapter 7, we specified a password system for an email hosting service, similar to the example used throughout Chapter 2. When a user joins the service, he/she is required to define a new numerical password. We restricted our attention to the part of the system that defines when a password is of an acceptable length. This password number must be 4 digits long (and so within the range of 0 and 9999). If the number chosen is not of the right length, the system will output a response message asking the user to select a new number within the correct range. If the number is within the correct range, then the system outputs a response message to this effect.

An initial specification of the system's password requirements, PwdCore, was given in terms of subspecifications of the natural numbers, booleans and strings, together with some axioms to model the domain (Example 7.6, p. 234, Chapter 7). To restrict the specification of the system to the relevant functionality, a final specification PwdSys was defined by hiding functions of PwdCore that we did not wish exposed.

We developed a theorem in SSL about PwdSys: given any input x of a password, there is always an appropriate response message to be output explaining whether the password is of the correct length or not (Section 7.3.6, pp. 245–247 of Chapter 7). This theorem is a truth about the specification PwdSys, given known properties about its required behavior. However, to build an implementation of the password checking system, we need to obtain a function for producing such a message for given passwords. In isolation, the theorem does not tell us what this function is.

We will now use our synthesis methods to extract such a password checking function which outputs an appropriate response for a given password number input.

We have seen how the proof of the theorem can be encoded as a term in the logical type theory (Chapter 8, Section 3.2.1, p. 260). The proof-term for the theorem involves a critical subterm because it involves proving the required property over PwdCore and then applying (hide) to show the property holds for PwdSys.

Consequently, we can extract an extended realizer from the proof of the theorem for PwdSys. To do this, we first need to obtain a modified realizer for the proof of the property over PwdCore.

We show how to extract a modified realizer from part of the proof about PwdCore, and then an extended realizer from the proof about PwdSys. We then show how the extended realizer can be used to build a consistent extension of PwdSys.

9.5.1 Extracting a modular realizer

The main part of the proof described in previous chapters involved a derivation of

$$\vdash \text{PwdCore} \diamond \forall x : nat \bullet \exists y : string \bullet ValidMsg(x, y) \tag{9.67}$$

The Skolem form of the formula $A = \forall x : nat \bullet \exists y : string \bullet ValidMsg(x, y)$ is

$$Sk(A) = \forall x : nat \bullet Valid(x, f_A(x))$$

Thus, by the definition of modular realizability, the theorem can be viewed as a specification of a function f_A that outputs an appropriate response message for a given password number.

The proof of this theorem can be encoded as a typed proof-term of the form

$$\vdash q^{\text{PwdCore} \diamond \forall x : nat \bullet \exists y : string \bullet ValidMsg(x,y)}$$

(see Section 7.3, pp. 245–247). where

$$\begin{aligned} q = {} & \mathsf{use}\ x : nat.\ \mathsf{case}\ p_4\ \mathsf{of} \\ & \quad \mathsf{inl}(u).\mathsf{show}(\text{'Password acceptable'}, \mathsf{app}(p_7, \mathsf{app}(p_6, p_5))), \\ & \quad\quad \mathsf{inr}(v).\mathsf{show}(\text{'Please choose a password in correct range'}, p_8) \end{aligned}$$

where p_4 is of the form

$$\mathsf{specific}(\mathsf{ext}_1(\mathsf{union}_1(BtoB \bullet (\mathsf{rec}([T, F], boolean, [(\mathsf{inr}(\mathsf{ax}(\textsc{NatBoolean}, T = T))), (\mathsf{inl}(\mathsf{ax}(\textsc{NatBoolean}, T = T)))])), \textsc{StringBool}), \langle SExt, AExt \rangle), inRange(x))$$

The proof-term encodes constructive information obtained from the ($\exists$-I) applications used within the proof — in particular, the witness string, y, for a valid message such that $ValidMsg(x, y)$ given a password number x, depending on the length of the password number.

By inspection, it can be seen that this proof-term is modular. We can apply Theorem 9.3.2 to obtain a function

$$\begin{aligned}
&\mathsf{extract}_{\mathsf{mod}}(q) = \\
&fn\ x : nat => \\
&\quad match\ rec([true, false], [inr(()), inl(())])inRange(x)\ with \\
&\qquad Inl(x_u)\ \ => \text{‘Password acceptable’}, \\
&\qquad Inr(x_v)\ \ => \text{‘Please choose a password in correct range’}
\end{aligned}$$

such that

$$\textsc{PwdCore} \models Sk(P)[\mathsf{extract}_{\mathsf{mod}}(q)/f_P]$$

This function has the required property. For instance, because $inRange(9999) = true$, we have that

$$\begin{aligned}
&\textsc{PwdCore} \models \mathsf{extract}_{\mathsf{mod}}(q)9999 = \\
&\qquad match\ (rec([true, false], [inr(()), inl(())])inRange(9999))\ with \\
&\qquad\quad Inl(x_u)\ \ => \text{‘Password acceptable’}, \\
&\qquad\quad Inr(x_v)\ \ => \text{‘Please choose a password in correct range’} \\
&\Rightarrow \textsc{PwdCore} \models \mathsf{extract}_{\mathsf{mod}}(q)9999 = \\
&\qquad match\ true\ with \\
&\qquad\quad Inl(x_u)\ \ => \text{‘Password acceptable’}, \\
&\qquad\quad Inr(x_v)\ \ => \text{‘Please choose a password in correct range’} \\
&\Rightarrow \textsc{PwdCore} \models \mathsf{extract}_{\mathsf{mod}}(q)9999 = \text{‘Password acceptable’}
\end{aligned} \tag{9.68}$$

The last inference holds because equality in our models preserves $\rhd_{SML}$ reduction and

$$\begin{aligned}
&match\ true\ with \\
&\quad Inl(x_u)\ \ => \text{‘Password acceptable’}, \\
&\quad Inr(x_v)\ \ => \text{‘Please choose a password in correct range’} \\
&\qquad\qquad\qquad\qquad \rhd_{SML} \text{‘Password acceptable’}
\end{aligned}$$

The inference (9.68) shows we have the correct property for input 9999 because

$$\textsc{PwdCore} \models ValidMsg(9999, \text{‘Password acceptable’})$$

and so

$$\textsc{PwdCore} \models ValidMsg(9999, \mathsf{extract}_{\mathsf{mod}}(q)9999)$$

9.5.2 Extracting an extended realizer

The conclusion of (9.67) specifies a function for PWDCORE, which, although specifying the password system, also exposes some extraneous functionality. To encapsulate this functionality, we apply the (hide) rule to (9.67), obtaining the final theorem about the specification PWDSYS

$$\vdash \text{PWDSYS} \diamond \forall x : nat \bullet \exists y : string \bullet ValidMsg(x, y) \tag{9.69}$$

The resulting theorem still describes a function that outputs an appropriate response message for a given password number, but for the specification PWDSYS of the encapsulated password system.

The proof of this theorem can be encoded as a typed proof-term of the form

$$\vdash p^{\text{PWDSYS}\diamond\forall x:nat\bullet\exists y:string\bullet ValidMsg(x,y)}$$

(see Section 7.3, pp. 245–247) where

$$\begin{aligned} p = \mathsf{hide}(&\mathsf{use}\ x : nat.\ \mathsf{case}\ p_4\ \mathsf{of} \\ &\quad \mathsf{inl}(u).\mathsf{show}(\text{'Password acceptable'}, \mathsf{app}(p_7, \mathsf{app}(p_6, p_5))), \\ &\quad \mathsf{inr}(v).\mathsf{show}(\text{'Please choose a password in correct range'}, p_8), \\ &\qquad \{ge, inRange\}) \end{aligned}$$

This proof-term is not modular, so we cannot extract a modular from the proof. However, by applying Theorem 9.4.2, we can obtain an extended realizer from p. This involves application of the process used in the proof of Lemma 9.4.2, to remove critical subterms of p. Essentially, we extend the specification PWDCORE with the function definition $\{f_A = \mathsf{extract}_{\mathsf{mod}}(q)\}$. By the nature of our extraction process, this results in a consistent extension,

$$NewSpec(\text{PWDSYS}, A, \mathsf{extract}_{\mathsf{mod}}(q))$$

This process gives us the proof-term

$$p' = \mathsf{unextract}(\text{PWDSYS}, A, \mathsf{extract}_{\mathsf{mod}}(q))$$

with the theorem $NewSpec(\text{PWDSYS}, A, \mathsf{extract}_{\mathsf{mod}}(q)) \diamond A$ as type.

The required extended realizer is then

$$\begin{aligned} &\mathsf{extract}_{\mathsf{mod}}(p) = \mathsf{extract}_{\mathsf{mod}}(p') = \\ &\qquad \mathsf{extract}_{\mathsf{mod}}(\mathsf{unextract}(\text{PWDSYS}, A, \mathsf{extract}_{\mathsf{mod}}(q))) = f_A \end{aligned}$$

which satisfies

$$NewSpec(\text{PWDSYS}, A, e) \diamond \forall x : nat \bullet Valid(x, \mathsf{extract}_{\mathsf{mod}}(p)(x))$$

where the specification $NewSpec(\text{PWDSYS}, A, \mathsf{extract}_{\mathsf{mod}}(q))$ is a consistent extension of PWDSYS. This is the required function which, given any input x of a password, outputs an appropriate response message explaining whether the password is of the correct length or not.

9.6 The Curry–Howard protocol for program synthesis

Our extraction map leads to an effective application of the Curry–Howard protocol for the synthesis of extended realizers from proofs of specifications. In this section, we show this.

9.6.1 Logical and computational type theories

In the Curry–Howard protocol of Chapter 3, we gave a general framework for program synthesis from proofs of specification that generalized state-of-the-art proofs-as-programs.

The protocol requires a logical type theory and a computational type theory. We take the logical type theory as the in $LTT(\mathsf{SSL})$ of Chapter 8 (identified as an LTT for SSL in Section 3.2.1 of Chapter 8, p. 81). We shall take our computational type theory to be SML (as defined in this chapter).

9.6.2 Conformance to the Curry–Howard protocol

The Curry–Howard protocol (Definition 3.2.5, Chapter 3, p. 87) holds between the $LTT(\mathsf{SSL})$ and SML, for the following reasons

1. There are extraction maps etype from formulae of $LTT(\mathsf{SSL})$ to the types of SML and $\mathsf{extract}_{\mathsf{SSL}}$ from proof-terms of $LTT(\mathsf{SSL})$ to programs of SML,

$$\mathsf{extract}_{\mathsf{SSL}} : PT(LTT(\mathsf{SSL})) \to Term(SML)$$
$$\mathsf{etype} : Formulae(\mathsf{SSL}) \to Type(SML)$$

   such that, given a proof $d \in PT(LTT(\mathsf{SSL}))$ such that

$$\vdash_{LTT(\mathsf{SSL})} d^{\mathrm{S}\mathrm{P} \diamond A}$$

   then $\mathsf{extract}_{\mathsf{SSL}}(d)$ is a lambda term of SML, is of type $\mathsf{etype}A$. The map etype was defined in Fig. 9.6. The map $\mathsf{extract}_{\mathsf{SSL}}$ is defined in the construction for the proof of Theorem 9.4.2, using the map $\mathsf{extract}_{\mathsf{SSL}}$, which was given in Fig. 9.7. The required typing property was shown in Theorem 9.4.3.
2. There is a realizability relation extr between programs and formulae, such that, for any proof

$$\vdash_{LTT(\mathsf{SSL})} p^{\mathrm{S}\mathrm{P} \diamond A}$$

   it is true that there is an extended realizer for $\mathrm{S}\mathrm{P} \diamond A$:

$$\mathsf{extract}_{\mathsf{SSL}}(p)\ \mathsf{extr}\ \mathrm{S}\mathrm{P} \diamond A$$

   The realizability relation was identified in Definition 9.2.5. The required property holds by Theorem 9.4.2.

9.6.3 Application of the protocol

Recalling the process of protocol application described in Chapter 3, Section 3.3, p. 87, we have successfully taken the required steps:

1. We defined a signature and a logical calculus that involves the signature, in Chapter 4. This involved deriving some properties that were orthogonal to the protocol process itself, but which were necessary for deriving the extraction theorem. Specifically, we provided a semantics for the calculus (in Chapter 4) and proved soundness (in Chapter 5).
2. We defined a logical type theory for the logical calculus in Chapter 5.
3. We identified a programming language and described it by means of a computational type theory in Chapter 4.
4. Finally, in this chapter, we completed the process by proving the Curry–Howard protocol to hold over the above domains.

9.7 Discussion

We have shown how to synthesize correct *SML* functions from proofs about *CASL* specifications. We have achieved one of the main goals of this part of the monograph: adapting proofs-as-programs to SSL, building upon the results of previous chapters. We applied the Curry–Howard protocol of Chapter 3 from Part II. A new notion of realizability was given between SSL specifications and formula pairs and *SML* functions. We then defined an extraction map from proofs in the logical type theory of Chapter 8 to realizing *SML* functions, which are terms of a computational type theory.

Additionally, we defined a method for incorporating extracted programs back into a *CASL* specification, in order to develop partly executable extensions.

In Chapter 10, we will extend our calculus and its logical type theory to accommodate parametrized specifications. We will show how the extraction results of this chapter can be extended to that augmented calculus.

Our method of defining consistent extensions involved extracting functions from a theorem and then adding them back into the theorem's specification. This leads to an intriguing possibility for program development: beginning with an abstract, nonexecutable specification, a system designer could repeatedly apply our method to derive a fully executable specification. Because our method always produces consistent extensions, the final specification can be viewed as a structured program that is provably correct. We will return to this question in more depth in Chapter 11.

10

Generic Specifications

In this chapter we extend our results to generic, parametrized specifications as they are treated in *CASL* [CoF01].

Generic (parametrized) specifications permit the abstraction of specifications. Abstraction is an important concept in structured development of systems, because it facilitates the encapsulation of system components that are applicable to a variety of problem domains.

Commonly, abstraction defines a system component with parameters to denote re-configurable parts. Then, by instantiation of the parameters, the abstract component can be made concrete to suit a particular problem domain. So a generic specification abstracts over a specification by parametrizing over sub-specifications. The parameters are the aspects of the specification that are open to change. The abstract generic specification is made more concrete by instantiating these parameters.

There are two main approaches to parametrization in the algebraic specification literature [Wir90, pp. 752–759]. The first method is by lambda abstraction, where a parametrized specification is taken as a mapping from argument specifications to a result specification. We do not deal with this approach in our work. The second approach, adopted by *CASL*, is by pushouts, where parametrization is a generalization of the notion of unions and extensions. A generic specification consists of a (fixed) main body in union with parameters that are open to change. Instantiation is taken as a kind of translation of parameters.

CASL defines generic specifications via the concept of a named specification — a structured specification that is given a name. Named specifications are an important idea on their own as they permit the reuse of complicated specifications by simply referring to a name, without the need to rewrite specifications in full. So we add new rules to SSL and its logical type theory, in order to reason with generic and named specifications, and show how these preserve the strong normalization and Church–Rosser properties, and finally show how to extract correct *SML* terms from proofs.

We proceed as follows:

- In Section 10.1 we summarize how generic and instantiated specifications are represented in *CASL* through the concept of named specifications.
- Section 10.2 adds new rules to SSL to deal with named, generic, and instantiated specifications. We discuss semantic issues, and show that soundness is unaffected by our changes.
- Section 10.3 adds corresponding type inference rules to LTT(SSL). We discuss proof-theory issues, and show that the normalization and Church–Rosser properties are unaffected.
- Finally, section 10.4 shows that we can still extract extended realizers from proofs in the extended system.

Throughout this chapter we illustrate our techniques by two simple examples (one, ultimately deriving from Sannella and Tarlecki in [TS89], of a specification for a warehouse parametrized by a specification of a catalogue of parts stored in the warehouse).

In the next chapter, we shall take extended versions of SSL and LTT(SSL) and show how to produce executable refinements of specifications. We will then give a method for refining non-generic specifications and generic specifications into provably correct *SML* modules and functors, respectively.

10.1 Generic and instantiated specifications

In this section, we define named, generic and instantiated specifications as treated in *CASL* [CoF01].

We define these constructions as structured specification expressions from the collection **CSpec** defined in Section 7.2 of Chapter 7. This will enable us to use the maps

$$Sig : \mathbf{CSpec} \to \mathbf{CSig}$$

and

$$Mod : \mathbf{CSpec} \to \{M \subseteq Mod(\Sigma) \mid \Sigma \in \mathbf{CSig}\}$$

to give the visible signature and semantics of our new constructions.

Example 10.1 *(Warehouse catalogue: domain considerations).* For the purposes of illustrating our ideas, we consider a very simple example, to be used throughout this chapter and the next.

We shall specify and reason about a generic system for a warehouse that houses parts. The warehouse supplies clients with replacements for faulty parts. The warehouse keeps track of the parts by means of a catalogue that contains a list of each part's possible replacements, indexed by the part's name. If a client has a faulty part, the warehouse system's catalogue can be used to locate a replacement, using the faulty part's name.

The system may be instantiated to work for a range of particular domains. For instance, we may wish to use the system for the car manufacturing industry, where warehouse parts are car parts; or for the textile industry, where parts are garments. Different warehouses can use different catalogues and involve different types of part.

We shall therefore define the warehouse system as generic over a specification of a catalogue and parts. To specify the essential elements of a catalogue, we will require a specification of lists. Lists themselves will be given as a generic specification.

10.1.1 Named and generic specifications

In the language of *CASL*, a generic specification is given through the notion of a *named* specification, that is, a specification name with a definition for it. Generic specifications are named specifications with parameter specifications.

A named specification is a specification with an associated name by which it may be referred to in other specifications. In this way, named specifications permit the reuse of specifications.

Definition 10.1.1 (Named and generic specifications). A named specification is written

$$\begin{array}{l}\textbf{spec}\ \textsc{Sn}[\textsc{Sp}\_1]\ldots[\textsc{Sp}\_n]\quad\textbf{given}\quad\textsc{Sp}''\_1,\ldots,\textsc{Sp}''\_m=\\ \quad\textsc{Sp}\\ \textbf{end}\end{array}\tag{10.1}$$

where the $\textsc{Sp}\_i$ $(i=1,\ldots,n)$ are specifications, called the *parameters* for the specification $\textsc{Sn}$, and the $\textsc{Sp}''\_j$ $(j=1,\ldots,m)$ are specifications, called the *imports* for $\textsc{Sn}$. The specification $\textsc{Sp}$ is called the *body* of $\textsc{Sn}$. If $n=0$, there are no parameter specifications $\textsc{Sp}\_i$, and $\textsc{Sn}$ is said to be a *non-generic* definition, otherwise it is called *generic*.

To keep a record of named specifications and what they refer to, *CASL* assumes a *global environment*, consisting of a mutable set of named specification declarations. When a named specification for $\textsc{Sn}$ is defined, the global environment is extended to include the definition, provided a definition for $\textsc{Sn}$ has not been given previously. We shall adopt the following notation to deal with the global environment.

Definition 10.1.2. We write Global to denote the global environment, and then $\mathsf{Global} := \mathsf{Global} \cup \{\mathrm{D}\}$ means that the global environment is extended by the named specification definition D. We write $\textsc{Sn} \in \mathsf{Global}$ to mean that $\textsc{Sn}$ is defined in Global. Therefore $\textsc{Sn} \notin \mathsf{Global}$ otherwise.

We understand the well-formedness and semantics of a named specification as given by the imports extended by the union of the parameter specifications and then extended by the body. That is, given $\textsc{Sn} \in \mathsf{Global}$ of the form (10.1) in Definition 10.1.1, we define the visible signature of $\textsc{Sn}$ to be

$$\begin{array}{l}\mathit{Sig}(\textsc{Sn}) = \mathit{Sig}(\{\textsc{Sp}''\_1\ \textbf{and}\ \ldots\ \textbf{and}\ \textsc{Sp}''\_m\}\\ \qquad\qquad\qquad\qquad\textbf{then}\ \{\textsc{Sp}\_1\ \textbf{and}\ \ldots\ \textbf{and}\ \textsc{Sp}\_n\}\ \textbf{then}\ \textsc{Sp})\end{array}$$

Similarly, the semantics of SN is

$$Mod(\textsc{Sn}) = Mod(\{\textsc{Sp}''\_1 \textbf{ and } \ldots \textbf{ and } \textsc{Sp}''\_m\} \\ \textbf{then } \{\textsc{Sp}\_1 \textbf{ and } \ldots \textbf{ and } \textsc{Sp}\_n\} \textbf{ then } \textsc{Sp})$$

Remark 10.1. Note that, on their own, generic specifications are not considered specification expressions — we are not permitted to apply union, translations, hidings or extension to generic specifications. In order to build a specification expression using generic specifications, each parameter has to be instantiated in all references to a specification name SP.

Remark 10.2. The declared parameters show just which parts of the generic specification are intended to vary between different references to it. The imports, by contrast, are fixed, and common to the parameters, body, and arguments. This illustrates the difference between declaring parameters and leaving them implicit in an extension of the form used to provide the semantics for generic specifications.

Remark 10.3. A non-generic specification is a specification expression and may be used to construct new specification expressions. In this case, we simply view the non-generic specification SN $\in$ Global

$$\begin{array}{l} \textbf{spec } \textsc{Sn} \quad \textbf{given} \quad \textsc{Sp}''\_1, \ldots, \textsc{Sp}''\_m = \\ \quad \textsc{Sp} \\ \textbf{end} \end{array} \tag{10.2}$$

as shorthand for the expression

$$\{\textsc{Sp}''\_1 \textbf{ and } \ldots \textbf{ and } \textsc{Sp}''\_m\} \textbf{ then } \textsc{Sp}$$

Example 10.2 *(Warehouse catalogue: generic lists).* Because a catalogue is a list of parts, we need a specification of lists LISTS: which we shall give first as a generic specification.

First we assume a specification of the natural numbers, NAT. For convenience we often omit irrelevant axioms, denoting them by an ellipsis ... . We shall build LISTS as a generic specification of lists of elements of sort *Elem* that contains an operation *hd* that returns the head of a list, on top of NAT:

$$\begin{array}{l} \textbf{spec } \textsc{Lists}[\{\textbf{sorts } Elem\}] \textbf{ given } \textsc{Nat} = \\ \textbf{sorts } List(Elem) \\ \textbf{ops } nil : List(Elem); cons : Elem \times List(Elem) \to List(Elem); \\ hd : List(Elem) \to Elem; size : List(Elem) \to nat \\ \textbf{preds } \in : Elem \times List(Elem) \\ \textbf{axioms } \forall k : List(Elem) \bullet size(k) > 0 \Rightarrow hd(k) \in k \\ \ldots \end{array}$$

We assume LISTS is a specification name in the global environment.

10.1.2 Instantiation

A non-generic specification may be used as a structured specification in place of its definition body. However, a generic specification must be instantiated in order to be used as a structured specification expression. *CASL* follows a pushout approach to instantiation, consisting of a process of applying morphism to "fit" a generic specification's parameters with instantiating specifications and then taking the pushout of the result.

Instantiation is done through *fitting arguments* ([CoF01], section 6.2.2). A fitting argument for a parameter specification consists of an instantiating specification and a symbol mapping that defines a morphism between the signatures of the two specifications.

Definition 10.1.3 (Fitting argument). Take two specifications $\textsc{Sp}$ and $\textsc{Sp}'$. Given a well-formed symbol mapping $SM : Sig(\textsc{Sp}) \to Sig(\textsc{Sp})'$ such that $M|_{SM} \in Mod(\textsc{Sp})$ for every $M \in Mod(\textsc{Sp}')$, we have that

$$\textsc{Sp}' \ \textbf{fit} \ SM$$

is a well-defined *fitting argument* for $\textsc{Sp}$. We call SM the *fitting morphism* for this fitting argument.

The form of an instantiated specification is as follows.

Definition 10.1.4 (Instantiated specifications). Instantiation of a generic specification with name $\textsc{Sn}$, defined in Global, is written

$$\textsc{Sn}[FA_1]\ldots[FA_n]$$

where $FA_1, \ldots, FA_n$ are well-defined fitting arguments from the parameters of $\textsc{Sn}$

$$\textsc{Sp}\_1 \ldots \textsc{Sp}\_n$$

to a set of *instantiating* specifications

$$\textsc{Sp}'\_1 \ldots \textsc{Sp}'\_n$$

Remark 10.4. The instantiation is valid provided that $\textsc{Sn}$ has been defined in the global environment.

To understand the semantics of instantiation, we required the following definitions.

Definition 10.1.5 (Well-definedness of instantiations). We take the following instantiation of a generic specification with name $\textsc{Sn}$ defined in Global:

$$\textsc{Sn}[FA_1]\ldots[FA_n]$$

where

$$\begin{array}{l} \textbf{spec}\ \textsc{Sn}[\textsc{Sp}\_1]\ldots[\textsc{Sp}\_n]\quad \textbf{given}\quad \textsc{Sp}''\_1,\ldots,\textsc{Sp}''\_m = \\ \quad \textsc{Sp} \\ \textbf{end} \end{array} \tag{10.3}$$

and $\textsc{Sp}'\_1,\ldots,\textsc{Sp}'\_n$ are the argument specifications for the fitting arguments.

Let $\textsc{Sp}^*$ stand for the imports and parameters extending body of $\textsc{Sn}$:

$$\{\textsc{Sp}''\_1\ \textbf{and}\ \ldots\ \textbf{and}\ \textsc{Sp}''\_m\}\ \textbf{then}\ \{\textsc{Sp}\_1\ \textbf{and}\ \ldots\ \textbf{and}\ \textsc{Sp}\_n\}\ \textbf{then}\ \textsc{Sp} \tag{10.4}$$

We define a morphism over $\textsc{Sn}$,

$$FM(\textsc{Sn}[FA_1]\ldots[FA_n])$$

to be the morphism formed by the the fitting arguments extended to a morphism applicable to the signature of $\textsc{Sp}^*$. When there is no ambiguity about the instantiated specification referred to, we shall simply write FM for this morphism.

We define $Inst(\textsc{Sn}[FA_1]\ldots[FA_n])$ to be

$$\textsc{Sp}^*\ \textbf{with}\ FM\ \textbf{then}\ \{\textsc{Sp}'\_1\ \textbf{and}\ \ldots\ \textbf{and}\ \textsc{Sp}'\_n\} \tag{10.5}$$

The instantiation $\textsc{Sn}[FA_1]\ldots[FA_n]$ is *well-defined* only when $Sig(Inst(\textsc{Sn}[FA_1]\ldots[FA_n]))$ is a pushout of the body and argument signatures of $\textsc{Sn}$.

This requirement is formalized as follows. Let

$$\begin{aligned} \Sigma_a &= Sig(\{\textsc{Sp}''\_1\ \textbf{and}\ \ldots\ \textbf{and}\ \textsc{Sp}''\_m\}\ \textbf{then} \\ &\qquad \{\textsc{Sp}\_1\ \textbf{and}\ \ldots\ \textbf{and}\ \textsc{Sp}\_n\}) \\ \Sigma_b &= Sig(\textsc{Sp}^*) \\ \Sigma_p &= Sig(\{\textsc{Sp}''\_1\ \textbf{and}\ \ldots\ \textbf{and}\ \textsc{Sp}''\_m\}\ \textbf{then} \\ &\qquad \{\textsc{Sp}\_1'\ \textbf{and}\ \ldots\ \textbf{and}\ \textsc{Sp}\_n'\}) \\ \Sigma_i &= Sig(Inst(\textsc{Sn}[FA_1]\ldots[FA_n])) \end{aligned}$$

Let FA denote the morphism formed from the fitting arguments extended with identities over imports to now map Σ_a to Σ_p.

The instantiation is *well-defined* only when FM can be used as a morphism in the pushout diagram:

$$\begin{array}{ccc} \Sigma_a & \xrightarrow{i_1} & \Sigma_b \\ {\scriptstyle FA}\downarrow & & \downarrow{\scriptstyle FM} \\ \Sigma_p & \xrightarrow{i_2} & \Sigma_i \end{array}$$

where i_1 and i_2 are morphisms from Σ_a to Σ_b and Σ_p to Σ_i, respectively. (For our purposes, we may additionally require that these morphisms be injections.)

Remark 10.5. The requirement on well-definedness in the previous definition means that, if the translated body

$$\{\text{SP}^* \textbf{ with } FM\}$$

and the union of the argument specifications

$$\{\text{SP}'\_1 \textbf{ and } \ldots \textbf{ and } \text{SP}'\_n\}$$

share any symbols, these symbols must be the result of applying FM to symbols shared with

$$\{\text{SP}''\_1 \textbf{ and } \ldots \textbf{ and } \text{SP}''\_m\} \textbf{ then } \{\text{SP}\_1 \textbf{ and } \ldots \textbf{ and } \text{SP}\_n\}$$

We understand the signature of an intantiation as

$$Sig(\text{SN}[FA_1]\ldots[FA_n]) = Sig(Inst(\text{SN}[FA_1]\ldots[FA_n]))$$

and the semantics of an instantiation to be

$$Mod(\text{SN}[FA_1]\ldots[FA_n]) = Mod(Inst(\text{SN}[FA_1]\ldots[FA_n]))$$

Remark 10.6. The difference between imports and parameters in a generic specification is shown by the semantics for instantiation. The symbols and axioms of imports may be used by the parameters in a generic specification — and may also be used by the arguments in an instantiation. However, they are fixed — not permitted to change according to the fitting arguments. In contrast, the parameters denote specifications that can be changed according to the fitting arguments. In this way, they denote the parts of an abstract generic specification that may vary according to the concrete application by the instantiation mechanism.

Example 10.3 *(Warehouse catalogue: instantiating lists to specify the catalogue and the warehouse).* We instantiate and extend LISTS to give the specification of the general form of a catalogue. We use lOR for the operation that provides a list of replacements for a part, and Rep for the predicate meaning that one part can be replaced with another, with their obvious semantics.

$$\textbf{spec } \text{CATALOGUE} = \text{LISTS}[Elem \mapsto Part] \textbf{ then } \text{SP}$$

where SP stands for

sorts $Catalogue$
ops $myCat : Catalogue; lOR : Part \to List(Part)$
preds $Rep : Part \times Part \times Catalogue; In : Part \times Catalogue$
axioms $\forall i, e : Part \bullet e \in lOR(i) \wedge In(e, myCat) \Rightarrow Rep(e, i, myCat);$
$\forall i : Part \bullet size(lOR(i)) > 0$

The specification of the generic warehouse system is then

$$\begin{aligned}&\textbf{spec } \textsc{Warehouse}[\textsc{Catalogue}] =\\&\textbf{ops } rep : Part \to Part\\&\textbf{axioms } \forall i : Part \bullet size(lOR(i)) > 0 \Rightarrow In(rep(i), myCat)\\&\qquad\qquad\qquad\qquad\qquad\qquad\qquad\Rightarrow Rep(rep(i), i, myCat)\end{aligned}$$

Given a faulty part name as input, the function *rep* uses the catalogue to obtain a replacement part, if it exists.

Observe that the warehouse specification uses the general form of the catalogue as a parameter. This general catalogue contains enough information for us to adequately define the behavior of replacement searches in the catalogue. To make the warehouse specific for a given problem domain we need only provide an appropriate definition of the catalogue specification, such that a fitting argument can be given to the general form of catalogue.

When we refer to this specification, we shall sometimes denote the body of the generic specification by BodyWare.

We assume Warehouse and Catalogue are specification names in the global environment.

10.2 Extensions to logical calculus

We now extend the SSL calculus to cover generic specifications. Recall how, in the case of structured specifications, we defined structural rules in SSL corresponding to specification building operations. These rules involved changing the specification for a known theorem $\vdash \textsc{Sp} \diamond A$ to construct a new theorem about a new specification. We proceed similarly, defining rules for construction using generic and named specifications.

We are concerned with two kinds of construction done using these specifications: the definition of named (possibly generic) specifications and the instantiation of generic specifications. Correspondingly, we give the following rules

- A rule for defining named, and consequently parametrized, specifications (Defn).
- Two rules for instantiating a parametrized specification by fitting arguments (Fit_1) and (Fit_2).

In addition, we will extend the axiom rule of SSL to allow us to use visible axioms from named specifications.

The new rules are shown in Fig. 10.1, p. 350.

10.2.1 New specification/formula pairs

We extend the calculus to permit named (possibly generic) specifications. That is, we now reason with pairs

$$\text{Sp} \diamond A$$

where Sp is any specification expression (including non-generic named specifications and instantiated specifications) or a generic named specification. That is, conclusions of our theorems vary over elements of $Pairs(\mathsf{SSL})$, redefined below.

Definition 10.2.1 (Pairs for new rules).

$$Pairs(\mathsf{SSL}) = \left\{ \text{Sp} \diamond F \;\middle|\; \begin{array}{l} \text{Sp is a specification expression or a generic} \\ \text{specification and } F \text{ is a formula from} \\ \bigcup_{\Sigma \in \mathbf{CSig}} WFF(\Sigma, Var) \end{array} \right\}$$

10.2.2 Global environment

In *CASL*, the definition of a named specific specification extends the global environment of available specification definitions. We formalize this by making the following metalogical assumption. We assume that the mutable global environment state Global is defined at any stage of a proof. The state does not change in any of the previously defined rules of SSL. However, upon application of the (Defn) rule, the Global state is expanded to include a new specification definition that may be used in subsequent stages of the proof.

10.2.3 Definition rule

The definition rule (Defn), see Fig. 10.1, p. 350, corresponds to the introduction of a new named specification. The formula is not affected, but the global environment must now include the new name. Note that the actual final specification obtained in proving a given formula will depend on the order in which the applications of (Defn) are introduced.

10.2.4 Fitting rules

The fitting rules[1] (Fit_1) and (Fit_2) in Fig. 10.1, p. 350, corresponds to instantiating the specification parameters $\text{Sp\_}1, \ldots, \text{Sp\_}n$ by the specifications $\text{Sp}'\text{\_}1, \ldots, \text{Sp}'\text{\_}n$. The morphism for the instantiation $FM(\text{Sn}[FA_1] \ldots [FA_n])$ is as defined in Definition 10.1.5.

[1] At WADT2001, Sannella pointed out that one can use the earlier rules to simulate the effects of (Fit). While this is true our aim has always been to accommodate our system to actual practice. Including the (Fit) rule explicitly allows the user to use the standard apparatus of *CASL* directly.

Let $\textsc{Sp}^*$ stand for the imports and parameters extending body of a named specification $\textsc{Sn}$:
$\{\textsc{Sp}''\_1$ **and** ... **and** $\textsc{Sp}''\_m\}$ **then** $\{\textsc{Sp}\_1$ **and** ... **and** $\textsc{Sp}\_n\}$ **then** $\textsc{Sp}$

$$\frac{\Gamma \vdash \textsc{Sp}^* \diamond A}{\Gamma \vdash \textsc{Sn} \diamond A} \text{ (Defn)}$$

$\mathsf{Global} := \mathsf{Global} \cup$
$\{$**spec**$\textsc{Sn}[\textsc{Sp}\_1] \ldots [\textsc{Sp}\_n]$ **given** $\textsc{Sp}''\_1, \ldots \textsc{Sp}''\_m =$
$\textsc{Sp}$
end$\}$
provided that the resulting Global contains only one definition of $\textsc{Sn}$.

$$\frac{\Gamma \vdash \textsc{Sn} \diamond B \quad \textsc{Sn} \in \mathsf{Global}}{FM`\Gamma \vdash \textsc{Sn}[FA_1] \ldots [FA_n] \diamond FM(B)} \text{ (Fit}_1\text{)}$$

where FM is the pushout morphism for the FA_i, as described in Definition 10.1.5.

$$\frac{\Gamma \vdash \{\textsc{Sp}'\_1 \textbf{ and } \ldots \textbf{ and } \textsc{Sp}'\_n\} \diamond B \quad \textsc{Sn} \in \mathsf{Global}}{\Gamma \vdash \textsc{Sn}[FA_1] \ldots [FA_n] \diamond B} \text{ (Fit}_2\text{)}$$

where each $\textsc{Sp}'\_i$ is the argument specification for FA_i.

$$\frac{\textsc{Sn} \in \mathsf{Global}}{\emptyset \vdash \textsc{Sn} \diamond A} \text{ (Ax* I)}$$

where $A \in \mathit{Axioms}(\textsc{Sn})$

Fig. 10.1. The new rules of SSL.

10.2.5 New axiom introduction rule

The third rule of Fig. 10.1, p. 350, allows us to use an axiom from the named specification, even if the specification is generic. It extends the axiom rule but we now have to ensure that $\textsc{Sn}$ is in Global.

We require the following notation to speak about axioms of named specifications.

Definition 10.2.2. We write $\mathit{Axioms}(\textsc{Sp})$ to denote the set of visible axioms in a structured specification $\textsc{Sp}$. If $\textsc{Sn}$ names a generic specification, we write $\mathit{Axioms}(\textsc{Sn})$ to denote the set of visible axioms in its body and imports. We call the instantiation $\textsc{Sn}[FA_1] \ldots [FA_n]$ an *unevaluated instantiation*. We call its expansion, given by the semantics above, an *evaluated instantiation*.

Remark 10.7. The axiom rule permits us to introduce axioms from a named specification. After the introduction of an axiom, the full range of logical rules is available for further reasoning about a named specification (generic or not).

Note that these rules will not change the specification, but can be used to derive new properties about it.

If the specification is not generic, we can also use any structural rule to build new specifications and theorems from this named specification.

However, in *CASL*, a generic specification name is not considered to be a specification expression to be used with the structuring operations (hiding, extension, translation or unions). Consequently, the structural rules for hiding, extension, translation and unions cannot be used with generic specifications. The only structural rules available to a generic specification are the (Fit_i) ($i = 1, 2$).

10.2.6 Soundness

Soundness of SSL with the new rules of Fig. 10.1 follows from the original proof, due to the straightforward semantics of generic and instantiated specifications.

Theorem 10.2.3. *The extended system of logical and structural rules is sound.*

Proof. The proof is a simple extension of the proofs of Theorem 7.4.2 given in Chapter 7.

Soundness of a proof ending in (Defn) follows because $Mod(\textsc{Sn}) = Mod(\textsc{Sp}^*)$, in

$$\frac{\Gamma \vdash \textsc{Sp}^* \diamond A \quad \textsc{Sn} \notin \mathsf{Global}}{\Gamma \vdash \textsc{Sn} \diamond A} \text{(Defn)}$$

$$\mathsf{Global} := \mathsf{Global} \cup$$
$$\{\mathbf{spec}\textsc{Sn}[\textsc{Sp}\_1]\ldots[\textsc{Sp}\_n] \quad \mathbf{given} \quad \textsc{Sp}''\_1, \ldots \textsc{Sp}''\_m = \textsc{Sp} \quad \mathbf{end}\}$$

For soundness of a proof ending in (Fit_1),

$$\frac{\Gamma \vdash \textsc{Sn} \diamond B \quad \textsc{Sn} \in \mathsf{Global}}{FM`\Gamma \vdash \textsc{Sn}[FA_1]\ldots[FA_n] \diamond FM(B)} (\mathrm{Fit}_1)$$

take any

$$M \in Mod(\textsc{Sn}[FA_1]\ldots[FA_n]) = Mod(Inst(\textsc{Sn}[FA_1]\ldots[FA_n])) =$$
$$Mod(\textsc{Sp}^* \ \mathbf{with}\ FM \ \mathbf{then}\ \{\textsc{Sp}'\_1 \ \mathbf{and}\ \ldots \ \mathbf{and}\ \textsc{Sp}'\_n\})$$

By the IH, $N \models B$ for any

$$N \in Mod(\textsc{Sn}) = Mod(\textsc{Sp}^*)$$

By definition the semantics for translation and extension, it follows that $M \models FM(B)$.

We proceed similarly for the case of a proof ending in(Fit_2).

The soundness of (Ax^* I) is trivial. □

10.3 Extensions to the logical type theory

It is a simple task to extend the logical type theory for SSL to represent proofs that use the new rules.

In this section we add new proof-terms and type inference rules corresponding to the new rules. Types of $LTT(\mathsf{SSL})$ now vary over the pairs used in the new rules, consisting of formulae with named and instantiated specifications. The Curry–Howard isomorphism is preserved, so that correct typing of a term corresponds to a valid proof according to the rules of SSL.

We do not change the reduction relation over the resulting lambda calculus so the Church–Rosser and proof normalization theorems are trivially preserved. Further reductions are possible to eliminate redundancies in our proofs. This is done by removing named and instantiated specifications with equivalent structured specifications. We briefly discuss these reductions at the end of this section.

10.3.1 New proof-terms and typing rules

We inductively extend the set of proof-terms of $LTT(\mathsf{SSL})$ by means of four new proof-term constructors:

$\mathsf{name}(p, \text{SN}; \text{SP\_1}, \ldots, \text{SP\_}n; \text{SP}''\text{\_1}, \ldots, \text{SP}''\text{\_}m)$	naming
$\mathsf{instantiate}(p, \text{SN}; FM)$	instantiation
$\mathsf{ax}(\text{SN}, x)$	axiom

where SN a specification name, $\text{SP\_1}, \ldots, \text{SP\_}n$, $\text{SP}''\text{\_1}, \ldots, \text{SP}''\text{\_}m$ are specification expressions, FM is a symbol mapping and p is a proof-term.

We type these rules with specification/formula pairs according to the rules of Fig. 10.2.

It is easy to see that a well-typed proof-term uniquely determines the form of an SSL proof of its type, and vice versa. That is, the Curry–Howard isomorphism (Theorem 5.2.5 of Chapter 8) is preserved by the new type inference rules.

Theorem 10.3.1.
The Curry–Howard isomorphism:

- *Given a natural deduction proof D of $\vdash_{\mathsf{SSL}} \text{SP} \diamond A$, we can construct a well-typed term $f^{\text{SP}\diamond A}$.*
- *Given a well-typed term $f^{\text{SP}\diamond A}$, we can construct a natural deduction proof D of*

$$\begin{array}{c}\vdots\\ \vdash \text{SP} \diamond A\end{array}$$

Recall three functions for determining proof information from a proof-term, defined on page 259: given a proof-term d with a derivation

$$\Gamma \vdash_{LTT(\mathsf{SSL})} d^{\text{SP}\diamond F}$$

we can compute the following data from d

Let $\textsc{Sp}^*$ stand for the imports and parameters extending body of a named specification $\textsc{Sn}$:

$$\{\textsc{Sp}''\_1 \textbf{ and } \ldots \textbf{ and } \textsc{Sp}''\_m\} \textbf{ then } \{\textsc{Sp}\_1 \textbf{ and } \ldots \textbf{ and } \textsc{Sp}\_n\} \textbf{ then } \textsc{Sp}$$

$$\frac{\Gamma \vdash p^{\textsc{Sp}^* \diamond A}}{\Gamma \vdash \mathsf{name}(p, \textsc{Sn}; \textsc{Sp}\_1, \ldots, \textsc{Sp}\_n; \textsc{Sp}''\_1, \ldots, \textsc{Sp}''\_m)^{\textsc{Sn} \diamond A}} \text{ (Defn)}$$

$\mathsf{Global} := \mathsf{Global} \cup$
$\{\textbf{spec}\textsc{Sn}[\textsc{Sp}\_1] \ldots [\textsc{Sp}\_n] \quad \textbf{given} \quad \textsc{Sp}''\_1, \ldots \textsc{Sp}''\_m =$
$\textsc{Sp}$
$\textbf{end}\}$

provided that the resulting Global contains only one definition of $\textsc{Sn}$.

$$\frac{\Gamma \vdash p^{\textsc{Sn} \diamond B} \quad \textsc{Sn} \in \mathsf{Global}}{FM`\Gamma \vdash \mathsf{instantiate}(p, \textsc{Sn}; FM)^{\textsc{Sn}[FA_1] \ldots [FA_n] \diamond FM \bullet B}} \text{ (Fit}_1\text{)}$$

where FM is a fitting morphism for the FA_i.

$$\frac{\Gamma \vdash p^{\{\textsc{Sp}'\_1 \textbf{ and } \ldots \textbf{ and } \textsc{Sp}'\_n \diamond B} \quad \textsc{Sn} \in \mathsf{Global}}{\Gamma \vdash \mathsf{instantiate}_2(p, FA_1, \ldots, FA_n)^{\textsc{Sn}[\mathtt{FA_1}] \ldots [\mathtt{FA_n}] \diamond B}} \text{ (Fit}_2\text{)}$$

where each $\textsc{Sp}'\_i$ is the argument specification for FA_i.

$$\frac{\textsc{Sn} \in \mathsf{Global} \quad \{A\} \in \mathit{Axioms}(\textsc{Sn})}{\emptyset \vdash \mathsf{ax}(\textsc{Sn}, A)^{\textsc{Sn} \diamond A}} \text{ (Ax}^* \text{ I)}$$

Fig. 10.2. The type inference rules.

1. the current context $\mathrm{con}(d)$
2. the specification $\mathrm{sp}(d)$ for which d is a derivation,
3. the derived formula, $\mathrm{for}(d)$

These functions are easily extended to our new terms, as in Fig. 10.3.

10.3.2 Proof-term reductions

We do not add any additional rules to define the $\rhd_{\mathsf{SSL}}$ reduction relation over proof-terms. Consequently, strong normalization and the Church–Rosser property follow easily.

Theorem 10.3.2 (Strong Normalization and the Church–Rosser property). *The extended calculus is strongly normalizing and satisfies the Church–Rosser property.*

$$\begin{aligned}
\mathrm{sp}(\mathsf{name}(p, \textsc{Sn}; \textsc{Sp_1}, \ldots, \textsc{Sp_}n; \textsc{Sp}''\textsc{_1}, \ldots, \textsc{Sp}''\textsc{_}m)) &= \textsc{Sn} \\
\mathrm{sp}(\mathsf{instantiate}(p, \textsc{Sn}; FM)) &= \textsc{Sn}[FA_1] \ldots [FA_n] \\
\mathrm{sp}(ax(\textsc{Sn}, x)) &= \textsc{Sn}
\end{aligned}$$

$$\begin{aligned}
\mathrm{for}(\mathsf{instantiate}(p, \textsc{Sn}; FM)) &= FM \bullet \mathrm{for}(p) \\
\text{If } \{a : A\} \in Axioms(\textsc{Sn}) \quad \mathrm{for}(ax(\textsc{Sn}, a)) &= A \\
\mathrm{for}(\mathsf{name}(p, \textsc{Sn}; \textsc{Sp_1}, \ldots, \textsc{Sp_}n; \textsc{Sp}''\textsc{_1}, \ldots, \textsc{Sp}''\textsc{_}m)) &= \mathrm{for}(p)
\end{aligned}$$

$$\begin{aligned}
\mathrm{con}(\mathsf{instantiate}(p, \textsc{Sn}; FM)) &= FM \bullet \mathrm{con}(p) \\
\mathrm{con}(ax(\textsc{Sn}, a)) &= \emptyset \\
\mathrm{con}(\mathsf{name}(p, \textsc{Sn}; \textsc{Sp_1}, \ldots, \textsc{Sp_}n; \textsc{Sp}''\textsc{_1}, \ldots, \textsc{Sp}''\textsc{_}m)) &= \mathrm{con}(p)
\end{aligned}$$

$FA_1, ...FA_n$ are the fitting argument specifications which can be recovered from the fitting morphism FM.

Fig. 10.3. Extensions of the sp, for and con functions to new proof-terms.

Proof. Strong normalization follows trivially from the original proof, as the new terms can be treated similarly to the other neutral terms. Church–Rosser follows because the new terms do not introduce any critical pairs. □

10.3.3 Extended logical type theory

We extend the logical type theory of Chapter 8 as follows:

$$LTT(\mathsf{SSL}) =$$
$$\langle PT(\mathsf{SSL}), Pairs(\mathsf{SSL}), (.)^{(.)}, \vdash_{LTT(\mathsf{SSL})}, PTR(LTT(\mathsf{SSL})), \triangleright_{\mathsf{SSL}}, NR(\mathsf{SSL})\rangle$$

consisting of:

- a set of extended proof-terms $PT(\mathsf{SSL})$,
- a set of types, taken as the extended pairs $Pairs(\mathsf{SSL})$,
- a typing relation $(.)^{(.)}$ between proof-terms and types, so that if $p \in PT(\mathsf{SSL})$ has type $(\textsc{Sp} \diamond F) \in Pairs(\mathsf{SSL})$, and we write $p^{\textsc{Sp} \diamond F}$,
- a type inference relation given by the original rules of $\vdash_{LTT(\mathsf{SSL})}$ extended by those of Fig. 10.2, and
- a reduction relation $\triangleright_{\mathsf{SSL}}$ is unchanged.

10.3.4 Proof-term simplifications

In [PCW02] Poernomo, Crossley and Wirsing presented some additional reductions for simplifying proofs that involve the new rules for generic specifications. These reductions involved matching definition and fitting rules, and then mapping instantiated specifications to a semantically equivalent specification expression that does not involve a named specification. This changes the associated conclusion specification, but provides a potential means for simplifying proofs.

We can define these reductions via transformations over proof-terms. However, application of these transformations will change the resulting type, corresponding to a change in the conclusion specification. Consequently, we do not include them in our reduction rules for $\rhd_{\mathsf{SSL}}$, but instead permit them as optional transformations that can be applied by the prover if required.

Remark 10.8. In this way, these reductions are similar to those described in Section 8.2 of Chapter 8 (p. 270). These reductions result in changing the specification of the conclusion and so were not used in our normalization strategy.

We wish to reduce proof redundancies in which a generic specification is defined (by an application of rule (Defn)) and then immediately instantiated (by an application of the rule (Fit)). The form of such a proof is

$$\dfrac{\dfrac{\begin{array}{c}\vdots\\ \Gamma \vdash d^{\mathrm{SP}^*\diamond A}\end{array}}{\Gamma \vdash \mathsf{name}(d, \mathrm{SN}; \mathrm{SP}\_1, \ldots, \mathrm{SP}\_n; \mathrm{SP}''\_1, \ldots, \mathrm{SP}''\_n)^{\mathrm{SN}\diamond A}}\;(\mathrm{Defn})}{FM`\Gamma \vdash \mathsf{instantiate}(d', \mathrm{SN}; FM)^{\mathrm{SN}[FA_1]\ldots[FA_n]\diamond FM\bullet A}}\;(\mathrm{Fit}_2) \tag{10.6}$$

where d' is

$$\mathsf{name}(d, \mathrm{SN}; \mathrm{SP}\_1, \ldots, \mathrm{SP}\_n; \mathrm{SP}\ ''\_1, \ldots, \mathrm{SP}''\_n)$$

and

$$\begin{array}{ll}\mathsf{Global} := \mathsf{Global} \cup & \{\mathbf{spec}\mathrm{SN}[\mathrm{SP}\_1]\ldots[\mathrm{SP}\_n] \quad \mathbf{given} \quad \mathrm{SP}''\_1, \ldots \mathrm{SP}''\_m = \\ & \quad \mathrm{SP} \\ & \quad \mathbf{end}\}\end{array}$$

where SP^* is of the form given in Definition 10.1.5.

So, such a proof can be transformed to a proof (that does not use (Defn) or (Fit_i) $(= 1, 2)$):

$$\dfrac{\dfrac{\begin{array}{c}\vdots\\ \Gamma d^{\mathrm{SP}^*\diamond A}\end{array}}{FM`\Gamma d_2'^{\,\mathrm{SP}^*\ \mathbf{with}\ FM\diamond FM\bullet A}}\;(\mathrm{trans})}{FM`\Gamma \vdash d_3'^{\,\{\mathrm{SP}^*\ \mathbf{with}\ FM\}\ \mathbf{and}\ \{\mathrm{SP}'\_1\ \mathbf{and}\ \ldots\ \mathbf{and}\ \mathrm{SP}'\_\mathrm{N}\}\diamond FM\bullet A}}\;(\mathrm{union}_1) \tag{10.7}$$

where $d_2' = FM \bullet d_1$ and $d_3' = union_1(d_2', \{\mathrm{SP}'\_1\ \mathbf{and}\ \ldots\ \mathbf{and}\ \mathrm{SP}'\_n\})$.

We justify this transformation by the semantics for instantiation: The conclusion of the original proof (10.6) is equivalent to the conclusion of the transformed proof (10.7). To see this, first observe that the formulae are identical. Also, the unreduced proof (10.6) concludes with the specification $\text{SN}[FA_1]\ldots[FA_n]$. By the semantics for instantiation, this specification denotes the concluding specification of the reduced proof (10.7). That is, the reduction transforms the concluding specification from an *unevaluated* instantiation to an equivalent, *evaluated* instantiation.

We also note that the premise of the original proof (10.6) is identical to the premise of the transformed proof (10.7). Thus the transformation yields a proof that does not use (Defn) or (Fit), yet proves the same conclusion from the same premise.

This process may be formalized by defining a transformation mapping $\succ$ over proof-terms as follows:

$$\mathsf{instantiate}(\mathsf{name}(p, \text{SN}; \text{SP\_1}, \ldots, \text{SP\_}n; \text{SP}''\text{\_1}, \ldots, \text{SP}''\text{\_}m), \text{SN}; FM) \succ INST_1$$

where $INST_1$ is the proof-term

$$union_1(FM \bullet p), \{\text{SP}''\ \text{\_1} \ \textbf{and} \ \ldots \ \textbf{and} \ \text{SP}''\text{\_}m\}) \ \textbf{and} \ \text{SP}''\text{\_}m\}\})))$$

We assume that all values in Global in the original proof remain defined in the transformed proof, including any definitions made in discarded occurrences of (Defn). We are required to make this assumption, because it is possible that these definitions are used in other proofs. For example, in the proofs:

$$\cfrac{\cdots}{\Gamma_1 \vdash \mathsf{ax}(\text{SN}, x)^{\text{SN}\diamond C}} \ (\text{Axiom})$$
$$\vdots$$
$$\Gamma_2 \vdash d^{\text{SN}\diamond A}$$

and

$$\vdots$$
$$\cfrac{\cfrac{\Gamma_3 \vdash c_3^{\text{SP}^*\diamond C}}{\Gamma_3 \vdash \mathsf{name}(c_3, \text{SN}; \text{SP\_1}, \ldots, \text{SP\_}n; \text{SP}''\text{\_1}, \ldots, \text{SP}''\text{\_}m)^{\text{SN}\diamond C}} \ (\text{Defn})}{FM`\Gamma_3 \vdash FM \bullet \mathsf{name}(c_3, \text{SN}; \text{SP\_1}, \ldots, \text{SP\_}n; \text{SP}''\text{\_1}, \ldots, \text{SP}''\text{\_}m)^{\text{SN}[FA]\diamond FM\bullet C}} \ (\text{Fit})$$
$$\vdots$$
$$\Gamma \vdash b^{\text{SP\_2}\diamond B}$$

the lower proof defines SN to be the name of a generic specification over the specification body SP. That proof may be transformed, because the definition is immediately followed by an instantiation. However, the definition of SN must still remain in Global, because it is required by the upper proof. We also wish to reduce redundancies in proofs where a generic specification is defined in Global, an axiom of the generic specification is introduced, and the generic specification is then instantiated (by an application of the rule (Fit)). The form of such a proof is

$$\dfrac{\dfrac{\textsc{Sn} \in \mathsf{Global} \quad \{a : A\} \in Axioms(\textsc{Sn})}{\Gamma \vdash \mathsf{ax}(\textsc{Sn}, x)^{\textsc{Sn} \diamond A}} \text{(Ax* I)}}{FM`\Gamma \vdash \mathsf{instantiate}(ax(\textsc{Sn}, x), \textsc{Sn}; FM)^{\textsc{Sn}[FA_1] \dots [FA_n] \diamond FM \bullet A}} \text{(Fit)} \qquad (10.8)$$

A must be an axiom of the body of SN, its parameters, or one of its imports. We only present the first case here as they are very similar. Such a proof can be reduced to a proof with the body SP of SN instead of SN and which does not use (Fit):

$$\dfrac{\dfrac{\dfrac{\dfrac{A \in Axioms(\textsc{Sp})}{\Gamma \vdash \mathsf{ax}(\textsc{Sp}, a)^{\textsc{Sp} \diamond A}} \text{(Ax* I)}}{\Gamma \vdash d_1^{\{\textsc{Sp}''\_1 \textbf{ and } \dots \textbf{ and } \textsc{Sp}''\_m\} \textbf{ then } \{\textsc{Sp}\_1 \textbf{ and } \dots \textbf{ and } \textsc{Sp}\_n\} \textbf{ then } \textsc{Sp} \diamond A}} \text{(ext}_2\text{)}}{FM`\Gamma \vdash FM \bullet d_1^{\textsc{Sp}^* \textbf{ with } FM \diamond FM \bullet A}} \text{(trans)}}{FM`\Gamma \vdash d_3'^{\{\textsc{Sp}^* \textbf{ with } FM\} \textbf{ and } \{\textsc{Sp}'\_1 \textbf{ and } \dots \textbf{ and } \textsc{Sp}'\_\textsc{n}\} \diamond FM \bullet A}} \text{(union}_1\text{)} \qquad (10.9)$$

where

$$d_1 = \mathsf{ext}_1(ax(\textsc{Sp}, x), \textsc{Sp}''\_1 \textbf{ and } \dots \textbf{ and } \textsc{Sp}''\_m, \textsc{Sp}\_1 \textbf{ and } \dots \textbf{ and } \textsc{Sp}\_n)$$
$$d_3 = \mathsf{ext}_2(FM \bullet d_1, \textsc{Sp}'\_1 \textbf{ and } \dots \textbf{ and } \textsc{Sp}'\_\textsc{n})$$

The justification for this reduction is similar to that given for (Defn)/(Fit) pairs. The reduction may be formalized by extending the reduction relation $\succ$ on proof-terms as follows:

$$\mathsf{instantiate}(Axiom(\textsc{Sn}, x), \textsc{Sn}; FM) \succ INST_2$$

where $INST_2$ is the proof-term

$$union_1(\{\textsc{Sp}\_1 \textbf{ and } \dots \textbf{ and } \textsc{Sp}\_n\}, FM \bullet \mathsf{ext}_2(p, \{\textsc{Sp}''\_1 \textbf{ and } \dots \textbf{ and } \textsc{Sp}''\_m\} \textbf{ then } \{\textsc{Sp}\_1 \textbf{ and } \dots \textsc{Sp}\_n\}, Axiom(\textsc{Sp})))$$

Remark 10.9. There are changes to all specification labels that follow either of the reductions above. This is because, unlike the normalizing for structural rules given in [WCP98], the proposed reductions result in conclusions with different (but equivalent) specification labels. (Both reductions result in a change from unevaluated instantiations to evaluated instantiations.)[2]

[2] It is possible to make further reductions on the structure of a proof, by first moving applications of (Defn) down proofs and then matching with applications of (Fit) to apply the above reduction. We also observe that an application of (Defn) and a logical rule (*), say, may be swapped, to give an equivalent proof. In fact the swapping is transitive over multiple occurrences of (∗). Often, if there are several applications of logical rules between the applications of the (Defn) and (Fit) rules, then (Defn) can be swapped over each rule, matched with (Fit) and discarded.

10.4 Structured proofs-as-programs revisited

In the previous chapter we adapted techniques for extracting programs from constructive proofs to produce executable refinements of specifications. This gave a method that used the (Sk) rule and unSkolemization to derive an executable refinement for a given specification. We now extend this to the extraction of *SML* programs and to our calculus extended by the new rules.

We need only extend our modular extraction map $\mathsf{extract}_{\mathsf{mod}}$ over the new proof-terms, as follows:

$$\mathsf{extract}_{\mathsf{mod}}(\mathsf{instantiate}(p, \mathrm{SN}; FM)) = FM \bullet \mathsf{extract}_{\mathsf{mod}}(p)$$
$$\mathsf{extract}_{\mathsf{mod}}(\mathsf{instantiate}_2(\mathrm{SN}, FA_1, \ldots, FA_n, p)) = \mathsf{extract}_{\mathsf{mod}}(p)$$
$$\mathsf{extract}_{\mathsf{mod}}(\mathsf{name}(p, \mathrm{SN}; \mathrm{SP\_1}, \ldots, \mathrm{SP\_}n; \mathrm{SP''\_1}, \ldots, \mathrm{SP''\_}m)) = \mathsf{extract}_{\mathsf{mod}}(p)$$

Extraction over proof-terms for (Ax* I) is the same as for (Ax I) (see Fig. 9.7, p. 310), mapping to (), because we assume all axioms are Harrop.

The definition of *modular* proof-terms is unchanged for the new proof-terms. Extraction of modular realizers from modular proofs still holds.

Theorem 10.4.1 (Extraction of modular realizers). *Take any set of typed proof-terms* $\Gamma = \{u_1^{G_1}, \ldots, u_n^{G_n}\}$. *We define* Γ' *to the corresponding set of Skolemized formulae*

$$\Gamma' = \{Sk(G_1)[x_{u_1}/f_{G_1}], \ldots, Sk(G_n)[x_{u_n}/f_{G_n}]\}$$

If there is a well-formed modular *proof-term p for the proof of* $\mathrm{SP} \diamond P$

$$\Gamma \vdash_{LTT(\mathsf{SSL})} p^{\mathrm{SP} \diamond P}$$

then there is a proof

$$\Gamma' \vdash_{\mathsf{SSL}} \mathrm{SP} \diamond Sk(P)[\mathsf{extract}_{\mathsf{mod}}(p)/f_P]$$

Proof. We extend the proof of Theorem 9.3.2 with additional cases to deal with the new rules. If the proof ends in an axiom rule, then we have the result trivially, as the conclusion formula must be Harrop.

We deal with the new rules similarly to the structural rules.

Case: (Fit$_1$). Assume that the proof term $p^{\mathrm{SP} \diamond P}$ is of the form $\mathsf{instantiate}(d, \mathrm{SN}; FM)^{\mathrm{SN}[\mathrm{FA}_1]\ldots[\mathrm{FA}_n] \diamond FM \bullet B}$ derived from a proof of the form

$$\frac{\Gamma \vdash d^{\mathrm{SN} \diamond B} \quad \mathrm{SN} \in \mathsf{Global}}{FM`\Gamma \vdash \mathsf{instantiate}(d, \mathrm{SN}; FM)^{\mathrm{SN}[FA_1]\ldots[FA_n] \diamond FM \bullet B}} \ (\mathrm{Fit}_1)$$

By the IH, there is a proof

$$\Gamma' \vdash_{\mathsf{SSL}} \mathrm{SN} \diamond Sk(A)[\mathsf{extract}_{\mathsf{mod}}(d)/f_A] \tag{10.10}$$

But, by definition, $\mathsf{extract}(p)$ is $FM(\mathsf{extract}_{\mathsf{mod}}(d))$, and so (10.10) is the same as writing

$$\Gamma' \vdash_{\mathsf{SSL}} \text{SP\_1} \diamond Sk(A)[FM^{-1}(\mathsf{extract}_{\mathsf{mod}}(p))/f_A] \tag{10.11}$$

By applying (Fit_1) to (10.11), we obtain

$$FM`\Gamma' \vdash_{\mathsf{SSL}} \text{SN}[FA_1]\ldots[FA_n] \diamond Sk(A)[\mathsf{extract}_{\mathsf{mod}}(p)/f_A]$$

as required.

The case of (Fit_2) is similar.

Case: (Defn). Assume $p^{\text{SP}\diamond P}$ is of the form $\mathsf{name}(d, \text{SN}; \text{SP\_1}, \ldots, \text{SP\_}n; \text{SP}''\text{\_1}, \ldots, \text{SP}''\text{\_}m)^{\text{SN}\diamond A}$ derived from a proof of the form

$$\frac{\Gamma \vdash d^{\text{SP}^*\diamond A}}{\Gamma \vdash \mathsf{name}(d, \text{SN}; \text{SP\_1}, \ldots, \text{SP\_}n; \text{SP}''\text{\_1}, \ldots, \text{SP}''\text{\_}m)^{\text{SN}\diamond A}}\ (\text{Defn})$$

with the Global environment defined by

$$\begin{array}{l} \mathsf{Global} := \mathsf{Global} \cup \{\mathbf{spec}\text{SN}[\text{SP\_1}]\ldots[\text{SP\_}n] \quad \mathbf{given} \quad \text{SP}''\text{\_1}, \ldots \text{SP}''\text{\_}m = \} \\ \qquad\qquad\qquad \text{SP} \\ \qquad\qquad\quad \mathbf{end} \end{array}$$

where this is an extension of Global if SN had not been previously added. By the IH,

$$\Gamma' \vdash \text{SP}^* \diamond Sk(A)[\mathsf{extract}_{\mathsf{mod}}(d)/f_A]$$

But $\mathsf{extract}_{\mathsf{mod}}(d) = \mathsf{extract}_{\mathsf{mod}}(p)$, so we can apply (Defn) to obtain

$$\Gamma' \vdash \text{SN} \diamond Sk(A)[\mathsf{extract}_{\mathsf{mod}}(d)/f_A]$$

as required. □

Remark 10.10. The extraction map "ignores" the occurrence of the name proof-term for the (Defn) rule and also the $\mathsf{instantiate}_2$ proof-term for (Fit_2). So, extracting a program from a proof that ends in an application of (Defn) is the same as extracting a program from a proof without this application.

Our reasons are similar to those for the definition over structural unions. The new rules do not affect the computational nature of the extracted term, and, because the extracted term uses operations from the premise specification that are a subset of the operations available to the conclusion specification, we can still reason about the term using the conclusion specification.

Remark 10.11. The extraction map over the proof-term for (Fit_1) applies the fitting morphism FM, for a similar reason to the renaming structural rule. That is, the constructive content is unchanged by the rule, but the morphism must be applied in order to reason about the extract term using the new specification.

The extension of a specification expression by the definition $f_A = e$,

$$NewSpec(\text{SP}, A, e)$$

is defined as before (Theorem 9.4.1 of the previous chapter, page 329). Note that, because this is an extension of a *specification expression* $\textsc{Sp}$, $\textsc{Sp}$ is never a generic named specification.

By Theorem 10.4.1, the following rules are a conservative extension of SSL and LTT(SSL) (in the sense of Lemma 9.4.1 of the previous chapter, page 331):

$$\frac{}{\vdash NewSpec(\textsc{Sp}, A, e) \diamond A}\ (\text{Sk}) \tag{10.12}$$

for a modular realizer e of $\textsc{Sp} \diamond A$, and corresponding type inference rule

$$\frac{}{\vdash \mathsf{unextract}(NewSpec(\textsc{Sp}, A, e), A, e)^{NewSpec(\textsc{Sp},A,e)\diamond A}}\ (\text{Sk})$$

with

$$\mathsf{extract}_{\mathsf{mod}}(\mathsf{unextract}(\textsc{Sp}, A, e)) = f_A$$

As before, we extract extended realizers from any proof by first transforming the proof into a modular proof by repeated applications of (Sk) to eliminate hidden terms that are used as witnesses for existential statements.

There is a complication because the (Sk) rule cannot be applied to proofs that involve generic specifications. This is not a problem, because of the following lemma, which tells us that, if the proof preceding the application of (Defn) is modular, then the proof following the rule is modular as well.

Lemma 10.4.1. *Take any proof-term $\emptyset \vdash d^{\textsc{Sp}\diamond A}$. Assume d contains at least one generic subterm and that each* generic *subterm of d of the form*

$$\mathsf{name}(p, \textsc{Sn}; \textsc{Sp}\_1, \ldots, \textsc{Sp}\_n; \textsc{Sp}''\_1, \ldots, \textsc{Sp}''\_m)$$

is such that p is modular.

Then d is modular.

Proof. By a simple induction on the derivation of d — the induction is simple and is omitted here.

The idea is as follows. Non-modular proof-terms must contain proof-terms corresponding to (hide).

However, because each generic subterm t of the form

$$\mathsf{name}(p, \textsc{Sn}; \textsc{Sp}\_1, \ldots, \textsc{Sp}\_n; \textsc{Sp}''\_1, \ldots, \textsc{Sp}''\_m)$$

has a generic specification in its type, it cannot be used as a premise for any structural rule. So d will not contain any subterms of the form

$$\mathsf{hide}(e, SL)$$

with t as a subterm of e. □

This lemma means that elimination of modular proof-terms by our process of (Sk) application need only consider subterms that do not involve name. So we have the following lemma.

Lemma 10.4.2. *Given any term* $\emptyset \vdash d^{\mathrm{SP} \diamond A}$, *there is a term* $\vdash \mathsf{modular}(d)^{\mathrm{SP}' \diamond A}$ *such that* $\mathsf{modular}(d)$ *contains no non-modular subterms and* $\mathrm{SP} \gg \mathrm{SP}'$.

Proof. We give a recursive definition of $\mathsf{modular}(d)$ using a depth-first traversal of the proof tree for d. (Recall that d is a proof-term and therefore represents a natural deduction proof tree.)

Let $n(t)$ be the total number of non-modular subterms in the proof-term t. We define a terminating sequence of proof terms $d = d_0, ..., d_k = \mathsf{modular}(d)$. Given d_i, we determine d_{i+1} as follows.

Case 1. If d_i does not contain any non-modular subterm, then $d_{i+1} = \mathsf{modular}(d) = d_i$ (viz, the sequence terminates).

Case 2. Otherwise, normalize d_i to give a proof-term d'. As long as d_i has no assumptions, the normalized proof-term d' will contain no subterms of the form $\mathsf{hide}((\lambda x : A.p), \Sigma)$.

1. Take the leftmost innermost non-modular subterm of the form $t = \mathsf{hide}(e : B, \Sigma)$ in d'. That is, take the first non-modular subterm t in d' which does not contain any non-modular subterms. So, e itself contains no non-modular subterms.
2. Extract a modular realizer from e to yield a new *SML* term $f = \mathsf{extract}_{\mathsf{mod}}(e)$. Let $\mathrm{SP} \equiv \mathit{NewSpec}(\mathit{sp}(e), A, \mathsf{extract}_{\mathsf{mod}}(e))$. Then, in t, replace all occurrences of e by $\mathsf{unextract}(\mathrm{SP}, A, \mathsf{extract}_{\mathsf{mod}}(e))$, to give

$$t' = t[\mathsf{unextract}(\mathrm{SP}, A, \mathsf{extract}_{\mathsf{mod}}(e))/e]$$

   The proof-term $\mathsf{unextract}(\mathrm{SP}, A, \mathsf{extract}_{\mathsf{mod}}(e))$ is well-typed.
   Note that t' proves the same formula as t by the typing of $\mathsf{unextract}$ and Lemma 9.4.1, but with a refined specification (adding the definition of $f_A = \mathsf{extract}_{\mathsf{mod}}(e)$). The term t' is modular because hidden symbols now do not hide the function f_A, which can be used in $\mathsf{extract}_{\mathsf{mod}}(t')$.
3. Replace t by t' in d', to give $d_{i+1} = d'[t'/t]$. Then d_{i+1} has at least one less non-modular subterm than d_i, and proves the same theorem as d_i.

Since $n(d_{i+1}) < n(d_i)$, this process yields a k such that $n(d_k) = 0$. Then we take $\mathsf{modular}(d) = d_k$, which is a proof-term with no non-modular subterms.

Note that SP' is a conservatively correct extension of SP by Lemma 9.4.1. So the final specification will be a conservatively correct refinement of $\mathrm{sp}(d)$. □

As a consequence, we can always apply the above theorem to extract modular realizers from proofs that involve (Defn) — provided the proofs that precede (Defn) are modular.

We extract extended realizers by the same process described by Theorem 9.4.2, Section 9.4 of Chapter 9, but now using Lemma 10.4.2 to remove modular subterms. This gives us the required theorem.

Theorem 10.4.2 (Extraction of extended realizers). *There is an extraction map* $\mathsf{extract}_{\mathsf{SSL}}$ *from proof-terms to SML terms such that, given any proof* $\vdash d^{\mathrm{SP} \diamond A}$ *it is the case that* $\mathsf{extract}_{\mathsf{SSL}}(d)$ *is an extended realizer of* $\mathrm{SP} \diamond A$.

It is easy to see that the new rules for SSL, the new proof-terms and typing rules for LTT(SSL) and the extraction map $\mathsf{extract}_{\mathsf{SSL}}$ preserves satisfaction of the Curry–Howard protocol, as identified to hold for the original system SSL in Section 9.6 of Chapter 9.

10.5 Example: warehouse specification

We now illustrate the extended calculus and extraction theorem by continuing our warehouse example. We shall derive the following theorem about WAREHOUSE:

$$\vdash \text{WAREHOUSE} \diamond \forall i : Part \bullet \exists y : Part \bullet size(lOR(i)) > 0 \Rightarrow (In(y, myCat) \Rightarrow Rep(y, i, myCat)) \quad (10.13)$$

This theorem states that, if the list of replacements for a part i is of size greater than zero, then there is a replacement part y in the warehouse catalogue that can replace i in the catalogue. The theorem can be considered a specification of a realizing function f

$$\vdash \text{WAREHOUSE} \diamond \forall i : Part \bullet size(lOR(i)) > 0 \Rightarrow (In(f(i), myCat) \Rightarrow Rep(f(i), i, myCat))$$

We will use the extended calculus to simultaneously derive the formula and construct the specification WAREHOUSE. Using the Curry–Howard isomorphism we represent this proof as a proof-term in the extended logical type theory from which we shall then extract a realizing function.

For convenience we shall omit the contexts and Global.

We begin with the axiom for $\in$ from the generic specification LISTS:

$$\vdash \mathsf{ax}(\text{LISTS}, l)^{\text{LISTS} \diamond \forall k : List(Elem) \bullet size(k) > 0 \Rightarrow hd(k) \in k}$$

We instantiate the specification to obtain a theorem about $\in$ for elements of $List(Parts)$ by applying the rule (Fit$_1$), with the fitting argument $[Elem \mapsto Parts]$, which yields the fitting morphism FM:

$$\vdash \mathsf{instantiate}(\mathsf{ax}(\text{LISTS}, l), \text{LISTS}; FM)^{\text{LISTS}[Elem \mapsto Parts] \diamond \forall k : List(Parts) \bullet size(k) > 0 \Rightarrow hd(k) \in k}$$

We next apply (ext$_1$) to this obtaining the same theorem over CATALOGUE and (trivially) the (Defn) rule to get

$$\vdash \mathsf{name}(\mathsf{ext}_1(\mathtt{instantiate}(\mathsf{ax}(\text{LISTS}, l), \text{LISTS}; FM), \text{BODYCAT}))(lOr(i)) \\ (\mathsf{ax}(\text{CATALOGUE}, c), \text{CATALOGUE};;)^{\text{CATALOGUE} \diamond \forall k : List(Parts) \bullet size(k) > 0 \Rightarrow hd(k) \in k}$$

and call this proof-term q_4. We apply ($\forall$-E) to this theorem with $lOR(i)$ for k and the formula obtained from the CATALOGUE axiom by ($\forall$-E) to get:

$$\vdash \mathsf{ax}(\text{CATALOGUE}, c) lOR(i)^{\text{CATALOGUE}\diamond size(lOR(i))>0}$$

Then these two theorems we apply ($\Rightarrow$-E) obtaining the formula

$$\vdash q_5^{\text{CATALOGUE}\diamond hd(lOR(i))\in lOR(i)} \tag{10.14}$$

where $q_5 = (q_4(lOr(i)))((\mathsf{ax}(\text{CATALOGUE}, c)lOR(i)))$.

Now we use the axiom for lOR given by WAREHOUSE, labelling it with w. Applying ($\forall$-E) twice, with $hd(lOR(i))$ for e, we obtain:

$$\vdash q_6^{\text{CATALOGUE}\diamond hd(lOR(i))\in lOR(i)\Rightarrow(In(hd(lOR(i)),myCat)\Rightarrow Rep(hd(lOR(i)),i,myCat))}$$

where $q_6 = (\mathsf{ax}(\text{WAREHOUSE}, w)\, i)(hd(lOR(i))))$.

Applying ($\Rightarrow$-E) to this formula and (10.14), we get

$$\vdash q_6 q_5^{\text{CATALOGUE}\diamond In(hd(lOR(i)),myCat)\Rightarrow Rep(hd(lOR(i)),i,myCat)}$$

We apply (ext_1) to the axiom for lOR, extending CATALOGUE by the body of the specification WAREHOUSE, that is: BODYWARE. We then apply ($\exists$-I):

$$\vdash p_2^{\text{CATALOGUE } \mathbf{then} \text{ BODYWARE}\diamond\exists y:Part\bullet In(y,myCat)\Rightarrow Rep(y,i,myCat)}$$

where $p_2 = \mathsf{show}(hd(lOR(i)), \mathsf{ext}_1(\mathsf{app}(q_6, q_5), \text{BODYWARE}))$.

Applying ($\forall$-I) over the sequent gives us the goal formula with proof-term p_3 but with the specification CATALOGUE **then** BODYWARE where

$$p_3 = \mathsf{use}\ i : Part.\ \mathsf{show}(hd(lOR(i)), \mathsf{ext}_2(\mathsf{app}(q_6, q_5), \text{CATALOGUE}))$$

We apply the (Defn) rule to the sequent, abstracting over CATALOGUE to obtain the specification WAREHOUSE, and our goal theorem

$$\vdash p_4^{\text{WAREHOUSE}\diamond\forall i:Part\bullet\exists y:Part\bullet In(y,myCat)\Rightarrow Rep(y,i,myCat)} \tag{10.15}$$

with

$$p_4 = \mathsf{name}(\mathsf{use}\ i : Part.\ (hd(lOR(i)),$$
$$\mathsf{ext}_2(q_6 q_5, \text{BODYWARE})), \text{WAREHOUSE}; \text{CATALOGUE})$$

Now applying **extract** to the proof-term p_4, gives the function

$$fn\ i => hd(lOR(i))$$

which returns the first element in the list of replacements for i. Here the function lOR is a parameter, coming from the particular catalogue chosen for the warehouse specification.

Theorem 10.4.1 ensures this function satisfies the goal formula, in the sense that

$$\text{WAREHOUSE} \models \forall i : Part \bullet In(\mathsf{extract}(p_4)(i), myCat) \Rightarrow Rep(\mathsf{extract}(p_4)(i), i, myCat)$$

is true for WAREHOUSE.

Observe that, because WAREHOUSE is a generic named specification, we cannot apply the (Sk) rule to extend it by this function. However, given a specification for CATALOGUE, we can instantiate WAREHOUSE and then apply (Sk) to extend the result with the function. For example, if we have a specification for catalogues of car parts, CARCATALOGUE with a fitting argument FA from CATALOGUE and CARCATALOGUE and corresponding morphism FM, we could apply (Fit_1) to (10.15), obtaining

$$\vdash p_5^{\textsc{Warehouse}[FA] \diamond \forall i:CarPart \bullet In(FM(\mathsf{extract}(p_4))(i), carCat) \Rightarrow Rep(FM(\mathsf{extract}(p_4))(i), i, carCat)} \tag{10.16}$$

with

$$\begin{aligned} p_5 = \mathtt{instantiate}_1(&\textsc{Warehouse}, \\ &\mathsf{name}(\mathsf{use}\ i : Part.\ (hd(lOR(i)), \\ &\mathsf{ext}_2(\mathsf{app}(q_6, q_5), \textsc{BodyWare})), \textsc{Warehouse}; \textsc{Catalogue})) \end{aligned}$$

We obtain the extension

$$NewSpec(\textsc{Warehouse}[FA], A, \mathsf{extract}(p_5))$$

where A is $\forall i : Part \bullet \exists y : CarPart \bullet In(y, carCat) \Rightarrow Rep(y, i, carCat)$ and

$$\mathsf{extract}(p_5) = FM(\mathsf{extract}(p_4))$$

This extension permits us to use the function $f_A = \mathsf{extract}(p_5)$ in further reasonings about the car parts warehouse.

Remark 10.12. Observe that the theorem (10.13) is an unSkolemized version of the axiom for *rep*:

$$\forall i : Part \bullet size(lOR(i)) \Rightarrow (In(rep(i), myCat) \Rightarrow Rep(rep(i), i, myCat))$$

In the next chapter we shall investigate the development of executable refinements from specifications using our extraction process. The idea is that, in certain cases, if we extract a function for the unSkolemized form of an axiom for a function, we can define the function to be equal to the realizer, producing an executable, consistent refinement. We shall return to this example and derive an executable refinement of WAREHOUSE.

10.6 Discussion

We have treated generic specifications for reasoning and program synthesis, presenting extensions to the structural rules of SSL that have permitted us to:

- reason about named and generic specifications by:

  - Defining named, possibly generic specifications and to conclude new truths about the result from known truths about the body of the named specification.
  - Instantiating generic specifications and concluding new truths about the instantiation from known truths about the generic specification.
- Extract extended realizers from proofs in our system, utilizing process of the Curry–Howard protocol.

Generic specifications provide a useful abstraction mechanism for system development. Our extensions are therefore valuable for reasoning and program synthesis within the context of abstractions of system elements. Further, the fact that our approach utilizes the Curry–Howard protocol represents another justification for use of the protocol in adapting the proof-as-programs paradigm to new logics and computational theories.

In the next chapter, we shall present an application of the process of extracting programs using proofs in SSL, to develop *SML* modules. The idea is that structured and generic specifications employ structuring and abstraction mechanisms that often parallel those used for developing structured (module-based) *SML* programs. By extracting *SML* terms for every function used in a specification, we shall show that it is possible to transform a *CASL* specification into a *SML* module or functor.

11

Structured Program Synthesis

This chapter discusses how our calculus and the structured proofs-as-programs results can be applied to give methods for *structured program synthesis*.

We will take structured programs to be executable *CASL* specifications — specifications in which every function symbol has an executable definition as a lambda term. The specification building operations are viewed as imposing architectural structure over executable programs. We consider executable *CASL* as an intermediate implementation language, with the possibility of further translation to a conventional structured programming language such as *SML* or C#.

We outline two complementary methods for the development of structured programs:

- We give a process for the construction from new structured programs from known structured programs, using the rules of SSL (including the Skolemization rule to define consistent extensions using extraction).
- A specification can be refined to a specification that retains the signature of the specification and includes all models of the original specification. We define a process for the systematic refinement of abstract (non-executable) specifications, repeatedly using SSL and extraction to obtain definitions of all functions of a signature.

These methods apply different stages of the software development process. The first method assumes we have a repository of structured programs, for use in implementation-level reasoning about and construction of new structured programs. The second method involves a higher-level of abstraction and is useful when we wish to develop structured programs from an abstract specification. The approaches are related, because the programs in the repository should be correct — and one way of guaranteeing their correctness is to construct them using our refinement methods.

One of the main aims of algebraic specification is to provide a formal basis to support the systematic development of correct programs from specifications by means of verified refinement steps. Our refinement technique is an important and novel use of our extraction to achieve this goal. In general, when refining a specification proving consistency can be very difficult. The great advantage of our process is that consistency is guaranteed — provided, of course, the original specification and the specifications used in a proof was consistent.

Also, our methods for reasoning about and constructing structured programs shows that a logic such as SSL can be useful for reasoning and synthesis at the implementation-level, as well as higher-levels, of program specification.

Because our methods involve structured programs, we must address their architectural design issues. A *software architecture* comprises the design decisions made about the structure and texture of software, defining how a system is implemented as a configuration of components (see, e.g., [Ran00, pp. 10–12]). In our context, components are basic executable specifications, and architectural configurations are defined by structured specification operators. Our methods require us to make design decisions about software architecture, using the language of *CASL*. We therefore need to be careful that our methods yield architectures that conform to accepted design standards of intelligibility, coherence and maintainability. This is possible because, in our methods, design decisions are made through interactive application of SSL structural rules.

We proceed as follows:

- In section 11.1 we define the notion of an executable specification, and discuss how such specifications can be thought of as comprising a structured programming language. We discuss the relation between implementation architectures and specification structuring operators. We then outline a simple process that uses SSL for reasoning about and synthesizing structured programs.
- In section 11.2 we show how to use SSL and program extraction for reasoning about and synthesizing structured programs.
- Section 11.3 reviews notions of refinement for non-generic and generic specifications and presents our method of refinement based on program extraction. We identify some problems concerning how final implementation architectural design can be affected by the refinement process, and provide how these problems can be avoided by considering implementation architectures earlier in the refinement process. We illustrate our refinement techniques, continuing with the warehouse example of Chapter 10.
- Conclusions are drawn in section 11.4.

11.1 Structured programs

Structured programming languages are used to build systems from components. As a minimal definition, a component is a cohesive (semantically related) grouping of functionalities. The main activities of structured programming are twofold:

- *Definition of basic, atomic components.* This activity occurs at a low-level of granularity, and is primarily concerned with the traditional programming task of function definition.
- *Configuration of components to form larger, compound components.* This activity is of a coarser-grained nature, focusing on the definition of usage relationships between components to construct larger components.

The second activity defines system architectures — that is, the the dependencies, encapsulation and hierarchies that form a component configuration and, ultimately, a component-based system.

There is a range of popular structured programming languages, some in industrial use, such C++, Java or Visual Basic, and some of a more theoretical nature, such as modular *SML*. By virtue of the fact that these languages are structured, their operational and denotation semantics are complex.

Rather than introduce the syntax and semantics of a separate structured programming language, in this chapter we shall simply consider a subset of *CASL* — executable *CASL* — consisting of specifications in which every function symbol has a unique equational definition. We show that this subset exhibits the features common to structured programming languages. Specifically, we show that

- executable *CASL* can be equipped with a simple operational semantics (11.1.1), and
- it facilitates the two activities of structured programming identified above: construction of basic components (11.1.2) and of compound component architectures (11.1.3).

Because these properties are satisfied, it is possible to extend the results of this chapter to a conventional structured programming language by defining a translation from executable *CASL*.

11.1.1 Executable *CASL*

Executable *CASL* consists of executable specifications. By an executable non-generic specification we simply mean a specification in which every function has an executable definition and every sort has free data type declaration (that is, it is given by constructor functions). A generic specification is executable if its body is executable.

We make these definitions formal as follows.

Definition 11.1.1 (Executable specification). Let $\text{S}\text{P\_A}$ be a well-formed, consistent specification in which every symbol is declared at most once.

Then,

1. If SP_A is non-generic then SP_A is called *executable* if, and only if,
   a) for each visible[1] sort symbol s used in SP_A, SP_A contains a basic or partial specification in which s has a free data type declaration.
   b) for each visible function symbol f of SP_A, there is a unique equational definition
   $$f = \mathsf{dec}(f) \in Axioms(\text{SP\_A})$$
   for some *SML* term $\mathsf{dec}(f)$
   c) every hidden subspecification in SP_A of the form
   $$\text{SP } \mathbf{hide} \ SL$$
   is such that SP is executable.
2. If SP_A is generic of the form $\mathbf{spec}\text{SP\_B}[\text{SP}_1]\ldots[\text{SP}_n]$ **given** $\text{SP}''\text{\_}1\ldots \text{SP}''\text{\_}m = \text{SP}$, then SP_A is called *executable* if, and only if, SP is executable.

Remark 11.1. Items 1b and 1c of the definition together entail that we have a unique equational definition of every (visible or hidden) function of the given specification.

When considering structured specifications as structured programs, we require a full operational semantics for functions. This semantics defines how a function application will evaluate to an answer, with respect to a given executable specification.

Recall that the lambda terms of our specifications are equipped with standard lambda reduction rules (see Fig. 9.3 in Section 9.1 of Chapter 9, p. 303). These rules are used within our logic to equate terms modulo lambda reduction. However, in isolation, these rules are not enough to define a satisfactory operational semantics, because they do not evaluate the application of a specification function symbol. Function application can occur in the lambda terms of our specifications. Because we assume function symbols always have an equational definition, we should be able to use this definition to evaluate such applications.

To provide a full operational semantics of an executable specification, we proceed as follows. We take the usual lambda reduction rules and extend them to replace references to functions with their equational definitions, according to the executable specification.

We define a reduction relation of the following form

$$\text{SP} \vdash f \hat{\triangleright} d$$

with the intended meaning that the term f evaluates to d in the context of component SP. This relation is given by the transitive closure of the following rules that define a one-step reduction relation $\triangleright$.

[1] Visible axioms and symbols are defined in Definition 7.2.2, Chapter 7, p. 236.

First, our reduction relation preserves $\rhd_{SML}$.

$$\boxed{f \ \rhd_{SML} \ d \text{ entails } \textsc{Sp} \vdash f \rhd d} \tag{11.1}$$

where $\hat{\rhd}_{SML}$ is defined as in Fig. 9.3 of Chapter 9, p. 303.

Then, we add a rule that tells us we can replace a function with its equational definition in an evaluation:

$$\boxed{\textsc{Sp} \vdash g[f/x] \ = \ g[\mathsf{dec}(f)/x]} \tag{11.2}$$

if

$$f = \mathsf{dec}(f) \in Axioms(\textsc{Sp})$$

for the *SML* term $\mathsf{dec}(f)$.

Note that repeated application of $\rhd$ need not give a terminating sequence of reductions. This occurs when two functions are mutually recursive without a base case. So we let $\hat{\rhd}$ denote the finite transitive closure of $\rhd$, if this closure exists, otherwise we let it be the null relation.

The following lemma tells us that any property that we can prove about a program g for the executable specification $\textsc{Sp}$, is also true of its final value.

Lemma 11.1.1. *If*

$$\textsc{Sp} \models P[g/x]$$

and

$$\textsc{Sp} \vdash g \ \hat{\rhd} \ k$$

then

$$\textsc{Sp} \models P[k/x]$$

Proof. This follows easily from the following facts

- we assume that the models of $\textsc{Sp}$ equate interpretations of terms that are reducable according to the lambda reduction $\rhd_{SML}$.
- For any function symbols g and f, given that

  $$f = \mathsf{dec}(f) \in Axioms(\textsc{Sp})$$

  it must be the case that interpretations of $g[f/x]$ and $g[\mathsf{dec}(f)/x]$ are equal under all models of $\textsc{Sp}$.

□

Remark 11.2. This lemma means that any statement that is provable about a program is also true of the program's evaluation, according to our operational semantics.

11.1.2 Specifications as software components

Software components are commonly defined as reusable entities that encapsulate a range of cohesive (semantically related) functionalities. Components can be basic (atomic building blocks) or compound (constructed out of smaller components).

A component enables clients to use its functionality by exposing an interface. The interface describes the functions offered. At its most simple level, this interface description takes the form of a signature list. At a more complicated level, the interface description may involve descriptive annotations, such as English documentation, behavioral description (using, for instance, a CCS process description or a UML activity graph), logical specifications of input/output behavior or non-functional property descriptions [Szy98].

The executable specifications form components in accordance with this definition. We can consider a specification's signature, together with its visible axioms, to form a component interface. The signature lists the functions available for a client and the axioms document this functionality.

Basic executable specifications are therefore the basic components of our structured programming language.

A compound component is a component that is built from smaller components, using the mechanisms of the structured programming language. In executable *CASL*, we consider the structured and generic executable specifications as compound components. The specification structuring operators and the naming and instantiation of generic specifications are our way of building larger components from smaller ones. The way in which a compound component is constructed, by the use of the structuring operators, comprises an architecture.

11.1.3 Architectures of executable specifications

An important area of software engineering research is concerned with understanding architectures of structured programs, for specification, construction, analysis, and maintenance [MT00, KMND00, PRS02].

A software architecture is a hierarchical configuration of components, expressing usage relations between components and their interfaces. In general, configuration can take many forms. We will consider four important patterns of configuration

- *Composition of components.* This involves using two or more components together to form a larger component that combines the components' functionality and interfaces.
- *Wrapping of component interfaces.* Functionality is redefined for use in new context.
- *Further encapsulation of interfaces.* Functionality is hidden by constraining exposure of interface description.

- *Abstraction and instantiation of components.* A configuration of components is parametrized by abstracting over required functionality that may then be provided by a specific component via instantiation.

Composition is by far the most important means of building a software architecture. For instance, in component based design, a compound component is often specified first — to conform to an industry standard API, for instance. It is then built by combining smaller components.

Wrapping and encapsulation are also important to rationalize and encapsulate aspects of the design in divide-and-conquer fashion. For instance, if we wish to define a component with three functions, from two components each with twenty functions, some form of wrapping and encapsulation is important.

Abstraction and instantiation are important for component reuse, providing a means of developing generic components that suit a range of problem domains.

The structure of an executable specification defines its architecture as a component, where:

- basic executable specifications are basic components,
- unions and extensions are the means of combining components,
- renaming permits wrapping and adaption of component interfaces,
- hiding provides encapsulation of functionality, by constraining the interface, available to a component user, and
- generic named specifications and instantiated specifications are the means of providing component abstractions.

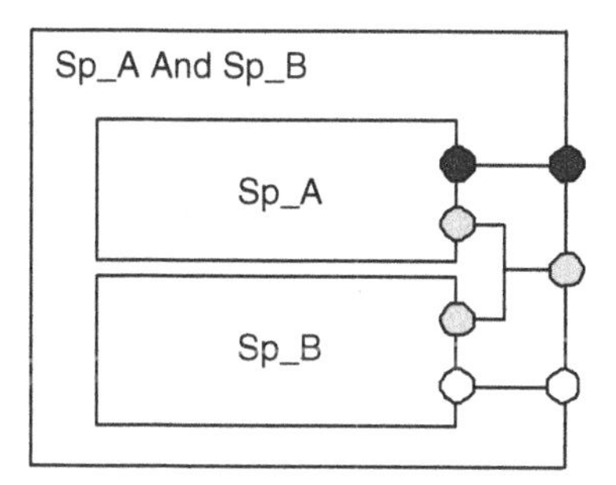

Fig. 11.1. The architecture corresponding to unions of two executable specifications.

We can visually represent how these operations define design decisions about component-based architectures. We represent components as boxes, with a list of circles denoting the component interface. A basic component is an empty box, while a compound component is a box that contains other components. A line between component interfaces denotes a usage relation.

Then, the implementation architecture resulting from unions of executable specifications is depicted as in Fig. 11.1. The two specifications are considered as components, with the union combining the components to form a compound component, and exposing amalgamated union signature as an interface. The final architecture is a compound component that is hierarchically composed from two smaller components, using their functionality to define a larger range of services. We consider the architecture of extensions to be similar.

The implementation architecture resulting from renaming of an executable specification is given in Fig. 11.2 The specification is considered as a component, with the renaming defining a wrapping of the component interface.

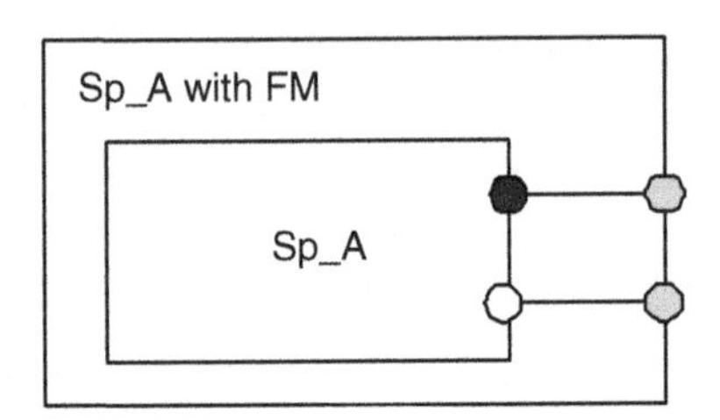

Fig. 11.2. An implementation architecture resulting from renaming of an executable specification.

The implementation architecture resulting from hiding of symbols in an executable specification is depicted in Fig. 11.3. The specification is considered as a component, with the hiding operator providing further encapsulation of functionality through hiding parts of the component interface.

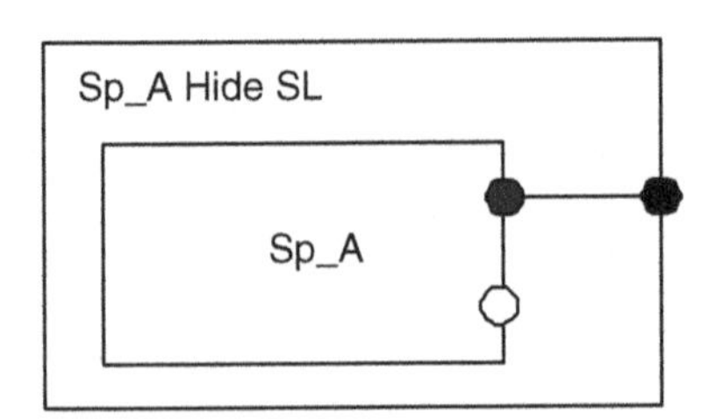

Fig. 11.3. An implementation architecture resulting from hiding of symbols for an executable specification.

In Fig. 11.4, the architectures resulting from (a) the definition and (b) instantiation of a generic executable specification. In Fig. 11.4 (a), the specification is considered as a template of a component, with the (possibly non-executable) parameter specifications representing parameters that need to be identified to make the template into a functioning architecture. In Fig. 11.4 (b), such a parameter argument has been found and is used to instantiate the architecture to form a functioning architecture.

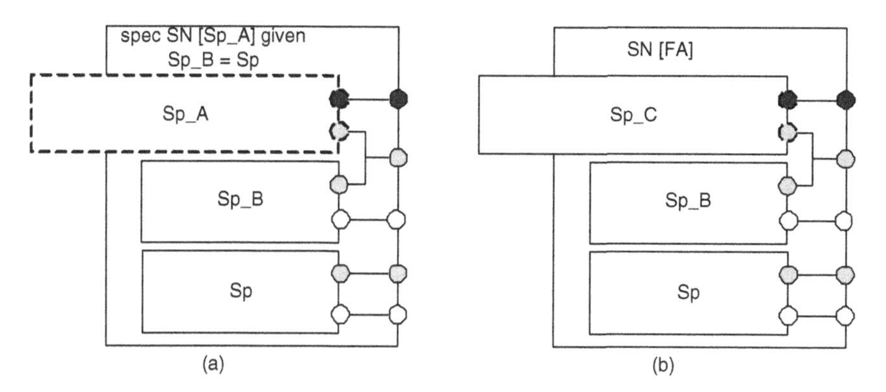

Fig. 11.4. Architectures corresponding to (a) the definition and (b) instantiation of a generic executable specification.

In designing a component-based system for extension, it is vital that architectures are designed carefully to address issues of optimal performance, reuse and maintainability. In general, this is achieved by careful use of design patterns for component composition, which are known to be *appropriate* for a problem domain. In practice, the appropriateness of a pattern is usually a judgment of the human designer, based on his/her experience of the domain — see [GHJV95], for instance. The field of software engineering is still too young for the fully automated design and configuration of appropriate software architectures.

For our purposes, this means that we require that our methods for construction of structured programs permit the formation of an architecture through *interactive* design decisions.

Remark 11.3. It is possible to define a further translation from executable *CASL* to a conventional structured programming language such as *SML* or C#, for compilation and integration with other software. Translation into these languages should preserve the architectural design decisions of executable specification components. Such a translation is possible — by virtue of the fact that these languages permit the architectural patterns we have discussed. For instance, it is possible to define a map from executable specifications to *SML* structures and functors. Essentially, we discard axioms, taking an equational functional definition as a function definition in a structure, and map specification signatures or *SML* signatures.

Using that translation, it is possible to see that the structural and abstraction operations for executable specifications correspond to patterns for composition, encapsulation, wrapping, abstraction and instantiation of modules and functors in *SML*.

11.2 Reasoning about structured programs

We can use SSL and program extraction for reasoning about and synthesizing structured programs. The idea follows simply because we consider executable *CASL* as our structured programming language. So, if we restrict our SSL to only reason about executable specifications, then our calculus and extraction results constitute a method for specification of, reasoning about, and synthesis of structured programs.

11.2.1 SSL for construction of components

By restricting our consideration to executable specifications only, the sequents of SSL are viewed as statements about components. So,

$$\vdash \text{SP} \diamond A$$

is understood as asserting that A is true of a component SP.

The logical rules of SSL are then taken as a means of deducing new truths about a component from its known properties. For instance, the ($\wedge$-I) rule permits us to prove the conjunction $(A \wedge B)$ about a component SP, given that we already have proofs of A and B from SP:

$$\frac{\Gamma_1 \vdash \text{SP} \diamond A \quad \Gamma_2 \vdash \text{SP} \diamond B}{\Gamma_1, \Gamma_2 \vdash \text{SP} \diamond (A \wedge B)} \;(\wedge\text{-I})$$

The structural rules are considered as a means of constructing new components from old ones, and deducing a truth about the result. For instance, the (union$_1$) rule

$$\frac{\Gamma \vdash \text{SP\_1} \diamond A}{\Gamma \vdash (\text{SP\_1 } \textbf{and} \text{ SP\_2}) \diamond A} \;(\text{union}_1)$$

tells us that, if A is true about a component SP_1, then A is also true about the component composition of SP_1 and SP_2.

Because the structural rules are used to build components, they are our means of designing new architectures. Under our view of specifications as components, these rules treat components as black-box. For instance, in the (union$_1$) rule above, we assume both specifications SP_1 and SP_2 have already been defined and are executable. The application of the rule combines components and reasons with their interface information (signature and logical axioms). However, the rule does not change or extend the interface or functionality of a component. In this sense, the structural rules are not concerned with the "inside" of a component — only with its interface.

To extend an architecture with additional required functionality, we use the rule (Sk) and extraction:

$$\frac{}{\vdash \mathsf{unextract}(\text{SP}, A, e)^{NewSpec(\text{SP},A,e) \diamond A}} \;(\text{Sk})$$

where e is a modular realizer of $\textsc{Sp} \diamond A$ (obtained by extraction from the proof of $\textsc{Sp} \diamond A$) and $NewSpec(\textsc{Sp}, A, e)$ is defined by

$$\textsc{Sp}\ \mathbf{then}\ \{\langle f_A : \mathsf{etype}(A)\}, \emptyset, \emptyset\rangle, \{f_A = e, Sk(A)\}\}$$

Under our view, extraction and (Sk) constructs a new component architecture,

$$NewSpec(\textsc{Sp}, A, e)$$

However, it also produces a new sub-component for use in this architecture,

$$\{\langle f_A : \mathsf{etype}(A)\}, \emptyset, \emptyset\rangle, \{f_A = e, Sk(A)\}\}$$

The sub-component involves a single function and has an interface that provides an axiomatization of this. This new component, when combined with $\textsc{Sp}$ to give $NewSpec(\textsc{Sp}, A, e)$, results in "correct" architecture. By "correct," we mean that given that $\textsc{Sp}$ is consistent, then the combination with the new component is also consistent (this is true by Theorem 9.4.1 of Chapter 9, p. 329).

Remark 11.4. Our calculus is interactive. Proofs can be done by hand, or by the use of encoding in interactive theorem prover, such as Isabelle [NPW02]. So the architectural design decisions made using the structural rules permit the architect to exercise his/her design experience to build an optimal architecture. On the other hand, use of the calculus ensures that the resulting architecture is correct with respect to a set of required properties given by the derived formula, or relatively consistent with respect to its derivation from other components.

11.2.2 Process of construction

Our methods suggest an approach to design and correct construction of software architectures.

Given a repository of correct components (executable, consistent specifications), we can use our calculus to combine these components to obtain new components that satisfy required properties, and then add the result back to the repository, for component reuse. This leads to the following process, depicted in Fig. 11.5:

Process 11.2.1. Take a repository consisting of a finite set of consistent, executable specifications (components). We extend the repository to include a new component by:

1. Repeatedly select components from the repository
2. Use SSL to simultaneously
    - construct a larger component from the selected components and
    - derive a required theorem about the result
3. Either
    - add new specification back to the repository, or

- use extraction and (Sk) rule to consistently extend the specification with a functional definition of the realizer of the derived theorem

4. Add the resulting specification back to the repository.

By the nature of the rules of SSL, the resulting extended repository is still consistent.

This process enables us to use SSL for component-based software development and, in particular, correct component reuse.

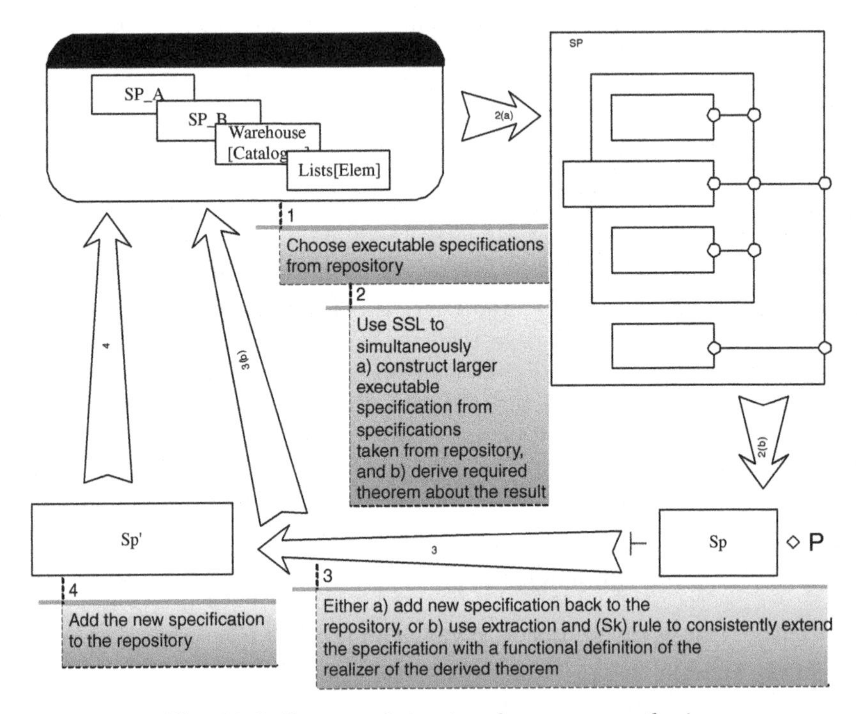

Fig. 11.5. Process of structured program synthesis

Example 11.1 *(Password checking system).* We consider how the process of structured program synthesis would work for the two main examples used in this part of the book.

Consider the password checking system example of the first three chapters of this part. In Chapter 7, we specified a password system for an email hosting service. An initial specification of the system's password requirements, PwdCore, was given in terms of subspecifications of the natural numbers, booleans and strings, together with some axioms to model the domain (Example 7.6, p. 234, Chapter 7). To restrict the specification of the system to relevant functionality, a final specification PwdSys was defined by hiding functions of PwdCore that we did not wish exposed.

For the purposes of illustration, we shall assume that PWDCORE and PWDSYS are executable specifications – all function symbols used in these specifications have executable definitions. This can be achieved simply by adding appropriate function definitions for the boolean, string and integer specifications.

We developed a theorem in SSL about PWDSYS:

$$\vdash \text{PWDSYS} \diamond \forall x : nat \bullet \exists y : string \bullet ValidMsg(x, y) \tag{11.3}$$

This specified that, given any input x of a password, there is always an appropriate response message to be output, explaining if the password is of the correct length or not (Section 7.3.6 of Chapter 7). We encoded our proof as a term in the logical type theory for SSL (Chapter 8, Section 3.2.1, p. 260). Finally, we used our synthesis techniques to extract a realizing function for the theorem (Chapter 9, Section 9.5).

Because the proof-term for the theorem involved a critical subterm, we extracted an extended realizer from the proof of the theorem for PWDSYS. To do this, we first need to obtain a modified realizer f for the proof of the required property over the subspecification PWDCORE, defined

$$\begin{aligned} fn\ x : nat => match\ rec([true, false], [inr(()), inl(())])inRange(x)\ with \\ Inl(x_u)\ => \text{`Password acceptable'}, \\ Inr(x_v)\ => \text{`Please choose a password in correct range'} \end{aligned}$$

such that

$$\text{PWDCORE} \models Sk(P)[f/f_P]$$

By application of the (Sk) rule, it is possible to obtain

$$\mathsf{unextract}(\text{PWDSYS}, A, \mathsf{extract}_{\mathsf{mod}}(q))^{NewSpec(\text{PWDSYS}, A, \mathsf{extract}_{\mathsf{mod}}(q)) \diamond A}$$

The new specification is a relatively consistent extension of PWDSYS:

$$\text{PWDSYS} \gg NewSpec(\text{PWDSYS}, A, f)$$

The application of the (Sk) rule corresponds to taking an implementation step in defining the password checking system, adding a correct function definition to the specification.

Now because we have assumed PWDCORE and PWDSYS are executable specifications, we have that $NewSpec(\text{PWDSYS}, A, f)$ is also executable. If PWDSYS is taken from a consistent repository of specifications, then, following our process, $NewSpec(\text{PWDSYS}, A, f)$ may be added back to the repository without affecting consistency.

11.3 Using extraction to obtain executable refinements

A specification can be refined to a specification, which retains the signature of the specification and includes all models of the original specification. Given certain kinds of specification expressions, we can obtain consistent refinements of specifications by the techniques of Chapter 9 (extended in Chapter 10).

Our refinement steps are given through the proof and extraction process. Briefly, our method of refinement is as follows. Given a (possibly parametrized) specification SP with universal axioms and given a function symbol f in the body of SP, we collect all the visible axioms for f into a formula, say $\forall x : s \bullet A(x, f(x))$. Then we build a modular proof of the unSkolemization

$$\mathrm{SP}' \diamond \forall x : s \bullet \exists y \bullet A(x, y)$$

using our calculus for some extension of SP. From this proof, we extract a lambda term as a modular realizer. We then extend the specification SP′, defining f to be equal to this term, and restrict the resulting signature to be the same as the original specification SP. By virtue of the fact that the lambda term is a modular realizer, this is a consistent refinement of SP with an executable definition of f. By repeating this process for all functions, we obtain a final executable refinement of the specification.

Our refinement method will not work for all specifications — we need to consider a subset, the *properly encapsulated* specifications. These are specifications where, given a function that is yet to be programmed, all axioms defining its behavior are visible.

Before explaining our method, we formalize the concept of refinement and proper encapsulation.

11.3.1 Refinements

The notion of refinement can be given as a partial order over specification expressions. One specification implements another if it shares the same signature and includes the model classes.

Definition 11.3.1 (Specification refinements). A specification SP_1 is said to be a *refinement* of a specification SP (written SP $\rightsquigarrow$ SP_1) if

- Sig(SP) = Sig(SP_1).
- all models of SP_1 (restricted to Sig(SP)) are also models of SP — that is, when every $C \in Mod(\mathrm{SP\_1})$ are such that $C|_{\mathrm{Sig(SP)}} \in Mod(\mathrm{SP})$.

We say a refinement SP $\rightsquigarrow$ SP_1 is relatively consistent if, assuming all Mod(SP) is consistent, then so is Mod(SP_1).

This definition can also be made to apply to generic named specifications, because Sig and Mod are defined over these specifications.

Note that, according to this definition, a refinement of a generic specification need not itself be generic. It is interesting to consider refinements of generic specifications that preserve the generic structure. This notion of *generic refinement* is now defined.

Definition 11.3.2 (Generic refinement). Let SN be the name of a generic specification defined by

$$\textbf{spec}\text{SN}[\text{SP\_1}]\ldots[\text{SP\_}n]\ \textbf{given}\text{SP}''\text{\_1}\ldots\text{SP}''\text{\_}m = \text{SP}$$

Now define a new generic specification SN′ by

$$\textbf{spec}\text{SN}'[\text{SP\_1}]\ldots[\text{SP\_}n]\ \textbf{given}\ \text{SP}''\text{\_1}\ldots\text{SP}''\text{\_}m = \text{SP}'$$

Then SN′ is said to be a *generic refinement* of SN if, and only if, for all fitting arguments $FA_1, \ldots, FA_n$, $\text{SN}'[FA_1]\ldots[FA_n]$ is a refinement of $\text{SN}[FA_1]\ldots[FA_n]$. We write $\text{SN}' \sqsubseteq \text{SN}$ in this case.

Lemma 11.3.1. *Specification building operators are monotonic with respect to* $\rightsquigarrow$. *That is, for any specification expressions* SP_1, SP_1, SP_2 *and* SP_2′, *if* SP_1 $\rightsquigarrow$ SP_1′ *and* SP_2 $\rightsquigarrow$ SP_2′, *then*

$$\begin{array}{c} \text{SP\_1}\ \textbf{and}\ \text{SP\_2} \rightsquigarrow \text{SP\_1}'\ \textbf{and}\ \text{SP\_2}' \\ \text{SP\_1}\ \textbf{then}\ \text{SP\_2} \rightsquigarrow \text{SP\_1}'\ \textbf{and}\ \text{SP\_2}' \\ \text{SP\_1}\ \textbf{hide}\ SL \rightsquigarrow \text{SP\_1}'\ \textbf{hide}\ SL \\ \text{SP\_1}\ \textbf{with}\ \rho \rightsquigarrow \text{SP\_1}'\ \textbf{with}\ \rho \end{array}$$

Proof. As in [Cen94, p. 172]. □

11.3.2 Proper encapsulation

A specification is *properly encapsulated* by hiding a symbol list when each visible function symbols has its axioms' visibility preserved, or else is already executable.

Definition 11.3.3 (Proper encapsulation). Let SP be a specification, and SL a symbol list. We say that SP is properly encapsulated by SL when, for each $f \in \text{Sig}(\text{SP}\ \textbf{hide}\ SL)$, either

1. each axiom in $Axioms(\text{SP})$ that involves f is also an axiom in $Axioms(\text{SP}\ \textbf{hide}\ SL)$, or
2. f has an executable definition in SP.

A specification is properly encapsulated if every subspecification of the form SP **hide** SL is such that SP is properly encapsulated by SL.

If the function has no definition, we should expose all information about the function's behavior for the purposes of refinement.

Remark 11.5. This definition restricts the hiding of functions and axioms. Given a properly encapsulated specification, all hidden functions have executable definitions. All visible functions have an axiomatization with no hidden axioms — we do not permit axioms for visible functions that involve hidden symbols.

11.3.3 Refinement method

Given a properly encapsulated specification SP, we can construct an executable specification SP_EXEC such that $SP \rightsquigarrow$ SP_EXEC.

Rather than attempting to achieve this in a single step, we proceed systematically in a stepwise fashion, incorporating more and more design and implementation decisions with each step. These include choosing between the options of behavior left open by the specification, between the algorithms that realize this behavior, between data representation schemes, etc. Each such decision is recorded as a separate step, typically consisting of a local modification to the specification.

So, developing structured programs from a specification then involves a sequence of such steps:

$$\text{SP\_0} \rightsquigarrow \text{SP\_1} \rightsquigarrow \text{SP\_N}$$

Here, SP = SP_0 is the original specification of requirements and each SP_I-1 $\rightsquigarrow$ SP_I for any $i = 1, \ldots, n$ is an individual refinement step. The aim is to reach a specification that is an exact description of an algebra. Our our case, the exact description is given by an executable specification.

Each of these steps involve deriving the definition of a function as constructive witness for the unSkolemized form of the axioms for a function, then adding the definition back into the specification to yield a consistent refinement. This process follows according to the proof of the following theorem.

Theorem 11.3.4. *Let* SP *be a specification expression that is properly encapsulated by SL with $Ax = Axioms(\text{SP})$ and $Sig(\text{SP}) = \langle S, TF, P \rangle$.*

Take a nonexecutable $f \in TF/SL$ and let D_f be the conjunction of every axiom in Ax that involves f.

Take a proof

$$\vdash d^{(\text{SP } \mathbf{then} \text{ SP}') \diamond A} \tag{11.4}$$

such that $Sk(A)[f/f_A] = D_f$, SP′ *is some specification expression and where* SP **then** SP′ *is a consistent implementation refinement of* SP*, and* SP′ *does not have f as a visible symbol.*

As the proof d must be modular, there is a modular realizer $\mathsf{extract}(d)$ *such that*

$$\text{SP } \mathbf{then} \text{ SP}' \diamond Sk(A)[\mathsf{extract}(d)/f_A]$$

Then

$$\text{SP} \rightsquigarrow (NewSpec((\text{SP } \mathbf{then} \text{ SP}'), A, e) \mathbf{ with } [f_A \mapsto f])$$

is a consistent refinement of SP *and is properly encapsulated.*

Proof. First note that, by the definition, $NewSpec(\text{SP } \mathbf{then} \text{ SP}', A, e)$ is a refinement of SP. So we need to show that this is a consistent refinement.

First, because we have an SSL proof (11.4), and we assume that all applications of structural rules in our proofs preserve consistency, we can assume that SP **then** SP′ is consistent.

Then, by Theorem 9.4.1 of Chapter 9 (p. 329), we have that

$$NewSpec(\text{SP } \mathbf{then} \text{ SP}', A, e) \tag{11.5}$$

is consistent. Assuming the normal form of this specification is

$$nf(NewSpec(\text{SP } \mathbf{then} \text{ SP}', A, e)) = \langle \Sigma, Ax \rangle \ \mathbf{hide}\ SL \tag{11.6}$$

we know that $\langle \Sigma, Ax \rangle$ **hide** SL is consistent (because normal forms preserve model classes), and contains more than the trivial model.

It remains to show that the renaming

$$NewSpec(\text{SP } \mathbf{then} \text{ SP}', A, e) \ \mathbf{with}\ [f_A \mapsto f] \tag{11.7}$$

is consistent. It is enough to show that

$$Mod(NewSpec(\text{SP } \mathbf{then} \text{ SP}', A, e) \ \mathbf{with}\ [f_A \mapsto f])$$

contains more than the trivial model.

Let ρ be the identity over all of Σ, except over f_A, where $\rho(f_A) = f$. It can be seen by (11.6), that the normal form of (11.7) must be

$$\begin{aligned} nf(NewSpec(\text{SP } \mathbf{then} \text{ SP}', A, e) \ \mathbf{with}\ [f_A \mapsto f]) &= \langle \rho(\Sigma), \rho(Ax) \rangle \ \mathbf{hide}\ SL \\ \langle \Sigma/f_A, Ax/\{D_f[f_A/f], f_A = e\} \cup f = e \rangle \ \mathbf{hide}\ SL & \end{aligned} \tag{11.8}$$

because we can assume that the hidden symbols SL do not contain f_A.

Now, as (11.5) is consistent, there is a non-trivial model

$$N \in Mod(\langle \Sigma, Ax \rangle \ \mathbf{hide}\ SL)$$

but, by the semantics of specifications with hide, this must be the reduct of a non-trivial model

$$M \in Mod(\langle \Sigma, Ax \rangle)$$

We construct a non-trivial model M' of $\langle \Sigma/f_A, Ax/\{D_f[f_A/f], f_A = e\} \cup f = e \rangle$ from M. We define M' from M so that

$$\begin{aligned} s^{M'} &= s^M \quad s \text{ a sort of } \Sigma/f_A \\ R^{M'} &= R^M \quad R \text{ a predicate of } \Sigma/f_A \\ g^{M'} &= \begin{cases} f_A^M & \text{if } g \text{ is function symbol } f \\ g^M & \text{if } g \text{ is any other function symbol} \end{cases} \end{aligned}$$

Now,

$$M \models Ax/\{D_f, D_f[f_A/f], f_A = e\}$$

but because $D_f, D_f[f_A/f], f_A = e$ are the only axioms in Ax that involve f and f_A, it can be seen that

$$M' \models Ax/\{D_f, D_f[f_A/f], f_A = e\} \tag{11.9}$$

But also, by definition of M', it can be seen that, as

$$M \models D_f[f_A/f], f_A = e$$

it must be the case that

$$M' \models D_f, f = e \tag{11.10}$$

So (11.9) and (11.10) show that there is a non-trivial model M' for the specification

$$\langle \Sigma/f_A, Ax/\{D_f[f_A/f], f_A = e\} \cup f = e\rangle$$

It follows from this that, by hiding the list of hidden symbols for the normal form (11.8), we have a non-trivial model for (11.7) as required. □

Remark 11.6. Observe that Theorem 11.3.4 will fail to hold if the specification to be refined, SP, is not properly encapsulated. Because we will use repeated applications of this theorem to obtain refinements, this means we must only consider a refinement process that begins with a properly encapsulated specification.

If SP is not properly encapsulated, the visible axioms for f, D_f, will not be all axioms in the normal form for SP, and so the normal form (11.6) in the proof

$$nf(NewSpec(\text{SP } \mathbf{then} \text{ SP}', A, e)) = \langle \Sigma, Ax\rangle \ \mathbf{hide}\ SL$$

will contain axioms in Ax for f in addition to D_f. So the fact that we know $D_f[f_A/f]$ is true for models of (11.6) is not enough to entail that (11.9) is true

$$M' \models Ax/\{D_f, D_f[f_A/f], f_A = e\}$$

for the model M' that takes f_A and f to denote the same object.

Remark 11.7. Note that implicit axioms obtained from schema are always preserved across functions. They need not be proved or unSkolemized. This is because, if I is an implicit axiom, then SP $\diamond$ $I[f/x]$, SP $\diamond$ $I[f_A/x]$, and SP $\diamond$ $I[\mathsf{extract}(p)/x]$ will all hold.

Remark 11.8. In Theorem 11.3.4, consistency of the refined specification is guaranteed, provided we assume consistency of the extension SP **then** SP′ has been verified during its derivation in the proof (11.4):

$$\vdash d^{(\text{SP } \mathbf{then} \text{ SP}')\diamond A}$$

If instead we prove the unSkolemized axioms A with respect to SP

$$\vdash d^{\text{SP}\diamond A}$$

then the refinement is consistent only with respect to the consistency of SP (not with respect to some extension). Thus, use of the extension introduces extra consistency obligations in the derivation, but enables the use of other structured specifications that may help in the derivation.

We use this theorem to define a refinement process for non-generic properly encapsulated specifications as follows.

Process 11.3.5. Let SP be a non-generic properly encapsulated specification.
Let R denote a finite set of executable specifications.
We define the following refinement process

$$\text{SP} = \text{SP\_0} \rightsquigarrow \text{SP\_1} \rightsquigarrow \text{SP\_N} = \text{SP\_EX}$$

where SP_EX is an executable specification

1. Let $i = 0$.
2. Take the first visible function symbol f in the specification SP_I that is not executable.
3. Let D_f be the conjunction of all the visible axioms for f, and let A be the unSkolemized form of D_f — so that $Sk(A) = D_f[f/f_D]$.
   Prove the theorem
   $$\vdash p^{(\text{SP } \mathbf{then} \text{ SP}')\diamond A} \tag{11.11}$$
   so that SP′ ∈ R is some specification expression where SP **then** SP′ is a consistent refinement of SP, and where SP′ does not have f as a visible symbol.
4. Use Theorem 11.3.4 to obtain a refinement specification
   $$\text{SP\_I+1} = (NewSpec((\text{SP } \mathbf{then} \text{ SP}'), A, e) \ \mathbf{with} \ [f_A \mapsto f])$$
   as a consistent refinement of SP_I.
   Then SP_I+1 has one less nonexecutable function than SP_I.
5. If SP_I+1 is executable, then $n = i + 1$ and we are done. Otherwise, let $i = i + 1$ and continue as in step 2.

The nature of this process guarantees that the refinement process will terminate, as each refinement will have strictly one less non-executable function symbol than the last, provided we can find proofs of the form given by (11.11).

We now extract programs for every function declared in the imports SP″_1, ..., SP″_m or in the body SP.

Process 11.3.6 (Executable refinements from a generic specification).
Suppose SN names a consistent, generic specification of the form

$$\mathbf{spec}\text{SN}[\text{SP\_1}]\ldots[\text{SP\_}n] \ \mathbf{given} \ \text{SP}''\text{\_1}\ldots\text{SP}''\text{\_}m = \text{SP},$$

where SP″_1 ... SP″_m are executable, all the sorts declared in the body of SP are basic or have free datatype declarations and the unSkolemized versions of all the axioms of SP can be proved using only executable specifications and the parameter specifications. Then we can obtain a consistent, executable refinement, SN_EXEC, that is a conservative extension of SN.

We sketch the process for producing a chain of refinements SN = SN_0 $\sqsubseteq$ SN_1 $\sqsubseteq$ ... SN_n = SN_EXEC. We shall assume that each SN_i is of the form

specSN_i[SP_1]...[SP_n] **given** SP$''$_1...SP$''$_m = SPBODY_i.

1. Given the properly encapsulated generic specification SN_i, take any function f in SPBODY_i for which there is no executable definition.
2. Prove the unSkolemization, A_f, of the conjunction, D_f, of all the visible axioms for f to give a proof

   $$\vdash \text{SP}'' \diamond A_f$$

   where

   SP$''$ = {SP$''$_1 **and** ... **and** ...SP$''$_n} **then**
   {SP_1 **and** ... **and** ...SP_m} **then** SPBODY_i **then** SP$'$

   and where SP$'$ $\in$ R is an executable specification from the repository.
3. Apply Theorem 11.3.4 to obtain a consistent refinement

   SP_I+1$'$ = ($NewSpec$((SP$''$ **then** SP$'$), A_f, e) **with** [$f_A \mapsto f$])

   because f_A does not occur in SP$''$, this specification has the same signature and model classes as

   SP_I+1$'''$ = SP$''$ **then** SP$'$ **then**
   $(\langle\langle\emptyset, f_A : \mathsf{etype}(A_f), \emptyset\rangle, \{D_f[f_A/f], f_A = e\}\rangle)$ **with** [$f_A \mapsto f$]
   = {SP$''$_1 **and** ... **and** ...SP$''$_n} **then**
   {SP_1 **and** ... **and** ...SP_m} **then** SP_I+1$''''$

   with

   SP_I+1$''''$ =
   SPBODY_i **then** SP$'$ **then**
   $(\langle\langle\emptyset, f_A : \mathsf{etype}(A_f), \emptyset\rangle, \{D_f[f_A/f], f_A = e\}\rangle)$ **with** [$f_A \mapsto f$]

4. Now set

   SN_$(i+1)$ = **spec**SN$'$_i[SP_1]...[SP_n] **given** SP$''$_1...SP$''$_m
   = SP_I+1$''''$

Then SN_$(i+1)$ is a refinement of SN_i which, by construction is a conservative extension. Each SN_$(i+1)$ contains one less non-executable function than the previous SN_i, so the chain of refinements must terminate at some finite stage r, say, so that SN_r = SN_EXEC, as required.

Example 11.2 *(Refinement of the warehouse specification).* Consider the warehouse example of the previous chapter, given by the generic specification WAREHOUSE,

$$\begin{aligned}&\textbf{spec } \mathrm{WAREHOUSE}[\mathrm{CATALOGUE}] =\\&\textbf{ops } rep : Part \to Part\\&\textbf{axioms } \forall i : Part \bullet size(lOR(i)) > 0 \Rightarrow In(rep(i), myCat)\\&\qquad\qquad\qquad\qquad\qquad\qquad \Rightarrow Rep(rep(i), i, myCat)\end{aligned}$$

Given a faulty part name as input, the function *rep* uses the catalogue to obtain a replacement part, if it exists.

In Section 10.5 of that chapter we derived the following:

$$\vdash p_3^{\mathrm{CATALOGUE}\ \textbf{then}\ \mathrm{BODYWARE}} \diamond A \tag{11.12}$$

with a proof-term of the form

$$p_3 = \mathsf{use}\ i : Part.\ \mathsf{show}(hd(lOR(i)), \mathsf{ext}_2(\mathsf{app}(q_6, q_5), \mathrm{CATALOGUE}))$$

where A is

$$\forall i : Part \bullet \exists y : Part \bullet In(y, myCat) \Rightarrow Rep(y, i, myCat)$$

This theorem is the unSkolemized form of the axiom for *rep*, D_{rep}

$$\forall i : Part \bullet size(lOR(i)) > 0 \Rightarrow In(rep(i), myCat) \Rightarrow Rep(rep(i), i, myCat)$$

in the warehouse specification body.

Because the proof-term is modular, there is a modular realizer of (11.12) such that

$$\begin{aligned}&\mathrm{CATALOGUE}\ \textbf{then}\ \mathrm{BODYWARE} \models\\&\quad \forall i : Part \bullet In(\mathsf{extract}(p_3)(i), myCat) \Rightarrow Rep(\mathsf{extract}(p_3)(i), i, myCat)\end{aligned}$$

We can apply Theorem 11.3.4 to obtain a refinement

$$\begin{aligned}&\mathrm{CATALOGUE}\ \textbf{then}\ \mathrm{BODYWARE} \rightsquigarrow\\&\qquad (NewSpec((\mathrm{CATALOGUE}\ \textbf{then}\\&\qquad \mathrm{BODYWARE}), A, \mathsf{extract}(p_3))\ \textbf{with}\ [f_A \mapsto rep])\end{aligned} \tag{11.13}$$

Because CATALOGUE **then** BODYWARE is properly encapsulate this is a consistent refinement.

Now, because *rep* is the only visible nonexecutable function symbol in the body of WAREHOUSE, the refinement (11.13) can be used in a single step application of Process 11.3.6 to obtain the generic specification refinement

$$\mathrm{WAREHOUSE}' \sqsubseteq \mathrm{WAREHOUSE}$$

where WAREHOUSE′ is defined

$$\textbf{spec}\ \text{WAREHOUSE}'[\text{CATALOGUE}] = \text{BODYWARE}\textbf{then}$$
$$\langle\langle\emptyset, f_A : \mathsf{etype}(A), \emptyset\rangle, D_{rep}[f_A/rep], f_A = \mathsf{extract}(p_3)\rangle \textbf{with}\ [f_A \mapsto rep]$$

This specification can be rewritten to the equivalent, easier to read executable refinement

spec WAREHOUSE[CATALOGUE] =
BODYCAT **then**
ops $f_A : Part \to Part$
axioms $\forall i : Part \bullet In(f_A(i), myCat) \Rightarrow Rep(f_A(i), i, myCat)$;
$f_A =$ `fn i => hd(listOfReplacements(i))`

11.3.4 Software development via specification refinement

Our methods yield a notion of provably correct systematic component development and reuse.

Our method of refinement is done with respect to a repository of structured programs. Once we refine a specification into a structured program, we can add it back to this repository. This repository consists of executable specifications, consistent with respect to their derivation by refinement. The expanded repository can then be used to the refinement of other abstract specifications.

A possible systematic development process results, of the form of Fig. 11.6. By virtue of the refinement process, components developed in this way are provably correct, in the sense of being relatively consistent with respect to the specifications from which they are refined.

11.4 Discussion

Our methods require us to consider the relation between the specification structure and required implementation *architecture* carefully. We showed that specification building operations correspond to architectural design decisions. In particular, we found that

- Our refinement process preserves and extends the structure of the original abstract specification in the architecture of final executable specification. So we must be careful with the abstract specification's structure, because decisions about this structure will affect the executable specification structure.
- Our use of the SSL calculus as in a method of constructing correct structured programs entails that the structural and Skolemization rules now correspond to architectural design decisions.

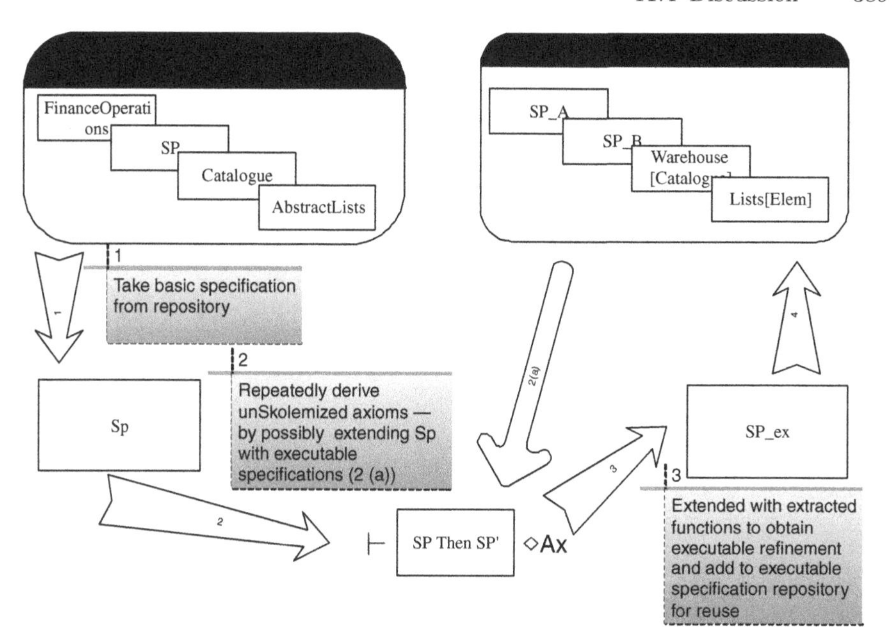

Fig. 11.6. Software development using the process of executable refinement (Process 11.3.6).

This is in contrast to usual approaches to specification refinement, where a structured specification is not considered to impose any architectural constraints on its final implementation. Architectural considerations are left as outside the scope of structured specifications.

Our techniques are compositional by virtue of the fact that they are grounded in a compositional proof system, SSL. SSL derives a proof about a structured specification in a modular fashion, using knowledge about subspecifications to derive knowledge about the composed specification. By employing extraction to extend and refine specifications from a proof, we compositionally employ known results about subspecifications. Thus, the divide-and-conquer approach of proof reuse in SSL corresponds to a form of knowledge reuse in construction of architectures.

Part V

Epilogue

12

Conclusions: Toward Constructive Logic as a Practical 4GL

As we said at the beginning, ultimately, we would like to solve problems by building well-structured, comprehensible, correct programs, solely through reasoning with domain knowledge. The discipline of the software engineer is still far from realizing this goal in an industrial setting. This monograph has explored research ideas that are a small step in this direction which may be useful in the long-term.

Traditional software development distinguishes between specification of requirements and implementation of requirements. Specifications are not formally included in the commonly used industrial third generation programming languages (3GLs).[1] Instead, a specification is written in a different system (such as English or predicate logic), and serves as a prescriptive goal to be achieved through implementation.

This distinction leads to a division between the tasks of specification and implementation. Although sometimes desirable, problems can result. In particular:

- The separation of tasks leads to an inefficient demarcation of development roles, between specification expert and programmer. A specification expert is needed to understand a problem, but is not usually skilled in providing a solution. Implementation is a task that requires knowledge of and experience with the programming language. A programmer is skilled in the technology to provide a solution, but, by virtue of this, usually lacks knowledge about

[1] We define 3GLs to be high-level procedural languages in which single instructions abstract away from several lines of machine code. Examples of 3GLs are FORTRAN, COBOL, C, C++, Java or C#. Programming a 3GL requires expertise that is specific to this task. In contrast, 4GLs are meant to abstract away from 3GLs, with the aim of enabling end-user programming that is, programming of a product by the person who will use the product [Sch99a, 440–443]. For example, a spreadsheet package may be thought of as a 4GL, which enables end-user programming for an accountant. 4GLs may be procedural, such as Visual Basic for Applications (VBA), or declarative, such as Microsoft Excel, SQL or Prolog.

the problem domain. The different skills sets of the two roles can impede communication and the process of development.
- Traditional implementation involves many issues that are orthogonal to the problem domain. It would be preferable to minimize these issues, to facilitate development that is driven entirely by specification, where the only skills required are knowledge of and the ability to reason about the problem to be solved.
- Because implementation is a separate task from specification, there is usually no formal guarantee that an implementation correctly satisfies its specification. For instance, ambiguities can arise in the transition from a specification to implementation.

To solve these problems, program construction should be driven primarily by knowledge of the problem domain, and less by orthogonal, language specific issues.

This is the goal of many approaches to program synthesis. The idea is that, given a specification, an automated or interactive process of synthesis yields a program that behaves according to the specification. In this way, program synthesis occurs at a higher level of abstraction than 3GL programming, forming a 4GL infrastructure that can treat specification declaration as a part of the programming task. Following Tyugu [Tyu88, p. 8], we distinguish three different approaches to program synthesis:

- *Transformational synthesis:* a program is derived stepwise from a specification by means of transformations or refinements. Refinement calculi [Dij76, MV93, Mor94, Bac80] achieve transformational synthesis through languages that mix non-executable specifications and programs. These calculi provide rules for refining non-executable specifications into executable terms that satisfy the specification. Repeated recursive application of rules over a term with non-executable subterms will eventually yield an executable term. Related techniques [HHS85] have been employed to obtain structured programs from both model-oriented specifications (such as B specifications [Abr96, pp. 501–550]), and from algebraic structured specifications (such as OBJ or CASL [CoF01]).
- *Deductive synthesis:* uses deduction of a proof of solvability of a problem and derives a program from the proof. Deductive synthesis can be interactive (semi-automated) or completely automated.
  Automated deductive synthesis is what occurs in high-level logic programming languages like Prolog and automatic theorem provers such as the Boyer–Moore prover [BM79] or OTTER [McC92].
  Interactive deductive synthesis often involves constructive logic, such as the systems described in [MW91] and [KBB93]. Of these constructive approaches, a subset is based on forms of constructive type theories [HN88, Tyu88, CMH86, CH88, PC01, CP01].
- *Inductive synthesis:* a program is based on a declaration of input-output requirements or examples of input-output pairs. Examples of methods that

fall into this category include inductive logic programming [Plo71, Mug92] and neural and belief networks [RN95, pp. 563–597].

This monograph has been concerned with deductive and transformational approaches. We have considered

- interactive deductive synthesis of imperative programs with side-effects (in Part III) and
- the combination of deductive and transformational approaches for structured program synthesis (in Part IV).

Our approaches generalize the deductive synthesis techniques commonly referred to as *proof-as-programs*, based on constructive logic, type theory, and the Curry–Howard isomorphism.

We have used a framework, called the Curry–Howard protocol, to identify how to adapt proofs-as-programs to new logics and programming paradigms. A strength of this framework is its apparent generality — we have demonstrated it successfully with two separate contexts (imperative programs and structured specifications). However, further use of the protocol will determine its ultimate success as a general framework.

Why is it interesting to adapt proofs-as-programs to the new contexts?

We argue that the construction of a program should be driven more by knowledge of the problem domain and less by orthogonal, language specific issues. This is the goal of fourth generation programming languages (4GLs). To move toward this goal, we have advocated the integration of specifications and, more generally, the ability to reason about domain knowledge into the process of program construction. Constructive logic permits us to define a system while hiding as much low level implementation detail as possible. Constructive logic does not require implementation knowledge, but only knowledge of, and the ability to reason about, the problem domain.

In contrast to declarative, fully automated program synthesis, such as SQL or Prolog, but similar to procedural languages such as C, Java or Microsoft's VBA, a constructive logic requires the development of a program to be guided incrementally by a system designer. This permits a degree of design freedom, which is important when developing complex system architectures for reuse and manageability. Our work defines a 4GL in the sense that, programs can be constructed purely through reasoning about domain knowledge. By the fact that we use adapt constructive synthesis, we can guarantee that our synthesized programs are correct with respect to their specification.

There already exist experimental 4GLs. However, most do not address two concerns that are important for industry uptake:

- Many industrial programming tasks are essentially imperative in nature. Common tasks usually involve state in some sense — for example, accessing and changing values in a database is fundamental to most industrial applications. We argue that imperative issues will never disappear as industrial concerns, no matter how high level the language used for development.

Therefore a useful 4GL must in some way incorporate state. Our work in imperative program synthesis in Part III addresses state issues, by extending 4GL proofs-as-programs ideas.

- Industrial strength programs are large and often difficult to maintain. Structured programming is an important means of developing and maintaining a system according to compositional, divide-and-conquer principles. It is therefore important that an industrial strength 4GL be intrinsically structured and compositional. The work of Part IV in synthesis, extension and refinement using structured algebraic specifications represents a step in this direction.

Our systems are by no means ready for industry use as they stand. Further work must involve examining efficiency of extracted programs and ease of use of our systems within a theorem proving environment.

However, we believe the results of this monograph are positive and, it is hoped, will stimulate further investigation. By adapting constructive synthesis, we have contributed some results toward the possibility of constructive logic as a 4GL and the goal of building complex, correct imperative and structured programs, solely through reasoning with domain knowledge.

Part VI

Appendix

A

Constructive Logic

This appendix presents some background information on constructive logic, type theory and the Curry–Howard isomorphism.

A.1 Constructive logic

Logic is the study of formal systems of deduction.

Classical logic is the most well-known formal system, formalizing the way in which mathematicians commonly reason. However, many other formal systems have been developed to reasoning about different domains.

Other formal systems include, amongst many others, modal logics [HC68], the temporal logic of actions (TLA) [Lam94], the Hoare logic [Hoa69] and linear logic [Gir87]. Each of these logics formalize some aspect of reasoning about a problem domain. For instance, modal logics are used to formally reason about possibility and necessity, while the TLA and Hoare Logic can be used to reason about the dynamic behavior of distributed and imperative programs, respectively.

Following classical logic, constructive (or intuitionistic) logic is one of the earliest formal systems of deduction. Constructive logic has its roots in the philosophy of Intuitionism, a position on the foundations of mathematics. Intuitionism was first described by Brouwer [Bro75, Bro81]. Its formalization with constructive logic was given by Heyting [Hey71]. One of its most important philosophical exponents was Dummett, with his anti-realist verificationist meaning theory [Dum77, Dum91].

Intuitionism restricts the ways in which mathematical reasoning should be done. It maintains that mathematics is dependent solely on the intuition of the creative subject (the mathematician). This means that the objects of mathematics are constructions of the subject, whose properties and meaning are given entirely by virtue of their construction. Consequently, for an intuitionist, a mathematical proof is permissible in as much as it encodes the constructions of the subject. For instance, the truth of the statement $(A \vee \neg A)$ is only known if we can provide a construction that shows either A or $\neg A$ to be true.

This is in contrast to the Platonic, realist view on mathematics, where objects and their meaning are considered to be external and objective, existing in their own right apart from subjective creation. In that view, we can reason about properties of an object that do not have to be constructed. Classical logic complements this Platonic view. The semantics of classical logic presupposes an external world in which statements are objectively true or false. Thus, in classical logic, the statement $(A \vee \neg A)$ is true, because either A is true or not.

For the most part, intuitionism gained acceptance only in philosophical circles. It was not adopted by mathematicians, who, on the whole, freely use non-constructive arguments when it suits them. However, constructive logic has found great application in the realm of computer science, where the notion of construction is fundamental (all programming is about constructing functions). Its application is largely due to the correspondence between constructive proofs and algorithms, resulting the Curry–Howard isomorphism property.[1]

A.1.1 Constructive evidence

Constructive logic is distinguished in that it requires constructive evidence of a formula's truth. There three important formal definitions of "constructive evidence":

1. a constructive proof,
2. a realizer, or
3. inhabitation of the formula seen as a type.

The Curry–Howard isomorphism is a property of constructive logic that shows that these three notions are coincident. We now briefly review these definitions and the isomorphism.

A.1.2 Constructive proofs

The notion of a constructive proof was first made precise by Kolmogorov and Heyting [Kol32, Hey71]. In the Brouwer–Heyting–Kolmogorov (BHK) explanation, provability of a compound formula is given in terms of the provability of the components of the formula. This may be defined as follows, assuming that we have a notion of constructive evidence for atomic formulae:

- A proof of an atomic formula A is constructive evidence that guarantees A.
- A proof of $(A \wedge B)$ is a construction that provides a proof of A and a proof of B.

[1] Also, from a philosophical perspective, it has been argued that constructive logic need not be tied to intuitionism, and can in fact complement weaker notions of constructivism, realist or Platonist views [Res00].

- A proof of $(A \vee B)$ is a construction that provides either a proof of A or a proof of B.
- A proof of $(A \Rightarrow B)$ is a construction that, when given any proof of A provides a proof of B.
- A proof of $\forall x \bullet A$ is construction that, when given a term a, provides a proof of $A[a/x]$.
- A proof of $\exists x \bullet A$ is a construction that provides a witness term a and a proof of $A[a/x]$.
- There is no proof of the absurdity, $\perp$.

In this way, constructive evidence to support the truth of a compound formula is to be found in the evidence that supports its components.

Gentzen devised the natural deduction calculus, a set of rules for building constructive proofs [Gen69]. For instance, the ($\vee$-I$_1$) rule of natural deduction tells us that we have a constructive proof of $(A \vee B)$, provided that we have a constructive proof of A:

$$\frac{\Gamma \vdash A}{\Gamma \vdash (A \vee B)} \; (\vee\text{-I}_1)$$

The ($\forall$-I) rule tells us that we have a proof of $\forall x \bullet A$ provided that we have a proof of $A[y/x]$ for free variable x in A:

$$\frac{\Gamma \vdash A[y/x]}{\Gamma \vdash \forall x \bullet A} \; (\forall\text{-I})$$

A.1.3 Realizability

The original notion of realizability was introduced by Kleene in [Kle45, Kle52] as a semantics for intuitionistic arithmetic. His idea was to show that natural numbers can be used to encode constructive evidence of formula truth.

Given formulae about the natural numbers, we can take the set of realizers to range over functional terms built from the naturals with pairs, projections (π_1 and π_2), disjoint unions (formed using inl and inr) and a unit constant ().[2] Then, we say that a formula is valid (realizable) when we can find a realizer for it, according to the following definition:

- () is realizer of an atomic formula A if A is true.
- p realizes $(A \wedge B)$ if $\pi_1(p)$ realizes A and $\pi_2(p)$ realizes B.
- p realizes $(A \vee B)$ if it is of the form $\mathsf{inl}(q)$ with q realizing A, or of the form $\mathsf{inr}(q)$ with q realizing B.
- p realizes $A \to B$, if it is a functional such that, given any realizer q of A, (pq) is a realizer of B.

[2] Kleene's original definition was for formulae, and involved functionals encoded as natural numbers. Later, more general definitions of realizability were developed using combinatorial algebra or the lambda calculus to define functionals.

- p is realizer of $\forall x \bullet A$ if it is a functional, such that, when given a number t, provides a realizer (pt) of $A[t/x]$.
- p is a realizer of $\exists x \bullet A$, such that, $\pi_1(p) = t$ for a witness number t and $\pi_2(p)$ realizes $A[t/x]$.
- There is no realizer of the contradiction, $\perp$.

When a realizer can be constructed for a formula, it is true. In this way, a realizer corresponds to Brouwer's notion of constructive evidence to support the formula.

This notion of realizability can be adapted to many-sorted and higher-order formulae. Since then, various variations on the idea have been developed — see Troelstra [Tro73, Tro98] for a detailed overview. In our work, we shall be concerned with realizability for many-sorted formulae with functional sorts.

A.1.4 The Curry–Howard isomorphism

A third notion of constructive validity is given by the Curry–Howard isomorphism. This property tells us that constructive natural deduction corresponds to a kind of type theory, where proofs correspond to terms, formulae to types, logical rules to type inference and proof normalization to term simplification. The original idea was first described by Curry in [Cur34] and extended to intuitionistic first order logic by Howard [How80].

Essentially, a constructive type theory corresponding to natural deduction for predicate logic is a typed lambda calculus with dependent product and sum types and disjoint unions. The rules of natural deduction then have corresponding type formation rules.

Example A.1. The ($\vee$-I$_1$) rule of natural deduction corresponds to a typing rule

$$\frac{\Gamma \vdash p^A}{\Gamma \vdash \mathsf{inl}(p)^{A \vee B}} \ \vee\text{-I}_1$$

which tells us that $\mathsf{inl}(p)$ is correctly typed with $A \vee B$, provided that p is typed with A. The formula $A \vee B$ is taken to be a disjoint union type.

Example A.2. For instance, the ($\forall$-I) rule of natural deduction for first order constructive logic with arithmetic corresponds to a typing rule

$$\frac{\Gamma \vdash p^{A[y/x]}}{\Gamma \vdash \lambda x.p^{\forall x \bullet A}} \ (\forall\text{-I})$$

that tells us that $\lambda x.p$ is correctly typed with $\forall x \bullet A$, provided that p is typed with $A[y/x]$. The formula $\forall x \bullet A$ is taken as a dependent product type, by virtue of the type inference rule corresponding to ($\forall$-E):

$$\frac{\Gamma \vdash p^{\forall x \bullet A}}{\Gamma \vdash \lambda x.(pa)^{A[a/x]}} \ (\forall\text{-E})$$

This is the elimination rule for dependent product types, because it shows that $\forall x \bullet A$ parametrizes the type A over possible instantiation by the number a.

We can view the isomorphism as a means of relating the BHK interpretation of logical connectives to a form of constructive realizability (similar to that given by Diller [Dil80]). This is achieved if we take realizability to be inhabitation of types by terms, as the lambda calculus enables definition of functionals.

A.2 Constructive type theories

First order and many-sorted logic have straightforward type theories, similar in form to that given by Howard [How80] for first order logic (see, for instance, [Sch99b, pp. 1–13]). Crossley and Shepherdson [CS93] provided a constructive type theory that is modular over sorts (datatypes such as natural numbers, booleans, lists, etc). We will also see that, by treating sorts as closed under functional constructors, we can provide a limited form of higher-order reasoning.

The Curry–Howard isomorphism also can be applied to a range of fully higher-order constructive type theories, each corresponding to a different form of constructive logic that permits predication over logical formulae.

The main motivation of these theories is to provide a single framework that unifies logical reasoning about computational objects with the typing of these objects. Typically, a programming language can be understood with a type system that defines the type of values for input and output of functions, and reduction rules for evaluating function application. By the Curry–Howard isomorphism, logic is also such a type system. Because logic reasons about computational entities, it is of foundational interest to examine how logical and computational domains can be unified within a single type theory.

For our purposes, unified higher-order theories are worth describing briefly, as they form the basis of two important program synthesis methods (*Nuprl* and *Coq*). As we shall later see, our work presents a different approach to program synthesis, which argues against unified type system for programming and reasoning.

A.2.1 Higher-order type theories

Higher-order type theories fall into two camps: the predicative theories and impredicative theories. The differences lie in impredicativity — the scope of quantification over types to form new types. Impredicativity is an important issue, because, if treated incorrectly, a paradoxical type system can arise. For example, Martin-Löf's original constructive type theory permitted a type of all types that contains itself as a term and is closed under quantification. Girard showed that this type theory entails that every type is inhabited and so corresponds to an inconsistent logic, exhibiting a form of Russell's paradox [Gir72].

A.2.2 Predicative type theories

The predicative type theories of Martin-Löf [ML75, ML84] restricts quantification according to hierarchies of type universes. Quantification over types from one universe forms a type of a higher universe, such that lower universe types cannot quantify over types of higher universes.

The idea is as follows. Basic computational datatypes (for example, integers and booleans) are defined in a base universe U_0. This universe is closed under logical formula types that predicate over the basic types. However, datatypes that define functions over U_0 or formulae that quantify over U_0 are necessarily types of the next universe U_1. The rest of the universe hierarchy is similarly constructed.

In these theories, datatypes and logical types can be represented within a single type theory and are treated similarly. Logical types are introduced by means of introduction and elimination rules corresponding to natural deduction rules. Datatypes are also introduced via construction typing rules for constructors and eliminated by typing rules for recursion operators.

These predicative theories are essentially functional programming languages, but with more powerful type systems. This is an advantage for the programmer, because, in particular, Martin-Löf describes an extensible methodology in which we are permitted to define new datatypes, simply by adding introduction, elimination, and reduction rules. This corresponds to programming practice of defining new abstract data types. However, these theories have a disadvantage for the logician or mathematician, as their syntax and usage differs significantly from more commonly used deductive systems (such as many-sorted logic).

A.2.3 Impredicative type theories

The impredicative type theories of, for instance, Girard [Gir72], Reynolds [Rey74] and Coquand [MLM90] permit quantification over types to form a type itself. For instance, in Coquand's calculus of constructions [MLM90], formulae types may be formed by universal quantification over any type including *Prop*, the type of all formulae types. Thus, $\forall P : Prop.P$ is a proposition of type *Prop*. To avoid paradox, these impredicative theories require that the type of formulae types is not itself a formulae type, but a different kind of type (so all formulae correspond to types, but not all types correspond to formulae).

In contrast to predicative theories, impredicative theories do not treat datatypes and logical propositions as the same — but rather encode the former using the latter. This results from the computational power of predicative theories, that permits an encoding of commonly used computational datatypes (see, e.g., [BB85]). For example, the natural numbers datatype can be represented using Church numerals, with the polymorphic type $\forall P : Prop.P \to (P \to P) \to P$.

While mathematically elegant, such encodings are unnatural and inefficient for programming. No software engineer would use Church numerals to write arithmetic functions, because the syntax is difficult to understand and evaluation is typically impractical (for example, Church-style numbers have no linear-time predecessor function).

Luo's extended calculus of constructions [Luo94] solves some practical problems with impredicative theories, unifying an impredicative logical type theory, for representing higher-order constructive logic, with a predicative computational type theory of datatypes, for representing objects in the logic.

A.2.4 Theorem proving

The isomorphism and constructive type theories have been used in many interactive theorem provers — for example, the Endinburgh Logical Framework [GMW79, HHP87], Nuprl [CMH86], Isabelle [NPW02] and the ALF system [MN94]. One of the earliest approaches to theorem proving with type theory was the Automath work [Bru70]. Amongst other advantages, type theory permits complex proof tactics and parametrized lemmata to be given simply as functions over terms, and the automation of proof simplification.

A.2.5 Disadvantages of unified type theories

Higher-order type theories aim to represent datatypes and logical formulae within the same system. It can be argued that, while useful from a foundational perspective, such approaches are not desirable in practice.

In general, programming and logical reasoning are two very different tasks. Implementations of higher-order type theories have existed for over 30 years now, but have largely failed to make an impact in the software development community — in contrast to, for instance, the Hoare Logic [Hoa69] or model-based refinement methods such as the B method [Abr96].

It is possible that this failure is due to the very nature of a unified system of program typing and logical reasoning. By definition, such a system aims to do two complicated tasks, and necessarily has a more complicated syntax and a steeper learning curve.

We hypothesize that, following the more successful approaches to formal software development, different languages for programming and logical reasoning should still be employed in practice. This motivates the work of Crossley and Shepherdson [CS93], which emphasized the importance of a logic that is commonly understood by people trained in formal reasoning for software development (such as first-order or many-sorted logic).

The idea of separating proofs is essential to the Curry–Howard protocol, described in Part II of this monograph. Parts III and IV of this monograph are applications of the protocol, and can be seen as a further argument for the separation of proofs from programs, to achieve a practical approach to constructive synthesis in new logical and programming contexts.

References

[Abr96] Jean-Raymond Abrial, *The B-Book: Assigning Programs to Meanings*, Cambridge University Press, 1996.

[AL97] Martín Abadi and Rustan Leino, *A logic of object-oriented programs*, TAPSOFT '97: Theory and Practice of Software Development, 7th International Joint Conference CAAP/FASE, Lille, France (Michel Bidoit and Max Dauchet, eds.), Lecture Notes in Computer Science (LNCS), vol. 1214, Springer-Verlag, Berlin, 1997, pp. 682–696.

[And93] Penny Anderson, *Program Derivation by Proof Transformation*, PhD thesis, Carnegie Mellon University, 1993.

[Bac80] Ralph-Johan Back, *Correctness Preserving Program Refinements: Proof Theory and Applications*, vol. 131, Mathematisch Centrum, Amsterdam, 1980.

[Bar84] Henk Barendregt, *The Lambda Calculus: Its Syntax and Semantics*, North-Holland, 1984.

[Bar90] ———, *Functional Programming and Lambda Calculus*, Handbook of Theoretical Computer Science: Volume B: Formal Models and Semantics (Jan van Leeuwen, ed.), Elsevier, 1990.

[BB85] Corrado Böhm and Alessandro Berarducci, *Automatic synthesis of typed λ-programs on term algebras*, Theoretical Computer Science **39** (1985), 135–154.

[BBHdP93] Nick Benton, Gavin M. Bierman, Martin Hyland, and Valeria de Paiva, *A term calculus for intuitionistic linear logic*, Typed Lambda Calculi and Applications, International Conference on Typed Lambda Calculi and Applications, TLCA '93, Utrecht, The Netherlands, March 16–18, 1993, Proceedings (Berlin) (Marc Bezem and Jan Friso Groote, eds.), Lecture Notes in Computer Science (LNCS), vol. 664, Springer-Verlag, 1993, pp. 75–90.

[BC85] Joseph Bates and Robert Constable, *Proofs as programs*, ACM Transactions on Programming Languages and Systems **7** (1985), no. 1, 113–136.

[BCH99] Michel Bidoit, María Victoria Cengarle, and Rolf Hennicker, *Proof systems for structured specifications and their refinements*, Algebraic Foundations of Systems Specification (Egidio Astesiano, Hans-Jörg Kreowski, and Bernd Krieg-Brückner, eds.), IFIP State-Of-The-Art Reports, Springer-Verlag, Berlin, 1999, pp. 385–433.

[BCR^{+}99] Patrick Bellot, Jean-Pierre Cottin, Bernard Robinet, D. Sarni, Jean Jeneutre, and E. Zarpas, *Prolegomena of a Logic of Causality and Dynamism*, Studia Logica **62** (1999), no. 1, 77–105.

[BG77] Rod Burstall and Joseph Goguen, *Putting theories together to make specifications*, Proceedings of the 5th International Joint Conference on Artificial Intelligence, vol. 31, Akademie-Verlag, Berlin, 1977, pp. 1045–1058.

[BG80] ______, *The semantics of CLEAR, a specification language*, Proceedings of the 1979 Copenhagen Winter School on Abstract Software Specification (Berlin) (Dines Bjorner, ed.), Lecture Notes in Computer Science (LNCS), vol. 86, Springer-Verlag, 1980, pp. 292–232.

[BM79] Robert S. Boyer and J. Strother Moore, *A Computational Logic*, Academic Press, New York, 1979.

[BR90] Patrick Bellot and Bernard Robinet, *Logical synthesis of imperative object-oriented programs*, Logic Programming Synthesis and Transformation, 8th International Workshop, LOPSTR'98, Manchester, UK, June 15–19, 1998, Proceedings (Berlin) (Pierre Flener, ed.), Lecture Notes in Computer Science (LNCS), vol. 1559, Springer-Verlag, 1990, pp. 316–318.

[Bro75] Luitzen Egbertus Jan Brouwer, *Collected Works I*, North-Holland, 1975.

[Bro81] ______, *Brouwer's Cambridge Lectures on Intuitionism*, Cambridge University Press, 1981.

[Bru70] Nicolaas Govert de Bruijn, *The Mathematical Language AUTOMATH, its usage and some of its extensions*, Proceedings of Symposium of Automatic Demonstration, INRIA, Versailles, 1968 (Berlin), Lecture Notes in Mathematics (LNM), vol. 125, Springer-Verlag, 1970, pp. 29–61.

[BS93] Ulrich Berger and Helmut Schwichtenberg, *Program development by proof transformation*, Proceedings of the NATO Advanced Study Institute on Proof and Computation (Helmut Schwichtenberg, ed.), 1993, pp. 1–45.

[BS95a] ______, *The greatest common divisor: A case study for program extraction from classical proofs*, Types for Proofs and Programs, International Workshop TYPES'95, Torino, Italy, June 5–8, 1995, Selected Papers (Berlin) (Stefano Beradi and Mario Coppo, eds.), Lecture Notes in Computer Science (LNCS), vol. 1158, Springer-Verlag, 1995, pp. 36–46.

[BS95b] ______, *Program extraction from classical proofs*, Logical and Computational Complexity, Selected Papers, Logic and Computational Complexity, International Workshop, LCC '94, Indianapolis, Indiana, USA, 13–16 October 1994 (Berlin) (Daniel Leivant, ed.), Lecture Notes in Computer Science (LNCS), vol. 960, Springer-Verlag, 1995, pp. 77–97.

[Cen94] Maria Victoria Cengarle, *Formal Specifications with Higher-Order Parametrization*, PhD thesis, Ludwig–Maximilians-Universität, München, 1994.

[CH88] Thierry Coquand and Gérard P. Huet, *The calculus of constructions*, Information and Computation **76** (1988), no. 2/3, 95–120.

[CMH86] Robert Constable, N. Mendler, and D. Howe, *Implementing Mathematics with the Nuprl Proof Development System*, Englewood Cliffs, NJ: Prentice-Hall, 1986, Updated edition available at `http://www.cs.cornell.edu/Info/Projects/NuPrl/book/doc.html` (Accessed May 2003).

[CoF01] CoFI Language Design Task Group on Language Design, *Casl, the Common Algebraic Specification Language (version 1.0.1), Summary, 25 March 2001*, March 2001, Available at `http://www.brics.dk/Projects/CoFI/Documents/CASL/Summary/` (Accessed May 2003).

[Coo78] Stephen A. Cook, *Soundness and Completeness of an Axiom System for Program Verification*, Siam Journal of Computing **7** (1978), no. 1, 70–90.

[Cou90] Patrick Cousot, *Methods and logics for proving programs*, Formal Models and Semantics: Volume B (Jan van Leeuwen, ed.), Elsevier and MIT Press, 1990, pp. 841–994.

[CP01] Johny Crossley and Iman Poernomo, `Fred`: *An approach to generating real, correct, reusable programs from proofs*, Journal of Universal Computer Science **7** (2001), 71–88.

[CPW00] John Crossley, Iman Poernomo, and Martin Wirsing, *Extraction of structured programs from specification proofs*, Recent Trends in Algebraic Development Techniques: 14th International Workshop, WADT '99, Lecture Notes in Computer Science (LNCS), vol. 1827, Springer-Verlag, 2000, pp. 419–437.

[CS93] John N. Crossley and John C. Shepherdson, *Extracting programs from proofs by an extension of the Curry–Howard process*, Logical Methods: Essays in honor of Anil Nerode (John N. Crossley, Jeffrey B. Remmel, Richard A. Shore, and Moss E. Sweedler, eds.), Birkhäuser, Boston, 1993, pp. 222–288.

[Cur34] Haskell Curry, *Functionality in combinatory logic*, Proceedings of the National Academy of Science of the USA, vol. 20, 1934, pp. 154–180.

[Dij76] E. W. Dijkstra, *A Discipline of Programming*, Prentice-Hall, Englewood Cliffs (N.J.), 1976.

[Dil80] Justus Diller, *Modified realization and the formulae–as–types notion*, To H. B. Curry: Essays on Combinatory logic, Lambda Calculus, and Formalism, Academic Press, New York, 1980, pp. 491–501.

[Dum77] Michael Dummett, *Elements of Intuitionism*, Oxford University Press, 1977.

[Dum91] ———, *The Logical Basis of Metaphysics*, Harvard University Press, 1991.

[FC92] Jordi Farrés-Casals, *Verification in ASL and Related Specification Languages*, PhD thesis, University of Edinburgh, 1992, Report CST-92-92.

[FD88] Kokichi Futatsugi and Razvan Diaconescu, *CafeOBJ Report: The Language, Proof Techniques and Methodologies for Object-Oriented Algebraic Specification*, World Scientific, 1988.

[Fef79] Solomon Feferman, *Constructive theories of functions and classes*, Logic Colloquium '78: Proceedings of the Colloquium held in Mons, August 1978 (Maurice Boffa, Dirk van Dalen, and Kenneth McAloon Boffa, eds.), Studies in Logic and the Foundations of Mathematics, vol. 97, North-Holland, 1979, pp. 159–224.

[FGJM85] Kokichi Futatsugi, Joseph Goguen, Jean-Pierre Jouannaud, and José Meseguer, *Principles of OBJ2*, Conference Record of the Twelfth Annual ACM Symposium on Principles of Programming Languages, New Orleans, Louisiana, January 1985, 1985, pp. 52–66.

[Fil99] J.-C. Filliâtre, *Preuve de programmes impératifs en théorie des types*, Thèse de doctorat, Université Paris-Sud, July 1999.

[Fil03] ———, *Verification of Non-Functional Programs using Interpretations in Type Theory*, Journal of Functional Programming **13** (2003), no. 4, 709–745.

[Flo67] Robert Floyd, *Assigning meanings to programs*, Mathematical Aspects of Computer Science (J. T. Schwartz, ed.), Proceedings of the Symposium on Applied Mathematics, vol. XIX, American Mathematical Society, 1967.

[Gab96] Dov M. Gabbay, *Labelled Deductive Systems*, Oxford University Press, 1996.

[Gen69] Gerhard Gentzen, *Investigations into logical deduction*, The Collected Papers of Gerhard Gentzen (M. E. Szabo, ed.), North-Holland, 1969, pp. 68–131.

[GHJV95] Erich Gamma, Richard Helm, Ralph Johnson, and John Vlissides, *Design Patterns: Elements of Reusable Object-Oriented Software*, Addison-Wesley, Reading, MA, 1995.

[Gir72] Jean-Yves Girard, *Interpretation Fonctionnelle et Elimination des Coupures de L'arithmetique D'ordre Superieur*, Thèse de doctorat, Universite Paris VII, 1972.

[Gir87] ———, *Linear Logic*, Theoretical Computer Science **50** (1987), 1–102.

[GLT89] Jean-Yves Girard, Yves Lafont, and Paul Taylor, *Proofs and types*, Cambridge University Press, Cambridge, 1989.

[GMW79] Michael Gordon, Robin Milner, and Christopher Wadsworth, *Edinburgh LCF: a Mechanized Logic of Computation*, Lecture Notes in Computer Science (LNCS), vol. 78, Springer-Verlag, Berlin, 1979.

[GN87] Michael R. Genesereth and Nils J. Nilsson, *Logical Foundations of Artificial Intelligence*, Morgan Kaufmann Publishers, 1987.

[Gri90] Timothy G. Griffin, *The formulae-as-types notion of control*, Conference Record of the 17th Annual ACM Symposium on Principles of Programming Languages, POPL'90, San Francisco, CA, USA, 17–19 Jan 1990, ACM Press, 1990, pp. 47–57.

[Gru93] T. R. Gruber, *A translation approach to portable ontology specifications*, Knowledge Acquisition **5** (1993), no. 2, 199–220.

[GS78] T. Gergely and M. Szöts, *On the incompleteness of proving partial correctness*, Acta Cybernetica **4** (1978), no. 1, 45–57.

[Gun93] Carl A. Gunter, *Semantics of Programming Languages*, MIT Press, 1993.

[GWM+00] Joseph Goguen, Timothy Winkler, Jose Meseguer, Kokichi Futatsugi, and Jean-Pierre Jouannaud, *Introducing OBJ3*, Software Engineering with OBJ: Algebraic Specification in Action, Kluwer Academic Publishers, Boston, 2000.

[Har60] Ronald Harrop, *Concerning formulas of the types $A \to B \vee C$, $Arightarrow(Ex)B(x)$ in intuitionistic formal systems*, Journal of Symbolic Logic **25** (1960), no. 1, 27–32.

[Har84] David Harel, *Dynamic Logic*, Handbook of Philosophical Logic (Dov Gabbay and F. Guenthner, eds.), Oxford University Press, 1984.

[Hay90] Susumu Hayashi, *An introduction to PX*, Logical Foundations of Functional Programming (Reading, MA) (Gerard Huet, ed.), Addison-Wesley, 1990.

[HC68] G. E. Hughes and M. J. Cresswell, *Introduction to Modal Logic*, Methuen, London, 1968.

[Hey71] Arend Heyting, *Intuitionism: An Introduction*, North-Holland, 1971.

[HH86] C.A.R. Hoare and Jifeng He, *The weakest prespecification*, Fundamenta of Informaticae **9** (1986), Part I: 51–84, Part II: 217–252.

[HHH+87] C.A.R. Hoare, Ian Hayes, Jifeng He, Carol C. Morgan, A. W. Roscoe, Jeff W. Sanders, I. H. Sorensen, J. Michael Spivey, and Bernard Sufrin, *Laws of programming*, Communications of the ACM **30** (1987), no. 8, 672–686.

[HHP87] Robert Harper, Furio Honsell, and Gordon Plotkin, *A framework for defining logics*, Proceedings of the 2nd Annual IEEE Symposium on Logic in Computer Science, LICS'87, Ithaca, NY, USA, 22–25 June 1987, IEEE Computer Society Press, 1987, pp. 194–204.

[HHS85] Jifeng He, C.A.R. Hoare, and Jeff W. Sanders, *Data Refinement Refined*, Oxford, 1985.

[HHS87] C.A.R. Hoare, Jifeng He, and Jeff W. Sanders, *Prespecification in data refinement*, Information Processing Letters **24** (1987), no. 2, 127–132.

[HKPM97] Gérard P. Huet, Gilles Kahn, and Christine Paulin-Mohring, *The Coq Proof assistant Reference Manual: Version 6.1*, Coq project research report RT-0203, Inria, 1997.

[HN88] Susumu Hayashi and Hiroshi Nakano, *PX, a Computational Logic*, Foundations of Computing, MIT Press, 1988, Electronic edition available at `http://www.shayashi.jp/PXbook.html` (Accessed May 2003).

[Hoa69] C. A. R. Hoare, *An axiomatic basis for computer programming*, Communications of the Association for Computing Machinery **12** (1969), 576–80.

[Hoa81] ______, *A calculus of total correctness for communicating processes*, Scientific Computational Programming **1** (1981), no. 1–2, 49–72.

[Hoa85] ______, *Communicating Sequential Processes*, Prentice-Hall, 1985.

[How80] William A. Howard, *The formulae-as-types notion of construction*, To H. B. Curry : Essays on Combinatory logic, Lambda calculus, and Formalism, Academic Press, London, New York, 1980, pp. 479–490.

[HS86] John Roger Hindley and Jonathan Seldin, *Introduction to Combinators and Lambda-Calculus*, Cambridge University Press, 1986.

[HS96] Martin Hofmanna and Donald Sannella, *On behavioural abstraction and behavioural satisfaction in higher-order logic*, Theoretical Computer Science **167** (1996), no. 1-2, 3–45.

[HWB97] Rolf Hennicker, Martin Wirsing, and Michel Bidoit, *Proof systems for structured specifications with observability operators*, Theoretical Computer Science **173** (1997), no. 2, 393–443.

[JPBC03] John S. Jeavons, Iman Poernomo, Bolis Basit, and John Crossley, *A layered approach to extracting programs from proofs with an application in Graph Theory*, Proceedings of the 7th and 8th Asian Logic Conferences (Rod Downey, Ding Decheng, Tung Shih Ping, Qiu Yu Hui, Mariko Yasugi, and Guohua Wu, eds.), Singapore University Press and World Scientific, 2003, pp. 193–222.

[KBB93] Ina Kraan, David A. Basin, and Alan Bundy, *Logic program synthesis via proof planning*, Logic Program Synthesis and Transformation, Proceedings of LOPSTR 92, International Workshop on Logic Program Synthesis and Transformation, University of Manchester, 2-3 July 1992 (Berlin) (Kung-Kiu Lau and Tim P. Clement, eds.), Workshops in Computing, Springer-Verlag, 1993, pp. 1–14.

[Kle45] Steven Cole Kleene, *On the interpretation of intuitionistic number theory*, Journal of Symbolic Logic (1945), no. 10, 109–124.

[Kle52] ______, *Introduction to Metamathematics*, North-Holland, 1952.

[KMND00] Jeff Kramer, Jeff Magee, Keng Ng, and Naranker Dulay, *Software architecture description*, Software Architecture for Product Families: Principles and Practice (Reading, MA), Addison-Wesley, 2000, pp. 31–64.

[Kol32] Andrei Nikolaevich Kolmogorov, *Zur deutung der intuitionistischen logik*, Mathematische Zeitschrift **35** (1932), 58–65.

[Koz97] Dexter Kozen, *Kleene algebra with tests*, ACM Transactions on Programming Languages and Systems **19** (1997), no. 3, 427–443.

[Kre59] George Kreisel, *Interpretation of analysis by means of constructive functionals of finite types*, Constructivity in Mathematics (Arend Heyting, ed.), North-Holland, Amsterdam, 1959, pp. 101–128.

[KST97] Stefan Kahrs, Donald Sannella, and Andrzej Tarlecki, *The definition of Extended ML: A gentle introduction*, Theoretical Computer Science **173** (1997), 445–484.

[Lam94] Leslie Lamport, *The Temporal Logic of Actions*, ACM Toplas **3** (1994), 872–923.

[Lei94] Daniel Leivant, *Higher order logic*, Handbook of Logic in Artificial Intelligence and Logic Programming (Dov M. Gabbay, C. J. Hogger, and J. A. Robinson, eds.), vol. 2, Oxford University Press, 1994, pp. 229–3211.

[Luo94] Zhaohui Luo, *Computation and Reasoning: A Type Theory for Computer Science*, Oxford University Press, 1994.

[McC92] William McCune, *Automated discovery of new axiomatizations of the left group and right group calculi*, Journal Of Automated Reasoning **9** (1992), no. 1, 1–24.

[Mey97] Bertrand Meyer, *Object-Oriented Software Construction*, Prentice-Hall, 1997.

[Mey00] ———, *Agents, iterators and introspection*, Technology paper, ISE Corporation, available at `http://archive.eiffel.com/doc/manuals/language/agent/page.html` (Accessed May 2003), May 2000.

[ML75] Per Martin-Löf, *An intuitionistic theory of types*, Logic Colloquium '73 (H. E. Rose and J. C. Shepherdson, eds.), North-Holland, Amsterdam, 1975, pp. 73–118.

[ML84] ———, *Intuitionistic Type Theory*, Bibliopolis, 1984.

[ML85] ———, *Constructive mathematics and computer programming*, Mathematical Logic and Programming Languages, Prentice-Hall, London, 1985, pp. 167–184.

[MLM90] Per Martin-Löf and Grigori Mints (eds.), *Inductively defined types*, Lecture Notes in Computer Science (LNCS), vol. 417, Berlin, Springer-Verlag, 1990.

[MN94] Lena Magnusson and Bengt Nordström, *The alf proof editor and its proof engine*, Types for Proofs and Programs (Berlin) (Henk Barendregt and Tobias Nipkow, eds.), Lecture Notes in Computer Science (LNCS), vol. 806, Springer-Verlag, 1994, pp. 213–237.

[Mor94] Carol C. Morgan, *Programming from Specifications*, Prentice-Hall, 1994.

[MT98] Mihhail Matskin and Enn Tyugu, *Strategies of Structural Synthesis of Programs*, Proceedings of the 12th IEEE International Conference Automated Software Engineering, IEEE Computer Society, 1998, pp. 305–306.

[MT00] Nenad Medvidovic and Richard N. Taylor, *A classification and comparison framework for software architecture description languages*, IEEE Transactions on Software Engineering **26** (2000), no. 1, 70–93.

[MTH90] Robin Milner, Mads Tofte, and Robert Harper, *The definition of Standard ML*, MIT Press, Cambridge, MA, 1990.

[Mug92] Stephen Muggleton, *Inductive Logic Programming*, Academic Press, New York, 1992.

[Mur91] Chetan Murthy, *An evaluation semantics for classical proofs*, Proceedings of the 6th Annual IEEE Symposium on Logic in Computer Science, LICS'91, Amsterdam, Netherlands, 15–18 July 1991, IEEE Computer Society Press, 1991, pp. 96–107.

[MV93] Carol C. Morgan and Trevor Vickers, *On the Refinement Calculus*, Springer-Verlag, Berlin, 1993.

[MW87] Zohar Manna and Richard J. Waldinger, *The deductive synthesis of imperative LISP programs*, National Conference on Artificial Intelligence, 1987, pp. 155–160.

[MW91] Manna and Waldinger, *Fundamentals of deductive program synthesis*, Combinatorial Algorithms on Words, NATO ISI Series, Springer-Verlag, Berlin, 1991.

[Nak94] Hiroshi Nakano, *The non-deterministic catch and throw mechanism and its subject reduction property*, Logic, Language, and Computation: Festschrift in Honor of Satoru Takasu (N. D. Jones, M. Hagiya, and M. Sato, eds.), vol. 792, Springer-Verlag, Berlin, 1994, pp. 61–72.

[NP83] Bengt Nordström and Kent Petersson, *Types and specifications*, Information Processing 83 (R.E.A. Mason, ed.), North-Holland, 1983.

[NPW02] Tobias Nipkow, Lawrence C. Paulson, and Markus Wenzel, *Isabelle/HOL*, Lecture Notes in Computer Science (LNCS), vol. 2283, Springer-Verlag, Berlin, 2002.

[NR68] Peter Naur and Brian Randell (eds.), *Software engineering: Report on a conference sponsored by the NATO science committee, garmisch, germany, 7th to 11th october, 1968*, Brussels 39 Belgium, Scientific Affairs Division, NATO, January 1968.

[Par93] Michel Parigot, *Classical proofs as programs*, Computational Logic and Proof Theory: Proceedings of the 3rd Kurt Gödel Colloquium, KGC'93, Brno, Czech Republic, 24–27 Aug 1993 (Georg Gottlob, Alexander Leitsch, and Daniele Mundici, eds.), Lecture Notes in Computer Science (LNCS), vol. 713, Springer-Verlag, Berlin, 1993, pp. 263–276.

[PC01] Iman Poernomo and John Crossley, *Protocols between programs and proofs*, Logic Based Program Synthesis and Transformation, 10th International Workshop, LOPSTR 2000 London, UK, July 24–28, 2000, Selected Papers (Berlin) (Kung-Kiu Lau, ed.), Lecture Notes in Computer Science (LNCS), vol. 2042, Springer-Verlag, 2001.

[PC03] ———, *The Curry–Howard isomorphism adapted for imperative program synthesis and reasoning*, Proceedings of the 7th and 8th Asian Logic Conferences (Rod Downey, Ding Decheng, Tung Shih Ping, Qiu Yu Hui, Mariko Yasugi, and Guohua Wu, eds.), Singapore University Press and World Scientific, 2003, pp. 343–377.

[PCW02] Iman Poernomo, John Crossley, and Martin Wirsing, *Programs, proofs and parametrized specifications*, Recent Trends in Algebraic Development Techniques, 15th International Workshop, WADT 2001, Joint with the CoFI WG Meeting, Genova, Italy, April 1–3, 2001, Selected Papers (Berlin) (Maura Cerioli and Gianna Reggio, eds.), Lecture Notes in Computer Science (LNCS), vol. 2267, Springer-Verlag, 2002, pp. 280–304.

[Pet96] Hannes Peterreins, *A Natural-deduction-like Calculus for Structured Specifications*, PhD thesis, Ludwig–Maximilians-Universität, München, 1996.

[Plo71] Gordon Plotkin, *Automatic methods of inductive inference*, PhD thesis, Edinburgh University, 1971.

[PM89] Christine Paulin-Mohring, *Extracting F_ω's programs from proofs in the Calculus of Constructions*, Sixteenth Annual ACM Symposium on Principles of Programming Languages (Austin), ACM, January 1989, pp. 89–104.

[PM93] ———, *Inductive Definitions in the System Coq Rules and Properties*, Typed Lambda Calculi and Applications, International Conference on Typed Lambda Calculi and Applications, TLCA '93, Utrecht, The Netherlands, March 16–18, 1993, Proceedings (Berlin) (Marc Bezem and Jan Friso Groote, eds.), Lecture Notes in Computer Science (LNCS), vol. 664, Springer-Verlag, 1993, pp. 328–345.

[PMW93] Christine Paulin-Mohring and Benjamin Werner, *Synthesis of ML programs in the system Coq*, Journal of Symbolic Computation **15** (1993), no. 5/6, 607–640.

[Poe99] Iman Poernomo, *A labelled deduction system with Curry-Howard terms for extracting functional programs with state*, Proceedings of the Workshop on Automated Reasoning, Automated Reasoning Day 1999 (Matthias Fuchs, ed.), 17–18 April 1999.

[Poe03a] ———, *Proofs-as-imperative-programs: Application to synthesis of contracts*, Perspectives of Systems Informatics, 5th International Andrei Ershov Memorial Conference, PSI 2003, Akademgorodok, Novosibirsk, Russia, July 9-12, 2003, Revised Papers (Berlin) (Manfred Broy and Alexandre Zamulin, eds.), Lecture Notes in Computer Science (LNCS), vol. 2890, Springer-Verlag, 2003.

[Poe03b] ———, *Variations on a Theme of Curry and Howard: The Curry-Howard isomorphism and the proofs-as-programs paradigm adapted to imperative and structured program synthesis*, PhD thesis, Monash University, 2003.

[PRS02] Iman Poernomo, Ralf Reussner, and Heinz Schmidt, *Architectures of Enterprise Systems: Modelling Transactional Contexts*, Proceedings of the First IFIP/ACM Working Conference on Component Deployment (CD 2002) (Berlin), Lecture Notes in Computer Science (LNCS), vol. 2370, Springer-Verlag, 2002, pp. 233–243.

[Ran00] Alexander Ran, *Ares conceptual framework for software architecture*, Software Architecture for Product Families: Principles and Practice (Reading, MA), Addison-Wesley, 2000, pp. 1–29.

[Res00] Greg Restall, *Constructive logic for all*, Available at `http://consequently.org/papers/constructive.pdf` (Accessed May 2003), 2000.

[Rey74] J. C. Reynolds, *Towards a Theory of Type Structure*, Programming Symposium, Paris (Berlin), Lecture Notes in Computer Science (LNCS), vol. 19, Springer-Verlag, 1974, pp. 408–425.

[RN95] Stuart Russell and Peter Norvig, *Artificial intelligence: A modern approach*, Prentice-Hall, 1995.

[Sas86] James T. Sasaki, *Extracting Efficient Code From Constructive Proofs*, PhD thesis, Cornell University, May 1986, Available as Cornell University Computer Science Department Technical Report TR 86–757, p. 317.

[Sat97] Masahiko Sato, *Intuitionistic and classical natural deduction systems with the catch and the throw rules*, Theoretical Computer Science **175** (1997), no. 1, 75–92.

[SB83] Donald Sannella and Rod Burstall, *Structured theories in lcf*, Proceedings of the 8th Colloqium on Trees in Algebra and Programming (CAAP)

(Berlin), Lecture Notes in Computer Science (LNCS), vol. 159, Springer-Verlag, 1983, pp. 377–391.

[Sch82] Helmut Schwichtenberg, *On Martin-Löf's theory of types*, Atti Degli Incontri di Logica Mathematica, 1982, pp. 299–325.

[Sch85] ______, *A normal form for natural deductions in a type theory with realizing terms*, Atti del Congresso Logica e Filosfia della Scienza, oggi. San Gimignano, December 7–11, 1983 (Ettore Casari, ed.), CLUEB, Bologna, Italy, 1985, pp. 95–138.

[Sch99a] Stephen Schach, *Classical and Object-oriented Software Engineering with UML and Java*, 4th edition ed., McGraw-Hill, 1999.

[Sch99b] Helmut Schwichtenberg, *Classical proofs and programs*, Notes for the Marktoberdorf Logic Summer School '99, available at `www.mathematik.uni-muenchen.de/~schwicht/` (Accessed May 2003), 1999.

[SI98a] Jamie Stark and Andrew Ireland, *Invariant discovery via failed proof attempts*, Logic Program Synthesis and Transformation, 8th International Workshop, LOPSTR'98, Manchester, UK, June 1998, Selected Papers (Berlin) (Pierre Flener, ed.), Lecture Notes in Computer Science (LNCS), vol. 1559, Springer-Verlag, 1998, pp. 271–288.

[SI98b] ______, *Towards automatic imperative program synthesis through proof planning*, Proceedings of the 14th IEEE International Conference on Automated Software, IEEE Computer Society, 1998, pp. 44–51.

[Sim00] Harold Simmons, *Derivation and Computation: Taking the Curry–Howard Isomorphism Seriously*, Cambridge University Press, 2000.

[Smi93] Douglas R. Smith, *Constructing specification morphisms*, Journal of Symbolic Computation **15** (1993), no. 5 & 6, 571–606.

[SST92] Donald Sannella, , S. Sokolowski, and Andrzej Tarlecki, *Toward formal development of programs from algebraic specifications: parameterisation revisited*, Acta Informatica **29** (1992), 689–736.

[ST88a] Donald Sannella and Andrzej Tarlecki, *Specifications in an arbitrary institution*, Information and Computation **76** (1988), 165–210.

[ST88b] ______, *Toward formal development of programs from algebraic specifications: implementation revisited*, Acta Informatica **25** (1988), no. 3, 233–281.

[ST97] ______, *Essential concepts of algebraic specification and program development*, Formal Aspects of Computing **9** (1997), 229–269.

[SW83] Donald Sannella and Martin Wirsing, *A kernel language for algebraic specification and implementation*, Fundamentals of Computation Theory, Proceedings of the 1983 International FCT-Conference, Borgholm, Sweden, August 21-27, 1983 (Berlin) (Marek Karpinski, ed.), Lecture Notes in Computer Science (LNCS), vol. 158, Springer-Verlag, 1983, pp. 413–427.

[Szy98] Clemens Szyperski, *Component Software: Beyond Object-Oriented Programming*, Addison-Wesley, Reading, MA, 1998.

[Tro73] A. S. Troelstra (ed.), *Metamathematical Investigation of Intuitionistic Arithmetic and Analysis*, Lecture Notes in Mathematics, vol. 344, Springer-Verlag, Berlin, 1973.

[Tro98] ______, *Realizability*, Handbook of Proof Theory (Samuel R. Buss, ed.), Elsevier, 1998, pp. 407–473.

[TS89] Andrzej Tarlecki and Donald Sannella, *Toward formal development of ML programs: Foundations and methodology*, TAPSOFT'89: Proceedings of the International Joint Conference on Theory and Practice of Software Development, Barcelona, Spain, March 13–17, 1989, Volume 2: Advanced Seminar on Foundations of Innovative Software Development II and Colloquium on Current Issues in Programming Languages (CCIPL) (Berlin) (Josep Díaz and Fernando Orejas, eds.), Lecture Notes in Computer Science (LNCS), vol. 352, Springer-Verlag, 1989, pp. 375–389.

[Tyu88] Enn Tyugu, *Knowledge-based programming*, Addison-Wesley, Reading, MA, 1988, Transation of Концептуальное Программирование, Nauka, USSR, 1984.

[Wan78] Mitchell Wand, *A new incompleteness result for Hoare's system*, Journal of the ACM **25** (1978), no. 1, 168–175.

[WCP98] Martin Wirsing, John Crossley, and Hannes Peterreins, *Proof normalization of structured algebraic specifications is convergent*, Workshop on Algebraic Development Techniques, Proceedings of the Twelfth International Workshop on Recent Trends in Algebraic Development Techniques (Berlin) (J. Fiaderio, ed.), Lecture Notes in Computer Science, vol. 1589, Springer-Verlag, 1998, pp. 326–340.

[Wir82] Martin Wirsing, *Structured algebraic specifications*, Proceedings of the AFCET Symposium on Mathematics for Computer Science (Bernard Robinet, ed.), 1982, pp. 93–108.

[Wir86] ———, *Structured algebraic specifications: a kernel language*, Theoretical Computer Science **43** (1986), 123–250.

[Wir90] ———, *Algebraic specification*, Handbook of Theoretical Computer Science (Amsterdam, New York) (Jan van Leeuwen, ed.), vol. B, Elsevier, 1990, pp. 675–788.

[Wir91] ———, *Structured specifications: Syntax, semantics and proof calculus*, Logic and Algebra of Specification (Berlin) (Friedrich L. Bauer, Wilfried Brauer, and Helmut Schwichtenberg, eds.), NATO ASI Series F94: Computer and System, Springer-Verlag, 1991, pp. 411–442.

[WK98] Jos Warmer and Anneke Kleppe, *The Object Constraint Language: Precise modeling with UML*, Addison-Wesley, Reading, MA, 1998.

Index